Naturalized Fishes of the World

Naturalized Fishes of the World

By
Christopher Lever

Illustrated by Martin Camm

ACADEMIC PRESS
Harcourt Brace & Company, Publishers
SAN DIEGO   LONDON   BOSTON
NEW YORK   SYDNEY   TOKYO   TORONTO

ACADEMIC PRESS INC
525B Street, Suite 1900, San Diego,
California 92101-4495, USA
http://www.apnet.com

ACADEMIC PRESS LIMITED
24–28 Oval Road
LONDON NW1 7DX
http://www.hbuk.co.uk/ap/

This book is printed on acid-free paper

A catalogue record for this book is available from the British Library

ISBN 0-12-444745-7

Typeset by Phoenix Photosetting, Chatham, Kent
Printed in Great Britain by The Bath Press, Bath

96 97 98 99 00 01 EB 9 8 7 6 5 4 3 2 1

CONTENTS

PREFACE

The object of this book is to describe when, where, why, how, and by whom the various alien inland freshwater fishes now living in a wild state throughout the world were introduced, how they subsequently became naturalized, their present status in the countries to which they have been introduced, and their ecological (and in some cases socio-economic) impact on the native biota (and local economies). The criteria for inclusion of a species are that it should have been imported from its natural range to a new country either deliberately or accidentally by human agency, and that it should currently be established in the wild in self-maintaining and self-perpetuating populations unsupported by, and independent of, man.

Each species account is a monograph on an individual fish; at the end of some of them are brief details of the artificial distribution of allied, related or associated species. In general, only those species that have become successfully naturalized are discussed. The inclusion of those that failed to become established would have been of doubtful value unless the precise reasons (usually a combination of various factors) for each such failure were known, and would inevitably have resulted in numerous omissions. Exceptions have been made in the case of species that have succeeded in becoming established in some regions but have failed elsewhere, and when initial failures have been followed by eventual successes. Natural immigrants from one country to another have normally been included only if they have been augmented by later accidental or deliberate introductions by man, or in the case of an established alien spreading naturally to a new country. Because of the accepted definitions of an 'Introduced' and a 'Naturalized' species (see below), only international transfers are discussed in detail. It is recognized that this may be regarded as of doubtful ecological validity, since the movement of species between different drainages and river basins within the same country may be of considerably greater ecological significance than those between neighbouring countries. Intranational transfers, however, do not comply with the agreed definitions of 'Introduced' and 'Naturalized'. Translocations of a species from one part of a country to another part of the same country are listed under 'Translocations', but in general have only been included if such a species has become successfully naturalized elsewhere.

Terminology

The definitions of various terms used in the text and associated with the introduction of living organisms are as follows (for other definitions see Glossary):

Acclimatization. Living in the wild in an alien environment or climate with the support of and dependent (e.g. for food and/or shelter) on man.

Adventive. An introduced species which is not as yet established in the wild.

Alien. See Exotic.

Allochthonous. See Exotic.

Autochthonous. See Native.

Colonization. See Naturalization.

Establishment. See Naturalization.

Exotic. A species native to an area outside of, or foreign to, the national geographic area under discussion. An introduced species.

Feral. A species that has reverted to the wild from domestication (the term 'Feral' should never be used to describe the naturalization of a wild species).

Indigene. See Native.

Introduction. The deliberate or accidental release of a species into a country in which it is not known to have occurred within historic times. The movement by man, whether deliberate or accidental, of living organisms to a new location outside their recent national geographic range.

Invasive. An introduced species (not necessarily one that has had a negative ecological impact).

Native. A species that is a member of the natural biotic community.

Naturalization. The introduction of species to countries where they are not indigenous, but in which they may flourish under the same conditions as those that are native. More particularly, the establishment of self-maintaining and self-perpetuating populations unsupported by and independent of man of an introduced species in a free-living state in the wild.

Reintroduction. The deliberate release by man of a species into a national geographic area in which it was indigenous in historic times but where it subsequently became extinct.

Restocking. The deliberate release by man of a species into an area where it already occurs, with the intention of augmenting the existing population of that species.

Stocking. See Translocation.

Translocation. The deliberate or accidental movement by man of a species from an area where it is established, as either native or alien, to another area within the same national geographic range.

Transplantation. See Translocation.

Distribution of Species

Zoogeographically, the world was divided in 1858 by P.L. Sclater into six great faunal regions: the Palaearctic (Europe, northern Africa south to the central Sahara, and northern Asia); the Ethiopian (Africa south of the Palaearctic, and southern Arabia); the Oriental (tropical Asia and western Indonesia); the Nearctic (North America south to central Mexico); the Neotropical (central and South America, southern Mexico, and the Antilles); and the Australasian (Australia, Tasmania, New Zealand, Papua New Guinea, and some Pacific and eastern Indonesian islands). The division between the Australasian and Oriental regions is a deep trough in the ocean bed between Papua New Guinea and Borneo; this division is defined by (Alfred Russel) Wallace's Line, which extends south of the Philippines between Borneo and Celebes (Sulawesi) and between the islands of Bali and Lombok. The

Palaearctic and Nearctic regions together form the Holarctic region. The tropical region (not one of the great faunal regions) lies between the tropics of Cancer and Capricorn (i.e. between 23°N and 23°S of the equator).

Geographically, the world is divided into six continents which are sociopolitically subdivided into countries or nations: Europe includes western Turkey and the European former USSR; Asia stretches from eastern Turkey to the Far East; Africa includes the island of Madagascar (the Malagasy Republic) and some off-lying islands; North America comprises Canada, the United States, Mexico, and some adjacent islands; South America includes all mainland south of North America, including Central America (not a separate continental region); Australasia (and Oceania, which is not strictly a continental region) includes those countries in the Australasian faunal region plus oceanic islands, especially those in the Pacific and Indian Oceans.[1]

The system adopted here is based on these geographical and socio-economic boundaries rather than on the less precise division into faunal regions with the following exceptions: Africa includes only the continental land mass and Madagascar; Australasia includes only Australia, New Zealand, and Papua New Guinea; for zoogeographic reasons, Oceania includes most of the world's offshore island groups, regardless of locality.

Geographic/Political Nomenclature

Where these apply the latest name is normally used, with the previous name given in parentheses. In the case of countries such as the USSR and Yugoslavia, the prefix 'the former' has been inserted.

British County Names

In 1974, the names and boundaries of several counties in England and Wales, and in 1975 those of some in Scotland, were changed. As most of the events described herein antedate these changes, and because it would be confusing to use both old and new names, the old names and boundaries have been adhered to throughout. (Britain consists of England,

1. For more details see the *Universal Decimal Classification*, 4th international edition (Publication No. 179, British Standards Institution, London, 1972). Gr. 9.

Scotland, Wales and Northern Ireland; the British Isles comprises those countries plus the Republic of Ireland).

Unsuccessful Introductions

The term 'Unsuccessful Introduction' in the text means only that the species under discussion has failed to establish a viable naturalized population; it may, however, be successfully acclimatized, established in aquaculture, or have at one time become naturalized but subsequently died out.

Vernacular Names

The vernacular name appearing immediately after the scientific name in the headings to the species accounts is the one most commonly used; vernacular names in parentheses are the main alternative and/or local names. Where vernacular names are omitted, no generally accepted one exists.

Sources

Various bibliographies exist on the international transfer of (mainly inland) aquatic organisms (principally fish) and the ecological and socio-economic consequences of such transfers; of these, Rosenthal (1978) is perhaps the most comprehensive. For more recent references, a computer-search was made of *Biological Abstracts*, and a trawl carried out on the Internet and of *Zoological Record*[2] to 1995. Finally, much up-to-date information was provided by the many correspondents whose names appear in the list of acknowledgements.

Naturalized British Fishes

More detailed information on naturalized British fishes will be found in *The Naturalized Animals of the British Isles* (Lever, 1977).

Ecological Impact

More detailed information on the ecological impact of naturalized fishes will be found in *Naturalized Animals: the ecology of successfully introduced species* (Lever, 1994).

Purposes and Ecological Impacts of Introductions

For further reading on the purposes of introductions to the wild of exotic inland freshwater fishes and their ecological impacts, the following references and bibliographies are recommended:

Mosquito control: Gerberich & Laird, 1968; Haas & Pal, 1984.
Aquatic weed control (general): Shireman, 1984; Petr, 1987b.
Aquatic weed control (by Grass Carp): Negonovskaya, 1980; Shireman & Smith, 1983; Chilton & Muoneke, 1992.
Vector control: Petr, 1987a.
General: Vooren, 1968; Rosenthal, 1978; Taylor *et al.*, 1984; Welcomme, 1984 & 1988; De Moor & Bruton, 1988; Lever, 1994.

---

2. Published annually by BIOSIS and the Zoological Society of London.

ACKNOWLEDGEMENTS

Without the generous co-operation of many individuals and organizations, both in Britain and overseas, who patiently and generously responded to my requests for information, this book would have been very much the poorer.

I am especially grateful to the librarians and staff of the various institutions in which I carried out my research, in particular the Natural History Museum in London (where I was generously granted photocopying facilities) and the Linnean Society of London. In the former, my thanks are due to the Head of the Department of Library Services, Mr R.E.R. Banks, the Deputy Head, Mr C. Mills, and Mrs A. Datta, Mrs C. Gokce, Mr P. Cooper, Miss S. McGee, Miss J. Lloyd-Davies, Miss S. Carthy, Mr N. Stead, Mrs J. Harvey, Mrs R. Lanstone, Miss L. Mitchell, Miss C. Pease and Mrs E. Duggan; and in the latter to the Librarian and Archivist, Miss G.L. Douglas.

Other institutions in which I conducted my research were the British Library; the Science Reference Library; and the libraries of the Ministry of Agriculture, Fisheries and Food, the Zoological Society of London and the Royal Geographical Society.

For the benefit of future students of naturalized animals, all the material resulting from my research on this book has been deposited with that on my previous books in the library of the Natural History Museum in London.

Individuals to whom I am indebted for information are: Dr N. Asmoradjo (Ministerie Van Lanbouw, Vesteelt en Visserij, Surinam); Mr W.J. Ayton (Welsh National Water Development Authority); Dr J. Barnett (Australian Nature Conservation Agency); Dr R.C. Barrio (Ministerio de la Industria Pesquera, Cuba); Dr D.M. Bartley (FAO, Italy); Dr P. Bridgewater (Australian Nature Conservation Agency); Dr M.S. Burgos Flores (Dirección General de Recursos Naturales Renovables, Honduras); Mr J.A. Burton (UK); Dr P.L. Cadwallader (Queensland Fisheries Management Authority, Australia); Mr A.S. Churchward (Severn–Trent Water Authority, UK); Dr D.W. Coble (University of Wisconsin, USA), Dr S. Contreras (Universidad Autónoma de Nuevo León, Mexico); Professor W.R. Courtenay (Florida Atlantic University, USA); Miss A. Crampton (Ecuador); Mr O.A. Crimmen (Natural History Museum, London, UK); Dr E.J. Crossman (Royal Ontario Museum, Canada); Dr B.T. Cunningham (Ministry of Agriculture and Fisheries, NZ); Dr S.J. de Groot (Netherlands Institute for Fisheries Research); Mr P. Donoughue (Jacaranda Wiley, Ltd, Australia); Dr L.G. Eldredge (Pacific Science Association, Hawaii); Dr B. Elvira (University of Madrid, Spain); Dr M.A. Escalante (Universidad Autónoma de Sinaloa, Mexico); Professor B.W. Fox (UK); Dr W. Frost (Freshwater Biology Association, UK); Dr E. Gelosi (Stabilimento Ittiogenico de Roma, Italy); Mr J. Gibson (Ministry of Trade, Industry Cooperatives & Consumer Protection, Belize); Dr A. Gill (Natural History Museum, London, UK); Dr Z. Gonzalez de Gutierrez (Departmento de Recursos Pesqueros, Dominican Republic); Professor P.H. Greenwood (formerly Natural History Museum, London, UK); Dr D.A. Hensley (University of Puerto Rico); Mr F. Herbert (Royal Geographical Society, UK); Dr P. Hickley (Severn–Trent Water Authority, UK); Dr J. Holcik (Slovak Academy of Agricultural Sciences, Czechoslovakia); Dr A. Holden (Freshwater Fisheries Laboratory, UK); Mr B. Holmberg (National Board of Fisheries, Sweden); Dr D.W. Howat (Celltech Ltd, UK); Dr P.B.N. Jackson (Rhodes University, SA); Dr K. Jensen (Directoratet for vilt og Ferskvannfisk, Norway); Dr D.J. Jude (Center for Great Lakes & Aquatic Sciences, USA); Mr R. Keam (Australia); Mr R. Kennedy (UK); Mr R.W.F. Kitchin (NZ); Professor J.R. Krebs (University of Oxford, UK); Dr H. Kuhlmann (Bundesforschungsanstalt für Fischerei, Germany); Dr P. Laurent (Station d'Hydrobiologie Lacustre de Thonon, France); Mr D.F. Leney (Surrey Trout Farm and United Fisheries, UK); Dr R.S.J. Linfield (Anglian Water Authority, UK); Dr L.N. Lloyd (University of Adelaide, Australia); Dr R.H. Lowe-McConnell (formerly Natural History Museum, London, UK); Mr D. Marlborough (British Ichthyological Society); Dr G.C. Martinez

(Departamento de Ecosistemas Acuáticos, Nicaragua); Dr R.P. Malaca (Dirección Nacional de Acuicultura, Panama); Dr J.A. McCann (National Fisheries Center, USA); Dr R.M. McDowall (National Institute of Water & Atmospheric Research Ltd, NZ); Dr R.J. McKay (Queensland Museum, Australia); Dr C.G. Menafra (Ministerio de Agricultura y Pesca, Uruguay); Dr E.L. Mills (Cornell University, USA); Mr J. Moreau (Institute National Polytechnique de Toulouse, France); Professor P.B. Moyle (University of California, USA); Mr A.K. Murray (Agriculture & Fisheries Department, Hong Kong); Dr H. Nanne (Ministerio de Agricultura y Ganaderia, Costa Rica); Miss R. Ntsoane (CSIR, SA); Dr J.R.J. Palacios (Centro de Acuicultura de Tezcatapec, Mexico); Dr A.I. Payne (Coventry Polytechnic, UK); Mr L.R. Peart (Berkshire Trout Farm, UK); Dr R. Pethiyagoda (Wildlife Heritage Trust, Sri Lanka); Dr R.S.V. Pullin (ICLARM, Philippines); Dr T. Rahman (Freshwater Fisheries Research Station, Bangladesh); Dr C. Sanchez (Ministerio de Agricultura y Ganaderia, El Salvador); Dr C.D. Sandgren (Center for Great Lakes and Aquatic Studies, USA); Dr J.H. Selgeby (Ashland Biological Station, USA); Miss P. Shelley (Natural History Book Service Ltd, UK); Dr J.V. Shireman (University of Florida, USA); Mr R. South (Editor, *Fish & Game New Zealand*); Dr J.R. Stauffer (Pennsylvania State University, USA); Dr J.N. Taylor (Florida Atlantic University, USA); Dr T. Tohiyat (Ministry of Agriculture, Malaysia); Dr P. Tunnainen (Finnish Game & Fisheries Research Institute); Dr A.E. Del Valle (Dirección de Ecologia Animal, Argentina); Dr J.M. Valdevieso (Ministerio de Pesqueria, Peru); Professor D.F.E. Thys Van Den Audenaerde (Musée Royal de l'Afrique Centrale, Belgium); Miss S. Van Rooyen (Foundation for Research Development, SA); Dr R.L. Welcomme (FAO, Italy); Dr D.B. Wingate (Department of Agriculture, Fisheries and Parks, Bermuda); Dr H.E. Wood (Ministry of Agriculture, Lands & Fisheries, Trinidad); Dr A.H. Zumker (Organisatie ter Verbetering Van de Binnen Visserij, Netherlands).

I have to thank Martin Camm for providing the illustrations; the staff at Academic Press – in particular Andrew Richford and Shammima Cowan for their help and advice throughout; and my copy editor Maggie Beveridge.

I am especially grateful to Mr A.C. Wheeler, formerly of the British Museum (Natural History), for kindly reading and commenting on my manuscript, for answering various queries, and for pointing out my various errors of commission and omission. Responsibility for any that remain, however, is of course mine alone.

Finally, my thanks are due to Mrs R. Rugman and Mrs P. Berry for their patience in deciphering and typing my well-nigh illegible manuscript.

Christopher Lever
Winkfield, Berkshire, 1996

INTRODUCTION

Since early times, man has attempted to adapt and shape the world in which he lives to suit his own perceived requirements. Often these needs are more imagined than real, but this has not deterred man in his wishes for change.

One of the ways in which man has sought to modify the natural environment is by the introduction of animals and plants throughout the world. Although major introductions of exotic fishes into countries outside their natural range are a relatively recent phenomenon, in Europe introductions of some species are believed to date from Roman times, when Common Carp *Cyprinus carpio* from the River Danube were reared in ponds in Italy and western and southern Greece (Thienemann, 1950; Balon, 1974, 1995). In the Middle Ages (*c.* AD 600–1500) mediaeval monks transported Common Carp and, to a lesser extent, Perch *Perca fluviatilis*, throughout Europe, where they were widely cultured in monastic fish-stews and elsewhere for fast-day food, from where they inevitably escaped into nearby lakes and rivers (Welcomme, 1984; Bartley & Subasinghe, 1995). From around the middle of the 19th century, when advances were made in the artificial breeding of salmonids, in particular Rainbow Trout *Oncorhynchus mykiss*, up to the outbreak of the Second World War, international transfers of fish species, especially for sporting purposes and the provision of an additional food supply, increased apace. Introductions for these purposes have sometimes been associated with those for sentimental or nostalgic reasons, especially by colonists to North America, the Antipodes, and elsewhere in the Victorian era. Although much of the spread of the Rainbow Trout stems from this period, many of the species introduced failed to survive due to unsuitable conditions; one such was the Tench *Tinca tinca* released in the waters of tropical Asia and Africa, although some populations did manage to become established in India, Indonesia, South Africa and elsewhere (Welcomme, 1984).

After the end of the Second World War the number of introductions of alien fish species increased still further, helped by the development of advanced artificial spawning techniques which made feasible the movement of species other than salmonids that were unlikely to reproduce naturally outside their native range (Welcomme, 1984; Bartley & Subasinghe, 1995). Current records, reviewed by Welcomme (1992), indicate that the volume of fish introductions reached a peak during the early 1960s, but that mean rates from 1950 to 1985 do not show any significant change, from which it might be concluded that during that period there was little or no diminution in the rate of the transfer between countries of fish species.

Welcomme (1992) found that after 1985 there was a sharp fall in the number of international introductions; this, he concluded, may have been due to either (i) the delay in the input of data into the records; (ii) a real decline in the volume of introductions, following the enactment in many parts of the world of disabling preventative legislation (such as in Britain the Salmon and Freshwater Fishes Act, 1975 and the Wildlife and Countryside Act, 1981); or (iii) the reaching of a saturation point whereby a species has been introduced to all potentially suitable host countries. Under European Community legislation, however, species continue to be introduced into member countries, and in Britain there is still an extensive trade in fish either under ornamental licences issued by the Ministry of Agriculture, Fisheries & Food, or illegally.

Trends in Introductions

Welcomme (1992) analysed the trends in introductions of exotic fishes. He found that transfers of European cyprinids (mainly Common Carp), poeciliids, centrarchids and salmonids, took place at a relatively early stage, and have since been succeeded by those of cichlids and Asian cyprinids. In the 1980s, species from an increasing number of other families were subject to international transfers to satisfy the growing demands of the tropical fish trade.

In Europe, North America, Australasia and Oceania, introductions of alien fishes peaked towards the end of the 19th century and again, after a lull, in the 1960s and 1970s. Introductions into Asia (the Middle East) and Africa reached a zenith in the 1950s, and have since declined. The rates of introductions to the rest of Asia and to South and Central America, however, have continued to increase, and these continents are now the principal hosts of exotic fish species (Welcomme, 1992).

The fact that European countries and many island states throughout the world are important recipients of alien fish species is ascribed by Welcomme (1992) to the fact that both have generally impoverished ichthyofaunas and that introductions have been made, with a variety of motives, to increase their ichthyological biodiversity. It is, however, as Welcomme points out, difficult similarly to explain the many introductions to Asia and to North, South and Central America, all of which have rich native fish faunas, and from where indigenous species have been widely exported.

The naturalization of some of the alien fish species that are able to reproduce successfully in the wild has had catastrophic consequences. Some introductions, on the other hand, have been extremely beneficial. Chile, for example, which also supports naturalized populations of all three species, is currently the second leading producer of farmed salmonids in the world, following the introduction of Coho Salmon *Oncorhynchus kisutch*, Atlantic Salmon *Salmo salar*, and Rainbow Trout, and the Chilean export salmonid aquaculture industry provides both valuable foreign exchange and also much-needed employment in rural communities (Bartley & Subasinghe, 1995). The effects of the most widely introduced species – the Common Carp, Nile Tilapia *Oreochromis niloticus*, Mozambique Tilapia, *O. mossambicus*, Rainbow Trout, Largemouth Bass *Micropterus salmoides*, Mosquitofish *Gambusia affinis* and Grass Carp *Ctenopharyngodon idella* (all of which have become virtually cosmopolitan within the limits of their thermal tolerance) have on the whole been mixed.

Motives for Introductions of Alien Fishes

Alien fishes have been introduced with a variety of motives, which have been classified by Welcomme (1984, 1988, 1992) as follows (with adaptations):

Aquaculture
Sport (and the provision of familiar species for nostalgic reasons by early colonists).
Improvement of wild stocks (by the establishment of new food fisheries; to fill a vacant ecological niche; as forage for other (often also alien) fishes; to restore fisheries; to establish a naturalized stock; to control stunted populations; and for species conservation).
Ornament (through introductions by commercial fish farms and private aquarists).
Biological control (of aquatic weeds, mosquito larvae, snails, phytoplankton blooms, and other fish).
Accidentally (by escapes (or releases) from aquaculture, aquaria, or of bait fish; by natural diffusion; through importation with other species; as eggs, juveniles, or adults in ballast water of sea-going vessels; and through live transportation for human consumption).

Welcomme (1988, 1992) has analysed the relative importance of the above categories. He found that introductions made for aquacultural purposes have always comprised a significant proportion of the total, and have steadily increased in importance; since the early 1970s such importations have accounted for well in excess of 50% of all introductions made. Introductions for sporting purposes formerly provided the second major motive for international transfers of fish species, but since the late 19th century this has steadily declined in importance, and since the early 1950s has ceased to be a significant motive. Between about 1870 and 1910, the improvement of wild stocks in lakes, ponds, rivers, streams and artificial impoundments was also of prime importance, and increased in volume between 1950 and 1979, when it overhauled sport and recreational fishing as the second most important motive.

The biological control of noxious 'pest' organisms by exotic fishes is a more recent phenomenon, which reached a peak in the 1920s and again between 1960 and 1979. Most of the fish introduced for ornamental purposes have been imported by private aquarists or by tropical aquarium fish farms, by whom many were either deliberately released or from whom they accidentally escaped into the wild.

Welcomme (1984, 1988, 1992), from whom much of the following is derived, has also analysed the composition of the more important of these categories.

Aquaculture

Introductions of fish species from one country to another before the 20th century mainly involved such freshwater salmonids as the Rainbow Trout (MacCrimmon, 1971), Brown Trout *Salmo trutta* (MacCrimmon & Marshall, 1968) and Brook Trout *Salvelinus fontinalis* (MacCrimmon & Scott Campbell, 1969); these were introduced into both temperate regions and to waters at high elevations in the tropics for rearing in aquaculture in conjunction with the establishment of sport fisheries and the provision of an additional supply of food. Since 1970, introductions of salmonids have been largely of anadromous species and forms for mariculture breeding in cages. Common Carp reached their acme of popularity between around 1910 and the outbreak of the Second World War, and were successively replaced by tilapias from about 1950–1979 and by Chinese carps between 1960 and 1980.

Alien species have played an important part in the development of aquaculture around the world. The rearing of Common Carp in monastic fish-stews in Europe for fast-day food in early times paved the way for present day intensive commercial aquaculture of Rainbow Trout in Europe, cage culture of Atlantic Salmon and Pacific Salmon *Oncorhynchus* spp. in Chile, the culture of various species of tilapia in Brazil, Cuba, Sri Lanka and Thailand, and that of Chinese carp species throughout the world, and have all involved the use of exotic species. Many alien fishes introduced for aquacultural purposes escape from confinement into natural waters, in some of which they succeeded in becoming naturalized.

Sport

International movements of fishes for sport fishing comprise those species (principally salmonids) valued for their fighting qualities (and for their flesh). The most widely disseminated of these are the Rainbow and Brown Trout (introduced mainly between about 1900 and 1930) and, among centrarchids, the Largemouth Bass introduced mainly between 1930 and 1950 (Robbins & MacCrimmon, 1974). Several other species of lesser importance have also been introduced to provide diversity for the recreational angler.

Improvement of Wild Stocks

The principal objective for the introduction of exotic fishes under this heading is the foundation of a new commercial or subsistence fishery, usually in ichthyologically impoverished regions such as temperate ones which suffered glaciation during the Ice Age (*c.* 60 000–20 000 years BP), islands east of Wallace's Line, or high elevation montane waters. Introductions have also been made into newly-created artificial habitats (man-made impoundments such as dams and reservoirs) in which autochthonous species were unable to establish themselves, and to replace such species that died out as a result of environmental changes; in the latter category are those salmonids introduced into the Great Lakes of North America.

The introduction of one species often results in the necessity of following it with the introduction of another. When piscivorous species, for example, are introduced into fish communities which have not evolved in the presence of such predators, the consequential decline or extinction of indigenous species has frequently rendered the introduction of replacement food fishes essential. Examples are provided by the introduction of sunfishes *Lepomis* spp. to mitigate the impact of introduced black bass *Micropterus* spp.; the introduction of tilapias or *Cichlasoma* spp. as forage for the Tucunare *Cichla ocellaris*; and the introduction to the Salton Sea in California of the Bairdiella *Bairdiella icistia* as forage for the Orangemouth Corvina *Cynoscion xanthulus*. Conversely, piscivorous aliens have been widely introduced to control, or if possible to eliminate, stunted populations, most often of tilapias and sunfishes. When introduced herbivores such as Common Carp or Grass Carp cause eutrophication, changing organisms responsible for primary production from macrophytes to phytoplankton, Silver Carp *Hypophthalmichthys molitrix* have been introduced to curb the excessive growth of algal blooms.

At least three species have been introduced outside their natural range as a means of conserving threatened populations – mostly without conspicuous success. Thus the Huchen *Hucho hucho* has been widely but unsuccessfully transferred from the Danube basin to countries of western Europe and with limited success to Morocco (Welcomme, 1988), while the Mohave Tui Chub *Gila bicolor mohavensis* and the Arroyo Chub *G. orcutti*, both natives of California, have been translocated elsewhere within the USA and introduced to Mexico, with uncertain results (Contreras & Escalante, 1984).

Several species have been successfully introduced to establish important new commercial (rather than sporting) fisheries. Examples include tilapias (mainly

Nile Tilapia) introduced to rivers and lakes in South East Asia and South and Central America; *Cichlasoma* spp. and *Micropterus* spp. in lakes in Central America; Argentine Pejerryes *Odontesthes bonariensis* and Rainbow Trout in Lake Titicaca (on the borders of Bolivia and Peru); and the Tawes *Barbus gonionotus* in rivers of the Philippines and Celebes.

Ornament

Fishes introduced for ornamental reasons fall into two main categories. In the first are species such as the Goldfish *Carassius auratus* which has been widely distributed outside its eastern European, central Asian, and Chinese natural range, for breeding in ornamental fish ponds, from which it has frequently escaped to become established in natural waters.

In the second category are the innumerable species of small, mostly tropical, species which have been widely dispersed around the world by the unfortunately flourishing aquarium fish trade (Lachner *et al.*, 1970). While the majority of these species are unable to survive under normal conditions in temperate climates (which in general prevail in most of the countries to which they have been introduced), some have been able, albeit usually temporarily, to establish breeding populations in artificially heated waters discharged from factories or power stations and in naturally warm waters such as thermal hot springs. Populations of the Guppy *Poecilia reticulata* and the Redbelly Tilapia *Tilapia zillii* reported by Wheeler & Maitland (1973) and Lever (1977) to be established in artificially heated waters in England provide examples in the former category. The diversity of species reported by McAllister (1969) in the hot springs of the Banff National Park in Alberta, Canada, is the prime example in the latter category.

In the tropics, where the increasing fashion for ornamental fishes has led to their widespread culture for exportation to Brazil, Colombia, Malaysia, Peru, Singapore, Thailand and the USA (Florida), the situation is much more serious. Escapes from tropical aquarium fish farms in the host nations have been frequent, and are almost certainly responsible (possibly augmented by deliberate release of unwanted pet fish by private aquarists) for the many small tropical exotics established in Colombia, the Hawaiian Islands, and the continental USA (principally Florida). Because many of these escapes and releases are not reported to the authorities, the number of such naturalized species is probably far greater than has been recorded.

Biological Control

For many years exotic fish have been used for the biological control of unwanted (pest) aquatic organisms, for which they have numerous advantages compared with the alternative of chemical control with its concomitant environmental dangers.

Early attempts in the 1920s concentrated on the control of mosquito larvae, for which Sailfin Mollies *Poecilia latipinna* and Guppies were among the species most frequently used. By far the most important, however, was the aptly named Mosquitofish *Gambusia affinis*. Nowadays, however, in many countries the Mosquitofish is gradually being replaced by native larvivorous fishes (Haas & Pal, 1984), and any further significant spread of Mosquitofish seems unlikely.

The introduction of molluscivorous fishes has been proposed for the control of the aquatic snail vector of schistosomiasis (bilharzia). In the 1960s, the cichlid *Astatoreochromis alluaudi* was introduced for this purpose from Uganda with some success to aquaculture ponds and irrigation canals in Cameroun, the Central African Republic, the Congo, Zaire and Zambia (Jhingran & Gopalakrishnan, 1974). In spite of the presence of numerous mollusc-eating fish species, control in larger water bodies (both natural and artificial) seems virtually ineffective, though little research seems to have been done on the subject.

In the 1960s and 1970s, exotic herbivorous fishes began to be used to control aquatic vegetation, and Shireman (1984) outlined a range of specialized phytophagous habits that could be utilized to form sophisticated herbivorous fish communities. Several species, of which the Grass Carp is the prime example, have been introduced outside their natural range to control excessive growths of aquatic weeds; others include the Mozambique Tilapia, the Redbreast Tilapia *Tilapia rendalli* and the Redbelly Tilapia. The Silver Carp has been widely introduced specifically to control excessive growths of phytoplankton.

Accident

The numerous cases of escapes or releases of species from captivity that have resulted in the establishment of naturalized populations have almost invariably proved ecologically and/or economically detrimental. Exceptions are provided by (i) the Giant Gourami *Osphronemus gouramy* which, introduced to Madagascar in an unsuccessful attempt to rear it in aquaculture, now makes a valuable contribution to catches in the coastal Canal des Pangalenes (see Glossary); (ii) the Tawes is fished for commercially in Celebes; and (iii) the Oriental Weatherfish *Misgurnus anguillicaudatus* is the basis of a fairly valuable fishery in upland rivers in the Philippines.

After escapes or releases from captivity, natural diffusion via freshwater waterways is probably the most common means whereby alien species have spread to new countries. Thus, for example, Common Carp introduced to Brazil travelled down the Uruguay River into Uruguay and Argentina, and Courtenay and Hensley (1979) showed how quickly a species, the Spotted Tilapia *Tilapia mariae* in southern Florida, can spread in an alien environment. Similarly, the Eastern Mudminnow *Umbra pygmaea* and North American catfishes *Ictalurus* spp. spread by diffusion in the lowland waters of northern Europe, while numerous small cyprinids introduced to waters in the Danube basin have spread throughout that river system. Expansion of an alien's range by natural diffusion is particularly dangerous in the case of euryhaline species, such as for example the Mozambique Tilapia, which is capable of spreading along coastal waters and can even survive in brackish or saltwaters.

Another fairly common accidental means by which alien fishes are introduced outside their natural range is when species, most frequently cyprinids, are inadvertently transported with the juveniles of Common Carp or Grass Carp; in this way the Stone Moroko *Pseudorasbora parva* appeared in the River Danube, the Bleak *Alburnus alburnus* and Silver Bream *Blicca bjoerkna* in Cyprus, and several cyprinids and eleotrids in the waters of Tashkent, Uzbekistan, in the former USSR. Between 1935 and 1938, a *Haplochromis* sp. was accidentally introduced with Nile Perch from Lake Bunyoni in Uganda to Lakes Bulera and Luhondo in neighbouring Rwanda (De Vos *et al.*, 1990).

Ecological Impact of Naturalized Fishes

The ecological effect of naturalized fishes on native aquatic communities has been summarized by Taylor *et al.* (1984) as follows:

Habitat alterations (through the removal of vegetation (by consumption, uprooting, and increased turbidity) and the degradation of water quality (by siltation, substrate erosion, and eutrophication)).
Introduction of parasites, pathogens, and diseases
Trophic alterations (through forage supplementation, competition for food, and predation).
Genetic degradation (through hybridization).
Spatial alterations (through aggression and overcrowding).
Environmental effects
Socio-economic effects

Particularly with naturalized fishes, it is often difficult, or in some cases well-nigh impossible, accurately to assign individual causes to specific impacts, because frequently more than one is involved. Although the effects of introductions are, in general, hard to predict, exotic fishes are most likely to become naturalized (i) in a mild climate; (ii) in disturbed or man-made habitats such as reservoirs and canals; and (iii) in communities with a low species diversity.

Habitat Alterations

Alterations in habitat composition by naturalized fishes involve principally the displacement of aquatic vegetation and the degradation of water quality: the former can be brought about by the consumption of plant material by herbivorous species, by the uprooting of macrophytes through digging for food or nesting sites, and by roiling and organic enrichment which increase turbidity and thus reduce light penetration and photosynthesis. Modification of aquatic plant communities can significantly affect native fishes and other animals. Plants provide a variety of facilities, such as (i) oxygen, substrate and nutrients for invertebrate species that may be an important prey resource for fishes; (ii) nesting sites; (iii) cover for early life stages of various native animals and for adult fishes; and (iv) feeding places for waterbirds and aquatic mammals (Taylor *et al.*, 1984). Grass Carp have been widely introduced to control aquatic vegetation, but by feeding selectively on tender species may enhance the development of tougher plants, which can prove an even greater problem. Moreover, the removal of underwater vegetation in which phytophilous fishes spawn may inhibit their breeding. Nevertheless, habitat alterations may sometimes be beneficial, in that they can lead to the colonization by exotics of waters that were unsuitable for native species (Welcomme, 1988).

Reduction of macrophytes may promote increased turbidity (to the disadvantage of native species accustomed to living in clear water) through wave-mediated erosion and the continual mixing of silt previously stabilized by the roots of aquatic vegetation (Boyd, 1971). The roiling of shelving shore areas by bottom-feeding fishes such as the Common Carp and the nesting and spawning activities of shoaling species can result in a deterioration in water quality, albeit perhaps only temporarily, especially in a reduction in clarity (Taylor *et al.*, 1984). This shades out macrophytes, chokes benthic invertebrate species, and contributes to accelerated eutrophication through the speedier recycling of phosphate, as a result of which the composition and abundance of native fish fauna may be changed (Welcomme, 1988). These effects are seldom observed where Carp are long-established, probably because the problems usually coincide with population 'explosions' that so frequently follow the naturalization of an alien species (McClaren, 1980). Accumulations and subsequent decay of organic waste material may promote phytoplankton bloom, which in turn can result in imbalances in mineral and nutrient cycling in the ecosystem (Hestand & Carter, 1978). However, Welcomme (1984) believed that by ingesting organisms from the phosphate-rich substratum and subsequently excreting phosphate, carp help to mobilize this nutrient in a more soluble format.

Increased turbidity can disrupt the reproduction process of native fishes (e.g. by causing the desertion of nests, the elimination of spawning, and the loss of eggs) and other aquatic organisms and result in physiological stress which may affect normal respiratory and secretory functions. Concern about the destructive impact that naturalized fishes can have on water quality is confirmed by numerous records in the literature that document a correlation between habitat degradation (usually a consequence of siltation and turbidity) and a reduction in the numbers and diversity of species in native freshwater fish faunas (Taylor *et al.*, 1984).

All these effects can, of course, occur in the absence of exotic species, but are likely to increase or appear when they are present.

Introduction of Parasites, Pathogens and Diseases

In contrast to other ecological effects, the importation of parasites, pathogens and diseases can be made via exotic fish never intended for release into the wild. Thus, nematode parasites of the genus *Anguillicola*, which is endemic to Australian and Asiatic *Anguilla* spp., have been introduced into Europe with oriental eels intended for human consumption and not for stocking purposes. A second introduced disease, probably originating in ungutted Pacific salmon carcasses, is infectious haematopoietic necrosis (Welcomme, 1984, 1988).

Many diseases of salmonids that infect hatchery-reared fish and also occur in wild populations have been imported. Rainbow Trout from western North America have carried furunculosis to Europe and South America. In the same way, Common Carp from the former Yugoslavia were responsible for the appearance in the 1930s of infectious dropsy in cyprinids in western Europe. More recently, the North American Fathead Minnow *Pimephales promelas* has been proved to have introduced *Yersinia ruckeri*, which is the causative agent of redmouth disease to parts of northern Europe. Langdon (1990) reported on the disease risks associated with the introduction of exotic fish species, with special reference to Australia.

The pathogens of several diseases, such as infectious pancreatic necrosis, infectious haematopoietic necrosis, and bacterial kidney disease, can be transmitted vertically via the gametes, so unfertilized ova, sperm, eggs and embryos, as well as adult fish, are all potential vectors.

Some pathogens, such as the coccidian protozoa *Eimeria* and *Goussia*, are considered group specific, but many so regarded are in fact not. Thus the introduction to Victoria in the 1970s of Goldfish from Japan infected with *Aeromonas salmonicida* in turn infected captive and wild Australian Goldfish and Koi Carp (an ornamental variety of the Common Carp) with the bacterium of Goldfish ulcer disease, which Carson & Handlinger (1988) and Whittington & Cullis (1988) have shown to be extremely pathogenic for salmonids.

The importation of pathogens that are not group specific is the greater risk associated with the introduction of exotic species (Langdon, 1990). Thus, a virus of unknown origin, epizootic haematopoietic necrosis, which in the wild affects Perch, in 1986 began to infect reared Rainbow Trout in Australia; Langdon (1989) has shown it to be highly pathogenic for such autochthonous species as Silver Perch *Bidyanus bidyanus*, Mountain Galaxias *Galaxias olidus*, Macquarie Perch *Macquaria australasica* and, to a lesser extent, Murray Cod *Maccullochella macquariensis* and introduced Mosquitofish. Other pathogens that have been introduced to wild native fishes by imported aliens include (i) *Lernaea cyprinacea*, now of frequent occurrence in various species; (ii) *Chilodonella cyprini* and perhaps *C. hexasticha*, which kill Bream *Abramis brama*, Largemouth Bass and Murray Cod; and (iii) *Costia*, *Ichthyophthirius* and *Trichodina* spp. (Langdon, 1988).

Pathogens are frequently more serious in atypical hosts, and thus occur when such hosts come in contact with typical hosts. Examples include whirling disease in Rainbow Trout, caused by *Myxosoma cerebralis*, a normally harmless parasite of Brown Trout *Salmo trutta*, and proliferative kidney disease of salmonids. The myxosporean *Triangula percae*, which causes curvature of the spine in Australian Perch *Perca fluviatilis*, is unknown in the species' European range (Langdon, 1990).

To try to combat the dangers posed by pathogens imported to Australia with alien fishes, Langdon recommended rigorous health testing and certification, preferably before fish are moved; at the same time he conceded that such precautions would at best only reduce, rather than eliminate, the risk incurred.

Trophic Alterations

Naturalized fishes can alter trophic relationships in aquatic communities in at least three different ways, all of which may cause changes in the populations of native species. First, their presence may significantly increase the amount of prey available to native predators. Second, the feeding habits of naturalized fishes can reduce the amount of forage available to native species through a dietary overlap; if food supplies are low, this can result in exploitation competition which may cause changes in diet, growth, condition and breeding success among native species. Grazing by naturalized herbivorous fishes may result in the reduction or clearance of weed beds which support an aquatic community that is a source of food for native species. Finally, naturalized predatory fishes can profoundly affect the population dynamics of indigenous prey species (Taylor *et al.*, 1984).

The reduction in the population of an autochthonous species can sometimes be difficult to attribute with certainty to predation or competition from an exotic, and on occasion both influences may act in concert; Welcomme (1984, 1988) provides several examples. Thus, Largemouth Bass introduced to Lago de Patzcuaro and other lakes in Mexico have greatly reduced the numbers of a popular food fish, the Blanco de Patzcuaro *Chirostoma estor*, and of several species of Goodeidae, and of native fish populations in Lakes Atitlan and Calderas in Guatemala; elsewhere, the same exotic has seriously affected a number of (alien) cichlid species in Lake Naivasha, Kenya, as well as Bleak *Alburnus alburnus*, Northern Pike *Esox lucius*, Perch, and naturalized sunfishes (*Lepomis* spp.) in the oligotrophic glacial lakes of northern Italy; in some Zimbabwean waters Rhodesian Mountain Catfish *Amphilius platychir* have been locally eradicated. The disappearance of Northern Pike and Perch may have been as a result of competition; the impact of predation by Largemouth Bass in other waters has probably been avoided by the coetaneous introduction of forage fish such as *Lepomis* spp. (Welcomme, 1988).

In Lake Lanao in the Philippines, the naturalized goby *Glossogobius* sp., Walking Catfish *Clarias batrachus* and Largemouth Bass all prey on native endemic cyprinids (Payne, 1987).

As Welcomme (1984, 1988) points out, salmonids – especially Brown and Rainbow Trout – have, partly because of their exceptionally wide naturalized distribution, one of the worst records for damaging native species of fish. Rainbow Trout have been at least partly responsible for the reduction of indigenous salmonids in Lake Ohrid, in the former Yugoslavia (Nijssen & de Groot, 1974), of the rare Maluti Minnow *Oreodaemion quathlambae* in Lesotho, of *Schizothorax* spp. in Himalayan rivers in Nepal, and with Brown Trout have been implicated in the decline of the endemic Dwarf Inanga *Galaxias gracilis* and New Zealand Grayling *Prototroctes oxyrhynchus* (McDowall, 1984), and other galaxiids in Australia. Although Brown Trout have not had as severe an ecological impact as Rainbow Trout, McDowall (1984) attributes the disappearance of the Giant Kokopu *G. argentus* and Dwarf Galaxias *G. divergens* from some New Zealand waters to the presence of Brown Trout. Jackson (1960) found that Brown and

Rainbow Trout and Largemouth Bass had eradicated Haarder *Trachyistoma euronotus* and Cape Kurper *Sandelia capensis* in some inland South African waters.

In Lake Titicaca in the Peruvian Andes trout – principally the Rainbow Trout – and the Pejerrey *Odontesthes bonariensis* have together contributed to the decline of tooth carp species (*Orestias* spp., especially *cuvieri* and *pentlandi*) and catfish (*Trichomycterus rivulatus* and *T. dispar*). The exact cause of the decline was obscured by the possibility of predation, the pressure of overfishing on both naturalized and native species, and competition for food (Welcomme, 1988). Subsequently, the trout populatiids were themselves controlled by interspecific competition with the autochthonous species for benthic invertebrates which were their mutual food, and *Orestias* spp. still comprise 5640 t (94%) of the 6000 t of fish caught annually in Titicaca (Welcomme, 1988).

According to Deacon *et al.* (1964), exploitation competition for food following the introduction of the Guppy into the southwestern USA was probably responsible for the decline of the cyprinodont White River Springfish *Crenichthys baileyi*, as it has been for that of local cyprinodonts in Lake Victoria, Uganda. Welcomme (1967) attributed the decline of *Tilapia variabilis* in Lake Victoria to interference competition for nursery beaches with introduced Redbelly Tilapia *T. zillii*.

Sometimes, the apparent initial impact of an introduced predatory fish subsequently needs reassessing; thus Tucanares *Cichla ocellaris* introduced in the Chagres River in Panama later spread into Gatún Lake, where they were reported by Zaret & Paine (1973) to have eradicated the Silverside *Melaniris chagrensis*, four characin species and two poeciliids; piscivorous birds also disappeared, and a sudden resurgence of human malaria in the area was attributed to the spread of Tucanares, which prey on the small fish that used to eat mosquito larvae (Payne, 1987). According to Welcomme (1988), however, all the fish species subsequently recolonized the lake from refugia. Similarly, the significance of Nile Perch *Lates niloticus* in Lake Victoria may need to be reappraised.

Predation by large piscivorous fishes on adult prey may be matched by predation of smaller species on juveniles, and some cyprinodonts such as Mosquitofish are reported to feed on fish eggs. Although the significance of this has not been quantified nor harm to any species recorded, Welcomme (1988) reported

concern about possible depletion of stocks from such predation.

Interspecific exploitation competition between naturalized and native species can result in the exclusion of the latter; Welcomme (1988) draws attention to the cases of *Opsariichthys uncirostris*, *Hypseleotris swinhonis*, *Hemiculter leucisculus*, *H. eigenmanni*, *Percottus glehni*, *Pseudorasbora parva* and Grass Carp accidentally introduced to waters in Tashkent in the former USSR. Similarly, the Jaguar Guapote *Cichlasoma managuense* has removed local predators in El Salvador, and in some southern states in the USA autochthonous species have been displaced by naturalized tilapiine cichlids (Welcomme, 1988).

The quality of fish stocks can also deteriorate by 'stunting', described by Welcomme (1988) as 'a process whereby the population of a species expands rapidly, producing large numbers of individuals which mature and breed at a much reduced size'. This phenomenon occurs both in the wild and in captivity, and severely diminishes the sporting or commercial value of the species concerned. Species known to be subject to 'stunting' include the Bleak, Perch and Silver Bream *Blicca bjoerkna* in Cyprus; Redbreast Sunfish *Lepomis auritus*, Pumpkinseed *L. gibbosus* and catfishes (*Ictalurus* spp.) in France, The Netherlands and Italy; the Stone Moroko *Pseudorasbora parva* in the Danube basin and parts of the former USSR; and especially the Goldfish and Common Carp in India, South Africa, the USA and Australia (Welcomme, 1984). Other species known to suffer from 'stunting' are Green Sunfish *L. cyanellus*, Bluegill *L. macrochirus*, Redbreast Sunfish, Redbreast Tilapia, Nile Tilapia and Mozambique Tilapia; the last-named 'stunts' the most readily, and its eradication has been attempted – usually unsuccessfully – in many waters (Welcomme, 1988). 'Stunting' among tilapiine cichlids has greatly reduced their value for establishing wild commercial fish stocks and for raising in captivity; 'stunted' populations can overwhelm existing ones, and may even cause a shortage of oxygen, and for the same reasons prevent the introduction of new and more suitable species (Welcomme, 1988).

Genetic Deterioration Through Hybridization

Fishes (poikilotherms) have, in general, a greater potential for successful hybridization without sterility than mammals and birds (homoiotherms), and may produce long-lasting hybrids in the wild. Naturalized species may thus interbreed with either native congeners or with other exotics, especially, in the latter case, where one of the aliens is rare in comparison with the other. Under the pressures exerted through introduction, normal behaviour patterns may be abandoned and hybrids arise from species or genera that do not normally interbreed (Welcomme, 1988). Thus Maciolek (1984) refers to hybridization between the Largemouth Bass and the Bluegill in the Hawaiian Islands. If either of the two species involved possesses especially desirable characteristics, these will in all probability be weakened or even lost (Taylor *et al.*, 1984).

In Africa and South America especially, a considerable amount of hybridization, in particular among tilapiine cichlids, has taken place. Although there may frequently be an imbalance of sexes among hybrid progeny, hybridization does take place among native species in the wild, and because these hybrids are fertile, new genotypes are formed. Thus in Lake Itasy, Madagascar, parental populations of Nile and Longfin Tilapia *Oreochromis macrochir* were all but replaced by a hybrid known locally as '*tilapia trois quarts*'. Subsequently, Nile Tilapia established an equilibrium with the new hybrid, but Longfin Tilapia have virtually disappeared (Welcomme, 1984).

Other cichlid genera, such as *Cichlasoma*, and indeed other families, also readily interbreed under natural conditions. De Groot (1985), for example, describes the widespread interbreeding in the Netherlands of various strains of the Common Carp (Cyprinidae), and Welcomme (1988) refers to the concern that has been evinced in Europe about the translocation of hatchery-reared Salmonidae that are genetically inferior to wild-bred stock with which, if they escaped into the wild, they might interbreed. Similar fear has been expressed about the escape of captive-reared Atlantic Salmon from their sea-loch pens in Scotland, and especially of the threat to wild Atlantic Salmon from the proposed introduction to Britain of the Coho Salmon *Oncorhynchus kisutch* (Cooke, 1977). Among the Poeciliidae, the introduction of the Green Swordtail *Xiphophorus helleri*, Southern Platy *X. maculatus*, and Variegated Platy *X. variatus* to Mexico resulted in the genetic overwhelming of the autochthonous Monterey Platy *X. couchianus* which is now classified as endangered (Welcomme, 1988). When closely related species,

whose specific integrity is maintained by geographic barriers, are involved, there is always a danger of hybridization. Thus in many of the glacial lakes of Europe the identity of local stocks has been confused by the naturalization of various *Coregonus* species (ciscoes and whitefish) (Welcomme, 1984).

Spatial Alterations

In abnormally high numbers, interactions between native and naturalized fishes may be of significance to the distribution, density and existence of the former. Interference competition arising from aggression and territoriality may cause indigenes to be displaced from favoured microhabitats such as feeding grounds, nesting sites and refugia (Taylor *et al.*, 1984). In exceptional cases, overcrowding by naturalized species may inhibit breeding among harassed natives, as for example in the case of sunfishes (*Lepomis* spp.), and Largemouth Bass and other indigenous fishes (Swingle, 1957). Other naturalized species that can cause retarded growth and development and diminished fecundity in autochthonous populations include the Common Carp and tilapias.

Environmental Effects

As well as the various effects of naturalized fishes on the environment described above, Welcomme (1984) suggests that, vice versa, the environment can be of significance to introduced species even when hydrological and climatic conditions appear suitable. It is axiomatic that the more diverse the autochthonous fish community and the more complex the limnological ecosystem into which an alien species is introduced, the less will be its immediate significance. There is, for example, as Welcomme points out, a conspicuous lack of success in exotics becoming naturalized in species-rich tropical waters; he cites the cases of Nile Tilapia in Thailand, and Common Carp in competition with native tilapias in Madagascar as examples of exotics that initially apparently flourished but eventually disappeared. Payne (1987), however, points out that between 1931 and 1973 Mozambique Tilapia and Tawes *Barbus* (*Puntius*) *javanicus*/*gonionotus* were successfully introduced to species-rich Ranu Lamongan, a lake in Indonesia, where they now support a thriving fishery without having had any apparent ecological effects. Tilapia have also proved economically invaluable elsewhere in South East Asia, most notably in Sri

Lanka (but see below under Socio-economic Effects), where before their introduction a mature man-made lake yielded an annual average of only 1.07 kg of fish per hectare; after the establishment of tilapia the annual yield rose to 445 kg per hectare. Overall, by the mid- to late 1980s artificial waters in Sri Lanka produced a yearly average of 918 kg of fish per hectare, 99% of which consisted of naturalized tilapia which, because they feed on plankton, do not have to compete with native species (Payne, 1987). The most successful naturalized fishes are usually established where indigenous fish communities are either comparatively fragile or are composed of relatively few species, or which are already under the influence of overfishing or environmental disturbance (Welcomme, 1984).

This last factor could also be responsible for the success of such introductions as Common Carp and tilapiine cichlids in man-modified habitats such as reservoirs or in rivers downstream of dams, where native fishes have been unable to adapt to the new conditions, rather than the alteration of the ecosystem by the introduced fish to suit themselves (Welcomme, 1984). This suggests that some exotics may remain quiescent until such time as the autochthonous fish community becomes weakened by excessive fishing, pollution or disturbance of the environment by man. When an exotic becomes quickly naturalized and rapidly increases, Welcomme suggests that in many cases this is because initially it was able to make use of some hitherto unexploited food resource; once the alien comes into equilibrium with the ecosystem, the population declines and eventually the species merges into the native biota.

Socio-economic Effects

In addition to the ecological impact of naturalized fishes outlined above, some species have also on occasion been of socio-economic significance; this is especially so when a naturalized species not favoured for human consumption replaces a popular food species. Welcomme (1988) cites the example of Mozambique Tilapia successfully established in reservoirs in India, where although they did not supplant

a native species they were not accepted by the human population. This has been a not infrequent occurrence with some tilapiine cichlids, whose small size and poor flavour makes them widely unacceptable. Ravichandra Reddy *et al.* (1990), however, say that *O. mossambicus* in inland waters of India has contributed significantly to the development of freshwater fisheries, but in a few reservoirs has also been responsible for a decrease in the number of some endemic carp species.

Where local economies are dependent on the rearing of fish in captivity for human consumption, their financial vulnerability is considerable. Disease is probably the principal danger to artificially bred fish stocks, but other problems, due to the genetic integrity of the stock, also exist. Aquaculturists are largely dependent on exchanging genetic material of their stock by means of translocations and introductions, and care must be taken to minimize the inherent risks (Welcomme, 1988).

The influence of Nile Perch introduced to Lake Victoria, East Africa, on native fisheries is discussed in the species' account.

Of the 221 species of introduced fish listed by Welcomme (1988) only 165 (75%) have become successfully naturalized. In fact the percentage is probably considerably lower, since there must be numerous cases of failed introductions that have not been recorded.

There is, as Welcomme (1984) rightly says, a clear distinction between the criteria employed by industrialized and undeveloped societies in deciding whether or not to introduce a new species. While the former mainly consider environmental and conservation factors, the latter not unreasonably are concerned primarily with the production of enough food for the human population. The problem with such a view, of course, is that short-term benefit may in time be replaced by longer-term damage. More careful examination of case histories of the introduction of individual species elsewhere, and their behaviour and ecology in their natural range, could help to determine whether short-term gain should be sacrificed for the sake of long-term ecological and socio-economic stability.

SYNBRANCHIDAE

Swamp Eel (Ricefield Eel; Cuchia)
Monopterus albus

The Swamp Eel is a predatory fish well able to tolerate the deoxygenated waters of swamps, ditches and paddyfields (Welcomme, 1988).

Natural Distribution

Eastern Asia from northern China to Malaya.

Naturalized Distribution

Oceania: Hawaiian Islands (Oahu).

Oceania

Hawaiian Islands

As a highly valued food fish, the Swamp Eel was successfully introduced before 1900 from the Far East to Oahu in the Hawaiian Islands, where it is now well established in a number of ponds (Fowler, 1925; Brock, 1952, 1960; Herald, 1961; Kanayama, 1968; Maciolek & Timbol, 1980; Maciolek, 1984). According to the last-named author, Swamp Eels have had a negligible ecological impact in those waters on Oahu where they have become naturalized.

Arthington (1986) lists the Swamp Eel as among those species 'that have established self-maintaining populations in Australian inland waters', though she concedes that 'only a few specimens have ever been recorded'.

CLUPEIDAE

Threadfin Shad
Dorosoma petenense

The Threadfin Shad is a small fish, averaging around 50 mm in length, with broadly based habitat requirements in the temperate waters of eastern North America. Although the species' tendency to population explosions in reservoirs has caused it to be regarded with disfavour, it has proved useful as an introduced forage fish (Welcomme, 1988).

Natural Distribution

Ohio River in Kentucky and southern Indiana, west and south to Oklahoma, Texas and Florida, along the coast of the Gulf of Mexico to northern Guatemala and Belize (Burgess, 1980).

Naturalized Distribution

Central America: Puerto Rico. *Oceania:* Hawaiian Islands (Kauai and Oahu).

Central America
Puerto Rico

On 30 May 1963, 40 adult Threadfin Shad in spawning condition were flown from Atlanta, Georgia, to Ramey Field in Puerto Rico, where they were stocked in Guajataca reservoir and where fry and juveniles were found in the following February. Since then, Threadfin Shad have been successfully introduced in Cidra, Carite, Garzas and La Plata reservoirs. Introduced to Puerto Rico as a food fish for the alien Largemouth Bass *Micropterus salmoides*, Threadfin Shad have considerably improved the condition of the predator species (Erdman, 1984; Wetherbee, 1989).

Oceania
Hawaiian Islands

The Threadfin Shad was first introduced to the Hawaiian Islands in 1958, when a small number were flown from the Chino Fish Station in California to Oahu, where they were released in a hatchery pond at Nuuanu; 53 of these fish were subsequently liberated in a quarry pond on the campus of the University of Hawaii. Later, a large number of fish were imported to Hawaii from the Puddingstone Reservoir near Chino to be used as live bait for tuna fishing. In August 1959, a further 3000 Shad from the same source were released in various places on Kauai, Oahu and Maui (Brock, 1960; Hida & Thomson, 1962; Kanayama, 1968; Randall, 1960, 1980; Randall & Kanayama, 1972; Maciolek, 1984). The last-named author records the Threadfin Shad as being of negligible ecological importance where it is naturalized on Kauai and Oahu.

Tanganyika Sardine (Kapenta (Kiswahili and Shona); Lumpu, Ndagala (Kinjiruwanda))
Limnothrissa miodon

Limnothrissa miodon forms dense pelagic populations in Lake Tanganyika, where it is endemic, which provide a valuable commercial fishery (Welcomme, 1988).

Natural Distribution

Lake Tanganyika (East Africa).

Naturalized Distribution

Africa: Rwanda (Lake Kivu); Zaire (Lake Kivu); Zambia (Lake Kariba); Zimbabwe (Lake Kariba); Mozambique (Cahora Bassa Reservoir).

Africa

Rwanda; Zaire

Between 1952 and 1954, a study of the fishes of Lake Kivu, between Rwanda and Zaire, was undertaken by a team of Belgian ichthyologists, who found an absence of pelagic species feeding on the abundant plankton. To fill this empty niche, a Belgian agronomist, A. Collart, introduced thousands of fry of plankton-eating *Stolothrissa tanganyicae* and *Limnothrissa miodon* from Lake Tanganyika to Lake Kivu between 1958 and 1960. (The former species was also introduced to the Tanzanian waters of Lake Nyasa (Moreau *et al.*, 1988).) From 1974 to 1976, the presence of *miodon* was recorded from various littoral waters of Lake Kivu, and in the latter year scientists from the University of Liége in Belgium confirmed the species' successful establishment (Frank, 1977). *S. tanganyicae* seems to have disappeared.

Ecological and Socio-economic Impacts Before the introduction of Tanganyika Sardines to Lake Kivu the zooplankton community was composed of large and heavily pigmented (thus easily visible to predators) Copepoda (crustaceans) and Cladocera (water fleas) (Dumont, 1986). By 1981, just 20 years after the Sardine's introduction, this zooplankton community had changed completely (Reyntjens, 1982; quoted by De Iongh *et al.*, 1995). At least eight rotifer species (worms) now occurred, plus four small Cladocera species, not hitherto noted. The cyclopoid species recorded previously were still present, but their average sizes had fallen by almost 50%. *Daphnia* water fleas had disappeared, but a ciliate, *Stentor coeruleus*, usually a sessile, littoral species, was abundant, as were other protozoa. One of the Cladocera, *Alona* cf. *cambouei*, hitherto recorded only in the coastal weed-beds, had become pelagic. Although the zooplankton community had diversified, the biomass and size had both decreased. These changes were interpreted by De Iongh *et al.* (1995) as a consequence of heavy size-selective predation of the plankton by Sardines.

The socio-economic benefits accruing from the introduction of Sardines to Lake Kivu are two-fold; first, because the unemployment rate is extremely high in Rwanda, a densely populated country whose economy is based entirely on agriculture and agroindustries, labour-intensive exploitation and marketing is beneficial; and second, because the harvest of fish provides the people of Rwanda with a valuable source of protein. In December 1979, the United Nations Development Programme and the Rwanda government started a commercial artisanal fishery project at Gisenyi at the northern end of Lake Kivu (Spliethoff *et al.*, 1983; Dumont, 1986). Within a few years a viable fishery had developed, and in Rwandese waters the number of fishing units increased from six in 1980 to 144 in 1991. Recorded Rwandese catches increased from 65 t in 1981 to 370 t in 1987 (De Iongh *et al.*, 1995). The postwar situation has yet to be determined.

Zambia; Zimbabwe

Before the construction of the 5400 km² artificial Lake Kariba, between Zambia and Zimbabwe, various recommendations were made regarding stocking. One of these was to utilize the open waters of the lake by the introduction of a pelagic planktivorous species, since it was correctly assumed that none of the riverine fish would disperse into this niche. Between 1964 and 1968, 360 000 *Limnothrissa miodon* fingerlings were imported by the Zambian government from Lake Tanganyika to the lake, where they soon became naturalized. Experimental fishing, which began in Zambian waters in 1969–1970, was, however, generally disappointing, due almost certainly to the low stocking rate (1 t of *miodon* requires around 1 million fish). Experimental fishing which commenced in Zimbabwean waters in 1971 was more encouraging, and a flourishing commercial fishery was soon established, which Welcomme (1988) reported to yield around 15000 t a year.

The Tanganyika Sardine appears an ideal fish for an ecosystem such as that of Lake Kariba since, being a short-lived but highly fecund and rapidly growing species, it is able advantageously to exploit seasonally abundant nutrients carried into the lake by feeder tributaries (Bell-Cross & Bell-Cross, 1971; Frank, 1977; Marshall, 1979; Thys van den Audenaerde, 1992).

The Tanganyika Sardine has also been successfully introduced into Lake Bangweulo in Zambia (Moreau *et al.*, 1988).

Ecological and Socio-economic Impacts   In Lake Kariba, predatory fish species considerably increased following the introduction of Sardines. The numbers of Tigerfish *Hydrocynus vittatus*, in particular, rose from 5 to 10% of the gill net catch, and the population of the predator appears to be at least partially dependent on the abundance of Sardines. Several other fish species in Lake Kariba are known to be partly piscivorous, and some, such as *Heterobranchus longifilis*, probably feed on juvenile Sardines in shallow water (Marshall, 1995).

The population of fish-eating birds on Lake Kariba is comparatively low, and the White-winged Black Tern *Chlidonias leucoptera* is the only species able to take live Sardines, which it does by following the nocturnal fishing boats. The Grey-headed Gull *Larus cir-rocephalus poiocephalus* has taken to scavenging Sardines on the rocks where they are sun-dried, and has now become a breeding resident on Kariba where it was previously recorded only as a vagrant (Marshall, 1995).

The most noticeable impact of zooplanktivorous fish on zooplankton communities is predation on larger species which decrease in numbers and may even be eradicated, while smaller species increase. The total biomass also decreases. In Lake Kariba, the numbers of diaptomids and larger cladocerans fell rapidly after the introduction of Sardines, and *Chaoborus* seems to have died out. Rotifers and nauplii larvae became increasingly important constituents of the zooplanktonic community (Marshall, 1991, 1995).

Marshall and Junor (1981) suggested that Sardines in Lake Kariba may have contributed to the decline of a fern, *Salvinia molesta*, whose floating mats retain large amounts of nutrients which were released when the mats collapsed between 1970 and 1975.

Karenge and Kolding (1995) could find no evidence that the introduction of Sardines to Lake Kariba had had any negative impact on inshore fish populations, nor any signs of over-fishing.

Mozambique

From Lake Kariba, Tanganyika Sardines spread naturally down the Zambezi River to colonize the Cahora Bassa Reservoir in Mozambique in 1975 (Welcomme, 1988).

Ecological Impact   Vostradovsky (1986) has recorded predation on Sardines in Cahora Bassa by Tigerfish *Hydrocynus vittata* and Butter Catfish *Schilbe intermedius*, the populations of which have probably benefited from the Sardine's colonization.

OSTEOGLOSSIDAE

Bony Tongue
Heterotis niloticus

Heterotis niloticus, the sole representative of its genus, occupies the lower reaches of rivers and their flood-plains. The species' ability to survive in severely deoxygenated waters through the possession of auxiliary branchial air-breathing organs, together with its rapid rate of growth, have made it a popular species for aquaculture in Africa. Escapes from captivity have become established in the wild, and form the basis for local fisheries (Welcomme, 1988).

Natural Distribution

Basins of the Senegal, Niger, Chari (Cameroun and Chad) and Nile rivers.

Naturalized Distribution

Africa: Central African Republic; Gabon; Ivory Coast; Madagascar; Zaire.

Africa

Central African Republic

Heterotis niloticus was introduced from the Chari River in Cameroun to the Central African Republic in 1956, where it is said by Moreau (1979) to be a 'highly appreciated' source of food in the Oubangui River.

Gabon

In 1955, this species was imported from Cameroun to Gabon, where it is naturalized in the Ogoué River (Moreau *et al.*, 1988).

Ivory Coast

In 1958, *Heterotis* was successfully imported from Cameroun to the Ivory Coast, where it was also introduced with success in 1962 and 1971 respectively to Ayaini (Ayame) and Kossou reservoirs (Moreau *et al.*, 1988).

Madagascar

Kiener (1963), Moreau (1979), Moreau *et al.* (1988) and Reinthal and Stiassny (1991) record the introduction of *Heterotis* to Madagascar in 1963, where it has become naturalized at both high elevations and in the Canal des Pangalenes (see Glossary).

Zaire

Heterotis was imported to Zaire (Belgian Congo) in 1950 and again in 1966, and has become established in the wild in the Congo River and in Lake Toumba (Moreau *et al.*, 1988).

SALMONIDAE

Huchen (Danube Salmon)
Hucho hucho

The Huchen, which lives permanently in fresh water and does not migrate to the sea, is a highly valued game fish whose population has greatly declined in recent years due to pollution of the Danube basin (Welcomme, 1988).

Natural Distribution

Rivers of the Danube basin.

Naturalized Distribution

Europe: Spain. *Africa:* Morocco.

Europe

Spain

In 1968 Huchen from Czechoslovakia were introduced to Spain for angling purposes by the National Fisheries Service; they are currently established only in the drainage of the River Douro (Elvira, 1995a & 1995b).

Africa

Morocco

According to Welcomme (1988), the Huchen has been 'successfully introduced [from the Danube] into one locality in the Atlas Mountains near Meknes'. The date of this introduction is unrecorded.

Unsuccessful Introductions

Several unsuccessful attempts have been made to naturalize the Huchen, in both European river basins and in North America; these include in a tributary of the middle reaches of the River Rhône between 1957 and 1960, where although spawning occurred the species failed to become established (Vooren, 1972); in the Rhine in Germany in the 1950s (Schindler, 1957); in the Thames in England in the early 20th century (Wheeler & Maitland, 1973; Wheeler, 1974); to Belgium from the former Yugoslavia in 1954 and 1960 to control *Chondrostoma nasus*; and to Sweden from the same source in 1963 (Welcomme, 1981, 1988); to the United States in the early 1870s (Hessel, 1874); and to Canada from Czechoslovakia in 1968 (Crossman, 1984).

Golden Trout
Oncorhynchus aguabonita

The Golden Trout is an extremely local species which has been widely transplanted within the USA, where outside its natural range it has hybridized with Rainbow Trout *Oncorhynchus mykiss* (q.v.) and Cutthroat Trout *O. clarki* (Ferguson, 1915; Baughman, 1985).

Natural Distribution

Southern Pacific montane streams of the USA.

Naturalized Distribution

North America: Canada.

North America
Canada

In 1959, Golden Trout ova from Wyoming were imported to Alberta, where fry were subsequently released in South Fork, Three Isle, Gap and Galatea Lakes at high altitudes in the Alberta Rocky Mountains, and later to the Red Deer River drainage in the same province. In 1960, eyed ova from California were placed in Golden Lake in Jasper National Park, Alberta, and in Kaufman Lake in Kootenay National Park, British Columbia (Crossman, 1984).

Several of the 1959 introductions proved successful, especially in the South Fork Lakes where the species is naturalized. The two 1960 plantings failed to establish populations in the two national parks, and there have been no further introductions (Crossman, 1984).

'This introduction' wrote Crossman (1984), 'has added a species to the sport fishery without apparently causing problems with native species at this time'.

Pink Salmon (Humpback Salmon)
Oncorhynchus gorbuscha

The Pink Salmon was formerly of lesser importance as a source of food than other Pacific salmon species, but it has gained in significance for commercial and game fisheries since the 1940s (Welcomme, 1988).

Natural Distribution

Western seaboard of North America and eastern seaboard of northern Asia.

Naturalized Distribution

Europe: ? Finland; ? Norway. **South America:** ? Argentina; ? Chile.

Europe
Finland; Norway

Nikolsky (1958, 1961), Hardy, 1961; Muentyan (1963) and Vooren (1972) record that in 1957 between 3 and 5 million Pink Salmon smolts, reared from eggs brought from Sakhalin Island in the then Soviet Far East, were released in the Kola River on the Barents Sea in northwestern Russia, where the species became established in the White and Barents Seas. From this area, Pink Salmon dispersed naturally into Finnish and northern Norwegian waters, and have even occurred in Scotland and Ireland. Their status in Scandinavia is uncertain; Berg (1961) says they 'are known to have ascended about 40 rivers in North-Norway (and perhaps others as well)'. Frost and Brown (1967) reported that 'many have been caught in Norwegian waters and a few in British estuaries and rivers . . . in the unlikely event of their becoming established in Britain they would not compete with our native salmon [*Salmo salar*] because of the difference in habits', a somewhat cavalier assumption. Welcomme (1981) confirmed that they were established in northern Norway, but said that it was unknown whether they were reproducing (Welcomme, 1988). However, Berg (1977) and Bakshtansky (1980) say they were only briefly successful in Scandinavia. (See also Welcomme, 1991; Mazzola, 1992.)

Argentina; Chile

Repeated attempts have been made to establish Pink (and other Pacific) Salmon in Chile and Argentina, where their present status is unknown (Barros, 1931; Thompson, 1940; Fuster de Plaza & Plaza, 1954; Ringuelet *et al.*, 1967; Campos Cereda, 1970; Wood, 1970; Baigun & Quiros, 1985; Zamorano, 1991).

Translocation

In 1958, Pink Salmon were translocated to Newfoundland, Canada, where they established a small population on the Atlantic coast (Welcomme, 1988).

Chum Salmon
Oncorhynchus keta

According to Welcomme (1988), although the Chum Salmon in its native range is one of the most important salmon for commercial fisheries, it has been introduced elsewhere less than other Pacific salmon species.

Natural Distribution

Western seaboard of North America and eastern seaboard of northern Asia.

Naturalized Distribution

Europe: ? Finland; ? Norway. *South America:* ? Chile.

Europe

Finland; Norway

Chum Salmon introduced in the 1960s from the eastern to western former USSR have dispersed into Scandinavian waters, where their present status is in doubt. Welcomme (1981) records them as having 'appeared in the Baltic and in some in-flowing rivers' of Finland (and possibly Norway), but whether they are reproducing is unknown (Welcomme, 1988).

South America

Chile

Since 1970, Chum Salmon from the USA and Japan have been introduced on several occasions to Chile to try to establish populations in southern rivers, but so far with uncertain results (Campos Cereda, 1970; Wood, 1970).

Coho Salmon (Silver Salmon)
Oncorhynchus kisutch

The Coho Salmon, one of the most valuable commercial fish of the Pacific seaboard, has been successfully translocated to the east coast of North America and, after initial failures (Dymond, 1955), to the Great Lakes (Welcomme, 1988).

Natural Distribution

Western seaboard of North America and eastern seaboard of northern Asia.

Naturalized Distribution

Europe: France. *South America:* ? Chile.

Europe
France

Coho Salmon first appeared in French rivers, where they have since reproduced, in 1974, having escaped from aquaculture ponds (Euzenat & Fournal, 1982).

In the same year, Coho Salmon were unsuccessfully introduced to Cyprus, Greece and Germany (the then Federal Republic) (Welcomme, 1981, 1988), and in 1973 to Italy (Rossi, 1991; Mazzola, 1992). In the early 1980s, several Coho Salmon were caught in The Netherlands and Belgium, having apparently been released privately in the Somme estuary in Picardy, France (De Groot, 1985).

South America
Chile

Between 1901 and 1910, a total of 225 040 Coho Salmon eggs and young were introduced from the USA to Chile (Barros, 1931; Davidson & Hutchinson, 1938). Other introductions followed (Mann, 1954; De Buen, 1959), and as a result of private releases since the late 1960s (Campos Cereda, 1970; Wood, 1970) Coho Salmon are reproducing in captivity in the region of Chiloë and Lake Llanqihue (Welcomme, 1981, 1988), and possibly in the wild.

Unsuccessful Introductions

Between 1901 and 1910, some 377 180 Coho Salmon eggs and young from the USA were introduced to Argentina (Davidson & Hutchinson, 1938), where it is believed that reproduction has taken place in captivity only (Welcomme, 1988). In 1965 and 1978, eyed eggs of Coho Salmon were unsuccessfully introduced from Seattle, Washington, to Hokkaido in Japan (Umeda *et al.*, 1981; Chiba *et al.*, 1989). In the latter year, 11 000 Coho Salmon eggs were imported from the USA to Kerguelen in the South Atlantic; 6000 survived the voyage, of which 5000 hatched, but the species failed to become established (Davaine & Beall, 1982a).

Rainbow Trout
Oncorhynchus mykiss

Since 1874, the natural distribution of the Rainbow Trout has been so extended that it is now virtually cosmopolitan, and is one of the most highly valued freshwater game and aquaculture species. Environmental factors believed to be of primary importance in the success or failure of introduced populations are water temperature, rainfall and the occurrence of suitable spawning redds; seasonal water temperatures below 13°C are essential for self-propagation. The species' present range extends from near sea level to over 4500 m (MacCrimmon, 1971).

Any further spread of the naturalized distribution of the Rainbow Trout appears unlikely, except perhaps in northeastern Asia. However, an extension of range within the species' present distribution is probable, through the utilization of new impoundments constructed for water supply and flood control in many parts of the world (MacCrimmon, 1971).

The extensive world distribution of the Rainbow Trout can be explained by a combination of biological, social and economic factors. First, the species is well adapted to aquaculture in a variety of conditions. Second, it is widely held in high esteem as a game fish, both as a small fish in streams and ponds and as the much larger anadromous 'steelhead' variety. Third, it is highly regarded as a food species, and the farming of Rainbow Trout, both for direct consumption and for stocking purposes, has made it a valuable economic asset (MacCrimmon, 1972).

Natural Distribution

Coastal drainage of North America, mainly west of the Continental Divide, from the Kuskokwim River in Alaska (not north of 64°N) southward to around 25°S in the Río Presidia, Durango, Mexico (MacCrimmon, 1971).

Naturalized Distribution

Europe: ? Albania; Austria; British Isles; Bulgaria; Czechoslovakia; ? Denmark; France; Germany; Greece; Hungary; ? Italy; Luxembourg; ? The Netherlands; Portugal; Romania; Spain; Sweden; Switzerland; the former Yugoslavia. *Asia:* ? Afghanistan; India; ? Iran; Israel; Japan; ? Korea; ? Lebanon; Sri Lanka; ? Taiwan; *Africa:* ? Ethiopia; Kenya; Lesotho; Madagascar; Malawi; South Africa; Sudan; Swaziland; Tanzania; ? Tunisia; Uganda; Zambia; Zimbabwe. *South and Central America:* Argentina; Bolivia; Brazil; Chile; Colombia; Costa Rica; Ecuador; Panama; Peru;

Venezuela. *Australasia:* Australia; New Zealand; Papua New Guinea. *Oceania:* Hawaiian Islands; Réunion.

Europe

Albania

Although Rainbow Trout were reported to occur in montane streams in Albania in the late 1960s, their origin and date of introduction are unknown. They might possibly be derived from dispersals from the north Albanian Alps (via the former Yugoslavia) or the Pindus Mountains (via Greece) (MacCrimmon, 1971).

Rainbow Trout culture was first recorded in Albania in March 1971 in an aquaculture station at Tushemishta near Lake Ohrid. Broodstock and eyed eggs were obtained from Vrutok near Gostivar, Macedonia (in the former Yugoslavia), where Rainbow Trout, originally imported from the United States Bureau of Sport Fisheries and Wildlife's Iowa Hatchery, were being reared (Anon., 1971). MacCrimmon (1971) records Rainbow Trout in Albania as 'probably' self-sustaining.

Austria

In 1885, H. Köttl began to breed Rainbow Trout, obtained from Germany (Welcomme, 1981, 1988, 1991), in ponds at Neukirchen am Walde and in the Vöckla River (Von Pirko, 1910). By the late 1960s, Rainbow Trout were widespread in the provinces of Lower Austria, Burgenland, Upper Austria, Styria, Carinthia, Salzburg, the Tyrol and Vorarlberg, except

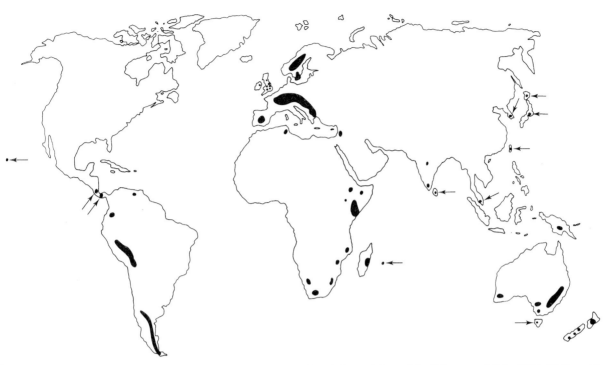

Naturalized distribution of Rainbow Trout *Oncorhynchus mykiss* (adapted from MacCrimmon, 1971, 1972; McDowall, 1990b).

in high montane streams occupied by Brown Trout *Salmo trutta* (MacCrimmon, 1971).

British Isles

The first shipment of Rainbow Trout eggs to England, numbering around 3000, arrived at the National Fish Culture Association at Delaford Park, Iver, Buckinghamshire, on 14 February 1884 (Clark, 1885). These were supplemented a year later by a further 15 000. Quantities of the resulting fish were despatched to various aquaculturists in England, and some were sent to Sir James Maitland's hatchery at Howietoun in Stirlingshire, Scotland, where by 1887 they were reported to be breeding.

In almost every winter from 1888 to 1905, Rainbow Trout eggs from the USA arrived in Britain, and were sent to hatcheries at Bridgnorth in Shropshire and Malvern Wells in Worcestershire (England); to Innishannon, Co. Cork (Republic of Ireland) in 1899–1901, where Moreton Trewen planted them in the River Bandon and contructed hatcheries on the river for stocking other nearby rivers and ponds (Worthington, 1941); to Ballymena in Co. Antrim in

1900 and 5 years later to Lough Shure on Arranmore Island off the coast of Co. Donegal (Northern Ireland) (Frost, 1940); and to Loch Uisg and several small lochans near Lochbuie on the Isle of Mull and to Inverness-shire (Scotland). In 1890, Rainbow Trout eggs arrived in England from S. Jaffé's hatchery in Osnabrück, Germany (see below); these are believed to have been the earliest importations not directly from North America.

No Rainbow Trout eggs seem to have been imported between 1905 and 1931 but, in the latter year and again in 1938–1939, D.F. Leney of the Surrey Trout Farm and United Fisheries imported a total of 100 000 eggs through the United States Bureau of Fisheries and Wildlife and from the government hatchery at White Sulphur Springs, West Virginia.

In his survey of Rainbow Trout in Britain, Worthington (1941) recorded the species as occurring in between 50 and 55 waters in the British Isles, of which only one was in Ireland. In a follow-up survey 33 years later, Frost (1974) found there was a total of 491 waters in the British Isles in which Rainbow Trout occurred, of which 462 were in

England, Scotland and Wales, 21 in Northern Ireland, and 8 in the Republic; she recorded a further 67 waters that had recently held Rainbow Trout, but in which their then status was uncertain. Frost's survey also revealed that in the decade after 1940 few Rainbow Trout had been introduced into British waters, and none into those of southern Ireland. A few introductions were made between 1950 and 1960, since when the number of waters stocked with Rainbow Trout has steadily increased – a pattern that continues to this day, reflecting the ever-increasing demand for trout fishing in Britain.

Frost's enquiry showed that of the various waters mentioned by Worthington (1941) as containing self-perpetuating populations of Rainbow Trout, only five – Blagdon reservoir in Somerset, the River Wye in Derbyshire, Lough Shure, the River Misbourne in Buckinghamshire and the River Wey in Surrey – certainly contained a breeding stock some 30 years later. Only Lough Shure and the Derbyshire Wye held self-maintaining populations unassisted by artificial stocking in both 1941 and 1974. Frost's survey further revealed that although Rainbow Trout were breeding in some 40 waters in the British Isles, only six – the Derbyshire Wye and a tributary, the Lathkill; a tributary of the Leigh Brook in Worcestershire; the River Chess in Hertfordshire; Lough Shure; and Lough na Leibe, Ballymote, Co. Sligo – held stocks that she considered to be self-perpetuating (for further details see Frost (1974) and Lever (1977)).

Since Frost's enquiry, some other breeding populations of Rainbow Trout have been discovered in the British Isles; for example, in Northern Ireland Rainbow Trout imported from Denmark were distributed between 1964 and 1968 from the Movanagher Ministry Fish Farm to 17 lakes, and have established self-maintaining populations in Shaws Lake, Co. Armagh, and in Lough Eyes in Co. Fermanagh. In the Irish Republic, Rainbow Trout are self-perpetuating in White Lake, Castlepollard, Co. Westmeath, as well as in Loughs Shure and Na Leibe, and there are reports of spawning in the Woodenbridge River, Co. Wicklow, and the Golden Grove brook and the Little Brosna River in Co. Tipperary (Wheeler & Maitland, 1973; Wheeler, 1974). It is of interest to note that nearly all the waters in the British Isles where Rainbow Trout breed successfully are spring-fed and alkaline.

The reasons for the comparative failure of Rainbow Trout to establish more viable populations in the British Isles, despite widespread breeding, remain obscure. One reason may be interspecific competition between the fry of Rainbow Trout and those of Brown Trout; the latter hatch earlier in the year than the former, and are thus able to dominate and harass them in the search for food and territory at a critical period in their development. Many of those Rainbow Trout that do manage to survive to become yearlings doubtless fall prey to larger cannibal Rainbow Trout and to Brown Trout; thus heavy artificial stocking of large fish, which is frequently accompanied by over-fishing, may be another factor inhibiting the establishment of more self-maintaining populations of Rainbow Trout in the British Isles.

Bulgaria

Rainbow Trout eggs were first introduced to Bulgaria from Germany in 1897–1898 by Tarand-Saxon Linke, by whom they were hatched at a fish farm at Kustendil.

In 1924, a state aquaculture centre was opened near Samokov for rearing both Brook Trout *Salvelinus fontinalis* and Rainbow Trout, and a year later 40 000 fingerling Rainbow Trout were placed in the Samokovska Bistritza River. In 1926, some Rila lakes, including Saragol and Grunchar, and most of the high-altitude montane rivers such as Vladeiska, Iskrezka, Blato (Sofia District) and Stara Reka, Chetirideset Izvora, Tschaia (Plovdiv District) were stocked with Rainbow Trout.

By the late 1960s, most of the high mountain rivers and lakes in the Rila, Rodopa, Pirin and Stara Planina ranges had been planted with Rainbow Trout, which it was believed might eventually displace native species (MacCrimmon, 1971).

Czechoslovakia

Rainbow Trout eggs were first imported to Czechoslovakia from Germany (according to Anon., 1971 from North America) in 1891, and were hatched at a trout farm at Litomyšl-Nedošín on the Loučná River in East Bohemia by S. Vacek. In 1918, eggs of the migratory 'steelhead' strain were brought to this hatchery from Wiesner's Trout Hatchery in Waltersdorf, Germany. Two years later, another hatchery, at Martin on the Turiec River, a tributary of the Váh in North Slovakia, began distributing Rainbow Trout. Subsequently, some Rainbow eggs were also imported to Czechoslovakia from Denmark.

Rainbow Trout populations in Czechoslovakia were largely wiped out by invading troops in 1944–1945, except for some in small montane streams. Restocking with eggs imported from Denmark took place between 1946 and 1949.

By the late 1960s, Rainbow Trout were established in most suitable waters throughout Czechoslovakia. In Slovakia, naturalized populations occurred in parts of the Okna, Torysa, Lučinka, Delňa, Hornád, Hnilec, Poprad, Slaná, Hron, Nikra and Váh rivers, and in Morské Oko, Evickino and Klinger Lakes, and in the Great Lake of Kolpachy of the Dunaj River basin. In Moravia, spawning took place in the upper reaches of the Olza, Ostravice and Moravice Rivers of the Odra basin, and in the Dyje, Morava, Svitava and Jihlava Rivers of the Dunaj basin. In Bohemia, Rainbow Trout occurred in the Stěnava River of the Odra basin and in such rivers as the Metnje, Tichá Orlice, Loučná, Doubravka, Lužická Nisa, Jizera, Ohře, Vuznice, Okrouhice, Nežárka and Blanice of the Labe River basin (MacCrimmon, 1971).

Denmark

Rainbow Trout are believed to have been first introduced to Denmark from Germany in about 1894. Subsequently, eggs were imported from the USA, Canada, New Zealand, Austria and France (Larsen, 1945, 1972). In Denmark, Rainbow Trout are used for aquaculture rather than for stocking purposes; escapes from captivity have, however, taken place, and wild populations are found in all the rivers of Jutland adjacent to trout farms (MacCrimmon, 1971; Jensen, 1987). According to Welcomme (1981, 1988), Rainbow Trout in Denmark are prone to disease, and may be responsible for the introduction of alien parasites.

France

In 1879, the Trocadéro Aquarium in Paris acquired Rainbow Trout eggs from the United States Fish Commission, which were hatched under the care of Dr J. de Bellesme who later released the fingerling fish (Bean, 1895). No further importations are on record until 1884, when 3000 eggs were despatched from the USA to C. Raveret-Wattel of La Société Zoologique d'Acclimatation in Paris (Clark, 1885; De Groot, 1985; Lever, 1992). In 1907, 10 000 Rainbow Trout eggs of the 'steelhead' strain were imported

from the USA to France, and by 1924 a total of 517 000 eggs had been imported.

In spite of the widespread releases, Rainbow Trout became established only in the basin of the Rhone (Roule, 1922) and in the provinces of Ariège (Ruisseau des Campels, near Aix-les-Thermes), the Pyrénées Orientales (in the Maurcillas) and Cantal (Vivier, 1955). Since then, however, Rainbow Trout have become naturalized in Alpine lakes such as Lac du Mont-Cenis in the Savoie, at an altitude of 2000 m, and in the Pyrénées in Lac d'Ilhéou near Cauterets at 1976 m, and in Lac de l'Oueil Nègre below Barrèges at 2339 m. They are also established in the upper drainage basins of the Gave d'Aspe and Gaube, and are said to breed in the Beaumes River of the Côte d'Or (MacCrimmon, 1971). They have failed to become established in Lake Annecy in eastern France (Hubault, 1955).

Germany

In 1882, 1200 eyed Rainbow Trout eggs were despatched from Northville Station, Michigan to Herr von Behr, president of the Deutsche Fischerei-Vereins (Giard, 1894; Walker & Patterson, 1898; Anon., 1909; Lever, 1977; De Groot, 1985), from where they were distributed to various hatcheries; the Breeding Institute of the Bavarian Fishing Club in Starnberg hatched 60 fry; Mr Haack, director of the Hüningen fish farm in Alsace (then in Germany), near Basel, received 400 eggs, and 30 months later 256 fish still survived in the hatchery (Hemsen, 1964).

Until 1928, Rainbow Trout eggs were periodically sent from (mostly) eastern North America to German aquaculturists. In 1896, S. Jaffé of Osnabrück acquired 10 000 'steelhead' strain eggs from California, the majority of which hatched successfully (Anon., 1897). Further shipments of 'steelheads' arrived in Germany in 1898 and 1902 (MacCrimmon, 1971).

By the late 1960s, Rainbow Trout were naturalized in Germany mainly in streams in the Erdinger Moos, the Dachauer Moos, the Freisinger Moos and the Isarkanal, in the rivers of Glonn, Mosach and Wielenbach, and in the basin of the Werra River in Thüringen (MacCrimmon, 1971).

Greece

Rainbow Trout were first introduced to Greece in January 1951, when a consignment of 200 000 eyed eggs from Switzerland was hatched at the new Louros Hatchery in Ioannina, from where they were planted in March in the Rivers Louros and Arachtos and in Lakes Jannina and Trichonis. In the following December, a further 500 000 eggs from the same source were hatched successfully, and in March 1952 fry were placed in Lakes Jannina, Trichonis, Viro (in the headwaters of the Louros), Ziro (Philippias), St Panteleimon (Ostrovo) in Macedonia, Zaravina, and in the Rivers Louros, Acheron, Calamas, Aoos, Voidomatis and Metsovitis (Arachtos) in Epire (Serbétis, 1954).

Between 1954 and 1969, a total of 7.5 million Rainbow Trout eggs was imported from Switzerland and reared in the hatcheries at Louros and Edessa, from where fish were distributed to various waters. By the late 1960s, Rainbow Trout were being fished for in Lakes Vegoritis, Ladonos and Tavropou, and the Rivers Voidomatis, Calamas, Louros, Arachtos, Acheron and Aliakmon in Epire (MacCrimmon, 1971).

Hungary

In 1882 and 1883 respectively, 443 and 434 Rainbow Trout, from an unnamed source, were imported to Count Palffy's Fish-Breeding Institute at Szomolány (subsequently Smolenice, Czechoslovakia) (Von dem Borne, 1885). In 1884, Mr Haack (see under Germany) despatched 500 Rainbow Trout from Hüningen to Szomolány at the request of the German Fishery Club (Hemsen, 1964). A further shipment was made in the following year, and subsequently Rainbow Trout began to be bred throughout Hungary; by 1910, 37 hatcheries had been set up by the Hungarian Forest Department, and by the outbreak of the First World War the number had increased to 200, which were supplied gratis by the government with eyed eggs. Between 1897 and 1917, a total of 26 383 000 eyed Rainbow Trout eggs were distributed throughout Hungary (Vásárhelyi, 1963).

After the War, most trout waters became depopulated and hatcheries were abandoned – the only surviving Rainbow Trout in Hungary being found in waters in the Bükk Mountains. By 1932, however, new aquaculture centres had been constructed, and introductions made in suitable waters. By the late 1960s, Rainbow Trout occurred in the Rivers Szinva, Eger, Hasznos, Gyöngyös and Rak, in others at Garadna and Szilvásvárad, and in Lakes Felsötárkány, Sikfökut and Hámor (MacCrimmon, 1971). Welcome

(1981, 1988), however, states that 'wild stocks exist [only] in two small streams'.

Italy

In 1895, Professor D. Vinciguerra recommended the introduction of Rainbow Trout to Sardinia (Cottiglia, 1968), and 2 years later 2-year-old fish were released in Lake Manate, where unsuccessful plantings continued until 1904 (Besana, 1910). In 1898, Vinciguerra received 20 000 eggs from the United States Fish Commission (Ravenel, 1898), and from that year onwards Rainbow Trout were liberated in both Sardinian waters and throughout the alpine regions of northern Italy.

Rainbow Trout were first introduced to waters in Liguria, near Genova, in 1911 (Arbocco, 1966), but may not have been planted in central and southern Italian waters until 1930 (MacCrimmon, 1971).

By the mid-1950s, Rainbow Trout had become naturalized in the headwaters of the Adige (Val Venosta), Toce (Val D'Ossola), Rodano (Vallese) and Pellice (Val Pellice), generally in waters above 1000 m (Sommani, 1957). In Liguria, they occurred a decade later in Lake Giacopiane and in the Rivers Magra, Vara, Entella, Centa, Arroscia, Pennavaira, Argentina and Roja, but were not self-perpetuating (MacCrimmon, 1971).

The introduction of Rainbow Trout in Italy has been most fruitful in the large reservoirs bordering the Appenines and the Alps. They have also been successful in Sardinian lakes near hatcheries, such as those at Gusana, Cughinadorza and Medio (Flumendosa), or in waters devoid of other competing salmonids (MacCrimmon, 1971). MacCrimmon (1971) and Welcomme (1981) describe their establishment in the wild in Italy as 'limited'.

Luxembourg

Rainbow Trout eggs from Denmark were first imported to Luxembourg in 1946; hatching was successful, and for at least the next two decades 300 000 eggs were brought in annually from the same source, and around 30 000 1- or 2-year-old fish were released every spring and autumn into the River Sûre, a manmade lake (Barrage Esch-Sûre), and into the Attert, Mamer, Ernz Blanche, Ernz Noire, Eisch and Syr streams; only in the Eisch and Syr are Rainbow Trout known to be self-maintaining (MacCrimmon, 1971; Rossi, 1991).

The Netherlands

According to De Groot (1985), the first Rainbow Trout introduced into The Netherlands are believed to have come from Germany; these fish were stocked (at an unrecorded date) as fry in Hasselbeek's fish farm at Zwaanspreng in Gelderland. In 1898, 1500 1-year-old fish were released in the moat of a fortress near Spaarndam north of Amsterdam (De Groot, 1985).

In 1923 and 1925, the Nederlandse Heidemaato-chappij-Arnhem (Organization for Soil Cultivation) acquired Rainbow Trout eggs from the USA; these were hatched in the Arnhem, Gulpen and Vaassen hatcheries, the fish being intended principally for direct consumption, though some were planted for angling purposes (MacCrimmon, 1971).

In 1968 and 1969, 42 500 Rainbow Trout were planted in the Veerse Meer (Steinmetz, 1968; Vooren, 1972), and others have been released in Lakes Vinkeveen, Nederhorst ten Berg and Zevenhoven (Vooren, 1968, 1972; Nijssen & De Groot, 1975; De Groot, 1985). Although MacCrimmon (1971) describes the Rainbow Trout as self-sustaining to a 'limited' extent in The Netherlands, Welcomme (1981, 1988) says it is reproducing only 'artificially'.

Portugal

In 1898 and 1899, Rainbow Trout eggs from the USA were consigned to Augusto Nobre at Vila do Condo in Portugal, and in the early 1900s others arrived from France. Later introductions of eggs from the USA were made in 1911, 1912, 1916 and the 1920s, but none resulted in naturalized populations (MacCrimmon, 1971).

After the Second World War, further importations of Rainbow Trout eggs from the USA were recommenced, and hatched fish have been stocked in streams north of the Douro River, and in still waters where some established populations occur (MacCrimmon, 1971).

Romania

Eyed Rainbow Trout eggs were imported to hatcheries in Transylvania (then part of Hungary) in 1885, and 5 years later to Moldavia, from where fish later escaped into nearby waters. Rainbow Trout were first deliberately released in Romanian rivers in Transylvania (in the Départements of Bihor, Arad and Clujl, and in the waters of the Somesul Rece and Somesul Cald) in 1895. Five years later the River Tarcau in northern Moldavia was stocked. In 1908, the rearing of Rainbow Trout eggs began in a newly constructed hatchery at Tarcau in Neamt, to which eggs were imported from the USA in 1925 (MacCrimmon, 1971).

By the late 1960s, there were 41 trout farms in Romania, and populations in the wild were established in Valea Bradului, Valea Minghetului (Baia Mare), Somesul Rece (Cluj), Sohodol (Craiova), Valea Minisului (Timisana), Birzava, Cerna (Banat), Vida (Beius) and Virghis (Odorhei). In Romania, Rainbow Trout do best in montane lakes between 800 and 2000 m asl (MacCrimmon, 1971).

Spain

Rainbow Trout eggs were originally imported to Spain from France in the 1890s, and these and early 20th century introductions were the basis for stocks reared in government hatcheries. The majority of populations in these hatcheries are, however, descended from eggs imported from the Mount Whitney, Mount Shasta and Hot Creek trout farms in California in 1965 (MacCrimmon, 1971).

Today, Rainbow Trout occur in most Spanish rivers which debouch into the Atlantic and Mediterranean; the species is naturalized in the Manzanares (Madrid), Iruelas (Avila), Cifuentes (Guadalajara), Riofrio (Granada), El Bosque (Cadiz) and the majority of the Huesca and Lerida rivers (MacCrimmon, 1971; Vooren, 1972). Elvira (1995a) lists it as occurring in the following drainages: North, Galicia, Douro, Tagus, Guadiana, Guadalquivir, South, Levant, Ebro and eastern Pyrenees.

Sweden

Rainbow Trout eggs were first imported to Sweden from Max von der Borne in Germany in the spring of 1892 by C. Hammarström for his private hatchery at Äggfors in Jämtland, from where fry were distributed to a number of lakes in Jämtland, Medelpad, Halsingland and Dalarna (Svärdson, 1968). During the next few years, further batches were imported from Germany, and the fry planted in lentic and lotic waters of southern Sweden, where eggs were first collected from naturally spawning Rainbow Trout in 1898; on hatching, the fish were kept in ponds in

Ängelsberg in Västmanland (Svärdson, 1968), from where they were subsequently planted in suitable waters throughout the country.

After Hammarström's death in 1911, interest in rearing Rainbow Trout declined, and the practice of stocking them was reviewed, as they survived in the wild only in waters devoid of other salmonids. After the mid-1950s, however, Rainbow Trout were again widely stocked in natural waters for 'put-and-take' angling (Svärdson, 1968).

One of the few places in Sweden where Rainbow Trout are naturalized is a small stream near Hedemora in Dalarna, which was originally stocked with fish from Denmark as long ago as 1927 and where Brown Trout *Salmo trutta* do not occur. Successful spawning has also been reported from a few localities in southern Sweden, and in some ponds with circulating water in Skåne (MacCrimmon, 1971). Welcomme (1981, 1988) reported naturalized populations as 'very limited'.

Switzerland

In 1888, 1890, 1891 and 1893, Rainbow Trout ova were sent to the Swiss government by the United States Fish Commission, but early attempts to establish the species in the wild in Switzerland failed (MacCrimmon, 1971).

In the 1920s, further Rainbow Trout were imported from the USA, and this time some naturalized populations resulted (MacCrimmon, 1971) MacCrimmon (1971) and Welcomme (1981, 1988) describe the species as self-sustaining in Switzerland on a 'limited' basis only.

Yugoslavia

In the former Yugoslavia, Rainbow Trout from Austria were first released by Professor I. Franke near Ljubljana in Slovenia (previously Laibach, Austria) between 1890 and 1892, and by M. Kulmer near Zagreb in Croatia in 1894. Other stock obtained directly from California was placed in waters near Sarajevo in Bosnia and Herzogovina by E. Schubert between 1902 and 1904. Until 1935, all apparently suitable waters (totalling around 60 000 ha) in the country were stocked with Rainbow Trout, but with little success (MacCrimmon, 1971).

By the late 1960s, Rainbow Trout could be fished for in Yugoslavia in the Radnovna River near Bled; in the Soča at Tolmin and its tributaries of Idrijca and Vipava; in the Lake of Bohinj; in the Sora River at Ljubljana; in the upper reaches of the Krka near Novo Mesto; in the Savinja at Celje; in the Bregana at Zagreb in the Sava valley; in the Gacka River at Otočac; in the River Lika and its feeders in the Ličko Polji plain at Gospić; in the Jadro River near Split; in the Trebišnjica River at Trebinje; in the Studenica tributaries; in the Stajnica at Jezerana; and in lakes in Plitvice. Rainbow Trout are naturalized in the Gacka, Stajnica and Trebišnjica Rivers and in the Plitvice lakes only, where in Lake Ohrid they have been implicated in the decline of the Lake Ohrid Trout *Salmo letnica.* (MacCrimmon, 1971).

Unsuccessful Introductions

Unsuccessful attempts have been made in Europe to naturalize Rainbow Trout also in Belgium, Cyprus, Finland, Iceland, Liechtenstein, Norway, Poland and the former USSR (MacCrimmon, 1971; Welcomme, 1981, 1988).

Asia

Afghanistan

Despite conflicting reports from MacCrimmon (1971, 1972) about the presence of Rainbow Trout in Afghanistan, Welcomme (1981, 1988) states they are 'present in fish culture stations and reputedly in north-flowing streams of the Hindu Kush'.

India

The earliest attempts to introduce Rainbow Trout from Scotland to waters in the Nilgiris Hills (2300 m asl) in Madras State in and after 1866 (Molesworth & Bryant, 1921; Natarajan & Ramachandra Menon, 1989; Sehgal, 1989) were unsuccessful. In 1906 or 1907, however, Henry Wilson of the government of Madras imported 100 1-year-old fish from Sri Lanka (then Ceylon), and subsequently also from Innsbruck (Austria), Germany and New Zealand (Jhingran, 1989; Shetty *et al.*, 1989; Sreenivasan, 1989), to a stream in Parson's Valley (Natarajan & Ramachandra Menon, 1989), which formed the original parental stock of Rainbow Trout in India (MacCrimmon, 1971). In the same year, trout ova and fingerlings were imported from S. Jaffé's hatchery in Osnabrück, Germany, and from the Auckland Acclimatization Society in New Zealand (Lever, 1992; McDowall, 1994b), and later from Howietoun in Stirlingshire, Scotland. By 1909, Rainbow Trout were established in the Rivers Mulkesti, Kaurmund, Mekod, Parson's Valley, Sandynulla, Yammakkal, Billithadahalla and Peermund (Sreenivasan, 1989), and in the Avalanche, Emerald, Krurumund, Meked and Pykara in the Nilgiris, where Brown Trout *Salmo trutta* had failed to survive, and 2 years later these waters were opened to anglers (MacCrimmon, 1971). By the late 1960s, Rainbow Trout were to be found in the Avalanche, Emerald Valley and Pykara river systems, but were supplemented annually by fry from the Avalanche Trout Farm. In 1968, an attempt was made to introduce Rainbow Trout to the Nilgiris from Japan (Natarajan & Ramachandra Menon, 1989; Sreenivasan, 1991).

In 1916, Rainbow Trout eggs from the government hatchery at Ooty Lake were hatched out at Chundavurrai, and the fry released in waters of the Kundale Valley and Nymakad (both around 1850 m) in Travancore; in 1932 a few hundred Rainbow Trout fry were brought from the Avalanche hatchery and released in ponds at Chittavurrai (about 2150 m): both these introductions ended in failure (Gopinath, 1942; MacKay, 1945).

In 1933, a hatchery was built at Arivikad (around 1850 m), and between 1933 and 1937 Lake Deviculam and Loch Finlay were repeatedly stocked. In the latter year the hatchery was closed, and the stock of about 170 fish was planted in waters of Hamilton's Plateau, where in December 1940 proof of breeding was obtained. However, as supplementary stocking was considered advisable, 10 000 ova were acquired from the Nuwara Eliya hatchery in Sri Lanka, 5000 of the resulting fry being placed in waters of the plateau. In January 1941, a second shipment of 5000 eggs from Sri Lanka was brought to Rajamaliay, from where 1300 fry were released at Devikulam and Chittavurrai (Gopinath, 1942; MacKay, 1945).

In the Garhwal Himalayas, Rainbow Trout were introduced in the mid-1980s by the State fisheries department to the Kaldyani Hatchery in the district of Uttarkashi and to the hatchery at Talwari in Charnoli district; the results of these introductions are uncertain (Johri & Prasad, 1989).

The first attempt to establish Rainbow Trout in Kashmir in 1904 failed because the eggs did not hatch (Mitchell, 1918). No further attempts were made until 1912, when ova from the Bristol Water Works at Blagdon, England, were successfully reared, and 3 years later Rainbow Trout were thriving in the Harwan Trout Farm (Mitchell, 1918). By the late 1960s, the species occurred in the Rivers Achabal, Kotsu, Panzat, Pahalgam, Gerinag, Kokernag, Kulgaam and Tehsil in Avnamalai and Kodaiknal, and in montane ranges of the Western Ghats, but again, as in the Nilgiris, the stock was supplemented annually by fry and fingerlings from hatcheries at Harwan and Achabal (MacCrimmon, 1971).

The Kulu Valley in Himachal Pradesh acquired its original stock of Rainbow Trout in 1919, when 5000 ova from Kashmir were brought to the Katrain Hatchery, from where fry were released in waters in Kulu (Jones & Sarojini, 1952). In 1923–1924, a further 10 000 eggs were sent to Kulu from Kangra, but in spite of regular stocking the fish failed to establish themselves (Singh & Kumar, 1989), and by 1942 aquaculture in both Kangra and Kulu was abandoned. In 1963, however, eyed ova from Kashmir hatched successfully in the Barot Trout Farm, and breeding began 3 years later, when fish from Kashmir were again transferred to the Katrain Hatchery. Rainbow Trout were successfully introduced to the Kulu Valley, Beas and Uhl Rivers in Himachal Pradesh where, however, they were once again stocked annually (MacCrimmon, 1971).

In the 1960s, Rainbow Trout were introduced in the River Manar in Kerala by a private angling association, where they became well established (MacCrimmon, 1971).

Welcomme (1981, 1988) says that in India Rainbow Trout are 'generally present above the 1200 m contour'.

Iran

Rainbow Trout were first introduced to Iran around 1966, where they were bred in private fish farms at Karadj and on the Jejerud River. Although most of the stock was reared for the market in Tehran, some was released in the Karadj reservoir (MacCrimmon, 1971).

Suitable waters for the species in Iran appear to be confined to those in the Elborz Mountains and to the Ligran-chay River in Azerbaijan Province. The distribution of Rainbow Trout in Iran is restricted to the southerly flowing Jejerud and Karadj Rivers and their associated reservoirs, and to the Mader-su and Namrud Rivers which debouch into the Caspian Sea (MacCrimmon, 1971).

Israel

The earliest attempt to introduce Rainbow Trout to Israel was made before 1935, when ova were hatched in Wadi el Kurn and the Ouja River (Hornel, 1935), but what became of them is unknown.

Rainbow Trout from Switzerland were introduced to Israel in 1947, and stocked in the River Dan in Upper Galilee where, although survival was poor, a naturalized population survived. In 1951, Rainbow Trout were being reared in hatcheries in northern Israel (Tal & Shelubsky, 1951), but this project, and a further introduction from Italy in 1968, was unsuccessful. Fish were again introduced from Italy in the following year for rearing on a farm in northern Israel, fed by the River Jordan (MacCrimmon, 1971).

In 1970, Rainbow Trout were accidentally introduced to Lake Kinneret, where they are said to be unable to reproduce naturally (Gophen *et al.*, 1983; Davidoff & Chervinski, 1984).

According to Ben-Tuvia (1981), 'a small population of rainbow trout is established in the upper reaches of the Jordan River but no data are available on its reproduction'. Welcomme (1988) says that 'one small self-sustaining population is maintained in ponds and a small naturally breeding population in the upper reaches of the River Jordan'.

Japan

In 1877, 10 000 Rainbow Trout ova of the *shasta* strain were despatched from the McCloud River by the California Fish Commission to S. Akekio in Tokyo (Wales, 1939; De Groot, 1985; Chiba *et al.*, 1989), where the hatched fish were released in the Yuki River, near Tokyo (Akekio, 1880).

In 1887, Rainbow Trout were freed in Lake Chuzenji near Nikko City in Tochigi; 6 years later Professor C. Sasaki of Tokyo received 10 000 Rainbow Trout eggs of the *irideus* strain (Worth, 1893), and between 1894 and 1896 the government purchased a total of 100 000 eyed 'steelhead' eggs from the USA. When the first fish were stocked is unrecorded, but 'steelhead' Rainbow Trout are said to have been placed in Lake Mashu and perhaps in some rivers on Hokkaido (MacCrimmon, 1971). Between 1907 and 1927, the United States Fish Commission sent a total of 1 290 000 Rainbow Trout eggs to Japan (MacCrimmon, 1971).

Rainbow Trout fingerlings have been widely stocked in Japan since 1955 (Okada, 1960; Chiba *et al.*, 1989), and by the late 1960s, they occurred in Lakes Chuzenji, Yumoko (Tochigi), Segenuma (Gunma) and Mashu, and in some rivers on Hokkaido; only Lake Yumoko continued to be stocked, and spawning redds were known to exist in the other lakes (MacCrimmon, 1971).

The current status of Rainbow Trout in Japan is uncertain; according to Kawanabe (1980) they are reproducing in the wild only in Hokkaido; Welcomme (1981, 1988) says that 'wild populations are established in some lakes', while Chiba *et al.* (1989) list them as 'being reproduced in certain experimental or natural ponds'.

Korea

In 1945, 120 000 Rainbow Trout eggs were imported to Korea, where they were reared in both the north and the south (MacCrimmon, 1972). Twenty years later, S.C. Chyung imported 10 000 'kamloops' strain Rainbow Trout to South Korea from the Coleman National Fish Hatchery in California, but all the resulting fry died. A year later a further shipment of 200 000 eggs was received from Montana, USA, and reared in the government hatchery in P'yŏngch'ang in Kangwŏn Province. Fingerlings from this stock were planted in several rivers in Kangwŏn, at Hak-Sa-Pyung, Back-Dam-Temple, Sam-Wha-Temple, Cho-Dan-Ri and Nak-Dong-Ri, where in 1969 the fish spawned for the first time (MacCrimmon, 1971).

Between 1967 and 1969, some 3 million eggs were shipped to Korea from the USA and Japan, the fish from which were stocked in private hatcheries in Kangwŏn.

According to Peng (1962), during the war Rainbow Trout survived in North Korea (Korean People's Democratic Republic) but died out in South Korea (Republic of Korea); their current status is uncertain.

Lebanon

In 1960, Rainbow Trout ova were imported from Brande, Denmark, to the experimental station at Anjar. Since 1966, fish have been placed in most of the river systems in Lebanon, but their present status is uncertain (MacCrimmon, 1971).

Sri Lanka

Rainbow Trout (from an apparently unrecorded source) were first introduced to Sri Lanka (then

Ceylon) in 1889 (MacCrimmon, 1971), but it was not until 1902 that breeding was discovered in the wild in waters on the Horton plains, 2195 m asl (Fowke, 1938). According to MacCrimmon (1971) and Welcomme (1981, 1988), Rainbow Trout are now 'breeding freely in waters above the 1220 m contour', principally in the Rivers Nuwara Eliya and Bulu Ellan, in Portswood Dam, in the Farr Lakes, Agra Oya, in the Gorge Valley rivers and in streams on the Horton plains (Wallis, 1969; Fernando, 1971).

Taiwan

Rainbow Trout were first introduced to Taiwan from Japan in 1957, when 3000 fertilized eggs were 'released in a reservoir' (Liao & Liu, 1989) or, according to MacCrimmon (1971) 'hatched and released in mountain streams', where all are believed to have perished. In 1959 and 1960, a total of 400 000 eyed ova were imported from the same source, but again all died (Liao & Liu, 1989) or, according to MacCrimmon (1971), were 'lost due to a typhoon'. In 1961, however, 50 000 eyed ova were imported, hatched and reared to maturity at Ma-Lin, a research station of the Lukang Branch of the Taiwan Fisheries Research Institute (Chen, 1976; Liao & Liu, 1989). From this hatchery, fingerlings have been introduced to high-altitude reservoirs at Taichin (Taichung) in 1963, at Wan Tai (Nanton) in 1964, and at Shihmen (Taoynam) in 1965. Adult fish have been caught, but it is not known with certainty whether the populations are self-supporting (MacCrimmon, 1971).

Unsuccessful Introductions

Unsuccessful attempts have been made in Asia to naturalize Rainbow Trout also in the People's Republic of China (Yo-Jun & He-Yi, 1989), Indonesia (Schuster, 1950; Muhammad Eidman, 1989), Iraq, Jordan (?), Syria, Thailand (Piyakarnchana, 1989; De Iongh & Van Zon, 1993), Turkey (Erencin *et al.*, 1972); Nepal, Pakistan (MacCrimmon, 1972), Malaysia (Ang *et al.*, 1989), the Philippines, and the USSR (MacCrimmon, 1971, 1972; Welcomme, 1981, 1988).

Africa

Ethiopia

The earliest introduction of Rainbow Trout to Ethiopia, believed to have come from Italy, was to Lake Aba Samuel in 1936 and 1940 (Tedla & Meskel,

1981). In March 1967, J.H. Blower of UNESCO imported some from Kenya, which were released in the Danka, a montane river near Goba, Bale, where Tedla and Meskel (1981) reported them to be 'performing well'. MacCrimmon (1971) said they were also planted in the Rivers Web and Dinchia in the same region. In 1973, some were transferred by F.H. Meskel and the Ministry of Agriculture to the River Sibilo, and in the following year others to the Rivers Chacha and Beressa and to Lake Wonchi in Shoa; they became established in the Sibilo, but their fate elsewhere was said by Tedla and Meskel (1981) to be uncertain, though Welcomme (1981, 1988) stated that 'some self-sustaining populations reputedly continue to exist in some highland streams', and MacCrimmon (1971) said that 'the potential for trout in Ethiopia is excellent'. Moreau *et al.* (1988) listed them as established in the Rivers Danka and Welo.

Kenya

Rainbow Trout obtained from the Jonkershoek Hatchery at Stellenbosch in South Africa were introduced to Kenya in 1910 by Commander Barry, RN, and became naturalized in the Nairobi River on the northern slopes of Mount Kenya. Nine years later, fish from this river were transferred to the Nanyuki River, also on Mount Kenya, where they likewise became established. In the same year, more eggs imported from Jonkershoek hatched successfully, and the fry established a population in streams at Molo and Londiani on the Mau Escarpment and, in 1921, in the Katamayu River in the Aberdare Mountains. A year later, Rainbow Trout were placed in a number of streams on the eastern and northeastern slopes of Mount Kenya (MacCrimmon, 1971).

By around 1927, Rainbow Trout had been planted in most apparently suitable Kenyan waters. They are presently naturalized in streams on Mount Kenya; in the Aberdare range; on the Mau Escarpment from Nakuru west to Kerichs; in the Cherangani Mountains; and on Mount Elgon. They are also established in the River Malewa, Lake Naivasha (Muchiri & Hickley, 1991). Although they can survive only at altitudes above 1500 m, breeding is seldom successful below 2000 m (MacCrimmon, 1971).

D.F. Smith built the first permanent trout hatchery in Kenya in 1947, and thereafter eggs have been imported on numerous occasions from the Surrey

Trout Farm in England and elsewhere to stock reservoirs and dams between 1500 and 2000 m (MacCrimmon, 1971).

Lesotho

Rainbow Trout from Cape Province, South Africa, were introduced to the Tsoelikane River in Lesotho (formerly Basutoland) in 1907 (Moreau *et al.*, 1988) and again in 1943 (Pike & Tedder, 1973), and in 1957 some were transferred from the Tlolohatse River to the Khubelu River (both in the Orange River system) (Liebenberg, 1967–1968). According to MacCrimmon (1971), 'the rainbow trout is self-sustaining in all streams in very high country in the Drakensberg Mountains, and only needs supplemental stocks in times of severe drought'. Rainbow Trout are currently present in the upper Orange system (the higher reaches of the Moremoholo, Senqu and Tsoelikane Rivers (Pike & Tedder, 1973; Gephard, 1977; Bruton & Merron, 1985; De Moor & Bruton, 1988).

Ecological Impact   In the Tsoelikane River, Rainbow Trout compete with the endangered native Maluti Minnow *Oreodaimon quathlambae* (Skelton, 1987), since the habitat preferences and diets of the two species are very similar, and possibly also prey on them in their early life stages (De Moor & Bruton, 1988). The range of Rainbow Trout in Lesotho is said to be restricted by excessive soil erosion which causes turbidity (Gephard, 1977).

Madagascar

Eyed Rainbow Trout ova from the River Isère near Vizille in France were placed in waters of Périnet (950 m asl) in the Malagasy Republic by M. Louvel in 1922, when a hatchery was also built at Manjakatompo in the Ankaratra, at an altitude of 1900 m (Kiener, 1963; Moreau, 1979; Reinthal & Stiassny, 1991).

Rainbow Trout in Madagascar are presently naturalized at Manjakatompo; in some rivers of Antenina; in tributaries of the Onire in Ambatolampy; and at Lakera in Fianarantso and Ialatsara in Ambohimahasoa, where they occur in 'cold, high altitude areas' (Welcomme, 1988) between 1700 and 2000 m.

Malawi

Rainbow Trout were first introduced to Malawi (then Nyasaland), from an apparently unrecorded source,

in 1907; these fish, and those from a subsequent importation in 1932, were used to stock streams and dams on the Zomba plateau (1220 m asl), where they became naturalized. From Zomba, Rainbow Trout were later translocated to streams on Mlanje Mountain at an altitude of around 1520 m, where they also became successfully established (MacCrimmon, 1971).

In 1951, fish from Jonkershoek, South Africa, and Sagana, Kenya, were reared in the Nchenachena hatchery (1077 m asl), where a second shipment arrived from Jonkershoek 3 years later. Rainbow Trout from the Nchenachena hatchery, which was abandoned before 1971, were placed in the Kaziweziwe River at Nchenachena and in streams and dams on the Nyika plateau, 2134 m asl, in 1953. Ten years later, waters on the Nyika plateau were successfully restocked with Rainbow Trout from Jonkershoek (MacCrimmon, 1971).

Rainbow Trout have been caught near Mlowe in Lake Malawi (in which epilimnetic temperatures range between 21 and 28.5°C), which they are able to reach from the Nyika plateau via the Chelinda, Kaziweziwe and/or Rukuru Rivers, but from which high waterfalls prevent their return. They have also been recorded from the Likangala River, which drains the Zomba plateau, near Lake Chilwa; this suggests that Rainbow Trout in Malawi can survive in waters as low as 500 m asl, but only reproduce at altitudes above 1000 m (MacCrimmon, 1971). Welcomme (1988) records them in Malawi as 'successfully introduced into cold mountain streams'.

South Africa

Rainbow Trout eggs from the UK were first imported to Cape Province, South Africa, in 1896, but none hatched successfully (MacCrimmon, 1971). However, eggs obtained in the following year from the same source were reared successfully at the Jonkershoek hatchery at Stellenbosch, from where fertilized eggs from mature fish were in 1899 sent to the Eastern Cape Province, Natal, and the Orange Free State (Anon., 1944).

In the Eastern Cape Province, the Frontier Acclimatization Society built a hatchery in the late 1890s on the banks of the Tyusha, a tributary of the Buffalo River, from which adult Rainbow Trout were subsequently introduced to the Rooikrantz dam on the Buffalo.

By 1930, a number of hatcheries had been established in various localities in South Africa, one of the first of which, at Pirie near Kingwilliamstown, was begun in the 1890s (De Moor & Bruton, 1988); others included those at Tedworth in Natal (1903) (Pike, 1980) and at Lydenbury in Transvaal (1915) (Du Plessis, 1961).

According to De Moor and Bruton (1988), (by whom the species is assigned to the genus *Parasalmo* (a junior synonym for *Oncorhynchus*) and from whom much of the ensuing information is taken), Rainbow Trout are regularly stocked in many river systems in South Africa, and the following account of their distribution (which, because many unrecorded stockings have taken place, is not complete) does not necessarily imply that naturalized populations have existed in all the localities mentioned, since data in the literature are often inconclusive.

Rainbow Trout were established in the Ibisi and Gungununu Rivers in East Griqualand by 1923, but had disappeared before 1948 (Harrison 1940, 1948). In 1932, they were said to be thriving in the Ingwangwana, Umzimkulu, Umzimouti, Ushiayke and Umtshezana Rivers in Natal (Day, 1932a,b). In the Transkei, Rainbow Trout were referred to by Hey (1926) as occurring in the Rivers Tsitsa, Little and Big Pots, Tina, Tinana, Mvenyani, Umzimhlava, Ginqoskoi, Ibisi, Ncweleni, Mnceba and Cancele in the Umzimvubu system, and also in the Gora and Manina Rivers of the Bashee system, and in the Engcobo of the Engwali system. Rainbow Trout were found in the main Umtata River until 1926 (Hey, 1926), and in the Potspruit and Sterkspruit (Lydenberg) by the mid-1920s (Du Plessis, 1961).

By 1949, Rainbow Trout were to be found in the western and southern Cape in the Liesbeeck River and lake; in Steenbras reservoir (where they did not reproduce); in the Rivers Eerste, Lourens, Upper Berg, Wemmer and Dwars; in the lower Breede, and in the Smalblaar, Holsloot, Dwars, Hex and Eland tributaries (Harrison, 1949).

In the eastern Cape and Transkei, Rainbow Trout had colonized the Rivers Tyume, Keiskammahoek, Upper Buffalo, Upper Kubusie, Langkloof, Kraai (near Barkly East), Indwe, Upper Tsomo, Upper Bashee, Umtata and Tsitsa, and tributaries near Maclear and Ugie; the Upper Tina, Luzi and Tinana Rivers (Mount Fletcher), and the Kenegha, Ceyata, Mabela and Mvenyana Rivers near Matatiele, as well as the nearby Umzimvubu and Krom Rivers (De

Moor & Bruton, 1988). In East Griqualand, Rainbow Trout were found in 1949 in the Rivers Umzimhlava (Mount Ayliff), Bizana (East Pondoland), and the Umzimhlava, Umzimvubu, Ibisi, Gungununu, Ndowana and Ingwangwana in Kokstad (Harrison, 1949).

De Moor and Bruton (1988) record that Rainbow Trout in South Africa are presently found in mountainous parts of the Cape and Natal (Crass, 1968, 1969), and in the eastern Transvaal. In the Cape, Rainbow Trout survive in most of the colder non-acidic waters of the south coast drainage system and in the Olifants system of the western Cape. They only occur in a few streams in the Eastern Cape, as most are too acid for their survival (Jubb, 1965).

According to De Moor and Bruton (1988), naturalized populations of Rainbow Trout are known to occur in South Africa in the following localities:

Western and *southern Cape.* In the Olifants system, where reproduction takes place in the main stream above the Leerkloof River and in the Leerkloof, Riet, Twee, Middeldeur, Kromme and Driehoek tributaries. Also in numerous agricultural impoundments in the Koue Bokkeveld. In the western Cape, Rainbow Trout are found in the Wemmershoek reservoir and Wemmer River (Harrison, 1966–1967), the Eerste, Lourens and Berg (upriver from La Motte), and also in the Dwars, Smalblaar, Molenaars, Elandspad, Holsloot, Witels, Elands and Jan du Toits Rivers, and the Bree River system.

Eastern Cape. In the Upper Tyume River (Gaigher, 1975a) and Bulk River dam (Port Elizabeth) (Donnelly, 1965).

Natal. In the Umtamvuna and Weza Rivers; in the Ndowana and Slang (in the upper Vaal system) and the Pivaan, Ingwangwana, Polela, Mnweni and Incandu Rivers (Crass, 1964); and in the upper reaches of the Umvoti, Umzimkulu, Umkomaas, Tugela, Buffalo and Phongolo Rivers (Crass, 1966).

Transvaal. Rainbow Trout have been established in rivers of the eastern and northeastern Transvaal since the 1920s (Le Roux & Steyn, 1968) at altitudes between 1220 and 1830 m, where Magoebaskloof, Sabie, Pilgrims' Rest, Lydenburg, Machadodorp, Dullstroom and Waterval Boven are the principal localities (Wallis, 1969).

Orange Free State. In the Upper Orange and Caledon Rivers (Skelton, 1986).

Populations are also known to occur in numerous other localities in South Africa, but since in most regular restocking occurs, their independent viability is in doubt.

Ecological Impact   Although the Rainbow Trout is not regarded as such a serious threat to native fishes as some other alien predators, such as the Smallmouth Bass *Micropterus dolomieui* (q.v.) and Largemouth Bass *M. salmoides* (q.v.) Skelton (1987) lists predation and/or interspecific competition by Rainbow Trout as one of the threats to the following South African Red Data Book fishes: *Barbus burgi, B. phlegethon, B. andrewi, B. treurensis, B. trevelyani, B. burchelli, B. tenuis, Sandelia bainsii,* and *Kneria auriculata.* According to Jackson (1960), Rainbow Trout have also eradicated local populations of Cape Kurper *S. capensis* and Haarder *Trachyistoma euronotus.* However, in most cases, as De Moor and Bruton (1988) point out, habitat destruction has been a more significant factor in the decline of these threatened indigenes than predation by, or competition with, Rainbow Trout; some exceptions are as follows.

In the Treur River in the Limpopo system, to which Rainbow Trout were introduced between 1957 and 1981, the combined impact of *mykiss* and *dolomieui* and the concomitant introduction of 'white spot disease' are believed to have contributed to the local extinction of *B. treurensis* (Pott, 1981).

In the Eerste River, where Rainbow Trout were introduced in 1901–1902, predation on *B. burgi* is believed to have been a major contributory factor in the extinction of that species in the 1930s or 1940s (De Moor & Bruton, 1988).

In the Buffalo River in the Eastern Cape, to which Rainbow Trout were introduced in 1901, a small relict population of *B. trevelyani* is severely threatened by *mykiss* predation. In the Tyume River, a combination of the impact of *mykiss, dolomieui*, water extraction, siltation and impoundment has forced *trevelyani* from optimum into marginal habitats (Gaigher, 1979).

Kleynhans (1984) believed that the introduction of Rainbow Trout in 1976–1977 to the Wilgekraalspruit, a tributary of the Crocodile River in the Incomati system of Lydenburg, was directly responsible for the disappearance there of *Kneria auriculata*: this supposition is supported by the fact that in 1985, after *mykiss* had itself vanished, *auriculata* reappeared.

Day (1932a,b) suggested that Rainbow Trout were less successful than Brown Trout *Salmo trutta* (q.v.) in colonizing many waters in Natal, probably due to preferential predation by *Barbus natalensis*: indigenous *Barbus* species move upriver with the coming of the spring rains; Rainbow Trout spawn after this dispersal whereas Brown Trout spawn beforehand, and the latter's ova are thus less susceptible to predation by *Barbus* species.

The introduction of alien Rainbow Trout and the translocation of the native Chubbyhead Barb *B. anoplus* has resulted in the colonization of new regions of the Natal uplands (especially Himeville and Nottingham Road) by several species of piscivorous birds, including kingfishers (Alcedininae), White-breasted Cormorants *Phalacrocorax lucidus*, Reed Cormorants *Haliëtor africanus*, African Darters *Anhinga rufa*, African Fish Eagles *Haliaeetus vocifer* and Ospreys *Pandion haliaetus* (Alletson, 1985).

De Moor and Bruton (1988: appendix I: 268–270) give in detail (with references) dates of initial stocking and earliest records of Rainbow Trout in various localities in South Africa.

Sudan

Rainbow Trout from a hatchery at Nyeri, Kenya, are believed to have been introduced to the Kinyeti River, near the Ugandan border with the Sudan in 1947. Two years later, fish hatched at the Kiganjo hatchery in Kenya, from eggs produced in Gloucestershire, England, were delivered as fingerlings to Juba in the southern Sudan, where they were placed in streams at Nagishot and Kitiri in the Eastern Equatorial Province (MacCrimmon, 1971), where Welcomme (1981) records them as 'established in highland waters'.

Swaziland

Moreau *et al.* (1988) list the Rainbow Trout as introduced in 1908 to Swaziland, where it became established in cold waters.

Tanzania

Rainbow Trout eggs from Scotland are believed to have been first imported to Tanzania (then Tanganyika) around 1924, and to have been hatched out in a small stream in the Usambara Mountains at an altitude of 1675 m; some of these fish were sent as fingerlings to streams on Mounts Meru and

Kilimanjaro (MacCrimmon, 1971). Three years later, the Mbusi River, near Lushoto in the western Usambara Mountains, was stocked. In 1931, fry were flown from Kenya to the Arusha region of Mount Meru, and in the same year Rainbow Trout became established in streams in the Mosho district of Mount Kilimanjaro. A further shipment of ova from Britain and South Africa arrived in Tanzania in the late 1930s, and from these Rainbow Trout became naturalized in highland streams in the Southern Provinces (MacCrimmon, 1971).

According to MacCrimmon (1971), all streams on the slopes of Kilimanjaro above 1370 m, with a water temperature of 15.6°C and a pH of 5, are suitable for Rainbow Trout; spawning occurs between May and July and again in December above 1670 m and at a temperature of between 11 and 12°C, and fish are stocked up to 2140 m where the temperature range is between 9 and 10°C. Rainbow Trout in Tanzania are currently found in waters on the Elton plateau, Mufindi and Mbeya in the southern provinces, on Mount Meru and the Pare Mountains in the north, and in the Usambaras in the coastal region (MacCrimmon, 1971).

Tunisia

Since 1965, Rainbow Trout from the then Federal Republic of Germany have been imported into Tunisia, where they are now found in rivers of the northern region near Ain Draham; whether populations are self-maintaining is unclear (MacCrimmon, 1971).

Uganda

In 1925, 2000 Rainbow Trout fry, derived from eggs imported to Uganda from the Solway Hatchery in Scotland, were successfully introduced to the River Suam on Mount Elgon (2440–4360 m). Five years later, the River Bukwa, at a similar altitude, was stocked with fingerlings transferred from the Suam. In 1952, 1960, 1962 and 1963, more fingerlings were placed in the Sipi River, where breeding was reported in the late 1960s. Introductions in the 1950s to the Rivers Nyaluti and Atar on Mount Elgon were unsuccessful (MacCrimmon, 1971).

In the Ruwenzori Mountains (the 'Mountains of the Moon') in southwestern Uganda, the Mpaga River (1680–5100 m) was first stocked in 1952 with 2000 eyed Rainbow Trout eggs from the Nyeri Trout Hatchery in Kenya. In the following year, the Mubuku and Nyamagasani Rivers were each stocked with 10 000 eyed ova from the Jonkershoek hatchery in South Africa, and in the same year Rainbow Trout were translocated from the Kinyeti River in the Sudan (see above) to the Aringa River on the Ugandan side of the Imatong Mountains. Between 1955 and 1958, the Murasegi, Kanyampara and Mbwa Rivers were also stocked (MacCrimmon, 1971). According to MacCrimmon (1971), Rainbow Trout in Uganda 'are known to be naturalized in rivers of Mount Elgon and the Ruwenzori Mountains'.

Zambia

Moreau *et al.* (1988) list Rainbow Trout, introduced from Malawi at an unrecorded date, as established in the wild in Zambia.

Zimbabwe

In 1910, 3000 Rainbow Trout ova from the Jonkershoek hatchery in South Africa were imported to Zimbabwe (then Southern Rhodesia), where 6000 fingerlings were later unsuccessfully planted in the Gwebi River, Darwendale (Toots, 1970). Further unsuccessful attempts to establish Rainbow Trout in Zimbabwe were made to waters in Matabeleland (Khami), Mashonaland (Cleveland, Mtepatepa and Borrowdale) and the Eastern Districts (Stapleford, Inyanga) (Toots, 1970), in 1916–1918, 1921, 1924 and 1927 (MacCrimmon, 1971). In 1929, however, 4000 Rainbow Trout ova from the hatchery at Pirie in South Africa were hatched at Penhalonga in the Stapleford Forest Reserve, and fry introduced in the Odzani River (where further plantings were made in 1930–1931) survived and became established. According to MacCrimmon (1971), 'Rainbow Trout are now naturalized in the highlands of the Inyanga and Chimanimani areas in some of the eastward-flowing rivers and tributaries, namely Pungwe, Gairesi, Haroni, and headwaters of the large Sabi catchment'.

Unsuccessful Introductions

Unsuccessful attempts have been made in Africa to establish Rainbow Trout also in Cameroun, Congo, Morocco and Mozambique (Moreau *et al.*, 1988) and Namibia (South West Africa) (Schrader, 1985).

South and Central America

Argentina

Because of the country's relative paucity of native fish species, at the end of the 19th century Dr Francisco P. Moreno, a scientist employed by the Argentinian government, decided to try to establish several species of exotic fish in the Patagonian cordillera. From 1900 to 1903 the Argentinian Ministry of Agriculture carried out feasibility studies regarding the possible introduction of various exotic species, and especially between 1904 and 1910 eyed ova of several salmonids (mainly *Salmo*, *Salvelinus* and *Oncorhynchus*) were imported from the USA, England and Germany, the fry from which were released in Andean rivers and lakes in the provinces of Nequén, Río Negro and Santa Cruz, and also in Alta Gracia and La Cumbre in the Province of Córdoba (Navas, 1987).

Only four species, the Rainbow Trout, Brown Trout, the landlocked ('*sebago*') variety of the Atlantic Salmon and the American Brook Trout (all q.v.), became established in suitable clear, well-oxygenated and cold rivers and lakes of the Patagonian cordillera, from Nequén to Santa Cruz and Tierra del Fuego. The Patagonian National Parks of Lanín, Nahuel Huapí, Lago Puelo, Los Alerces, Perito Moreno and Los Glaciares and that of Tierra del Fuego now support flourishing populations of these four species, which provide high-quality game fishing. Those introductions made in the Province of Córdoba and subsequently in streams of the Sierra de la Ventana, Buenos Aires, were unsuccessful.

In June 1904, some 70 000 Rainbow Trout eggs were despatched by ship from New York bound for the Nahuel Huapí Hatchery in Argentina; off the coast of Brazil, however, many started to hatch prematurely, and had to be dumped overboard; on 23 July the few remaining unhatched eggs, which it was decided could not reach the hatchery alive, were placed in Laguna La Grande (Tulian, 1910).

On 4 February 1905, a further shipment of fish ova from New York, which included 92 000 Rainbow Trout eggs, arrived in Buenos Aires, where most of the eggs were found to be dead or dying. On 10 February 1906, E.A. Tulian left New York for Argentina via England (where he picked up eggs of the Atlantic Salmon *Salmo salar*) with 25 000 Rainbow Trout ova, arriving in Buenos Aires on 17 March, when many of the trout eggs were found to be dying or dead (Tulian, 1910).

On 18 January 1908, yet another shipment of salmonid eggs, including 30 000 Rainbow Trout, left New York for Argentina, where once again a high percentage were found to have died. The surviving eggs were planted in tributaries of the Río Santa Cruz, which rises in Lago Argentino in the Andes at an altitude of around 8000 m (Tulian, 1910).

On 6 May 1908, Tulian left New York once again with around 300 000 'steelhead' Rainbow Trout eggs bound for Argentina via England, where an additional 50 000 Rainbow Trout eggs from Germany were acquired. This consignment arrived at La Cumbre hatchery in Argentina on 13 June. On this occasion the loss of eggs was very small (Tulian, 1910).

Between 1904 and 1923, a total of around 370 000 Rainbow Trout eggs was imported to Argentina from the USA, and in 1930 ova were translocated from Chile and the resulting fingerlings planted in waters in Córdoba (MacCrimmon, 1971). In 1969 a further shipment was acquired from Denmark.

Although Welcomme (1981, 1988) says that Rainbow Trout are not established in Argentina, MacCrimmon (1971) and Navas (1987) say they have become naturalized as described above.

Ecological Impact   Through interspecific competition for food and/or space and/or predation, naturalized salmonids (especially Rainbow Trout) in Argentina appear to have greatly depleted the populations of those few indigenous fish species with which they occur sympatrically. The reduction of native prey species has caused a corresponding decrease in the numbers of some indigenous mammals such as the Otter *Lutra provocax*, which seems unable to catch the alien salmonids. The three introduced trout, however, are preyed on by the native Canadian Otter *L. canadensis*, whose range extends to southern South America, and, according to Navas (1987), by the 'nutria europea *Lutra lutra*' (the European Otter).

Bolivia and Peru

In 1937, the Bolivian and Peruvian governments jointly set up the Chucuito Fish Culture Station in Puno, southern Peru, from where in 1939, 1941, 1942 or 1943 (accounts differ) Rainbow Trout were introduced to the waters of Lake Titicaca, which straddles the Bolivian/Peruvian border at an altitude of 3800 m asl, where the physicochemical constituents of the lake seem ideal for self-perpetuating Rainbow Trout populations and where by 1950 they were said to be 'abundant throughout the lake and its tributaries' (Everett, 1971, 1973). A commercial trout cannery was opened in 1961, but closed 9 years later, due apparently to a combination of overfishing, a decline in water level (Matsui, 1962; Everett, 1973) and 'bad environmental management' (Welcomme, 1988).

In 1948, Rainbow Trout translocated from Chile were imported to the Pongo Hatchery in Bolivia, from where Rainbow Trout have been planted in rivers and lakes in the Bolivian Andes, including the Cordillera Real, Cordillera Oriental, especially in the Tunari (near Cochamba) and rivers near Potosi, in many of which they have become naturalized.

In Peru, the importation of Rainbow Trout ova from the USA was first organized in 1928 by Señor Mitchell of the Compañia Metalurgica, which constructed a hatchery at Oroya on the banks of the Río Tishgo, from where Rainbow Trout were planted in suitable lentic and lotic waters of the Peruvian Andes in the districts of Junín and Pasco, especially in the Río Mantaro system (MacCrimmon, 1971).

In 1930, 50 mature Rainbow Trout were transferred from Oroya to the Ingenio Hatchery on the Río Chiapuquio, a tributary of the Mantaro in Huancayo. Plantings of Rainbow Trout from Oroya and Ingenio resulted in naturalized populations in central Andean waters, in the Mantaro system, in Lakes Junín (Chinchayacocha), Marcapomacocha, Punrun and elsewhere (MacCrimmon, 1971).

According to MacCrimmon (1971), 'Rainbow trout have been planted in all suitable accessible waters of the Andes above 2000 m altitude, where they have become self-sustaining'. In Titicaca they reach a weight of 12 kg and in Marcapomacocha (4500 m asl) 4.5 kg. Several other hatcheries have been established, from which 4–6 million Rainbow Trout are stocked annually (MacCrimmon, 1971).

Ecological Impact According to Welcomme (1981, 1988), Rainbow Trout in both Bolivia and Peru have 'eliminated many native local species including *Orestias* for which there was previously a fishery'. (See also under *Odontesthes bonariensis*.)

Brazil

According to Welcomme (1981), who gives no further details, Rainbow Trout from Britain were unsuc-

cessfully introduced to Brazil in 1913 and from Denmark in 1949. Welcomme (1988) also gives a date of 1942 for an introduction from an unnamed source. MacCrimmon (1971), however, says that Rainbow Trout from Denmark were first imported to Brazil in 1952, where they were placed in the Bocaina River, Río de Janeiro, where they became naturalized. In 1962, W.E. Ripley imported 200 000 ova from the Hot Creek Hatchery in California, and fingerlings were later stocked in the region around Campos de Jordão (Saõ Paulo), Nova Friburgo (Río de Janeiro) and upper Santa Catarina State where, according to MacCrimmon (1971) 'these plantings have resulted in naturalized populations'.

Chile

According to McClane (1965), Rainbow Trout from Germany were first introduced into Chile in 1905, where they were placed in waters to the south of Santiago.

Eigenmann (1927) recorded Rainbow Trout in Laguna del Inca and Río Blanco (Lautaro), and Oliver (1949) said they occurred in Río Bio-Bio and in the vicinity of Concepción; according to Mann (1954), they were abundant in rivers and lakes (all quoted by De Buen, 1959).

MacCrimmon (1971) said that Rainbow Trout were then naturalized more or less continuously in Chile between 31 and 51°S, with a discrete population in Río Loa, Antofagasta (23°S). Some 1.5 million fry were being stocked annually in Chile, where the Rainbow Trout is regarded as a very 'voracious' fish which displaces other salmonid species, and is the most widely distributed.

Colombia

'Steelhead' Rainbow Trout eggs were first unsuccessfully imported into Colombia from the USA in 1926. Eggs were incubated successfully in 1929 at Los Pozos (3400 m asl) near Lago de Tota, and fry were introduced to the lake from which, 3 years later, eggs were stripped from mature fish. Subsequently, Rainbow Trout became naturalized in Laguna de la Cocha near Pasto (MacCrimmon, 1971). MacCrimmon said that they were found in all stocked waters in the country with a water temperature not exceeding 17°C. Welcomme (1981) records the Rainbow Trout in Colombia as established in Lago de Tota and in 'some highland streams'. Because of climatic conditions in

Colombia, where Rainbow Trout have caused the local extinction of endemic parasitic catfishes of the genus *Trichomycterus*, the introduced fish can be induced to breed throughout the year, and their eggs are widely exported (Welcomme, 1981, 1988).

Costa Rica

Rainbow Trout were first imported into Costa Rica from the USA in 1927–1928. It was not, however, until October 1954, when Señor Jorge Zeledón introduced Rainbow Trout brought in from Mexico to a small man-made water in Finca la Georgina in the central Volcanic Range, that they became established: some of these fish managed to escape from the lake, where they formed a large population downstream in the Quebrada del Desengaño (MacCrimmon, 1971).

In 1956, Señor L. Cruz imported 50 000 Rainbow Trout eggs from the USA which he successfully placed in the Ríos Coton and Coto Brus, near the Panamanian border (McNally, 1959). Between 1959 and 1962, a total of 250 000 ova was obtained from the USA, the resulting fry being successfully introduced to the Ríos Macho, Pejiballe, Reventazón, Parrita, Humo, Poás, El Roble, La Paz and Playas (MacCrimmon, 1971). More were planted in the Ríos Durazno and Virilla, but were lost following eruptions of the Irazú volcano, and some adults were transferred by native people from the Parrita to the Río Savegre. Welcomme (1981, 1988) records Rainbow Trout in Costa Rica as 'well established in high altitude zones of all rivers', where they show 'very rapid growth'.

Ecuador

Rainbow Trout eggs from the USA were first unsuccessfully imported into Ecuador in 1928; in the following year, however, a further shipment arrived which was successfully hatched at El Sena and at Cotopaxi. In 1933, Dr P. Holst in the USA presented 100 000 Rainbow Trout ova to the Punyaru Hatchery on San Pablo Lake, and a year later the town of Tungurahua brought in more eggs from the USA, and planted the fry in nearby rivers and lakes. Between 1932 and 1939, Rainbow Trout became acclimatized in the Cotopaxi, Punyaru and Chirimachay hatcheries (MacCrimmon, 1971).

After 1951, Ecuador began to import Rainbow Trout from Chile, Colombia and Peru, as well as from the USA. From 1963 until at least 1971, between 8000 and 20 000 fish were introduced annually to suitable waters by the Ecuador government (MacCrimmon, 1971), and further fish were imported from Argentina in 1969.

According to MacCrimmon (1971), 'Rainbow trout are now naturalized in rivers, lakes, reservoirs, canals, and ponds of the mountains in the provinces of Carchi, Imbabura, Pichincha, Cotopaxi, Napo, Pastaza, Tungurahua, Chimborazo, Bolívar, Cañar, and Azuay'.

Panama

Rainbow Trout were first imported to the Río Chiriquí in western Panama, around 1500 m asl, in 1925 (Hildebrand, 1935b, 1938). This original introduction, supplemented by others in 1932–1933 and 1950 from the USA, resulted in the establishment of naturalized populations in both the Chiriquí and Caldera Rivers of El Volcan de Chiriquí (Walters, 1953). Further releases were made in 1968–1969 in various rivers and lakes in the Volcan/Boquete district (MacCrimmon, 1972). Welcomme (1981, 1988) records that in Panama Rainbow Trout 'breed freely above 1100 m'.

Venezuela

In 1935 and 1938 respectively, 50 000 and 100 000 Rainbow Trout ova were imported into Venezuela from the USA (MacCrimmon, 1972). After 1940, eggs were stripped from fish established in the central Andes (León, 1966).

Rainbow Trout are now naturalized in the central Venezuelan Andes in the states of Trujillo, Mérida and Tachira, at altitudes between 1700 and 4000 m asl (MacCrimmon, 1971).

Ecological Impact in western South America

According to Kear and Williams (1978) and Kear (1990), where they occur in the same fast-flowing waters in western South America, Rainbow Trout and other naturalized insectivorous fish have had a negative impact on native Torrent Ducks *Merganetta armata*. (See also under Brown Trout.)

Unsuccessful Introductions

Unsuccessful attempts have been made to naturalize Rainbow Trout in Central and South America in Guyana (British Guinea), Honduras, Paraguay, Puerto Rico (Nichols, 1929; Hildebrand, 1934;

Bonnet, 1941; Erdman, 1947, 1972, 1984; Iñigo, 1949; Soler, 1951), Uruguay and the West Indies (Cuba, the Dominican Republic and Trinidad) (MacCrimmon, 1971, 1972; Welcomme, 1981, 1988).

Australasia

Australia

New South Wales
Rainbow Trout were first introduced to New South Wales in 1894, when 3000 ova were imported from New Zealand to a hatchery at Prospect reservoir (Roughley, 1951; MacCrimmon, 1971; Jackson, 1981; McKay, 1984; Allen, 1989). By 1908, populations were well established in highland waters (Stead, 1908), and by 1922 Rainbow Trout were considered to be more successfully established than Brown Trout *Salmo trutta* (q.v.) (McCulloch, 1922). Rainbow Trout have done well in the tableland rivers of the Eastern Highlands of New South Wales, especially above 600 m, and sometimes as low as 300 m (Weatherley & Lake, 1967). They are established along the entire length of the Murrumbidgee River (including Lake Burrinjuck, an artificial water at the confluence of the Yass, Murrumbidgee and Goodradigbee Rivers (Burchmore & Battaglene, 1990)), in the Murray River to beyond the South Australian border, and in the Barwon and Darling Rivers as far downstream as Brewarrina (Lake, 1957; Weatherley & Lake, 1967; MacCrimmon, 1971; McDowall & Tilzey, 1980; Clements, 1988), below which summer temperatures up to 30°C limit their distribution (Weatherley & Lake, 1967; Cadwallader & Backhouse, 1983).

Queensland
Rainbow Trout were first imported to the hatchery at Killarney in 1896, from where between the following year and 1900 fingerlings were planted in streams between central and southern Queensland west and northwest of Brisbane; they only became successfully established, however, in Spring Creek, Killarney. In the early 20th century, Rainbow (and Brown) Trout were unsuccessfully transplanted from New South Wales to streams on the Atherton tablelands and in Lake Tinaroo in northern Queensland (MacCrimmon, 1971; McKay, 1978; Clements, 1988).

South Australia
Rainbow Trout were originally imported into South Australia from New South Wales in 1900, and fry were placed in the Sturt River of the Mount Lofty Range before the First World War (MacCrimmon, 1971). Between 1934 and 1936, fingerlings from the Ballarat Hatchery in Victoria were released in the Boughton, Torrens, Finniss and Angas Rivers (Morrissy, 1967, quoted by MacCrimmon, 1971). From 1951 to 1961, Rainbow Trout eggs from New Zealand, Tasmania, Victoria and Western Australia were hatched at Ovingham, a suburb of Adelaide; the fry were liberated in all permanent streams south of the Wakefield River, apart from the Glenelg River and Mosquito Creek (Morrissy, 1967). Rainbow Trout are still stocked in rivers and artificial impoundments of the Mount Lofty Range, and are commonest near Mount Lofty (particularly in the Light, Little Para and Sturt Rivers close to Adelaide, becoming less abundant further north and east) (MacCrimmon, 1971; Clements, 1988).

Tasmania
Rainbow Trout from New South Wales were unsuccessfully translocated to Tasmania in 1898 (Weatherley & Lake, 1967). In 1904, however, a man-made water, Lake Leake (640 m asl), was stocked with yearling fish, and 2 years later they entered a principal feeder, the Snow River, to spawn; ova from naturalized stocks on the Snow River and Great Lake (1030 m asl, which was stocked in 1910) on the Liawenee Canal were subsequently utilized to supply state hatcheries. After 1921, Rainbow Trout declined in Lake Leake and, during the Second World War, in Great Lake, due, it was thought, to predation by Brown Trout; between 1960 and 1969 some 57 000 mature Brown Trout were removed from these two lakes, and replaced with Rainbow Trout (Nicholls, 1957–1958). Today there are a large number of still-waters in Tasmania that contain Rainbow Trout (Lynch, 1970; MacCrimmon, 1971; Clements, 1988).

According to MacCrimmon (1971, 1972), 'self-sustaining populations of Rainbow Trout occur in most environmentally suitable streams of Tasmania', where Merrick and Schmida (1984), say that 'new techniques are being developed for increasing natural spawning success and juvenile survival'.

Victoria
Although the earliest recorded releases of Rainbow Trout in Victoria were made in the Croppers and Delatite Rivers in 1907, it seems probable that ova from New South Wales were first imported to the

state around 1898. Between 1907 and 1929 a further 92 rivers and streams were stocked, and in the latter year the first lake, Eaton's dam, was planted with Rainbow Trout; by 1948 another 14 natural and man-made lakes had been stocked (MacCrimmon, 1971; Cadwallader & Tilzey, 1980; Clements, 1988).

By the late 1960s, Rainbow Trout were naturalized in most of the upland tributaries of the Murray River basin, and in headwaters of the Glenelg, Tarwin, Agnes, Albert, Tarra, Mitchell, Timbarra, Snowy and Brodribb Rivers, and in south-flowing feeders of Lake Wellington, most of which are, none the less, restocked annually (Wharton, 1969; Crowl *et al.*, 1992). In rivers in northeastern Victoria, Rainbow Trout predominate over introduced Brown Trout; although rivers south of the Dividing Range are heavily stocked with Rainbow Trout, they are not abundant (MacCrimmon, 1971; Jackson & Davies, 1983).

Western Australia
In the late 1890s, several unsuccessful attempts were made to establish both Rainbow and Brown Trout in streams in southwest Western Australia. In 1930, eggs were imported from Victoria, a high percentage of which died *en route*. In the following year a second consignment of 20 000 eggs was more successful, many hatching in a fish farm at Pemberton. By the late 1960s, Rainbow Trout were established in rivers and impoundments along the coast from Albany to south of Perth (MacCrimmon, 1971; Clements, 1988).

Although Twyford (1991) says that Rainbow Trout in Australia 'have not become as common or widespread as the brown trout', McKay (1989) more accurately describes them as 'the most important salmonid stocked in Australia', where McDowall (1980) records them as occurring in 'high country from northern New South Wales to Victoria, in hills near Adelaide, South Australia, and in Tasmania'. (See also Tilzey, 1977, 1980.)

Ecological Impact   Where Rainbow and Brown Trout occur sympatrically in Australia the latter species tends to have a greater ecological impact than the former. For the effects of both species see under Brown Trout.

New Zealand

Scott *et al.* (1978) and McDowall (1990b), from whom much of the following account is taken, have attempted to unravel the complex story of the introduction of Rainbow Trout into New Zealand.

The earliest record of the importation of trout eggs to New Zealand from Lake Tahoe, California, was by the Auckland Acclimatization Society in 1878 (Druett, 1983; Lever, 1992; McDowall, 1994b), and although Thomson (1922) and Hefford (1926) and subsequent authorities assumed these to have been of *Oncorhynchus mykiss* (then *Salmo (Salvelinus) irideus*), they are now believed to have been more probably the Cutthroat Trout *O. clarki*, which failed to become established.

The first generally accepted successful importations of Rainbow Trout eggs to New Zealand were both made in 1883 (Ayson, 1910; Thomson, 1922, 1926; Hefford, 1926; Hobbs, 1948), again by the Auckland Acclimatization Society (Druett, 1983; Lever, 1992; McDowall, 1994b), but under the misapprehension that they were Brook Trout *Salvelinus fontinalis*. Scott *et al.* (1978) quote from the Society's Minute Book of 3 April 1883:

> The Secretary announced the arrival of 10,000 brook trout ova from San Francisco by the 'City of New York', from which 500 healthy young fish had been hatched, and also of 12,000 ova of the same fish by the 'Zealandia', about 5,000 or 6,000 of which appeared to be in good condition.

These two shipments resulted in 4000–5000 live fish, some of which were distributed in the Auckland area and elsewhere, while others were retained by the Society as breeding stock, their correct identity as 'steelhead' Rainbow Trout only being determined in 1887. As with the identity of the species imported, their place of origin was also originally incorrectly attributed by Ayson (1910) to the 'Baird Station on the McCloud River' and to the Russian River, both in California, whereas they almost certainly came from Sonoma Creek which flows into San Francisco Bay. Other shipments of Rainbow Trout eggs from the McCloud and Shasta Rivers were made by the Auckland Acclimatization Society and others in the late 1880s or 1890s and by the Auckland Society again in 1930 from Lake Almanor, California (Hobbs, 1937), the fish from which were subsequently planted in waters in both North and South Islands, in the latter especially in lakes in the Southern Alps. Recently, McDowall (1990b) refers to a suggestion that New Zealand has also been stocked with 'Gerard' Rainbow Trout, a large and rapidly growing variety originating in Kootenay Lake in British Columbia, Canada.

After their introduction to New Zealand, Rainbow Trout were reared and widely distributed by various acclimatization societies (Druett, 1983; Lever, 1992; McDowall, 1994b) such as Canterbury (first in 1885), Wellington (1891), Otago (1895), Southland (1900), Hawke's Bay (1900) and Westland (1907) (Thomson, 1922). Thomson estimated that by around 1920 at least 10 million had been released by the Auckland Society alone, including in the so-called 'thermal lakes' region of central Auckland, where in 1903 Lake Taupo was planted 'with outstanding success' (McDowall, 1990b, 1991).

Thereafter, Rainbow Trout continued to be released in many places throughout New Zealand from Auckland to Southland, rapidly becoming established in lakes of the interior such as those in central North Island and along the eastern slopes of the Southern Alps in South Island. Establishment in river systems was less successful, in spite of large scale and repeated plantings. Rainbow Trout do, however, survive in the upper reaches of some river systems not associated with lakes, such as the Waihou in the Thames region, large rivers such as the Mohaka, Ngaruroro, and Rangitikei in central North Island, and in several – mostly large – rivers in South Island, such as the Waimakariri, Rakaia, Rangitata and Waitaki, as well as some smaller ones such as the Waihou and Pelorus (McDowall, 1990a,b,c, 1994b).

Although more locally distributed than Brown Trout, Rainbow Trout in New Zealand are none the less 'widespread and locally abundant from Northland to Southland' (McDowall, 1979); despite reports to the contrary, there are no known populations of sea-migratory Rainbow Trout in New Zealand (McDowall, 1994b).

Ecological Impact As in Australia, trout in New Zealand were formerly considered of only minor ecological significance; according to McDowall (1968), the reason for the decline of stocks of native fish species (some of which seemed able to live sympatrically and even competed advantageously with introduced trout) are complex, and were only to a limited degree connected with the presence of the alien species (Wilson, 1995). Twenty years later, however, McDowall (1990a) considered that 'there is strong circumstantial evidence for serious impacts by these trouts on certain native fish species, primarily as a

result of predation on lacustrine populations, and also some indication of competitive displacement of other native fishes in streams'. Introduced Rainbow Trout may have contributed to the extinction before 1923 of the so-called New Zealand Grayling *Prototroctes oxyrhynchus;* to the decline in large lakes of the Koaro *Galaxias brevipennis* and in small coastal dune lakes of the freshwater crayfish *Paranephrops planifrons,* the crab *Halicarcinus lacustris* (Fish, 1966) and the Dwarf Inanga *G. gracilis.* In well-oxygenated mountain torrents, competition for food with Rainbow Trout may have contributed to the failure of the native Blue or Mountain Duck *Hymenolaimus malacorhynchus* to breed (Kear, 1972, 1990; Kear & Williams, 1978; Lever, 1994). (See also under Brown Trout.)

Papua New Guinea

Eyed Rainbow Trout ova were first imported into Papua New Guinea from New Zealand in 1952 by the Bulolo Gold Dredging Company, which successfully hatched 10 000 eggs in a raceway in the Upper Buane Creek, a tributary of the Bulolo River. The company continued to hatch eggs and release fingerlings in the Bulolo and its tributaries in the Wau area for several years, where the fish apparently survived until around 1964. Introductions by private individuals to provide sport for expatriate Europeans continued until 1959, but as on previous occasions no naturalized populations resulted (West & Glucksman, 1976).

In 1964, the Division of Fisheries imported 2000 Rainbow Trout fry from Jindabyne in New South Wales, which were planted in the Baiyer and Gumanch Rivers near Mount Hagen, where they reproduced and became established. Further small introductions were made in later years, and in 1971 a hatchery was opened at Mendi, from where by 1973 a total of 180 000 fish had been distributed throughout the highlands (West & Glucksman, 1976).

Rainbow Trout have been introduced in Papua New Guinea in the Eastern Highlands, Southern Highlands, Western Highlands, Enga, Chimbu and Morobe Districts, where, because they survive solely in cold montane streams above 2000 m, they pose no threat to the commercially valuable fish fauna of the warmer lower reaches (MacCrimmon, 1971, 1972; Glucksman *et al.*, 1976; West & Glucksman, 1976; Clements, 1988; Allen, 1991).

Ecological Impact In Papua New Guinea, competition for food with introduced Rainbow Trout in well-oxygenated upland torrents may have adversely affected the endemic Salvadori's Duck *Anas waigiuensis* (Kear, 1990; Kear & Williams, 1978; Lever, 1994). (See also under Brown Trout.)

Oceania

Hawaiian Islands

In June 1920, 34 500 Rainbow Trout fry and 4000 eggs (Needham & West, 1953) from the USA were released in waters on Kauai Island at Kokee, Kawaikinana, Kawaikoe, Mohihi and Waiakoali (Brock, 1952; MacCrimmon, 1971). Further shipments, totalling 203 000 eyed ova, were received from the United States Fish Commission between 1922 and 1928, and in 1940 and 1941 respectively fish were acquired from Washington and Portland, Oregon (Brock, 1952, 1960; MacCrimmon, 1971). Until 1951, Rainbow Trout were released in various waters on Kauai, Maui, Hawaii, Oahu and Molokai, but only at Kokee on Kauai did a naturalized population become established (Needham, 1949; Brock, 1952; MacCrimmon, 1971).

Before 1956, Rainbow Trout were introduced to waters in the Kohala Mountains (915–1220 m asl) on Hawaii, in 1957 near Hilo, and in 1959 in the Waipio Valley (MacCrimmon, 1971).

In eastern Maui, eyed Rainbow Trout eggs were placed in Waiakamadi and Haipuena streams (1220 m asl) in about 1960, where breeding was reported, and in 1962 Hanawi, Kopiliula, East and West Wailuaki and East Wailuanui (457 m asl) were similarly stocked with eyed ova imported from the Idaho National Fish Hatchery in the USA (MacCrimmon, 1971). In spite of all these introduc-tions, Rainbow Trout are currently naturalized in the Hawaiian Islands only at Kokee on Kauai (Maciolek, 1984).

Réunion

Rainbow Trout were first imported to Réunion in the Indian Ocean from the Mamjakatompo hatchery in Madagascar by M.J. Benoit in 1939–1940. A hatchery was constructed at Hellbourg (900 m asl) where, however, as the water temperature never fell below 14°C, the eggs hatched poorly, and stocks had to be continually replenished (MacCrimmon, 1971).

Rainbow Trout in Réunion are naturalized in the Marsouins River and its tributaries, mainly the Bras Cabot les Hauts; in the Bras de Ste-Suzanne, a tributary of the Bras de la Plaine; and in stretches of the Galets and Mât Rivers, although only the Marsouins and Bras de Ste-Suzanne are considered environmentally favourable (MacCrimmon, 1971).

Unsuccessful Introductions

Unsuccessful attempts to naturalize Rainbow Trout on islands in Oceania have been made on the Azores (Goubier *et al.*, 1983). Chatham (Skrynski, 1967), Madeira, Mauritius (MacCrimmon, 1971), Falklands (Arrowsmith & Pentelow, 1965), Kerguelen (Davaine & Beall, 1982a), Samoa (Phillipps, 1923; Fowler, 1925, 1932a) and Tahiti (Fowler, 1925; Maciolek, 1984).

Translocations

Rainbow Trout have been widely and successfully translocated within their native range in the USA, Canada and Mexico (MacCrimmon, 1971, 1972).

Sockeye Salmon; Kokanee (when land-locked); Blue-backed Salmon
Oncorhynchus nerka

The Sockeye Salmon, in its anadromous form, is one of the most abundant and important North American salmon. Although it may not reach the same size as other *Oncorhynchus* salmon (2 kg is a good weight), its red flesh makes it a valuable commercial species (McDowall, 1990b).

Natural Distribution

Western seaboard of North America and eastern seaboard of northern Asia.

Naturalized Distribution

Australasia: New Zealand.

Australasia

New Zealand

McDowall (1990b, 1994b), from whom much of the following account is derived, has traced the history of the introduction of the Sockeye Salmon to New Zealand, where it has become naturalized only in one river system in South Island.

The first shipment of eggs from Canada in 1900 all died. A second consignment of 500 000 ova (Ayson (1910) says 300 000 and Scott (1984) says 432 000) 2 years later, an intergovernmental gift (Flain, 1982), was more successful, around 160 000 (32%) arriving in good condition, and the present New Zealand population is all descended from this batch. The surviving ova were reared in the fish hatchery on the Hakataramea River in Otago, 5000 of the resulting 150 000 fry (Waugh, 1981) being planted in tributaries of the Waitaki River and 91 200 in streams flowing into the head of Lake Ohau (Scott, 1984), while 20 000 were retained for rearing-on (Waugh, 1981). In 1903 and 1904 respectively, 10 000 11-month-old fish and 6000 2.5-year-olds were liberated in the Hakataramea River. By the end of the latter year some 2000 3-year-old fish remained in the hatchery ponds, and these too were released (Waugh, 1981). Waugh (1981) and McDowall (1990b, 1994b) attribute the subsequent loss of interest in Sockeye Salmon in New Zealand to the introduction at the same time of the more viable Quinnat Salmon *O. tshawytscha* (q.v.).

The question of whether the New Zealand Sockeye Salmon population is derived from anadromous or land-locked Kokanee stock has for long been a matter of speculation. Since it was the declared intention of the New Zealand government, in introducing Pacific salmon species, to set up a salmon-canning industry, it has been assumed that the importation of the generally higher quality sea-run fish was intended, whereas New Zealand's population is entirely land-locked. Hardy (1983) has shown that

the 1902 shipment came from tributaries of Shushwap Lake on the Frazer River in British Columbia, Canada (the Scotch, Canoe, Salmon and Hatchery Creeks (Scott, 1984)), which support both sea-going and land-locked varieties. Scott (1984), from an examination of Canadian fisheries records, concluded that the eggs imported to New Zealand came from anadromous fish.

The early results of this introduction are unclear. Scott (1984) summarized what is believed to have happened: '. . . it is possible that in the first two decades after ova were introduced, some fish may have left the Waitaki and returned after a period of sea feeding; however, there is no unequivocal evidence that they did or that they have done so since'.

McDowall (1990b) agreed with Scott (1984) that the importation of 1902 was most probably of anadromous fish, some of which possessed non-migratory tendencies, which is quite typical of sea-going salmonids. Many of the young Sockeye Salmon in Lake Ohau, a deep glacial water fed by rivers flowing from the snowy Southern Alps, would have smolted and gone to sea, where some probably became disoriented and were unable to return to their natal water. A few probably did not migrate at all, but remained to maturity in fresh water, and the present population in New Zealand is undoubtedly descended from these non-migratory individuals. There would thus have been a rapid and powerful selection against migratory fish, resulting within two or three generations in an entirely land-locked stock. As Scott (1984) says, 'It is not surprising that the sockeye salmon should change in this way in New Zealand, since in its native range it is the one species of *Oncorhynchus* which is prone to form freshwater stocks'. (This was before the reassignment of *Salmo gairdneri* to *O. mykiss*).

The presence of Sockeye Salmon in New Zealand was apparently largely forgotten (to the extent that Druett (1983) claimed that '. . . this venture has to be considered a failure') until 1969, when a number were discovered in the spawning race below the Aviemore dam on the Waitaki River, having presumably dispersed downstream from Lake Ohau to Lake Waitaki. Sockeye Salmon in New Zealand are currently confined 'in substantial numbers' (Waugh, 1981) to waters in the Waitaki valley, such as Lakes Benmore, Waitaki, Ohau and Aviemore.

Summing up the status of the Sockeye Salmon in New Zealand, McDowall (1984) described it as 'fragile and presently subject to major changes because of planned modifications of river flow patterns in relation to hydroelectricity construction'. The map in McDowall (1990b) shows a recent release near the west coast of South Island, the success of which is 'still uncertain'. According to McDowall (1979), 'almost nothing is known about their interactions with the faunas of New Zealand lakes and rivers'.

Unsuccessful Introductions

Sockeyes have also been unsuccessfully introduced in Europe to Denmark, Finland and Sweden (Welcomme, 1981, 1988); in Asia to India (Natarajan & Ramachandra Menon, 1989; Shetty *et al.*, 1989; Sreenivasan, 1989, 1991; Yadav, 1993) and Japan (Tanaka & Shiraishi, 1970; Chiba *et al.*, 1989), and in South America to Argentina (Davidson & Hutchinson, 1938; Baigun & Quiros, 1985) and Chile (Barros, 1931; Davidson & Hutchinson, 1938; Mann, 1954; De Buen, 1959; Wood, 1970).

Translocations

Sockeye Salmon have been widely translocated within North America, although permanent populations have only become established outside their natural range in the Great Lakes (Welcomme, 1988).

Chinook Salmon (Quinnat Salmon (New Zealand); King Salmon; Spring Salmon; Tyee)
Oncorhynchus tshawytscha

The Chinook Salmon was the first of the various Pacific salmon species to be introduced outside its natural range, and attempts have since been made to establish it more widely than any other Pacific salmonid. Although it has been successfully translocated to the Great Lakes in North America, numerous attempts to establish it elsewhere, especially in Europe and Central and South America, have all ended in failure, and it is currently naturalized only in New Zealand (Welcomme, 1988). It is one of the largest salmonids, reaching a length of 1.5 m and a weight of 60 kg.

Natural Distribution

Western seaboard of North America and eastern seaboard of northern Asia.

Naturalized Distribution

Australasia: New Zealand.

Australasia

New Zealand

McDowall (1990b, 1994a,b) from whom much of the following account is taken, has traced the history of the introduction of the Chinook Salmon to New Zealand, where it is generally known as the Quinnat Salmon.

The earliest importations of Chinook Salmon to New Zealand were made in the 1870s. Thomson (1922) and Stokell (1955) record that in 1875 the Hawke's Bay Acclimatization Society acquired some Chinook eggs from Dr Spencer F. Baird, chairman of the United States Fish Commission and owner of the Baird Hatchery on the McCloud River (a tributary of the Sacramento) in California, USA. Owing to the lack of ice in which to pack the eggs to retard development, they began to hatch while crossing the Tasman Sea from Sydney, Australia, *en route* for Napier, and most of the fry were released in tributaries of the Waikato and Waihou Rivers south of Auckland.

In 1876, a large shipment of some 500 000 eggs was imported from the Baird Hatchery, jointly for the Auckland, Hawke's Bay, North Canterbury and Southland societies. The resulting fry were freed in various rivers from Auckland to Southland, including the Waikato, Wairoa, Hurunui, Ashley, Waimakariri, Avon, Rakaia, Rangitata and Oreti. Some were even planted in a stream in remote Preservation Inlet in southern Fiordland. In 1877, another shipment of

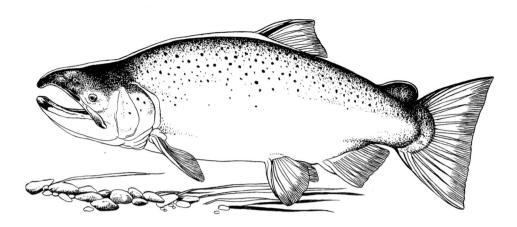

around 1 million ova was imported from the Baird Hatchery, which was distributed among the Auckland, Wellington, Nelson, Marlborough, Grey, Canterbury, Otago and Southland societies (McDowall, 1994b). Davidson and Hutchinson (1938) suggested the possibility of further importations of Chinook ova in the late 19th century, but confirmatory evidence appears to be lacking, as it also is for the contemporaneous return to New Zealand rivers of sea-run fish.

In 1889, Lake (*sic*) Ayson, recently appointed Chief Inspector of Fisheries in New Zealand, visited North America, where he was promised supplies of Pacific salmon ova from the Baird and Battle Creek Hatcheries in California, and from the Fraser River in British Columbia, Canada. Between 1901 and 1907, a total of 2.1 million Chinook Salmon eggs arrived in New Zealand from the Baird and other hatcheries in the Sacramento River basin (Flain, 1982), and possibly also from Canada. These were reared in a hatchery which had been constructed on the Hakataramea River, a tributary of the Waitaki, from where some 1.5 million young fish were planted in various portions of the Waitaki River system (McDowall, 1990b, 1994b). Large numbers were also released in the Hakataramea itself, and others were placed higher up the river in Lake Ohau (Waugh, 1981).

The earliest authenticated report of the return of sea-migratory Chinook Salmon to New Zealand waters came in 1906, when a 3- or 4-year-old gravid hen weighing 7.3 kg was caught in the Waitaki River in Otago; other fish were taken in the river later in the same year, and Chinook were reported to be spawning in the Hakataramea and in tributaries of the upper Waitaki such as the Maryburn, Grays and Haldane (McDowall, 1994b).

Ayson (1910) reported that in 1907 'quite a run of salmon came up the Waitaki River and spawned in several of its main tributary rivers. In the Hakataramea from 300 to 400 salmon spawned in the 2 miles [3.2 km] of river before it joins the Waitaki, and a number of these fish were caught and stripped and about 30 000 eggs put down to hatch'. (Stokell (1962), who says that 50 000 eggs were obtained in 1907 and that by 1923 1 510 000 had been secured, records their distribution as follows: Wairau River (600 000); Hokitika River (400 000); Upper Clutha (200 000); Queenstown Hatchery, Lake Wakatipu and the Clutha Drainage (150 000); and Tasmania (100 000), the balance of 60 000 being retained in the Hakataramea Hatchery; a few were also sent to the Seaforth–McKenzie River in Dusky Sound). A similar run took place in 1908, though on average the fish were considerably heavier (Anon., 1908), and runs of Chinook Salmon continue in the Waitaki to this day. Spawning was reported in the Tekapo, Maryburn, Ohau, Ahuriri and other tributaries of the Waitaki (Hobbs, 1937). Within a few years, Chinook had spread naturally northwards along the east coast of South Island, and had established sea-going runs in the Rangitata, Opihi, Ashburton, Rakaia, Waimakariri, Hurunui and Waiau Rivers (McDowall, 1990b).

Chinook Salmon in New Zealand are currently established primarily in rivers on the east coast of South Island, from the Clutha in Otago northwards to the Waiau in North Canterbury; the main runs occur in the Waimakariri, Rakaia, Rangitata and Waitaki Rivers, with smaller and fluctuating runs into other rivers such as the Wairau (Marlborough), Hurunui, Ashley, Ashburton, Opihi and Orari, and occasionally into the Waihao, Waikouaiti and Taieri.

Small stocks are also found in such west coast rivers as the Moeraki and Paringa, and from time to time also in the Arawata, Mohaka, Hollyford, Turnbull, Haast, Waiaoto, Manawatu, Wanganui, Taramakau, Grey and Ohinemaka, and some other rivers (Hardy, 1972; McDowall, 1987, 1990b, 1994b).

Very small runs have occurred in the Takaka River in Golden Bay, following efforts to establish a commercial run there (McDowall, 1984), and a few stragglers have been reported from the Aorere and Motueka Rivers (McDowall, 1990b, 1994b).

Several lakes on the eastern slopes of the Southern Alps (Heron, Sumner, Coleridge, MacGregor, Wakatipu, Hawea and Wanaka) support large stocks of Chinook Salmon, as do some on the west coast (Kaniere, Paringa, Moeraki, Mapourika and Okarito Lagoon, and possibly also Lake Ellery (Arawato River) (McDowall, 1990b, 1994b; Shutt, 1995; Wilson, 1995).

From time to time stray Chinook Salmon are reported from North Island (Lake Onoke/ Ruamahanga River, Rangitikei (where Hicks & Watson (1983) record that spawning occurred), Waiongoro, Tarawera, Tukituki and Esk) (McDowall, 1990b, 1994b; Shutt, 1995; Wilson, 1995).

Ecological Impact In the early years of their introduction to New Zealand, concern was expressed (by Thomson (1922) and others) about the suitability of Chinook Salmon, on account of their flavour, questionable sporting qualities, threat to young trout, and the risk of pollution from the bodies of 'spent' fish which had died after spawning. However according to McDowall (1990b),

> All these fears have proved groundless. There is nothing to suggest harmful interactions between trout and salmon, the quinnat has proved to be a very popular sporting fish, and pollution from rotting salmon bodies is negligible – some overseas studies even suggest that the nutrients released may contribute to the productivity of the very pure high-country spawning streams, and so be beneficial to the early growth of the young salmon. In addition, the salmon carcasses are a source of food for harriers (*Circus approximans*) [the Marsh Harrier *C. aeruginosus*] and even for cattle[!].

Australia

The Chinook Salmon was first introduced to Australia in 1877, but evidence on its present status

there is contradictory. Most authorities (e.g. Weatherley & Lake, 1967; Rogan, 1982; Cadwallader & Backhouse, 1983; Merrick & Schmida, 1984; McKay, 1984, 1989) agree with McDowall (1980) that 'self-sustaining populations not established in Australia, but for some years populations kept in lakes by release of hatchery-reared fish; about 150 000 released into 16 Victorian lakes in the 1930–40s; grew well only in Lakes Bullen Merri and Purrumbete ...'. Welcomme (1988), however, quoting MacKinnon (1987) says that the 'species has formed breeding populations in South Australia and Victoria ...'. Cadwallader (1995) states unequivocally that 'there is no evidence that Chinook Salmon have formed self-sustaining populations' in Australia.

Unsuccessful Introductions

Elsewhere, unsuccessful attempts to naturalize Chinook Salmon have been made in Europe in Denmark, France, (Royer, 1902) Ireland, Italy, Britain (?) (Welcomme, 1988), Germany (Mather, 1879; Von Behr, 1882), The Netherlands (Davidson & Hutchinson 1938; Vooren, 1972; De Groot, 1985), and Norway (Davidson & Hutchinson, 1938); in Africa in Madagascar (Kiener, 1963; Moreau, 1979; Reinthal & Stiassny, 1991); in Asia in India (Natarajan & Ramachandra Menon, 1989; Shetty *et al.*, 1989; Sreenivasan, 1989, 1991; Yadav, 1993) and Japan (Hikita, 1960); in Central and South America in Mexico (Davidson & Hutchinson, 1938), Nicaragua (Davidson & Hutchinson, 1938), Argentina (Davidson & Hutchinson, 1938; Thompson, 1940; Fuster de Plaza & Plaza, 1954; De Buen, 1959; Ringuelet *et al.*, 1967; Baigun & Quiros, 1985), Chile (Davidson & Hutchinson, 1938; Mann, 1954; De Buen, 1959; Campos Cereda, 1970; Wood, 1970; Zamorano, 1991); and in Oceania in the Hawaiian Islands (Davidson & Hutchinson, 1938; Needham & Welsh, 1953; Brock, 1960; Maciolek, 1984).

Translocations

As mentioned above, the Chinook Salmon has been successfully translocated outside its native range in North America to the Great Lakes.

Atlantic Salmon
Salmo salar

Numerous attempts have been made to naturalize the Atlantic Salmon outside its native range, where it is strictly anadromous; most of these have failed, and those that have succeeded have only resulted in land-locked populations (Welcomme, 1988).

Natural Distribution

Northwestern seaboard of Europe and northeastern seaboard of North America.

Naturalized Distribution

South America: Argentina; ? Chile. **Australasia:** Australia; New Zealand.

South America

Argentina

For the introduction to Argentina of the land-locked variety ('*sebago*') of the Atlantic Salmon, see under Rainbow Trout. (See also Marini, 1942.)

Chile

Atlantic Salmon were recorded as established in Chile by Quijada (1913), in Ríos Maullín, Bueno y Valdiria (Gotschlich, 1913) and Laguna del Inca (Eigenmann, 1927), but by the late 1940s they had become very rare or even extinct (Mann, 1954; De Buen, 1959).

Australasia

New Zealand

McDowall (1990b, 1994b), from whom much of the following account is derived, has traced the 'quite remarkable' story of the introduction of the Atlantic Salmon to New Zealand.

In the very early days of European settlement – certainly by 1858 – the Atlantic Salmon was singled out as a prime candidate for naturalization. That this was so is undoubtedly due to the fact that the vast majority of early settlers came from Britain, where *Salmo salar* is regarded as the game fish *par excellence*.

The first attempt to introduce Atlantic Salmon to New Zealand was made in 1864, when 600 young fish were put on board the *British Empire* in London bound for Canterbury, where in spite of elaborate preparations for their welfare on the ship all arrived dead (Thomson, 1922).

In 1868, a consignment of 100 000 Atlantic Salmon eggs, mostly from the River Tay in Perthshire, Scotland, but some from the Severn in Gloucestershire, England, arrived on the *Celestial Queen* in Port Chalmers, Otago, where from the surviving 8000 ova between 500 and 600 fry hatched suc-

cessfully (Thomson, 1922). Further shipments, the majority from the Severn and Ribble in England and the Tay, Firth and Tweed in Scotland, but some also from the Rhine in Germany and from the Miramichi River in Canada (Flain, 1982), were imported in 1869, 1871 (twice), 1873, 1875, 1876, 1878, 1881, 1884 (three times), 1887, 1889 (twice), 1895, 1901, 1905 (10 000 landlocked '*sebago*' Salmon ova from Maine, USA), 1907, 1908, 1909 and 1910–1911. Thomson (1922) estimated that nearly 5 million ova had been imported to New Zealand, and that around 2.7 million young fish had been released in the following rivers, tributaries and lakes of South Island; Waiau, Aparima, Mataura, Owaka, Clutha, Leith, Kakanui, Waitaki, Opihi, Temuka, Rangitata, Selwyn, Heathcote, Avon, Ashley, Perceval, Clarence, Hurunui, Nelson, Marlborough, Grey, Buller and Hokitata, and Lake Ada, Milford Sound: and on North Island in Ruamahanga, Hutt, Manawatu, Taranaki and Hawke's Bay. Few of the releases were successful (Druett, 1983).

The failure of Atlantic Salmon to become established in New Zealand was attributed by Whitney (1927) to predation on the smolts as they migrated downriver to the sea. A contributory factor was also probably that the releases, which occurred in rivers from Southland on South Island to Auckland on North Island, were all too frequently made in small batches, in violation of one of the cardinal rules of attempted naturalization. Realizing this, L. F. Ayson, the Chief Inspector of Fisheries, decided in the early 1900s to concentrate on stocking a single river, the Waiau, with large numbers of Atlantic Salmon, which he was certain would meet with the same success as had the Chinook Salmon *Oncorhynchus tshawytscha* (q.v.). Ayson's sentiments were echoed by the Association of New Zealand Acclimatisation Societies (Lever, 1992; McDowall, 1994b).

In 1916, a few Atlantic Salmon were caught in the mouth of the Waiau, including one that was believed to have spawned on several previous occasions, and in 1921 some smolts were caught in the estuary of the Waiau, proving that spawning had definitely occurred. By 1922, considerable numbers of Atlantic Salmon were being caught on rod-and-line in the Waiau, and a second spawning run was discovered in the Upokeroro River, which flows into Lake Te Anau. In 1923, the Marine Department began 'systematic stocking' of the Wanganui (McDowall, 1990b).

In 1928 and 1932 respectively, 42 000 and 40 000

Atlantic Salmon were placed in Lake Coleridge, Canterbury (Stokell, 1934), to try to determine the age/growth relationship in New Zealand fish, but they failed to become established (Godby, 1934). Liberations in the Clutha (150 000 in 1927) and Lake Wakatipu (6700 in 1932) were also unsuccessful (McDowall, 1990b).

As stocks in the Lake Te Anau system increased during the 1920s and early 1930s, debate arose as to whether New Zealand's Atlantic Salmon were seagoing or land-locked fish. A few were said to have been taken in the Wairaurahiri River, which flows into the sea at Te Waewae Bay, slightly west of the Waiau, suggesting that they must have migrated to sea and then returned to their natal water (Hutton, 1927). The fact that fish caught at the mouth of the Waiau were feeding implies that they were either residents or were moving downriver, rather than mature fish returning from the sea to spawn, and their generally poor condition, in comparison to sea-run European or North American stock, tends to confirm this belief (McDowall, 1990b).

The reasons why an anadromous population of Atlantic Salmon failed to develop in New Zealand are a matter of conjecture. According to McDowall (1990b), 'Probably the real cause is that once the small smolts went to sea and dispersed to feed there, they were unable to navigate in the strange new environment and could not find their way back to fresh water again'. But the new environment in which they found themselves would surely have seemed no more 'strange' to them than that of the Atlantic or North Sea in their native range.

Whatever the true reasons for the failure of a seagoing population of Atlantic Salmon to evolve in New Zealand, the stock naturalized there now is a land-locked one. For many years the fish had open access to the sea from Lake Te Anau, and some mature fish were taken by anglers in the estuary of the Waiau; since 1975, however, when the Mararoa dam was built across the Waiau downstream from Manapouri, the 'meagre' population has been perforce confined to fresh water (Gibbs, 1981; McDowall, 1990b).

The derivation of the Waiau/Manapouri stock is also a matter of speculation. It is known that most of the importations into New Zealand were of North Atlantic anadromous stocks (McDowall, 1990b), and it thus seems likely that for some unknown reason they lost their migratory instinct in their new home. This has happened in the case of such naturalized birds in Britain as the Canada Goose *Branta canadensis* and Mandarin Duck *Aix galericulata* (Lever, 1977, 1987).

Between 1923 and 1929, attempts were made to establish Atlantic Salmon in the Wanganui River, using a total of 4 million eggs (Hutchinson, 1975) obtained from the Upokeroro, but all were unsuccessful. More recently, in the early 1960s the Southland Acclimatization Society attempted to establish an anadromous population, using mainly fish from Sweden and Poland, on the basis that because Atlantic Salmon in the Baltic Sea do not migrate as far as those in the North Sea and Atlantic, they might form a sea-going stock in the southern hemisphere, but this project, too, failed (McDowall, 1990b, 1994b).

Summing up the present status of the species in New Zealand, McDowall (1984) said: 'Atlantic Salmon were once abundant and a sought-after game fish in the Te Anau – Manapouri system. . . . Today, few if any Atlantic Salmon are caught in Te Anau and Manapouri, but a relict population persists in Lakes Fergus and Gunn in the upper Eglinton valley'.

Australia

Atlantic Salmon were first introduced to Australia between 1864 and 1870, when some were released in Tasmania and Victoria (Allport, 1870a,b, 1875, 1880) and again to Tasmania in 1884 (Merrick & Schmida, 1984; McKay, 1984). They were introduced once more in 1963, when fish from Canada were planted in various still and flowing waters in New South Wales. Although Arthington (1989) lists the Atlantic

Salmon among those fish that have 'established self-maintaining populations in Australian inland waters', and Allen (1989) says it is 'present in the cool reaches of streams draining the southeastern coast, Murray–Darling system, and Tasmania', other authors (e.g. McDowall, 1980; McKay, 1984; Merrick & Schmida, 1984; Welcomme, 1988; Burchmore & Battaglene, 1990, say that it is present in at most two waters in New South Wales (Lake Jindabyne and Burrinjuck dam), where it is maintained solely by artificial stocking.

Unsuccessful Introductions

Atlantic Salmon have been unsuccessfully introduced in Europe to Cyprus and Greece (Welcomme, 1988); in Asia to Indonesia (Welcomme, 1988; Muhammad Eidman, 1989) and India (Jhingran, 1989; Shetty *et al.*, 1989); in Africa to South Africa (Welcomme, 1981; Bruton & Merron, 1985; De Moor & Bruton, 1988); in South America to Brazil (Welcomme, 1988); and in Oceania to the Falkland Islands (Arrowsmith & Pentelow, 1965) and Îles Kerguelen (Davaine & Beall, 1982b).

Translocations

Atlantic Salmon have been unsuccessfully translocated outside their natural range in North America to the Pacific states of the USA (Smith, 1896) and to British Columbia, Canada (Chapman, 1942; Carl & Clemens, 1953; Dymond, 1955; Carl & Guiguet, 1958).

Brown Trout
Salmo trutta

Brown Trout have been successfully introduced to at least two dozen countries and island groups outside their natural range, and now occur discontinuously in waters in all continents except Antarctica.

The principal limiting factor that determines the success or failure of introduced Brown Trout appears to be water temperature; the limits are 4–19.5°C for growth; 0–25 or 30°C for survival; and 0–15°C for egg

development. Other factors include the existence of suitable spawning redds, water quality, extremes of flow, food availability, and the degree of predation (Elliott, 1989). Welcomme (1988) says that where Rainbow Trout *Oncorhynchus mykiss* (q.v.) and Brown Trout live sympatrically, the two species tend to separate thermally, with the latter occupying the colder waters at higher elevations. The successful establish-

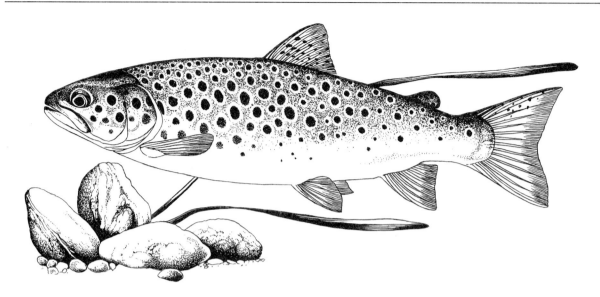

ment of the Brown Trout in the wild has been limited by its comparative lack of success in aquaculture; as a result, most introductions have been made for sporting purposes (Welcomme, 1988).

Natural Distribution

From Iceland and northern Europe southwards to the Mediterranean Sea, the islands of Sardinia, Sicily, Corsica, Cyprus and Crete, and North Africa, eastwards to the Black, Caspian and Aral Seas and their associated rivers, through Afghanistan into West Pakistan (Frost & Brown, 1967; MacCrimmon & Marshall, 1968; MacCrimmon *et al.*, 1970; Elliott, 1989).

(Note: this distribution includes that of the Sea Trout, an anadromous variety of the Brown Trout, which occurs in the British Isles, along the coast of western Europe between 40 and 65° N (including Iceland), and in the Black, Caspian and Aral Seas.)

Naturalized Distribution

Asia: India; Japan; Pakistan; ? Sri Lanka. **Africa:** Algeria; ? Ethiopia; Kenya; Lesotho; Madagascar; South Africa; Swaziland; ? Tanzania; ? Uganda; Zimbabwe. **North America:** Canada; USA **Central and South America:** Argentina; Bolivia; Chile; ? Ecuador; Panama; Peru; ? Venezuela. **Australasia:** Australia; New Zealand; Papua New Guinea. **Oceania:** Falkland Islands; ? Îles Kerguelen; ? Îles Crozet.

Unsuccessful European Reintroductions

In Europe, unsuccessful attempts have been made to reintroduce Brown Trout to The Netherlands, where they became extinct in 1945 (Vooren, 1972; Welcomme, 1981; De Groot, 1985).

For the likely early introduction by man of Brown Trout to high-mountain lakes in what is now the Austrian Tyrol see under Arctic Charr *Salvelinus alpinus* (Pechlaner, 1984).

Asia

India

The history of the introduction of Brown (and Rainbow) Trout into India is discussed by Howell (1916), Mitchell (1918), Molesworth and Bryant (1921), MacKay (1945), Jones and Sarojini (1952), MacCrimmon and Marshall (1968), Sehgal (1989) and others.

The first attempt to import Brown Trout to India was made in the Nilgiris Hills in Madras by H.S. Thomas in 1863, but all the eggs died *en route* (Day, 1876). In 1866, Francis Day (1876) imported 6000 ova of the 'Loch Leven' (Scotland) variety ('*levensis*'), obtained for him by Frank Buckland of the Acclimatization Society of the United Kingdom (Lever, 1992), and although the majority died shortly before their arrival in Ootacamund, the survivors hatched successfully. Further introductions were made to streams in the Nilgiris (e.g. Mulkerti,

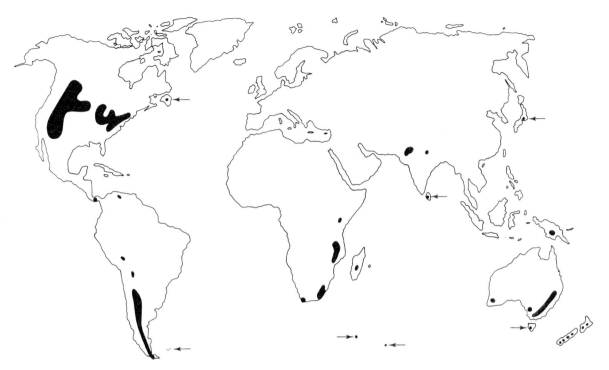

Naturalized distribution of Brown Trout *Salmo trutta* (adapted from Frost & Brown, 1967; MacCrimmon & Marshall, 1968; MacCrimmon *et al.*, 1970; Elliott, 1989; McDowall, 1990b).

Kaurmund, Mekod, Parsōn's Valley, Sandynulla, Yammakkal, Billithadahalla, Peermund and Avalanche (Sreenivasan, 1989)) between 1887 and 1906, some of which proved partially successful.

Kashmir acquired its first Brown Trout eggs in 1889, when ova imported from Europe hatched successfully, the young fish becoming well established in the Kashmir Valley according to MacCrimmon and Marshall (1968). Sehgal (1989), however, says that the earliest attempt to introduce Brown Trout to the Himalayas began in 1899, when a carpet merchant, Mr F. J. Mitchell, unsuccessfully imported a consignment of eyed eggs from England into Kashmir. In 1900 or 1901 (Yadav, 1993) more were imported as a gift from the Duke of Bedford to the Maharajah of Kashmir, but the entire shipment died on the voyage from England (Mitchell, 1918). Other consignments later in the same year and in 1901 from Sir James Maitland's hatchery at Howietoun in Stirlingshire, Scotland, were more successful, and the hatched fish became established in Mr Mitchell's ponds at Harwan in the Kashmir Valley in 1903 (MacCrimmon *et al.*, 1970), and subsequently in the River Kalapani near Abbottabad, and in Jammu, Gilgit, Himachal

Pradesh, Uttar Pradesh, Arunachal Pradesh, Nagaland and Meghalaya (Sehgal 1989).

Brown Trout were first successfully introduced to the River Beas in the Kulu Valley of the Punjab in 1909 by G.C.L. Howell (1916), who transferred some 23 000 eyed ova from Kashmir. Later importations were made in 1911 (Kangra), 1915 (Mahasu) and 1919 (Chamba) (MacCrimmon & Marshall, 1968), and Brown Trout are now established in the Beas and its tributaries and in other streams in the Kulu Valley and elsewhere, from where they have been translocated to waters in Chamba, in the Kangra Valley and the Simla Hills, Himachal Pradesh. Regular stocking of Brown Trout eggs in the upper reaches of the Ravi, Baner, Awa and Binun (Binwa) tributaries of the Beas, between 1912 and 1947, were relatively unsuccessful (Sehgal, 1989). The species' failure to become naturalized in the Kangra Valley is due largely to unsuitable riverine habitats.

In 1910, a consignment of 10 000 Brown Trout eggs was transferred from Kashmir to a hatchery at Kaldayani or Bhowali near Nainital in the Kumaon Hills of Uttar Pradesh in the Garhwal Himalayas; 2 years later a further batch was acquired from the same source, and fingerlings were stocked, with only partial success, in Lakes Nainital, Naguchiatal, Sathtal, Malwatal and others in the eastern Himalayas, and to the Pindar, Birehi and Asiganga Rivers (Sehgal, 1989). Attempts by Mr F. J. Mitchell to establish Brown Trout in the Darjeeling Himalayas in 1899 were similarly unsuccessful (Jhingran, 1989), as were others in Shillong and Arunachal Pradesh. According to Johri and Prasad (1989), in Uttar Pradesh Lake Deodital (3000 m asl) appears to be the only water in which Brown Trout are fully naturalized.

Between 1906 and 1916 (Gopinath, 1942), several shipments of Brown Trout ova arrived in Travancore in southern India from Sir James Maitland's hatchery at Howietoun in Stirlingshire, Scotland. Although the fry introduced into the waters of the Munnar High Range showed 'phenomenal growth' (Jones & Sarojini, 1952), they failed to breed until a hatchery was constructed at Rajmally, from where fish hatched out in 1932 were placed in the Eravikulam River in which they bred in the wild in 1937.

Today, Brown Trout in India occur in streams in Kashmir and Himachal Pradesh (Welcomme, 1981, 1988) where, however, they have a 'restricted distribution' (Shetty *et al.*, 1989). According to Singh and Kumar (1989) 'being a cold water fish, the distribution of trout [in Himachal Pradesh] is limited to the upper reaches of snowfed streams where the temperature seldom rises above 10°C'.

Japan

Worth (1893) documents the introduction of Brown Trout from the USA to Tokyo in 1892–1893, when Professor C. Sasaki acquired 10 000 'Loch Leven Trout *S. levensis*' eggs and 10 000 'German trout *Salmo fario*' eggs from the United States Fish Commission.

The only known naturalized population of Brown Trout in Japan today, where the species is not valued either for food or as a sporting fish, is in Chuzenji Ko Lake (Welcomme (1981, 1988) incorrectly gives Chezayi), Nikko City, Tochigi Prefecture (MacCrimmon & Marshall, 1968; Chiba *et al.*, 1989).

Pakistan

The first hatchery for Brown Trout in Pakistan was established at Shinu in the Kaghan Valley with eggs from the hatchery at Harwen in Kashmir (see above under India). Subsequently, releases of young fish in the neighbouring Kunhar River and its feeders resulted in flourishing naturalized populations, especially near Naran, including Saif-ul-Malak Lake (3050 m asl) (MacCrimmon *et al.*, 1970).

The transfer of Brown Trout from Gilgit in Kashmir to the Lutkoh River of Chitral in 1945 (Sehgal, 1989) established a thriving fishery there, and led to the construction of a hatchery in the neighbouring Bombrait Valley. Other hatcheries were subsequently built elsewhere, as a result of which Brown Trout fishing can now be found in many suitable streams above the 1190 m contour (MacCrimmon *et al.*, 1970; Welcomme, 1981, 1988).

Unsuccessful Introductions

Brown Trout have been unsuccessfully introduced in Asia to Indonesia (Schuster, 1950; Muhammad Eidman, 1989), Nepal (MacCrimmon *et al.*, 1970) and Sri Lanka (Fernando, 1971; De Silva, 1987).

Translocations

Brown Trout have been successfully translocated to several waters outside their natural range in the former Asian USSR (MacCrimmon *et al.*, 1970).

Africa

Algeria

Moreau *et al.* (1988) list the Brown Trout as an established game fish in Lac de Ghrib, Algeria, to which it was introduced from France at an unrecorded date.

Ethiopia

In 1967, Brown Trout from Kenya were introduced by J. H. Blower of UNESCO into an upland water, the River Webb, near Goba, in the Bale Administrative Region of Ethiopia, where they apparently became established (Tedla & Meskel, 1981). Moreau *et al.* (1988) list them as also occurring in the River Danka.

Kenya

Attempts to establish Brown Trout from England in Kenya between 1906 and 1910 ended in failure, but in 1921 populations became naturalized in rivers and streams in the Aberdare Mountains and on Mount Kenya at or above the 2000 m contour (Harrison *et al.*, 1963). Moreau *et al.* (1988) refer to an importation from the USA in 1948.

Brown Trout in Kenya have proved less adaptable to local conditions than Rainbow Trout *Oncorhynchus mykiss* (q.v.) and, as a result, the former are now confined to around eight rivers in the Aberdares and on Mount Kenya (MacCrimmon & Marshall, 1968), where according to Welcomme (1981) they are established 'in highland streams'.

Lesotho

Brown Trout were first introduced from Cape Province, South Africa, to Basutoland (now Lesotho) in 1904, where Harrison *et al.* (1963) reported them to be thriving locally in some large pools in the Langabeletsi, Quthing, Maseru, Sangabethu and Mokhotolong Rivers. Today, Brown Trout are established in most highland waters in Lesotho especially in the Tsoelikane River (Jubb, 1972) and the Senqu and Moremoholo Rivers (2740 m and 2560 m respectively asl) (Rondorf, 1976), and also in the Mokhotlong, Sangebethu and Khebelu Rivers (Gephard, 1977) in the Upper Orange River system, where their range is said to be restricted by excessive soil erosion (Gephard, 1977).

Madagascar

Brown Trout eggs first arrived at the Manjakatompo Station in the Ankaratra (1900+ m asl) in Madagascar in 1926, where they are now naturalized in montane streams above 1700 m (Kiener, 1961, 1963; MacCrimmon *et al.*, 1970; Moreau, 1979; Reinthal & Stiassny, 1991).

South Africa

De Moor and Bruton (1988) have traced the history of the introduction of Brown Trout to South Africa.

Although De Moor and Bruton (1988) state that Brown Trout were first imported to South Africa as a game fish in 1892, Hey (1947) says that the earliest importations of eggs from Britain to Cape Town and Natal were made in 1876, although these and subsequent introductions to Cape Town, Kingwilliamstown and the Drakensburg Mountains of Natal were unsuccessful.

In 1890, Brown Trout ova from Scotland were imported to the Boschfontein hatchery, Bolgowan, Natal (Pike, 1980). Two years later eggs from the Surrey Trout Farm in England were imported to the Cape, where the hatched fry were placed in a small river near Stellenbosch (Anon., 1962–1963). Many of the early accounts of importations of Brown Trout to the Cape refer to the 'Loch Leven *S. levensis*' variety, but these were later merged with *S. trutta* stocks at the Jonkershoek hatchery (Anon., 1962–1963). The first stripping of ova took place in 1897 (Harrison, 1940).

Brown Trout have subsequently been stocked on a regular basis in numerous river systems in South Africa, and the following account of their distribution, which is probably incomplete, does not necessarily indicate the presence of naturalized populations in the localities named, as references in the literature are sometimes ambiguous (De Moor & Bruton, 1988).

Before 1930, Brown Trout were established in the Mooi, Bushmans, Umgeni and Loteni Rivers in Natal, as well as the Umzimhlava River in East Griqualand (Bennion, 1923), who also refers to the existence of 'trout' in the Elandspruit River near Machadadorp in the eastern Transvaal; the Helpmekaar and Broederstroom River in north Transvaal; the Wemmer, Eerske and Hex Rivers in the Cape; rivers near Himeville, Natal; and the Wildebeeste and Little Pot Rivers of East Griqualand. Brown Trout also occurred in the White River (Bain's kloof), Cedarberg streams east of Uitkyk Pass in the Olifants system, the Buffalo, Keiskamma and Tyume Rivers in the eastern Cape, and in the Ginqiskodo River, Umzimkulu (De Moor & Bruton, 1988).

By 1949, Brown Trout were to be found in the Cape in Steenbras reservoir and the Lourens and Upper Berg Rivers, and in the Wemmer and Dwars (tributaries of the Upper Berg). In the Eastern Cape they occurred in the Smalblaar, Witte and Mitels Rivers in the Breede River system, and in the Keiskammahoek and upper reaches of the Buffalo River (De Moor & Bruton, 1988). 'Trout' of unspecified species were also reported by Harrison (1949) in tributaries in the upper reaches of the Olifants River system and in the Ndowana and Ingwangwana Rivers.

Today, Brown Trout are well established in montane regions of the Cape and Natal, and in the eastern Transvaal.

In *Natal*, they occur in the upper reaches of the Umkomaas, Umgeni, Umvoti, Tugela, Upper Bushmans (1200–2000 m asl), Buffalo and Phongolo Rivers (Crass, 1966), and also in the Midmar dam on the Umgeni River (Heeg, 1983) and in the Highmoor dam (Pike & Wright, 1972).

In the *Cape*, Brown Trout are found in the Breede River (Skelton, 1987). In the Buffalo River their numbers and distribution have declined, and they now occur only in the Maden dam (Jackson, 1982). They are also present in the Tyume River (Gaigher, 1975a), in the Cata River in the Keiskamma system, in the Gubu dam on the Kubusie River in the Kei system, and in small numbers in Lake Le Roux (De Moor & Bruton, 1988). Brown Trout are also found in the Keurbooms River and in the upper tributaries of the Caledon and Orange Rivers. In the Gwaayang River near George, where Brown Trout were said to be 'present but struggling' in the 1940s, they still survived in 1966 (Harrison, 1966a), but whether they exist there today is uncertain. They survive but do not reproduce in the Steenbras reservoir, and occur in the Wemmershoek River and reservoir (Louw, 1979). Naturalized populations are present in the White River (Bain's kloof) and in Mitchell's Pass in the Witels River (De Moor & Bruton, 1988).

In the Olifants catchment, Brown Trout have declined in numbers and range, and now only occur in the Riet, Twee, Middeldeur and Matjies Rivers (Kromme and Driekloof tributaries) upstream of the Leerkloof River, and in the Leerkloof itself where the population is said to be self-maintaining (De Moor & Bruton, 1988).

In the *Transvaal*, Brown Trout are naturalized in the Limpopo River system, in inlets to the Dap Naude and Ebenezer dams in the Letaba system, in the upper reaches of the Blyde River, and also in Dullstroom (Crocodile River in the Incomati system) (De Moor & Bruton, 1988).

South African Distribution   As elsewhere, Brown Trout in South Africa are restricted to cool, well-oxygenated highland waters, where their optimum habitat is the larger and slower-flowing streams with plenty of cover both above and below water. Brown Trout appear to be less tolerant than *O. mykiss* of adverse environmental conditions such as high temperatures, excessive water extraction, and high siltation (De Moor & Bruton, 1988), but seem to be adapted to survive at higher altitudes than the latter species.

Ecological Impact   Brown Trout are not considered as serious a threat to native South African

species as other introduced predators such as Largemouth Bass *Micropterus salmoides* (q.v.) and Smallmouth Bass *M. dolomieui* (q.v.), and since they are not as widely distributed as *O. mykiss* have not had as much impact as the North American species. Skelton (1987) includes predation and/or interspecific competition by Brown Trout as being one of the factors, albeit a minor one in view of the amount of habitat damage caused by man, contributing to the decline of such threatened native species as *Barbus bergi* (in the Berg, Eerste and Breede Rivers), *B. andrewi* (Berg and Breede), and *B. burchelli* (Breede).

The introduction of Brown Trout to the Umkomazana River in Natal was considered by Pike and Tedder (1973) as contributing (together with loss of habitat) to the decline and eventual extinction there of the native *Oreodaimon quathlambae*. The diet (principally aquatic insects such as mayflies and caddis flies) of the two species is very similar, so competition for food resources seems probable.

In research carried out in a moorland stream, Healey (1984) reported significant alterations in the species composition of aquatic invertebrates after the introduction of Brown Trout, with a fall in the number of species found and a restriction in their distribution, while some rarities have been eliminated.

De Moor and Bruton (1988: appendix II: 270–271) give in detail (with references) dates of initial stocking and earliest records of Brown Trout in various localities in South Africa.

Swaziland

According to Clay (1972), Brown Trout were introduced from South Africa to Swaziland in 1914 where Moreau *et al.* (1988) say they are reproducing.

Tanzania

Brown Trout were introduced to streams in the Takukuya area of Tanganyika (now Tanzania) in 1934, but repeated restocking has been considered necessary to maintain them, and naturalized populations are 'questionable' (Harrison *et al.*, 1963; MacCrimmon & Marshall, 1968).

Uganda

In 1930 Brown Trout were unsuccessfully introduced to waters in the Mount Elgon region of Uganda (MacCrimmon & Marshall, 1968). Two years later, however, a shipment of 50 000 ova from England arrived at a small hatchery on the River Mubuku in the foothills of the Ruwenzori Mountains (the 'Mountains of the Moon') in southwestern Uganda, where trout subsequently became established in the neighbouring River Ruimi (1220–5110 m asl) (MacCrimmon *et al.*, 1970). Whether they survive there today is unknown. Introductions to the Mubuku River and in 1952 to the Seti were unsuccessful (Twongo, 1988).

Zimbabwe

Toots (1970) quotes from a report by C. Dimmock in the *Rhodesia Agricultural Journal* (Vol. 8., No. 2) of December 1910, that:

> At the time when the acclimatization of trout is being attended to in North-West Rhodesia, perhaps an account of four years experiments in Southern Rhodesia will be of interest: 1907, 3,000; 1908, 4,000 and 1909, 4,000 Brown trout ova were received from Pirie Hatcheries, King Williams Town (South Africa) and hatched in Gwebi river in hatching boxes. In 1910, 3,000 Rainbow and 1,000 Brown trout ova were received from Stellenbosch. From this last batch 6,000 fingerlings were liberated in Gwebi River.

According to Toots (1970) this stocking was unsuccessful.

Between 1920 and 1940, large numbers of trout eggs were imported into Rhodesia (now Zimbabwe), where in spite of high mortality rates among both ova and fry, waters in Matabeleland (Khami), Mashonaland (Cleveland, Mtepatepa and Borrowdale) and the Eastern Districts (Stapleford, Inyanga) were stocked. It was not, however, until 1929, when the streams of Inyanga were planted, that Brown Trout became naturalized in Zimbabwe (Harrison *et al.*, 1963), where they are currently found mainly between 1500 and 2300 m asl.

Unsuccessful Introductions

Brown Trout have been unsuccessfully introduced in Africa to the Congo and Malawi (MacCrimmon *et al.*, 1970), and to Morocco (Moreau *et al.*, 1988).

North America

Canada

The history of the introduction of Brown Trout into Canada has been traced by Catt (1950), Dymond

(1955), MacCrimmon and Marshall (1968), Crossman (1984) and others.

Brown Trout are the most widely introduced alien fish in Canada, where 'although the dissemination of brown trout was slow and dependent on neighbouring American states, all Provinces but Prince Edward Island and Manitoba and the Northwest and Yukon territories ultimately experienced successful [i.e. naturalized] introductions' (MacCrimmon & Marshall, 1968).

Brown Trout were first introduced to *Newfoundland* in either 1884 (Frost, 1938; Dymond, 1955; Scott & Crossman, 1964) or 1886 (Andrews, 1966), when 100 000 ova of the 'Loch Leven *levensis*' variety were despatched from Sir James Maitland's hatchery at Howietoun in Stirlingshire, Scotland (Anon., 1887; Catt, 1950). Between 1886 and 1889, some 100 000 ova a year were imported (Andrews, 1966), and in 1892 ova of 'German' Brown Trout were introduced to Newfoundland (Dymond, 1955), where the resulting fry were subsequently released. In 1905–1906, some Brown Trout eggs were brought in from England, but thereafter introductions appear to have ceased (Crossman, 1984).

Today, Brown Trout are 'moderately widely distributed' in Newfoundland on the Avalon, Burin and Bonavista peninsulas, and around Trinity Bay (Crossman, 1984). The species' slow spread is believed to be due to lower than optimal environmental temperatures, and new waters have probably been occupied by sea-going populations (Sea Trout) which developed soon after the original plantings (Dymond, 1955; Crossman, 1984). Brown Trout in Newfoundland appear to be displacing indigenous salmonids and have become the dominant species in several lakes near St John's (Crossman, 1984).

Brown Trout, of both the 'Loch Leven' and 'German' types, were first transferred from New Brunswick and New Hampshire to *Nova Scotia* in 1925, where they were placed in three rivers in Guysborough County. In 1929–1930, fish of both varieties were again introduced, and freed in waters in Yarmouth, Queens and Lunenburg Counties. More were brought in in 1934, and releases were made in Pictou, Cumberland and Annapolis Counties (Crossman, 1984).

Brown Trout in Nova Scotia (where restocking has been abandoned) are presently naturalized in many watersheds, where they compete with native salmonids such as the Atlantic Salmon *Salmo salar* and Brook Trout *Salvelinus fontinalis* (Catt, 1950; Gilhen, 1974).

Brown Trout were first introduced to *New Brunswick* in 1921, when they were released in the Loch Lomond system, and later in waters in Saint John and King's or Charlotte Counties (Catt, 1950). They still survive, albeit in small numbers, in the Nashwaak River and possibly elsewhere, in the province, where stocking has been discontinued.

The earliest authenticated introduction of Brown Trout to *Quebec* dates from 1890, when 25 000 fry were imported from the Caledonia hatchery in New York (Catt, 1950), and released into Lac Brule near Ste Agathe, and probably in North River which flows into the lake. In 1951, 3000 fingerlings were introduced from Johns River, Vermont, to the Chateauguay River. More recently, artificially reared Brown Trout from lakes in the Atlas Mountains of Morocco (referred to by Crossman (1984) as of a subspecies 'macrostigma') were imported and unsuccessfully released in Lake Bernard in Mont Tremblant Park, and in Ruisseau Tremblant – a tributary of Lake Tremblant – or Ruisseau des Cascades in Terrebonne County (Crossman, 1984).

The Ste Agathe population, supported by regular restocking, is still extant, as are the eastern township populations, which are similarly sustained, from tributaries of the lower St Lawrence River beyond Quebec City to the border with the USA (Crossman, 1984).

The earliest recorded introduction of Brown Trout to *Ontario* was made in 1913, when fingerlings were liberated in the Speed River near Hespler in Waterloo County. Many other plantings of fish, believed to have come from hatcheries in Michigan, were made in Ontario between 1913 and 1918, and again in 1929 and 1930 when trout were released in the former year in Peterborough, Frontenac County's Muskoka District, and near Sudbury, and in the latter in seven lakes near Kenora. The stocking of Brown Trout in Ontario ceased in 1962 (by which time some 10 million had been introduced (MacCrimmon & Marshall, 1968)), but populations are established in Lakes Ontario, Erie and Huron, and their inflowing feeders (MacCrimmon & Marshall, 1968).

Dymond (1955), ascribes the success of the Brown Trout in Ontario in part to the fact that many of the waters there are too warm for the native Brook Trout *Salvelinus fontinalis*. Dymond found that in numerous waters in southern Ontario the Brown Trout was the only salmonid in the lower reaches, while Brook Trout survived in the upper and colder stretches. In the middle sections, both species occurred sympatrically.

According to MacCrimmon and Marshall (1968), Brown Trout were first introduced to *Manitoba* in 1943, where in the following year 20 000 fingerlings were released in West Blue Lake, in what is now the Duck Mountain Provincial Park; these fish failed to breed, as did others freed between 1951 and 1953 and more placed in Child's Lake in 1954. In spite of many subsequent efforts, no breeding populations of Brown Trout have become established in Manitoba, due in part it is believed to their failure to overwinter and to predation by Northern Pike *Esox lucius*.

Marshall and Johnson (1971) record that Brown Trout were introduced from the Banff National Park in Alberta to the Cypress Hills district of *Saskatchewan* in 1924, and more were brought in at later dates from Montana, Wisconsin, and again from Alberta. A total of 13 introductions in Cypress Hills between 1924 and 1930 resulted in naturalized populations in seven tributaries of the Missouri and Saskatchewan Rivers (Crossman, 1984). Attempts between 1931 and 1955 to establish Brown Trout in the Hudson Bay and North Battleford/Meadow Lake areas failed, probably because of unfavourable temperatures. The stocking of Brown Trout in Saskatchewan was restricted because it was considered an inferior game species to the native Brook Trout. Dymond (1955) says that the streams in Saskatchewan in which Brown Trout are established are devoid of native salmonids, perhaps because the water temperatures are too high.

Brown Trout were first imported from Montana to *Alberta* in 1924, when some were released in Lake Annette in the Jasper National Park and in the Raven River (in the Red Deer catchment). Fish believed to be of the 'levensis' strain gained access to Carrot Creek in the Bow River system in the following year (when a lorry transferring fish from Banff to eastern Alaska broke down), from where they eventually spread 140 km downriver in the Kananaskis River system (Nelson, 1965).

The population in Lake Annette was later deliberately eradicated, as it was regarded by anglers with disfavour. Through subsequent plantings, Brown Trout became established in much of western Alberta, particularly in the Athabasca, North

Saskatchewan, Red Deer, Bow and Milk Rivers drainages (Crossman, 1984). Miller (1949) ascribed the success of Brown Trout in Alberta in part to their earlier spawning than the indigenous and competing Cutthroat Trout *Oncorhynchus clarki*.

Brown Trout were first introduced to *British Columbia* in 1932, when fry reared from eyed ova brought from Lodge, Wisconsin, and from Manitoba were planted in Cowichan Lake on Vancouver Island. Between 1932 and 1935, a total of 547 496 (Neave & Carl, 1940) fry, fingerlings and adult fish were released in the Cowichan and Little Qualicum Rivers and in Cowichan Lake, where spawning took place in 1937 (Carl & Guiguet, 1958). From this established population, Brown Trout were in 1959 transferred into the Kettle River near Kootenay, apparently unsuccessfully, and more recently into the Adam River, 75 km north of the Campbell River on Vancouver Island, where a sea-going run of Sea Trout developed (Carl *et al.*, 1967). In 1935, fish bred from ova brought from Montana were placed in nine more waters on Vancouver Island, and in subsequent years stock was also imported from Washington State (Crossman, 1984). No more Brown Trout were introduced to British Columbia after about 1961, where the only known surviving population is that in the Cowichan River (especially above Skutz Falls) and Lake Cowichan and the Little Qualicum River. According to Dymond (1955), Brown Trout spawn in the Cowichan River with four species of native Pacific salmon (*Oncorhynchus*), the Rainbow Trout *O. mykiss*, and the Cutthroat Trout.

United States

The earliest importation of Brown Trout to the USA was made in the winter of 1882–1883, when a shipment of eggs from Von Behr, president of the Deutsche Fischeries Verein in Germany, was despatched to the Northville Hatchery in Michigan where it arrived on 18 February of the latter year; fry from these ova were released in the following April in the Père Marquette River, Michigan (Mather, 1889; Goode, 1903). In 1885, Von Behr sent a second consignment of 40 000 eggs to Northville and to the Caledonia Hatchery, New York, from which 28 000 fry were subsequently planted in waters of Long Island near the Hudson River (Laycock, 1966; Luton, 1985).

Further shipments of ova from Germany between 1884 and 1887 were reared at hatcheries in Cold Spring Harbor on Long Island and Caledonia; at Northville; at Central Station, Washington; and at Wytherville, Virginia (Smiley, 1884). Consignments of eggs from Germany were also sent to hatcheries in New Hampshire and Pennsylvania (Smiley, 1889a), and by 1892 Brown Trout ova had arrived at hatcheries in California (Smith, 1896).

In January of 1885, a shipment of 120 000 'Loch Leven' Brown Trout from the Howietoun Hatchery in Stirlingshire, Scotland (Smiley, 1889b), arrived at the Cold Spring Harbor Hatchery, from where in the same month 10 000 were sent to Wilmurt in the Adirondacks, New York; 5000 to Plymouth, New Hampshire (where fry were released in Sunapee Lake); 20 000 to Anamosa, Iowa (where fry were planted in West Okoboji Lake); 20 000 to St Paul, Minnesota; and 10 000 to Maine, where subfreezing temperatures killed 3000 and where fry that hatched from the remainder were placed in Branch Pond at Ellsworth. In April of the same year, 10 000 fry were sent to the Michigan Fish Commission; 6500 to Grand Rapids, Michigan; and 20 000 to the Crooked Lake Hatchery in northern Michigan; the balance of 7000 surviving fry was retained at the New York hatchery (Laycock, 1966). By 1887, Brown Trout were being reared in hatcheries extending from Maine, Maryland and Illinois in the east to California (Shelby, 1917; Moyle, 1976a,b) and Colorado in the west (MacCrimmon & Marshall, 1968).

Brown Trout are currently naturalized in the USA in the following states, with dates of earliest introduction where known: Arizona (1891–1892); Arkansas (1890); California (1892); Colorado (1887); Connecticut (1889–1890); Georgia; Idaho (1892–1893); Iowa (1885); ? Kentucky (1891–1892); Maine (1885); Maryland (1889–1890); Massachusetts (1887); Michigan (1883); Minnesota (1885); Montana (1889); Nebraska (1889); Nevada (1895); New Hampshire (1885); New Jersey (1889–1900); New Mexico (1893–1894); New York (1883); North Carolina (1887); Oregon (1897–1898); Pennsylvania (1886); Rhode Island (1894–1895); South Carolina (1887); South Dakota (1893–1894); Tennessee; Utah (1900); Vermont (1898–1899); Virginia (1890–1891); Washington (1884–1887); West Virginia (1925–1926); Wisconsin (1887); and Wyoming (1889–1890) (MacCrimmon & Marshall, 1968; MacCrimmon *et al.*, 1970; Courteney *et al.*, 1984, 1986).

North American Distribution   In both Canada and the USA, Brown Trout have tended to be released in waters considered to be marginal or unsuitable for native salmonids. The failure of many plantings to establish naturalized populations, and thus the discontinuous occurrence of the species across the continent, can be at least partially accounted for by the inimical habitats into which some captive-bred fish were introduced (Wiggins, 1950). Evidence for both the sympatric presence of Brown Trout with other salmonids and of the displacement of indigenous Brook Trout *Salvelinus fontinalis* and Rainbow Trout *Oncorhynchus mykiss* (Dymond, 1955) where such plantings have been made, suggests that were Brown Trout to have been released in Brook and Rainbow Trout waters from which they have been deliberately excluded, they would have become very much more widely naturalized (MacCrimmon & Marshall, 1968).

Ecological Impact   The ecological impact of Brown Trout in the USA has yet to be precisely assessed. They are said to have severely affected native Golden Trout *Salmo aquabonita* in California, Atlantic Salmon *S. salar* in northeastern states, and Brook Trout *Salvelinus fontinalis* in the Great Smoky Mountains National Park and neighbouring waters (Krueger & May, 1991; Courtenay, 1992; Courtenay & Moyle, 1992). On the other hand, they are highly valued as game fish. In short, as Courtenay comments, 'an introduction that has been beneficial in many areas can also be detrimental in some'.

Unsuccessful Introductions

Unsuccessful attempts have been made in North America to establish Brown Trout in Mexico, and in the Caribbean in Jamaica and Puerto Rico (Hildebrand, 1934; Bonnet, 1941; Erdman, 1947, 1972, 1984; Iñigo, 1949; MacCrimmon & Marshall, 1968; MacCrimmon *et al.*, 1970).

Central and South America

Argentina

Brown Trout were first introduced to Argentina from the USA in 1904, when some were released in rivers of Santa Cruz (De Plaza & De Plaza, 1949). (For the present distribution of Brown Trout in Argentina, see under Rainbow Trout.)

Bolivia

Brown Trout from Germany were released in waters in the Tunari range (Cordillera Oriental) in 1935, and according to Matsui (1962) were stocked (with Rainbow Trout q.v.) in Lake Titicaca, where Welcomme (1981) says they are less well established. They are now naturalized in upland waters of both the Oriental and Real ranges, where they hybridize with Rainbow Trout (MacCrimmon *et al.*, 1970).

Ecological Impact   According to Welcomme (1988), Brown Trout in Bolivia have had a negative impact on native *Orestias* populations.

Chile

Brown Trout ova were first imported from Hamburg, Germany, into Chile in 1905, where fry were subsequently widely planted in streams between 33°S and 41°S (MacCrimmon & Marshall, 1968). Quijada (1913) and Mann (1954) list the species as established in Chile, while Eigenmann (1927) refers to it in Lautaro, Puerto Montt, Rio Blanco. The currently naturalized distribution of Brown Trout in Chile is more or less continuous from 30°S (Coquimba) to 42°S (Puerto Montt), with only a few discrete populations elsewhere (MacCrimmon & Marshall, 1968). The dispersal of Brown Trout from government hatcheries, however, maintains populations between 19°S (Iquique) and 55°S (Tierra del Fuego), where (as in the Falkland Islands, q.v.) the species has become anadromous, and where some of the finest fly-fishing for Sea Trout in the world is found.

Ecuador

In 1950, 1957 and 1959, a total of 90 000 Brown Trout eggs from Chile, and in 1954 some from the USA, were imported into Ecuador, where they were placed in lakes and streams in the Provinces of Azuay and Cañar not already inhabited by Rainbow Trout (q.v.). In 1967 and 1968, further shipments of 100 000 and 350 000 eggs respectively were received from the USA, but the status of Brown Trout in Ecuador is uncertain (MacCrimmon *et al.*, 1970; Welcomme, 1981, 1988), although MacCrimmon and Marshall (1968), quoting Howard and Godfrey (1950) refer to 'the naturalization of Rainbow Trout in Ecuador in the 1930s'.

Panama

Although neither MacCrimmon and Marshall (1968) nor MacCrimmon *et al.* (1970) refer to the introduction of Brown Trout to Panama, Welcomme (1981, 1988) says that Brown Trout of unknown origin and imported at an unrecorded date are established 'in highland areas of western Panama'.

Peru

Although the date and origin of the first Brown Trout to be introduced to Peru is unknown, it was certainly not before 1928. The species is currently naturalized in lakes and rivers of the state of Puno at or above 2500 m (MacCrimmon & Marshall, 1968). Welcomme (1988) says of the species that 'a small stock is in Lake Titicaca [see under Bolivia] and other highland lakes'.

Venezuela

Léon (1966) records that Brown Trout were first imported into Venezuela in 1938, and that from 1941 ova were stripped from fish established in the wild. He also says that fingerlings were in the mid-1960s being released in rivers and lakes in the Andes Venezolanos region (Cordillera de Mérida) in the provinces of Trujillo, Mérida and Táchira, where their present status is uncertain.

South American Distribution

MacCrimmon and Marshall (1968) say that 'In South America, the naturalization of brown trout has occurred in areas typified by air temperatures reaching a maximum of 21°C in summer (January) and not exceeding 10°C in winter. The extensive elevated areas of the Cordillera of the Andes and Patagonia account for the presence of brown trout along the continental divide'.

Ecological Impact According to Kear and Williams (1978) and Kear (1990), where they occur sympatrically in fast-flowing waters in western South America, Brown Trout and other naturalized insectivorous fish have had a negative impact on indigenous Torrent Ducks *Merganetta armata*. (See also under Rainbow Trout.)

Unsuccessful Introductions Brown Trout have been introduced unsuccessfully in South America

apparently only in Colombia (MacCrimmon *et al.*, 1970; Welcomme, 1981, 1988).

Australasia

Australia

Roughley (1951), Weatherley and Lake (1967), MacCrimmon and Marshall (1968) and others have summarized the history of the introduction of Brown Trout to Australia.

The earliest successful introduction of Brown Trout to the antipodes was made, after four previous failures (McKay, 1984), in 1864, when eyed ova mainly from the River Wey near High Wycombe, Buckinghamshire, and the River Itchen, Bishopstoke, Hampshire, arrived in Tasmania from England on the clipper *Norfolk* (Youl, 1864; Arthur, 1878; Francis, 1879; Nicols, 1882; Seager, 1889; Stokell, 1955; Scott, 1964). Only around 200–300 eggs hatched, but from these a successful planting was made in the Plenty River, where a naturalized population became established. Although according to Scott (1964), 'this lot alone was the basis of the brown trout stocks in Tasmania . . .', Frost and Brown (1967) say that from a second importation of 1500 eggs in 1865 around 500 hatched successfully.

In the following year, the clipper *Lincolnshire* arrived in Tasmania with a shipment of 10 000 migratory Sea Trout ova from the River Tweed in Scotland, from which 1000 fry were subsequently hatched (Johnston, 1883; Scott, 1964). Most of these fish entered the smolt stage in September and October 1867, and were released in the Plenty River (Scott, 1964).

On the mainland of Australia, Ritchie (1988) has traced in detail the introduction of Brown Trout to Victoria.

On 18 August 1866, 1800 ova arrived from Tasmania on the *Southern Cross*, from which they were transferred to Sunbury; this shipment was followed by a second of 1300 eggs in August 1867, a third comprising 250 fry in the following November, and a fourth consisting of 1000 ova in August 1868; all these eggs and fry died before they could be released.

In December 1868, yet another consignment of fry from Tasmania arrived in Victoria, some of which were sent to Yan Yean and some to New Gisborne, where it is believed they were planted in Jackson's and Riddell's Creeks.

In 1869, ova from Tasmania were sent to C. H. Lyon at Ballan on the Upper Werribee River, where after hatching it was later found that 'some scoundrel had tampered with the box containing the trout fry . . . and taken a number of the fry therefrom'. In the same year, the Victoria Acclimatisation Society (Lever, 1992) received 100 Sea Trout which hatched at the Royal Park, from where fry were sent to Lal Lal, where they are believed to have been released in the West Moorabool River.

By about 1870 Brown Trout were beginning to be caught by anglers in Victoria, and in 1871–1872 the acclimatization society supplemented the existing stocks by distributing over 2612 fry in the following Victorian waters:

LOCATION	NUMBER
Watts River, Healesville	1500 +
Watts River, Fernshaw	200
Fyans Creek, Stawell	100
Creeks feeding Lake Colac	200
Bourke's Creek (Cardinia Creek Basin)	80
Riddell's Creek	260
Running Creek, Lillydale (Olinda Creek)	30
Boyd's Creek (Maribyrnong River basin)	60
Ferntree Gully Creek	12
Creek near Dandenong	170

In October 1872, when the first attempt was made to obtain fry from the wild, it was announced that 'many large fish are known to be in the [Riddell's] creek', providing strong circumstantial evidence that the fry released in 1868 had matured and were probably reproducing. In the same year, 430 fry were received from Tasmania, 380 of which were planted 'in a secluded area' in the upper reaches of the Coliban River and 50 in Kilmore reservoirs.

In 1872, the *Southern Cross* brought yet another consignment of 1500 Brown Trout ova from Hobart, Tasmania, and 200 Sea Trout eggs. The Sea Trout and some 300 Brown Trout ova remained in the Royal Park hatchery, while the balance of the Brown Trout eggs was distributed to Riddell's Creek, Tatong on the Broken River, and Ettrick. In 1871–1873, the Ballarat Fish Acclimatization Society (Lever, 1992) announced that it had successfully hatched and distributed Brown Trout in local waters.

Throughout the 1880s there were regular distributions of fry to numerous Victorian waters, including in 1881 Mitchell River, Gippsland; Cockatoo and

William Wallace Creeks, Gembrook; Stony Creek, Beaconsfield; Coliban River, Malmsbury; Hughes' Creek; Fisher's Creek, Narbethong; Taggerty River; Delatite River, and Kangaroo Flat reservoir. In the following year plantings were made in the Yea River, King Parrot Creek, Avon River, Tyer's River, Macalister River, Cockatoo Creek, Toorourrong reservoir, Rose River, Buffalo Creek, Tararalgon Creek, Morses Creek, Ovens River, and many others. In 1886, some 16 000 fry were released in various waters in Victoria, and in 1891 5000 were freed in the Jamieson River by Albert Le Souef.

At the Annual General Meeting of the Victoria Acclimatization Society in 1895 (Lever, 1992), it was reported that 'The Council are pleased to know that their labours in the cause of Pisciculture have been rewarded with success, as is evidenced by the frequent capture of English trout in many of the streams of the colony' (quoted by Ritchie, 1988). One final introduction of Brown Trout in Victoria was made by the Society in 1909, when 40 000 Brown Trout ova and 30 000 'Loch Leven' ova were hatched, the majority of the fry being placed in the government ponds in Studley Park.

From Victoria, Brown Trout were translocated to New South Wales in 1888, to southern Queensland (32 000 ova from New Zealand) in 1896, at an unknown date to South Australia, and before 1885 (Dannevig, 1885) to Western Australia.

Brown Trout are currently naturalized in Australia in cool, well-oxygenated upland waters above the 600-m contour from northern New South Wales and extreme southern Queensland (the Stanthorpe and Warwick districts) to the south coast of Victoria (where they are sometimes found at sea level) and at lower elevations (to 300 m) where there are extensive lakes and impoundments or where fast-flowing waters remain cool. According to Lake (1957), in the northerly areas of New South Wales, waters below 1200 m provide Brown Trout with a marginal or even lethal habitat in summer. The only apparent exceptions to the absence of Brown Trout at lower elevations in New South Wales are in the Clarence River near Grafton, the Shoalhaven River, and the Tuross River, where Weatherley and Lake (1967) ascribe their presence to small numbers being washed downriver during severe floods. Brown Trout are found in a few streams near Adelaide in South Australia, and in the Karri country near Pemberton in Western Australia.

Summing up, Weatherley and Lake (1967) say that with the exceptions referred to above, Brown Trout occur in Victoria and New South Wales in most of the main tributaries above 600 m of the Rivers McIntyre, Gwydir, Namoi, Macquarie, Lachlan, Murrumbidgee and Murray in the western watershed, and the Clarence, Macleay, Hastings, Manning, Hunter, Wollondilly, Shoalhaven, Tuross and Snowy in the eastern watershed.

In Tasmania, where temperatures tend to be lower, Brown Trout are abundant and widely distributed down to sea level in estuaries and inshore brackish waters.

(See also McDowall, 1980; Merrick & Schmida, 1984; Clements, 1988; Allen, 1989; McKay, 1989; Burchmore & Battaglene, 1990; Twyford, 1991; Crowl *et al.*, 1992).

Ecological Impact According to Jackson and Williams (1980) and Jackson (1981), with a few exceptions data on the interactions between trout, especially Brown Trout, and indigenous Australian fishes were up to those dates fragmentary and inconclusive.

Keam (1994) and Cadwallader (1995) have summarized what has recently been discovered about the ecological impact of trout in Australia. The principal significance of trout in Australian waters has undoubtedly been on populations of native galaxiids – a group of small fishes usually described incorrectly as 'minnows'. 'To the degree that generalizations are possible', wrote Keam, 'it is broadly clear that:

(a) In environments where galaxiids are abundant and a high proportion are easily secured, they are likely to form a major component of adult trout diet, leading to high growth rates and high average weight.

(b) In clear, open lakes and streams, this situation may not be sustainable in the longer term; decline may follow with great rapidity. In the case of some galaxiid species, populations may restabilise at a lower level in altered spatial distributions. Other galaxiid species or populations may decline past this point.

(c) In weedy, rocky or turbid lakes and streams offering adequate refuge and breeding habitat, galaxiid populations may decline to some extent in the presence of trout, but the species may have a high chance of sustainable coexistence. In

some waters management strategies may be important in enhancing galaxiid survival and trout food chains.

(d) The diets of adult galaxiids and juvenile trout overlap. In smaller streams which serve as trout spawning and nursery areas, galaxiids may be outcompeted for food and space; "in this light the non-migratory riverine species of galaxiids are likely to be the most vulnerable . . ." (Sloane & French, 1991). Additionally, under these circumstances juvenile galaxiids can be prone to direct predation by juvenile as well as older trout'.

Galaxiid species occurring within the distribution of naturalized trout and regarded by Keam (1994) to be both locally and, because their restricted distribution is the same as their national distribution, nationally at risk are the Barred or Brown Galaxias *Galaxias fuscus* in Victoria, and the Pedder Galaxias *G. pedderensis*, the Swan Galaxias *G. fontanus*, the Clarence Galaxias *G. johnstoni*, and the Saddled Galaxias *G. tanycephalus* in Tasmania. Galaxiids with a wider national distribution which Keam regarded as locally at risk are the Mountain Galaxias *G. olidus*, the Broadfinned or Climbing Galaxias *G. brevipinnus*, and the Spotted Galaxias or Spotted Mountain Trout *G. truttaceus* in Victoria. (See also, Tilzey, 1976).

Of the very few larger native fish species that coexist with trout, 'only in the case of trout cod [*Maccullochella macquariensis*] does the presently incomplete and inferential evidence suggest a possibly significant impact of trout' (Keam, 1994). Cadwallader and Backhouse (1983) said that 'it . . . appears that predation by and competition with trout has been one of the major factors in the decline of the . . . Trout cod'. Although Cadwallader and Gooley (1984) pointed out that overfishing and habitat degradation have probably contributed to the Trout Cod's decline, they added that areas of northeastern Victoria previously occupied by the species correspond with those presently inhabited by trout, and suggest that this is 'perhaps . . . more than coincidence', though at the same time indicating that supporting evidence is largely anecdotal.

To the above list of species affected by naturalized trout in Australia, Cadwallader (1995) added the Variegated Pygmy Perch *Nannoperca variegata* as 'vulnerable', the Yarra Pygmy Perch *Edelia obscura* and the Australian Grayling *Prototroctes maraena* as 'potentially threatened', and indicates that the position of the Macquarie Perch *Macquaria australasica* is 'indeterminate'.

'The numerous and widespread instances of fragmented galaxiid distribution patterns in the presence of trout (in a wide range of catchments, some in areas of natural vegetation, some in cleared areas)' wrote Cadwallader (1995), 'the more widespread distribution of these galaxiids in the absence of trout, and the observations of fragmentation of the galaxiid distribution pattern as trout progressively move upstream, provide a substantial body of evidence for an adverse impact of trout on these stream-dwelling galaxiids'. 'Predation has been the major cause of impact', Cadwallader continues, although 'direct competitive encounters may occur between juvenile trout and those species which occupy the same microhabitat and feed in the same manner on the same foods'. 'The impact of salmonids on the native fish fauna via the spread of pathogens', Cadwallader concludes, 'particularly via the widespread releases of hatchery-produced fish, is unknown. However, the occurrence of pathogens in salmonid hatchery stocks indicates that salmonids have undoubtedly played a major role in the spread of these disease organisms, irrespective of whether or not the pathogens were introduced in Australia with the salmonids'.

Cadwallader (1995) provides a comprehensive bibliography on the impacts of introduced salmonids on Australian fauna.

New Zealand

The earliest importation of Brown Trout to South Island, New Zealand, was made in 1867, when 800 ova were brought from Tasmania by A. M. Johnson for the recently founded Canterbury and Otago Acclimatization Societies (Arthur, 1879, 1882, 1884; Thomson, 1922; Scott, 1964; Lever, 1992; McDowall,

1990b, 1994b). In the following year, successful importations from Tasmania were made by the Southland, Otago, Canterbury and Nelson societies, and until at least 1875 repeated introductions of Brown Trout were made to New Zealand from the same source (Thomson, 1922; Hobbs, 1948; Druett, 1983). It was not until 1883 that the first importation of non-migratory Brown Trout was made from Europe (Scott, 1964), but for much of the remainder of the decade many introductions came directly from England, Scotland, Germany and Italy (McDowall, 1990b).

Scott (1964) outlines in detail the early introductions to New Zealand of migratory Sea Trout which were released as in the table below.

From these releases, naturalized populations of Sea Trout became established in southeastern South Island (Scott, 1964).

In 1870, the Auckland Acclimatization Society in North Island acquired 1000 Brown Trout eggs from Tasmania, from which only 60 fry hatched successfully (Thomson, 1922). Other early introductions to North Island recorded by Thomson were to Auckland again (1872–1874), Wellington (1874), Hawke's Bay (1876), and Wanganui (1877). Large numbers of trout hatcheries were built by private individuals and nearly every acclimatization society, and fish were released in river systems throughout both islands (Hobbs, 1937) so that, according to MacCrimmon and Marshall (1968), by the mid-1880s introductions to North Island had become as successful as those in South Island. Thomson (1922) estimated that by 1916 no fewer than 50 million Brown Trout had been planted by the Wellington, Canterbury, Westland, Otago and Southland Acclimatization Societies, and that between 1916 and 1921 a further 14 million had been released.

Today, both migratory and sedentary varieties of Brown Trout are widely distributed in most rivers of both North and South Islands south of around the

DATE	RIVER	NUMBER	ORIGIN	IMPORTER
1869–1870	Waiwera	?	River Hodder, England	Government of Otago
1871	Water of Leith	20	River Hodder, England	Government of Otago
1872	Shag	120	River Tweed, Scotland	Otago Acclimatization Society
1875–1876	Oreti	850	River Tweed, Scotland	Southland Acclimatization Society
1875–1876	Wyndham	250	River Tweed, Scotland	Southland Acclimatization Society
?	?	?	River Tweed, Scotland	Canterbury Acclimatization Society
—	Ova did not hatch	?	River Tweed, Scotland (1876)	Otago Acclimatization Society
?	?	?	Howietoun Hatchery, Scotland	A. M. Johnson

Coromandel Peninsula, and occasionally further north into Northland where, however, waters tend to be too warm for reproduction. On much of the east coast of North Island, from East Cape south to the Wairarapa, and in parts of the upper Rangitikei and Wanganui River catchments, Brown Trout are scarce because of unsuitable habitats. Elsewhere, apart from Stewart Island from which they are absent, they are almost universal, and are frequently much commoner than is apparent (McDowall, 1990b).

Ecological Impact   Evidence regarding the ecological impact of Brown Trout in New Zealand is fragmentary and confused. The problem has been addressed by, among others, McDowall (1984, 1990a,b,c, 1991), Crowl *et al.* (1992) and Wilson (1995).

For most native species, including some such as bullies (Eleotridae) which are heavily predated, there appears to be little or no evidence for significant detrimental impact (McDowall, 1990c). There are, however, several examples where impacts are considered to be important. Although in river estuaries there seems to be little evidence that predation by Brown Trout has noticeably depressed populations of either galaxiids or retropinnids (southern smelt), circumstantial evidence suggests that upstream the sympatric existence of Brown Trout and one of the largest amphidromous galaxiids, the Giant Kokopu *Galaxias argenteus*, is infrequent, implying competitive exclusion (McDowall, 1990c; Lever, 1994). There is similar evidence, albeit inconclusive, for the exclusion by Brown Trout juveniles of the smaller upland Dwarf Galaxias *G. divergens*, and by adults of the Common River Galaxias *G. vulgaris* (McDowall, 1990c).

In well-oxygenated mountain torrents, competition for food with Brown Trout may have contributed to the failure of the native Blue or Mountain Duck *Hymenolaimus malacorhynchus* to breed (Kear, 1972, 1990; Kear & Williams, 1978; Lever, 1994). (See also under Rainbow Trout.)

Summing up the position in New Zealand, McDowall (1990c) says, 'Although few explicit data are available, it appears that the addition of large salmonids has had harmful impacts on some indigenous species, particularly several galaxiids; these appear to result from predation and competitive exclusion ... it is difficult to separate the effects of human-induced habitat deterioration from interspecific interactions'.

Summing up the situation in the Antipodes, Crowl *et al.* (1992) conclude that in both Australia and New Zealand, Brown Trout appear to be more damaging than Rainbow Trout; that stream fish faunas seem to be less affected than those in lakes; that both species have been more successful in New Zealand than in Australia; and that both the introduction of Brown and Rainbow Trout and man-made land-use changes have impacted on native fish species. (See also under Rainbow Trout.)

Papua New Guinea

MacCrimmon *et al.* (1970), West and Glucksman (1976) and Glucksman *et al.* (1976) have summarized the history of the introduction of Brown Trout into Papua New Guinea.

Brown Trout were first imported in 1949, when Mr (later Sir) Edward Hallstrom (West & Glucksman incorrectly say Hallstorm), imported 20 000 fingerlings from Oberon, New South Wales, which he released unsuccessfully in the Rivers Arl and Wahgi and tributaries in the Nondugl area of the Western Highlands and in a pond at the Hallstrom Fauna Trust (Schuster, 1951).

In 1955, the Bulolo Gold Dredging Company imported 20 000 ova from New Zealand, which were released to provide sport-fishing for employees in the Bulolo River and its tributaries in the Wau area of Morobe (West & Glucksman, 1976).

The Division of Fisheries began importing Rainbow Trout (q.v.) in 1964, and Brown Trout in 1968, when around 1900 fry from Ballarat in Victoria were released with mixed results in Aunde Lake, Mt Wilhelm; in streams around Goroka; in the Mangani River, Mendi; in the Iaro River, Ialibu in the Erave/Purari watershed; and in the Nebilyer River, Mt Hagen (West & Glucksman, 1976), and also into the Asaro River in the Purari watershed; the Gumanch in the Wahgi/Purari watershed; and the Baiyer in the Sepik watershed (MacCrimmon *et al.*, 1970).

According to Glucksman *et al.* (1976), Brown Trout have been introduced in the following Districts of Papua New Guinea: Southern Highlands, Western Highlands, Chimbu, Eastern Highlands, and Morobe, including in the Gumanch and Baier systems (MacCrimmon *et al.*, 1970).

Ecological Impact   In well-oxygenated highland torrents, competition for food with introduced

Brown Trout may have adversely affected the endemic Salvadori's Duck *Anas waigiuensis* (Kear & Williams, 1978; Kear, 1990; Lever, 1994).

Oceania

Falkland Islands

Arrowsmith and Pentelow (1965), from whom the following account is derived, have recorded the history of the introduction of Brown Trout to the Falkland Islands.

The earliest attempts to establish Brown Trout in the Falklands were made during the Second World War, when small numbers of eyed ova were imported from Chile. In August 1947, a much larger shipment of 30 000 eggs was received in the islands as a gift from the government of Chile; these eggs were flown from a hatchery at Lautaro to Montevideo in Uruguay, from where they travelled by ship to Stanley, where they hatched successfully. Which streams were stocked with fry is uncertain, but it is believed that the majority were planted in the Moody and some probably into the Malo and Murrel Rivers on East Falkland: none were placed in waters on West Falkland.

In the same year, Dr J. E. Hamilton, a biologist employed by the government of the Falklands, obtained 10 000 Brown Trout eggs from D. F. Leney's Surrey Trout Farm in England, from which they were despatched by air to Montevideo and thence by ship to Stanley, where they arrived in January 1948. In each January of the succeeding 4 years (1949–1952) Brown Trout eggs were sent out to Stanley – 15 000 in 1949 and 10 000 in each of the other years. These eggs were stripped from fish in a pond near Haslemere in Surrey, and from others in Cobbinshaw Loch in the Pentland Hills south of Edinburgh, Scotland. No Sea Trout occur in these waters that have no outlets to the sea and hold only non-migratory Brown Trout. The first release in three rivers on West Falkland was made in 1950; fry were also planted in several land-locked ponds, but apparently without success.

The first rod-caught Brown Trout in the Falkland Islands were taken in 1954 by anglers fishing (with mutton (!) as bait) for *Aplochiton* (*Haplochiton*), freshwater and brackish southern smelts. Two years later, Brown Trout were first caught on a fly in the islands.

Most, if not all, the Brown Trout in the Falkland Islands, which now provide (together with Tierra del Fuego in Chile (q.v.) some of the finest fly-fishing for the species in the world, have taken to migrating to sea to feed, only returning to their natal rivers to spawn, and have thus become anadromous Sea Trout.

According to Welcomme (1981, 1988), who incorrectly gives the date of introduction as 1955, Brown Trout in the Falklands occur 'mainly in estuaries and coastal lagoons', where they 'may have led to a decline of native galaxiid species'.

Îles Kerguelen

Davaine and Beall (1982a) say that in 1955 115 000 Brown Trout ova from France were shipped to Îles Kerguelen in the Southern Ocean, of which 46 000 survived the voyage; from these only 743 fry hatched successfully, 721 of which were placed as 1-year-old fish in the Val Studer river system and 22 as 4-year-olds in the Rivière du Château; in 1961 a further 2000 alevins were added in the Val Studer.

In 1979, 30 000 anadromous Sea Trout eggs from the Baltic were imported to Îles Kerguelen, of which 29 200 arrived alive and from which 23 000 fish were safely hatched; these were planted as alevins in the Val de l'Acaena river system.

It is worth noting, as Davaine and Beall (1982a) point out, that the Brown Trout and Brook Trout *Salvelinus fontinalis* (q.v.) established in Îles Kerguelen are all descended from very small founder stocks. The former are naturalized in a variety of habitats, including lakes, ponds, streams, rivers and estuaries. Many of the first generation of fish hatched in the wild in the Rivière du Château migrated to the sea, and colonized the lower reaches of Rivières Norvégienne and Albatros which flow into the ocean in the Baie Norvégienne, establishing both non-migratory Brown Trout and anadromous Sea Trout subpopulations. (Lesel *et al.*, 1971; Davaine & Beall, 1982b).

MacCrimmon *et al.* (1970) and Welcomme (1981) say only that in Îles Kerguelen Brown Trout are 'resident in three small natural waters'.

Îles Crozet

In 1969 and 1972, a total of 95 400 Brown Trout eggs were despatched from France to Îles Crozet in the Southern Ocean, where 71 400 hatched successfully in the Rivières du Camp and Moby Dick and in Lac Perdu on Île de la Possession. Those placed in the colder and more oxygenated waters of the du Camp became locally established in very small numbers downstream of a small population of American Brook Trout *Salvelinus fontinalis* (q.v.) that had been introduced in 1969. A few Brown Trout have been taken in the Moby Dick where, however, they do not appear to reproduce (Davaine & Beall, 1982a).

Unsuccessful Introductions Elsewhere in Oceania, unsuccessful attempts to naturalize Brown Trout have been made in the Azores (Goubier *et al.*, 1983); the Chatham Islands (Skrynski, 1967; MacCrimmon & Marshall, 1968); Fiji (Welcomme, 1988); the Hawaiian Islands (Brock, 1960); and on Marion Island in the South Atlantic (Cooper *et al.*, 1992).

Arctic Char(r)
Salvelinus alpinus

The Arctic Charr is a highly variable fish, depending on its environment. In the north of its range it is anadromous, whereas further south it lives in cool clear mountain lakes where, because of their long isolation, populations have become dissimilar both from one another and from the migratory form. In the British Isles, where the entirely land-locked population is probably a relic of anadromous Post Ice Age fish, they occur in the Lake District of Cumbria, in many Scottish lochs, in a few lakes in north Wales, and in numerous Irish loughs. Owing to poor feeding conditions, land-locked individuals are generally much smaller than their migratory counterparts. In Loch Arkaig, Scotland, however, where they live under the cages of ranched salmon whose food they share, they reach a weight of 3.5 kg or more (A.C. Wheeler, pers. comm. 1995).

Natural Distribution

Circumpolar; Arctic and North Atlantic Oceans, northern Europe, the former USSR, and North America.

Naturalized Distribution

Europe: Austria; Cyprus; France; Germany; Italy; former Yugoslavia.

Europe

Austria

Pechlaner (1984), quoting historical evidence, argues persuasively that most, if not all, the numerous lakes in the Alps above the timber-line (the 'alpine horizon' of botanists) that contain populations of Arctic Charr must have been stocked by man rather than as a result of natural invasion. In particular, he shows that a specific set of montane lakes in the continental divide of the eastern Austrian Alps could not have been invaded naturally by the Arctic Charr they contain. These lakes, all with long-standing Charr populations, are the Vorderer Finstertaler See (2237 m asl), the Hinterer Finstertaler See (2256 m), the Mittlerer Plenderlesee (2317 m) and the Oberer Plenderlesee (2344 m), all in the Studai Mountains near Kuehtai; and the Berglersee (2466 m), the Wannenkarsee (2639 m), the Unterer Seekarsee (2655 m), the Laubkarsee (2681 m) and the Geislacher See (2702 m), all in the headwater region of the Oetz valley.

Pechlaner (1984) concludes that the extent of the glaciers during the last glacial advance in northern Europe (the Devensian, Weichselian or Würm period), which occurred approximately 100 000 to 18 000 years BP, confirms that the Charr in the lakes are postglacial immigrants. None of the lakes listed above existed as water bodies during the Egesen stadial (the early Dryas), part of the characteristic late-glacial sequence of climatic change following the ice advance of the Devensian, and could only have been formed after the end of the early Dryas around 10 000 years ago.

Pechlaner (1984) quotes evidence from mediaeval documents to show considerable stocking of fish in alpine lakes by the sovereigns of the Tyrol in the late 15th and early 16th centuries, specifically by Erzherzog Sigismund (died 1496) and Kaiser Maximilian I (died 1519), and subsequently by Erzherzog Ferdinand II (died 1595). Sigismund is specifically mentioned as having ordered the stocking with fish of five montane lakes, and four such lakes, described in a report dated 1500 as being without any fish, are 4 years later mentioned as containing both Brown Trout *Salmo trutta* (q.v.) and Arctic Charr.

Cyprus

Welcomme (1981, 1988) says that Arctic Charr from Britain were introduced to the island of Cyprus in 1970, 'where some isolated populations persist'.

France

Hubault (1955) states that Arctic Charr from the Lake of Geneva in Switzerland and from Scandinavia have been repeatedly stocked in lakes in eastern France, where stunted populations are now naturalized in Lakes Annecy and Leman in Haute Savoie, and in their inflowing tributaries.

Germany

According to Brenner (1984), in 1961 the 3.25-km² Sorpe reservoir in Nordheim Westfalen in the then Federal Republic of Germany was stocked with the fry and fingerlings of Arctic Charr hatched from eggs obtained from the Mondsee in Austria, and reared in the Landesanstalt für Fischerei in Albaum. In 1980, more Charr were planted in the 7.8-km² Rur reservoir. Both introductions appear to have been successful.

Italy

Arctic Charr were originally absent from the Italian peninsula, apart perhaps from some small montane lakes such as Molveno and Tovel in Trentino. Fry from the Zuggersee in Switzerland were successfully introduced into Lago Lugarno in 1895, and in 1910 fry from Lugarno were transplanted to Lago Maggiore.

Yugoslavia

Welcomme (1981, 1988) says that in 1928 and 1943 Arctic Charr from Switzerland and Austria respectively were successfully introduced to the former Yugoslavia, where at present 'populations are established in three deep lakes only'.

Unsuccessful Introductions

In Europe, Arctic Charr have been unsuccessfully introduced in Denmark (Jensen, 1987) and The Netherlands (Vooren, 1972; De Groot, 1985); in Australasia in New Zealand (McDowall, 1968; Clements, 1988); and in Africa in Morocco (Moreau *et al.*, 1988).

Translocations

Since early times, Arctic Charr (and other fish species) have been extensively translocated within Fennoscandia, originally by Laplanders in order to ensure a supply of food during their long annual migrations from the coastal forests to the mountains. More recently, in 1961 Arctic Charr were successfully translocated to Lake Pieskejaure in Swedish Lapland, where within 6 years they had eradicated their principal food source, the anostracan fairy shrimp *Polyartemia forcipata*; since then the Charr population has become stunted (Nilsson, 1972).

Brook Trout (Brook Char(r); Speckled Trout)
Salvelinus fontinalis

The discontinuity of the naturalized distribution of the Brook Trout (erroneously named a trout rather than a charr by early European immigrants to North America due to its similarity to the Brown Trout *Salmo trutta* (q.v.)) reflects the species' low tolerance of adverse environmental conditions rather than a lack of attempts to establish it more widely, although it is not as highly regarded for food or as a sporting asset elsewhere as it is in North America (MacCrimmon & Scott Campbell, 1969).

It is generally agreed that water temperature is the most important single factor governing the naturalized distribution of the Brook Trout; it seems able to survive in waters ranging from just above freezing to around 24°C (preferably not above 20°C), with an optimum range for growth of about 11–16°C and for breeding of between 9 and 11.5°C. Other factors affecting the ability of Brook Trout to become established outside their natural range

include latitude, altitude, precipitation, and geophysical and ecological conditions similar to those within their natural range (MacCrimmon & Scott Campbell, 1969).

It seems probable that, except perhaps in Asia, the current naturalized distribution of the Brook Trout is unlikely to be greatly extended.

Natural Distribution

Northeastern North America, from northern Quebec south of about 60°N and Labrador south of around 57°N, south through Newfoundland and Nova Scotia to about 40°N in Pennsylvania (with an extension south through central Virginia into extreme western North Carolina, eastern Tennessee and northern Georgia), westward to Wisconsin, Michigan and Manitoba.

Sea-going Brook Trout (known as 'coasters') occur mainly in the northern (Canadian) part of the species' natural distribution, and also in some rivers of Maine and Cape Cod.

Naturalized Distribution

Europe: Austria; ? Belgium; British Isles; Bulgaria; Czechoslovakia; Denmark; France; Germany; Greece; Italy; Norway; Poland; Romania; ? Spain; Sweden; Switzerland; ? Yugoslavia. **Asia:** Japan. **Africa:** Kenya; ? South Africa; Zimbabwe. **North America:** ? Mexico. **South and Central America:** Argentina; Bolivia; Chile; Peru; Venezuela. **Australasia:** Australia; New Zealand. **Oceania:** Falkland Islands; Îles Crozet; Îles Kerguelen.

Europe
Austria

Brook Trout, probably from Germany, were imported to Austria shortly after 1879 (Von Pirko, 1910), and are now naturalized in most Austrian provinces, including Vorarlberg, Tyrol, Salzburg, Upper and Lower Austria, Styria and Carinthia (MacCrimmon & Scott Campbell, 1969).

Belgium

According to Welcomme (1981, 1988), Brook Trout were introduced from the USA to Belgium in the 1890s, where the species only has a 'very limited' distribution, and 'due to its delicacy it is disappearing'. (See also Mulier (1900) and Poll (1949).)

British Isles

The first shipment of Brook Trout eggs from the USA to reach England arrived in the spring of 1869 (Vooren, 1972). Livingstone Stone, owner of the Cold Spring Trout Ponds at Charlestown, New Hampshire, USA, recorded that 'one lot was sent to England to Mr Frank Buckland, and was favourably noticed in the London *Times*'. This initial consignment was followed by a second 2 years later, when 10 000 eggs from Lake Huron fish were imported from the same source by John Parnaby, co-owner with J. J. Armistead of the Troutdale Fishery near Keswick, Cumberland (now Cumbria), England (Lever, 1977, 1992). In 1878, Sir James Maitland imported 10 000 Brook Trout eggs from the USA

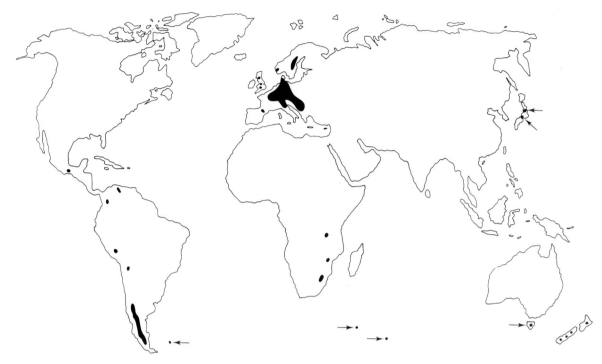

Naturalized distribution of Brook Trout *Salvelinus fontinalis* (adapted from MacCrimmon & Scott Campbell, 1969; McCrimmon *et al.*, 1971; McDowall, 1990b).

to his hatchery at Howietoun in Stirlingshire, Scotland.

These are generally accepted as the earliest introductions of Brook Trout to Britain (Ffennell, 1885). In a letter to *The Times* newspaper (28 October 1885), however, Mr Parker Gilmore claimed that in 1868 he had shipped to England from North America Brook Trout and Atlantic Salmon *Salmo salar* (q.v.) eggs and oyster spat, which were apparently distributed before his return and without his knowledge. He complained in his letter that 'the Prince of Wales had a pond stocked with *fontinalis*. When I visited it the person in charge was ignorant of my being the introducer, as was also His Royal Highness, until I informed him by letter . . . at the Zoological Gardens, I also saw *fontinalis* in a glass tank with another person's name given on the label as presenter and introducer . . . I have not only been a pecuniary loser, but also have never been credited with my work' (quoted by Lever, 1977).

Before the end of the 19th century, Brook Trout had been released in many waters in England and Scotland, including the following:

DATE	COUNTRY	LOCALITY
? *c.*1876	England	River Thames, Chertsey, Surrey
Before 1884	Scotland	Loch Uist and Lochbuie, Mull
Before 1886	England	Ponds at Tehidy, near Camborne, Cornwall
Before 1887	England	River Wey, Guildford, Surrey
1889	England	River Eden, Appleby, Westmorland
1880s	Scotland	Loch Brora and Kintradwell Burn, Sutherland
Before 1892	Scotland	Lochs in Cuilfail district above Kilmelford, near Oban, Argyll
1898	Scotland	River Buchart, Aberdeenshire

Before the turn of the century, Brook Trout were being taken by anglers in many British waters including, according to Armistead (1895), 'in salt water in some of our bays and estuaries', suggesting the existence of an anadromous population. By the early 1900s they had been planted in numerous waters in the Clyde watershed, in Loch Lomond, throughout Renfrewshire and Ayrshire (where they were flourishing in the Rivers Irvine and Ayr), and in waters in

Boland, Kilmarnock, Cessnock, Carmel and Alnwick (Walker, 1976).

Since then, despite repeated stocking in many localities, Brook Trout have succeeded in becoming naturalized in only a few widely scattered waters in England, the Scottish Highlands, and Wales (De Bunsen, 1962; Lever, 1977). They have apparently never been introduced into Ireland. (For further details of Brook Trout in the British Isles see Lever, 1977.)

Bulgaria

Brook Trout eyed ova were first unsuccessfully imported from Czechoslovakia to Samokov in western Bulgaria in 1930. In 1959, Brook Trout were planted in several landlocked mountain lakes devoid of Brown Trout, and since 1962 more have been released in another upland lake 1500 m asl (MacCrimmon & Scott Campbell, 1969). Welcomme (1981, 1988, 1991) indicates that Brook Trout are established in the wild in Bulgaria.

Czechoslovakia

Brook Trout were introduced to Black Lake in the Šumava Mountains of southwestern Bohemia before 1890. By the late 1960s, they were being released mainly in valley reservoirs of northern Bohemia and in suitable streams throughout Czechoslovakia. They became naturalized in the High Tatra Mountains of northern Slovakia (especially in Lake Štrbské Pleso); in middle and eastern Slovakia; in the Hruby-Jeseník Mountains of northern Moravia; in the Bohemian/Moravian uplands; and in the Krkonoše and Jizerské Mountains in northern Bohemia (MacCrimmon & Scott Campbell, 1969; Anon., 1971).

Denmark

Brook Trout are believed to have been originally introduced from Germany to trout hatcheries in central Jutland around the turn of the century, where escaped fish subsequently established free-living populations (Jensen, 1987). The species was deliberately released in the 1960s in the River Skjern å and in the Randers fjord, and in tributaries of the River Lindenborg å and in Hald Lake, in all of which populations became naturalized (MacCrimmon & Scott Campbell, 1969).

According to Welcomme (1981, 1988), Brook Trout from Germany were introduced to Denmark in 1895–1896 and 1902–1903, where they are presently 'well established in the upper reaches of some streams', but where 'interest in the species has declined since 1940'.

France

According to Bean (1895), Brook Trout were shipped by the United States Fish Commission to the Trocadéro Aquarium in Paris in 1879, from where they were later introduced in many waters in France, including the upper reaches of the River Seine, tributaries of the Rupt-de-Mad in Lorraine, and in streams in the Alps (Delachaux, 1901) and Voges Mountains, in none of which they became established (Vivier, 1955).

According to MacCrimmon and Scott Campbell (1969), Brook Trout became naturalized in Lacs Lutel, Domenon and Prénol, high altitude montane lakes in the Alps in the province of Dauphiné, and in other lentic and lotic waters between 1800 and 2300 m asl in the Alps in Savoie which are devoid of Brown Trout. The same authors also record naturalized Brook Trout populations in the upper drainage basins of two streams, the Aspe and Gaube, in the Pyrénées Mountains around 2400 m asl, and in ponds at Sturzelbronn and Mouterhouse near Biche on the Moselle. Hubault (1955) refers to their failure to become established in Lake Annecy in eastern France.

Welcomme (1981, 1988, 1991), however, who gives the earliest date of introduction as 1932, says only that there is 'one permanent population in a lake of the Eastern Pyrénées, but in regression'. (See also under Spain.)

Germany

In 1879, Livingstone Stone (see under British Isles) sent a consignment of Brook Trout ova from his Cold Spring Trout hatchery at Charlestown, New Hampshire, USA, to Friedrich von Behr of the German Fishery Association (Von Behr, 1883), by whom in 1885 eggs were transferred to a hatchery at Boitzenburg in Mecklenburg (Von dem Borne, 1885). Between 1884 and 1889, Germany imported 620 000 Brook Trout eggs, and produced 390 000 internally; in 1890, when German hatcheries yielded 750 000 eggs, the species was said to be established in several high altitude lakes and streams (such as the Lahn near Carlsbad) in Bavaria (De Groot, 1985).

By the late 1960s, MacCrimmon and Scott Campbell (1969) recorded naturalized populations of Brook Trout in Germany throughout the country, especially in Thüringen and the Werra River and its tributaries, and in high altitude mountain lakes, in some of which Brook Trout interbreed with native Brown Trout to produce a sterile hybrid (Vooren, 1972) known as 'Tigerfisch' (see also under Zimbabwe). Welcomme (1981, 1988) says that in Germany Brook Trout are, contrary to experiences elsewhere, 'generally regarded as desirable and a better quality sport fish than local alternatives'.

Italy

The earliest attempt to introduce Brook Trout to Italy was in 1891 (Welcomme (1981, 1988) incorrectly says 'after 1895'), when fry were released in Lake Idro in Lombardia (MacCrimmon & Scott Campbell, 1969). This planting, like one made in 1935 in streams in the Valley of Aosta, was unsuccessful. Soon after the Second World War, the Fishery Council of the Aosta Valley imported some Brook Trout ova from Denmark, fry from which were successfully placed in local waters where naturalized populations formed in at least two lakes at altitudes of 1940 and 2542 m asl in which potentially competing native salmonids do not occur (MacCrimmon & Scott Campbell, 1969). Welcomme (1981, 1988) records Brook Trout in Italy only as 'present in very few streams'. (See also Mazzola, 1992.)

Norway

Between 1870 and 1919, Brook Trout were unsuccessfully introduced on a number of occasions to various localities in Norway. Grande (1964), however, records that 1000 fry liberated in 1918 in a small tarn at Dyrdal, Øyfjell, in the Telemark district of southern Norway resulted in a naturalized population becoming established in local watercourses around 600–1000 m asl, where in the lower reaches they lived sympatrically with Brown Trout and where Welcomme (1981, 1988, 1991) confirms their existence.

Poland

Brook Trout eggs from Germany were imported in the last decade of the 19th century to Poland, where naturalized populations became established in a small number of mountain rivers and lakes, particularly near Biedrusko in a tributary of the Warta River, and in a stream near Jelenia Góra, in the Wroclaw *voivodship* of western and northwestern Poland. Owing to angling pressure and hybridization with Brown Trout, the population is only precariously established (MacCrimmon & Scott Campbell, 1969; Jasinski, 1981).

Romania

In the late 19th century, Brook Trout from Austria were released in three tributaries of the Somesul Mic River, Cluj (the Negrutza, Dumitreasa and Irisoara); naturalized populations became established in the Negrutza and in a tributary of the Crisul Repede River in the Bihor Mountains (in the western Carpathians); in the Gudea Mica, a tributary stream of the Upper Mures River in Transylvania; in the Putna, a tributary stream of the Moldova River in Moldavia; and in the Upper Bistritza River, Moldavia (MacCrimmon & Scott Campbell, 1969).

Spain

Brook Trout ova from Switzerland were first imported to the hatchery of the monastery of Piedra in 1934, but the resulting fish disappeared during the Civil War of 1936–1939. After the Second World War, fry from France were released in streams in the Pyrénées (see also under France) where, although MacCrimmon and Scott Campbell (1969) say that 'it is not known if these plantings have resulted in the establishment of naturalized populations', Welcomme (1981, 1988) says that the species has become established. Elvira (1995a) lists it as naturalized in the drainages of the Rivers Douro and Tagus.

Sweden

Eyed Brook Trout ova from Germany were first imported to the province of Jämtland in the winter of 1891–1892 (Svärdson, 1964). Until 1907, Brook Trout were stocked in around 120 Swedish waters, mainly in Jämtland, but also in Halsingland and North Dalarna (Svärdson, 1964), where some naturalized populations became established. Alm (1920) recorded Brook Trout to have been successfully introduced in some lakes, tarns and rivers in Örebro, Kopparberg and Jämtland, while Hanström and Johnels (1962) found populations established in similar waters in Norrland.

MacCrimmon and Scott Campbell (1969) were informed of self-maintaining populations of Brook Trout in most Swedish provinces, where they are most abundant in Västergötland, Småland, Närke, Västmanland, Dalarna, Hälsingland, Medlepad, Jämtland, Åugermanland, Västerbotten, Norrbotten and Lappland; woodland waters of the last three provinces seem particularly suitable, especially in Norrbotten where the annual precipitation is below 35 cm (50 cm of rainfall per annum seems to delimit the Brook Trout's European distribution). Naturalized populations in Sweden generally occur only below 200 m.

Switzerland

Goll (1887) records that in 1883 Brook Trout eggs from the Germany Fishery Association were sent to Vaud, Switzerland, where they were hatched at the Roveray Fish Cultural hatchery near Allaman. In 1885–1886, Frederick Mather of the Cold Spring Harbor Hatchery in New Hampshire, USA, sent 10 000 Brook Trout eggs to the Swiss government (Clark, 1887), which succeeded in establishing a few naturalized populations in various localities. The Swiss Fisheries Department, however, told MacCrimmon and Scott Campbell (1969) that because of continuing loss of suitable spawning redds, intensive stocking of both native and exotic salmonids is essential.

Yugoslavia

According to Welcomme (1981, 1988), Brook Trout from Austria were first introduced to the former Yugoslavia in 1892, where MacCrimmon and Scott Campbell (1969) said they were then naturalized in some waters of Slovenia and near the source of the River Bosna at Sarajevo, where they lived sympatrically with Brown Trout. Welcomme (1981) said that there were 'few permanent populations' and (1988) that 'very few remnant populations occur', suggesting that the species may soon be, if it is not already, extinct.

Unsuccessful Introductions

Brook Trout have been unsuccessfully introduced in Europe in Finland, Hungary and the former USSR (MacCrimmon & Scott Campbell, 1969); in The Netherlands (Mulier, 1900; Nijssen & De Groot, 1975; De Groot, 1985); and in Cyprus and Greece (Welcomme, 1981, 1988).

Asia

Japan

Brook Trout were first successfully introduced to Japan in 1901, when ova from the USA were introduced to the Nikko hatchery, the resulting fry being liberated in the Yugawa River in Nikko City on central Honshu. Brook Trout are currently naturalized in Kotoku Numa, Yuniko Lake and the Yugawa (all near Nikko) in the Tochigi Prefecture, and in the small Myozinike Lake in the Nagano Prefecture (Okada, 1960; MacCrimmon & Scott Campbell, 1969; Welcomme, 1981, 1988; Chiba *et al.*, 1989).

Unsuccessful Introductions

Brook Trout have been introduced in Asia without success in India (MacCrimmon & Scott Campbell, 1969) and the Lebanon (MacCrimmon *et al.*, 1971).

Africa

Kenya

Brook Trout ova from England were first introduced to Kenya in 1949, but the resulting fry, planted in waters around 2000 m asl on the slopes of Mount Kenya and in the Aberdare Mountains, all died (Harrison *et al.*, 1963). In 1961, ova were imported from the Paradise Brook Trout Company of Pennsylvania, USA, the hatched fingerlings being released high in the Aberdares in a large reservoir and in some streams, in one of the latter of which they succeeded in becoming naturalized (MacCrimmon & Scott Campbell, 1969; Welcomme, 1981, 1988).

South Africa

Le Roux and Steyn (1968) refer to the presence of Brook Trout in mountain headwaters of a tributary of the Olifants River in the eastern Transvaal. More recently, however, Welcomme (1981, 1988) says they are 'maintained only in one hatchery' in South Africa and Bruton and Merron (1985), Bruton and Van As (1986) and De Moor and Bruton (1988) make no mention of their establishment in the Republic.

Zimbabwe

In 1955, both Brook Trout and so-called 'Tiger Trout' (a Brook Trout × Brown Trout *Salmo trutta* hybrid) were stocked in Zimbabwean (then Rhodesian) waters, in some of which in the Inyanga Mountains Brook Trout became established (Jubb, 1961), but where according to Welcomme (1981, 1988), they are 'not widespread'. (See also Turnbull-Kemp, 1957, and under Germany.)

Unsuccessful Introductions

In Africa, introduced Brook Trout have failed to become established in Nyasaland (Malawi) and Tanganyika (Tanzania) (MacCrimmon & Scott Campbell, 1969), and in Morocco (Moreau *et al.*, 1988).

North America

Mexico

According to Jara (1945), 1500 Brook Trout reared from ova imported from the USA to the El Zarco Fish Station in the State of Mexico, were released in the Río Tendido near Fortin, Veracruz in 1939. By 1945, a total of 91 838 Brook Trout had been distributed to suitable waters in six states in Mexico, where MacCrimmon *et al.* (1971) were informed that they occurred principally only in the states of Puebla and Mexico, mainly above the 3000 m contour (MacCrimmon & Scott Campbell, 1969).

Contreras and Escalante (1984), however, say that the date of release 'was possibly mid-1800s' – perhaps a literal error. The presence of Brook Trout was assumed in the Valley of Mexico by Alvarez and Navarro (1957), in Río Yaqui by Hendrickson *et al.* (1980), and in Cuitzitán Michoacán by Rosas (1976b) (all quoted by Contreras & Escalante (1984)), and elsewhere. Welcomme (1981, 1988), on the other hand, says that introductions of Brook Trout to Mexico have been 'apparently unsuccessful'.

Translocations

Within North America, Brook Trout have been successfully translocated outside their natural range to Alaska, Arizona, California, Colorado, Idaho, Montana, Nebraska, Nevada, New Mexico, Oregon, South Dakota, Utah, Washington and Wyoming, and unsuccessfully to Delaware, Illinois, Indiana, Kentucky, North Dakota and Ohio (MacCrimmon & Scott Campbell, 1969). In Canada, they have been translocated with success to Alberta, British Columbia and Saskatchewan (MacCrimmon & Scott Campbell, 1969).

South and Central America

Argentina

Brook Trout were first introduced to South America in 1904, when 100 000 eggs from the USA were despatched to a fish farm on the shores of Lago Nahuel Huapí in Argentina (Thompson, 1940), where 3 years later fry were planted in a number of apparently suitable waters in the provinces of Córdoba, Buenos Aires, Tucuman, Salta and San Luis (Tulian, 1910). Between 1904 and 1908, E. A. Tulian, of the Argentinian Ministry of Agriculture, personally supervised the importation of a total of 535 000 Brook Trout eggs from the USA to Argentina, where by 1931 a further 52 000 had been introduced and fry and fingerlings had been successfully released in Patagonian lakes and in some of those in the Parque Nacional Nahuel Huapí, where according to Thompson (1940) Brook Trout had become the most abundant salmonid.

By the late 1940s, the species was widespread and common in the Nahuel Huapí, Lanín and Los Alerces National Parks (De Plaza & De Plaza, 1949). MacCrimmon and Scott Campbell (1969) and Welcomme (1981, 1988) reported Brook Trout to be naturalized in nearly all suitable rivers and large lakes (General Paz, Fontana and Cardiel) in the Patagonian steppe. (See also under Rainbow Trout.)

Bolivia

In 1948, Brook Trout were imported from Chile to the Pongo hatchery in Bolivia, from which a number of suitable rivers in the Cordillera Occidental and Altiplano were stocked, where some naturalized populations have become established (MacCrimmon & Scott Campbell, 1969; Welcomme, 1981, 1988).

Chile

Prior to 1927, Brook Trout from Lago Nahuel Huapí in Argentina were introduced into Chile, where they were placed in the Laguna del Inca near Portillo (2700 m asl), in the upper reaches of the Río Blanco, Aconagua, and in the middle reaches of the Río Cautín (Eigenmann, 1927; Golusda, 1927) from which, however, Mann (1954) said they seemed to have disappeared. (See also De Buen, 1959.)

MacCrimmon and Scott Campbell (1969), on the other hand, who say that Brook Trout eggs of unknown origin were first introduced to Chile in

1935–1936 and were planted at an unrecorded location, reported naturalized populations mainly in the province of Aconagua, and in the following rivers of the high Cordillera mountain range: Colorado, Estero Ojos de Aqua, Estero Pinquenes, Estero Juncal, Estero La Polvareda and Estero Los Leones. Fly-fishing for Brook Trout in Chile occurs in the Río Negro near Peulla, in the Río Petrohue which flows from Lago Todos los Santos, and in the Ríos Manso and Puelo which flow out of Argentina (Heusser, 1964), who says that some Brook Trout survive alongside Brown and Rainbow Trout (q.v.) in Laguna del Inca.

Peru

Brook Trout from the USA were first imported to the Chucuito Fish Culture Station at Puno on the west bank of Lake Titicaca in extreme southern Peru in 1950 (MacCrimmon *et al.*, 1971) where, together with Brown and Rainbow Trout (q.v.), they have become successfully established, as they have also in surrounding rivers and lakes above the 2500-m contour (MacCrimmon & Scott Campbell, 1969; Welcomme, 1981, 1988).

Venezuela

Brook Trout from the USA were originally introduced to Venezuela in 1937, where further introductions followed until at least 1942. Plantings were later made in various suitable waters in the Venezuelan Andes, and Brook Trout are now considered to be naturalized in the Motatan, Chama and Santo Domingo Rivers in the State of Mérida, and in numerous small lakes in the Paramos of the Andes (MacCrimmon & Scott Campbell, 1969).

Unsuccessful Introductions

Attempts to naturalize Brook Trout in South America have failed in Colombia and Ecuador (MacCrimmon & Scott Campbell, 1969).

Australasia

Australia

Brook Trout were first introduced to Tasmania in 1883, where they failed to become established. Further introductions followed, most recently from Canada in 1962, and naturalized populations are presently established in a few waters such as Clarence Lagoon (Merrick & Schmida, 1984; Clements, 1988), where the populations are regularly augmented by additional stocking. Since the 1970s, Brook Trout have been stocked in streams on the montane tablelands of New South Wales and in South Australia, in both of which they have failed to establish viable populations; in the former state, Brook Trout continue to be stocked to provide put-and-take angling (Cadwallader, 1995).

New Zealand

Thomson (1922) records that in March 1877, A.M. Johnson of Opawa, Christchurch, received the first consignment of Brook Trout ova to reach New Zealand, when a 'considerable stock' arrived from New York via California, USA. From these eggs a large stock of fish was obtained, which was sold to several acclimatization societies in various parts of the country (Lever, 1992; McDowall, 1994b). In the same year, the Auckland Society acquired 5000 ova direct from San Francisco, California, and other shipments arrived from the USA in 1883 and 1884, and from the Solway Fisheries in Scotland in 1887 (Thomson, 1922).

Plantings from these introductions were made around Auckland, Wellington and Christchurch, and from fish reared in Christchurch in Otago, Southland and Taranaki (McDowall, 1990b, 1994b). In Lake Emily a population of Brook Trout became established from releases made in 1932 and 1938, and others were naturalized in the Hatepe hydro lakes after liberations there in 1952 (McDowall, 1990b).

Although Brook Trout were released thereafter in many places in New Zealand, particularly in eastern South Island, they apparently failed to catch anglers' imagination. Since the mid-1980s, however, they seem to have attracted renewed interest, and further releases have been made in Lake Tikitapu in Rotorua; in Lake Opouahi in Hawke's Bay; in the Red Lagoons in the upper Waitaki; in Lake Dispute near Queenstown; and in Lake Henry near Te Anau, all of which, according to McDowall (1990b) will probably depend on restocking since their catchments are devoid of suitable spawning redds.

'In New Zealand' wrote McDowall (1990b), 'the brook char is found quite widely, although it is relatively little known. It occurs in some streams flowing into Lake Rotorua, a tributary of the upper Waikato (the Tahunaatara Stream), and in the Hatepe hydro lakes in the Hinemaiaia Stream, a tributary of Lake Taupo. It is also present in the headwaters of the Aorangi and Moawhanga Rivers (in the upper reaches of the Rangitikei River) and Lake Opouhai (Hawke's Bay). In the South Island it is surprisingly widespread in many inland Canterbury, Otago and Southland river systems, as well as reported from the upper Buller catchment. Lake Emily, in the upper Ashburton River catchment in inland Canterbury, has larger brook char than other waters in New Zealand.' (See also Ayson, 1910; Stokell, 1951; McDowall, 1979; Welcomme, 1981, 1988; Clements, 1988).

Unsuccessful Introduction   Brook Trout have been unsuccessfully introduced in Australasia to Papua New Guinea (Glucksman *et al.*, 1976; West & Glucksman, 1976).

Oceania

Falkland Islands

The earliest attempts to naturalize trout species in the Falkland Islands were made during the Second World War, when eyed ova of Brown Trout *Salmo trutta* q.v., Rainbow Trout *Oncorhynchus mykiss* (q.v.), and Brook Trout were received from Chile. The Rainbow Trout died out, but a few small Brook Trout became established in one water, the Moody Brook (Arrowsmith & Pentelow, 1965). Whether, like the Brown Trout, they became anadromous, is unknown.

Îles Kerguelen

In 1961, 25 000 Brook Trout eggs from France were shipped to Îles Kerguelen in the Southern Ocean, where 6000 arrived alive, 2548 of which later hatched successfully; the resulting fish were subsequently distributed as follows: 500 alevins in the Rivière du Château; 2000 alevins in Val Studer; and 2000 alevins and 48 yearlings in the Rivière du Sud (Lesel *et al.*, 1971).

As with Brown Trout *Salmo trutta* (q.v.), it is noteworthy that Brook Trout in Îles Kerguelen are descended from a very small founder stock, which became established in a wide range of habitats, including lakes, ponds, streams, rivers and estuaries (Davaine & Beall, 1982a). They now occur, with Brown Trout, in the upper reaches of the Val Studer and on their own in the Rivière du Château.

In 1973, some Brook Trout were caught in the Rivières Norvégienne and Albatros, but it was not until early 1977 that two small populations seemed to have become established in these rivers upstream of the areas colonized by Brown Trout. Apart from two individuals caught in 1973 and 1978, Brook Trout in Îles Kerguelen have not exhibited any sea-going instinct, and it thus seems probable that the two rivers were colonized either via a network of underground channels through the peat bogs or, possibly, though given the distance involved less likely, through deliberate transportation by anglers (Davaine & Beall, 1982a).

Îles Crozet

In 1969, 25 000 eyed Brook Trout eggs from France were despatched to Îles Crozet, also in the Southern Ocean, where all but 100 arrived safely. Of these, 18 900 were planted in the Rivière du Camp and 4000 in Rivière Moby Dick on L'Île de la Possession (Lesel *et al.*, 1971). Those placed in cold and well-oxygenated waters at high elevations in the Rivière du Camp hatched successfully, whereas those planted lower down all died. Davaine and Beall (1982a) reported a small population of Brook Trout to be established in the Rivière du Camp, where Brown Trout occur in even smaller numbers downstream.

Unsuccessful Introductions

In Oceania, attempts to naturalize Brook Trout in the Hawaiian Islands were unsuccessful (Maciolek, 1984). (See also Needham & Welsh, 1953; Brock, 1960; Randall, 1960, 1980; Randall & Kanayama, 1972.)

Lake Trout (Lake Char(r); Mackinaw)
Salvelinus namaycush

A number of attempts have been made to take advantage of the Lake Trout's tolerance of cold and deep waters by introducing it to high latitude and high altitude locations in Europe, South America and New Zealand. Apart from a few discrete populations in large deep lakes, however, none have been successful in establishing naturalized populations (Welcomme, 1988).

Natural Distribution

North America, from western Alaska to eastern Canada, southwards from the Arctic into southern Canada in the west and New York and the New England states of the USA in the east.

Naturalized Distribution

Europe: Germany, Switzerland. *South America:* Argentina. *Australasia:* New Zealand.

Europe

Germany

According to De Groot (1985), Lake Trout eggs from North America were shipped to Germany in the late 19th century as follows: 40 000 in 1881; 100 000 in 1883; 25 000 in 1884; and 50 000 in 1885. In 1882, fish, presumably hatched from eggs imported in the previous year, were introduced to German waters, but with what result is unrecorded.

Welcomme (1981, 1988) refers to an introduction from the USA that is not mentioned by De Groot, and also to introductions in 1969 and 1978 to the then Federal Republic (West Germany), where he says that Lake Trout are established only 'very locally in southern Germany'.

Switzerland

North American Lake Trout were first introduced to alpine lakes in Switzerland in 1888 (Delachaux, 1901; Vivier, 1955), where in the 1950s a breeding population became established (Heinz & Lorenz, 1955), especially in Lake Arnensee (Grimås & Nilsson, 1962), and other high-altitude mountain lakes such as Sägisthal, Fully and Barberine (Thienemann, 1950; Vivier, 1955; Muus & Dahlstrøm, 1968), in at least some of which they still survive (Welcomme, 1981, 1988).

Unsuccessful Introductions

Lake Trout have been unsuccessfully introduced in Europe to Britain (Von dem Borne, 1890), Denmark (Jensen, 1987), Finland (Nilsson & Svärdson, 1968),

France (Kreitman, 1929; Vooren, 1972; De Groot, 1985), and Sweden (Nilsson & Svärdson, 1968).

South America

Argentina

Tulian (1910) and Fuster de Plaza and Plaza (1954) have summarized what is known about the introduction of Lake Trout to Argentina.

In March 1904, E.A. Tulian, then with the Argentinian Ministry of Agriculture, arrived at Lago Nahuel Huapí with a consignment of eggs from New York, among which were 53 000 Lake Trout ova, of which only 5% were lost in transit. After hatching, most of the Lake Trout were released in Lakes Nahuel Huapí, Gutierrez, Traful and Correntosa, in the last two of which Tulian (1910) said they were still to be found 4 years later.

In February 1905, Tulian imported another shipment of fish eggs from New York, including 224 000 of Lake Trout, many of which were hatched at Nahuel Huapí and some in a hatchery at Alta Gracia in the province of Córdoba. A year later, another batch of fish eggs arrived from New York, with 80 000 Lake Trout ova among them, nearly all of which were landed safely. In 1908, yet another shipment from New York included 75 000 Lake Trout ova, making a grand total since 1904 of 432 000. The eggs imported in 1908 were hatched in Santa Cruz, the Lake Trout fry being released in the very deep Lago Argentina in southern Argentina and in some neighbouring waters, which are fed by several small snow- and ice-melt rivers and streams which rise in the Andes. Today, a few Lake Trout in poor condition survive only in Lago Argentina (Fuster de Plaza & Plaza, 1954; Welcomme, 1981, 1988). (See also Navas, 1987, and under Rainbow Trout.)

Unsuccessful Introductions

In South America, attempts to naturalize Lake Trout in Bolivia (Welcomme, 1988), Chile (Mann, 1954; De Buen, 1959) and Peru (Welcomme, 1981, 1988) all failed.

Australasia

New Zealand

In 1906, L.F. Ayson imported 'a case' (McDowall (1994b) says 50 000) of Lake Trout eggs from Northville, Michigan, USA (Donne, 1927) to New Zealand, where they were hatched at the Christchurch Acclimatization Society's station, from where 4000 fry were liberated in Lakes Pearson and Grassmere, two small waters in north Canterbury. Another batch of 4000 was intended for release in Lake Kanieri, but 'owing to the carelessness of the curator then in charge of the hatchery they were all lost' (Thomson, 1922). Those in the two Canterbury lakes were said to have done well, and by 1916 fish up to 4.5 kg were being caught (Stokell, 1951; Clements, 1988). (In North America, Lake Trout have been known to reach a weight of 46 kg.)

For many years few attempts were made to manage the Lake Pearson population (that in Lake Grassmere is believed to have died out many years ago), but eventually in 1977 and 1980 the New Zealand Wildlife Service translocated fish from Pearson to their hatchery at Wanaka as a founder breeding stock (McDowall, 1990b, 1994b).

McDowall (1984) summed up the present status of the Lake Trout in New Zealand thus: 'Present only in Lake Pearson in subalpine Canterbury, status threatened, exploited only incidentally by anglers seeking other species; largely unrecognized . . . Perhaps the most precarious [alien] species . . . Lake Pearson is a shallow basin lake and the survival of lake trout there for seventy-five years is a surprise. They are small and usually in poor condition, reaching a kilogram or a little more. The current strategy with lake trout is to establish and maintain a hatchery brood stock and at present the hatchery fish are surviving. The status of the wild lake population is hard to assess but is regarded as endangered'. (See also Druett, 1983; Clements, 1988.)

Unsuccessful Introductions

According to Shetty *et al.* (1989), *Salvelinus namaycush* × *S. fontinalis* hybrids ('Splake Trout') have been unsuccessfully introduced from Canada to India, and according to Chiba *et al.*, 1989, from Canada to Japan. Davaine and Beall (1982b) record the unsuccessful introduction of Lake Trout to Îles Kerguelen in the Southern Ocean.

Translocations

In North America, the Lake Trout has been widely translocated outside its natural range in both the western and eastern USA (Smith, 1896; Chapman, 1942; McDowall, 1990b).

COREGONIDAE

Lake Whitefish
Coregonus clupeaformis

The Lake Whitefish is an important commercial species in oligotrophic lakes throughout its native range (Welcomme, 1988).

Natural Distribution

Northern USA and Canada.

Naturalized Distribution

South America: Chile.

South America

Chile

In 1930, the Lake Whitefish was introduced from an unrecorded source to the province of Concepción in central Chile (Oliver, 1949), where it is established

and breeding (Welcomme, 1988). (See also De Buen, 1959.)

Unsuccessful Introductions

In Europe, the Lake Whitefish has been unsuccessfully introduced to Britain, France, Germany, The Netherlands and Switzerland (Hubault, 1955; Vooren, 1972; De Groot, 1985); in Asia to Japan (Chiba *et al.*, 1989); in South America to Argentina (Tulian, 1910; Baigun & Quiros, 1985); and in Australasia to New Zealand (Ayson, 1910).

Translocations

In North America the Lake Whitefish has been widely transplanted outside its natural range, particularly in the Pacific northwest (Smith, 1896; Chapman, 1942).

Common Whitefish (Powan (Scotland); Schelly (England); Gwyniad (Wales))
Coregonus lavaretus

Coregonus lavaretus is, like *Salvelinus alpinus*, believed to be a relic of a more widespread migratory species of the last Ice Age, and isolated lake populations have evolved their own individual and distinctive characteristics. It is an important commercial fish in oligotrophic lakes throughout its native range. It is a facultative anadrome, and Welcomme (1988) suggests that its establishment in the River Danube implies that it may also occur in other countries bordering that river from which it has so far not been recorded. (Coregonid nomenclature is still extremely uncertain, and is further confused by the many early introductions (A.C. Wheeler, pers. comm. 1995).)

Natural Distribution

The Baltic and North Sea basins; Britain, northern Europe, the northern USSR, and the Swiss Alps.

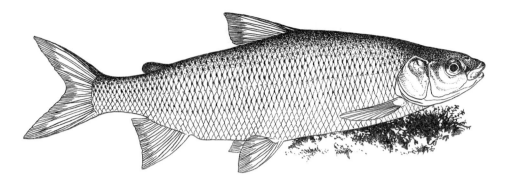

Naturalized Distribution

Europe: Belgium; Czechoslovakia; ? Germany; Italy.

Europe

Belgium

Welcomme (1988) says that populations of the Common Whitefish and the Broad Whitefish *C. nasus*, a native of the Arctic USSR and eastern Scandinavia, are established and breeding in Lake Bütgenbach some 55 km southeast of Liège in extreme eastern Belgium near the German border.

Czechoslovakia

Two subspecies of *C. lavaretus* have been introduced to Czechoslovakia, the Marina *C. l. maraena* from lakes in northern Germany in 1882, and *C. l. wartmani* from Austria at an unknown date (Anon., 1971). Welcomme (1988) says that Common Whitefish are established and reproducing in the main channel of the River Danube, and that they may well colonize Czechoslovakian reservoirs. They may also have spread undetected to other countries through which the Danube flows.

Germany

Common Whitefish introduced from the former USSR to the then Federal Republic (West Germany) to upgrade stocks in alpine lakes hybridized with local *Coregonus* species (Welcomme, 1981, 1988, 1991). Whether any pure *lavaretus* remain is unknown.

Italy

Common Whitefish first introduced to Italy in 1880 (Dottrens, 1955) (to Lago Maggiore in 1891 (Grimaldi, 1972)) to establish a naturalized population for commercial fishing, are today 'widespread in lakes in central and northern Italy' (Welcomme, 1988).

Unsuccessful Introductions

Hubault (1955) and Vooren (1972) have summarized what is known of introductions of *C. lavaretus* elsewhere in Europe.

Originally, many Alpine lakes had their own *Coregonus* species or subspecies (Thienemann, 1950). During the past two centuries many introductions and transplantations have been made, and it seems likely that today few such lakes have 'pure-bred' *Coregonus* species.

Between 1880 and 1887, several Swiss lakes, including Lake Constance, received stocks of *C. l. maraena* from northern Prussia (Besana, 1915). Lake Geneva has been planted with fry from several nearby lakes, particularly with *C. l. primigenius/jurassica* from Lake Neuchatel, which has largely replaced the endemic race which had been much reduced in numbers by over-fishing. Fry of *maraena* from Estonia were unsuccessfully released in Lake Geneva on numerous occasions between 1928 and 1932. In Lake Annecy in France, *C. lavaretus* from Lake Bourget was released in 1892, 1905, 1906 and 1911, and *C. schinzii* from Lake Constance in 1888 (Dottrens, 1955). *C. lavaretus* from Lake Bourget was also successful in Lake Aiguebelette, where the species is, however, 'dying out owing to the eutrophy of the lake' (Hubault, 1955).

Welcomme (1988) refers to presumed unsuccessful introductions of Common Whitefish in Europe in Greece, The Netherlands and former Yugoslavia.

Chiba *et al.* (1989) mention failed attempts with three subspecies (*maraena, ludoga* and *baeri*) in Japan.

Peled

Coregonus peled

In its native range the Peled is an important commercial species which has been the subject of numerous transplantations. It favours coldwater rivers and lakes, but is tolerant of fairly high temperatures in summer (Welcomme, 1988).

Natural Distribution

Northern USSR, and the Baltic Sea.

Naturalized Distribution

Europe: Czechoslovakia; Poland.

Europe

Czechoslovakia

Balon (1974) records the presence of small breeding populations of Peleds in the Czechoslovakian basin of the Danube River, where Welcomme (1981, 1988), who gives the dates of introduction as 1970 and 1979 and the source as the former USSR, suggests the species may have spread undetected to other Danube riverine countries.

Poland

In or around 1960, Peleds from the former USSR were introduced to improve commercial fisheries in Poland, where they are now established and breeding 'in a few lakes' (Welcomme, 1981).

Unsuccessful Introductions

Welcomme (1981, 1988, 1991) records unsuccessful attempts in Europe to naturalize Peleds in Finland and Yugoslavia, while in Asia Chiba *et al.* (1989) and Yo-Jun and He-Yi (1989) mention failures in Japan and China respectively.

Translocations

Mukhachev (1965) describes one of the most successful transplantations of Peleds within their native range to water basins of the Chelybinsk district east of the southern Ural Mountains in the former USSR.

OSMERIDAE

Wakasagi (Japanese Smelt)
Hypomesus nipponensis

The Wakasagi is a small (15 cm) species which frequents freshwater lakes and brackish lagoons. In Japan it is regarded as a luxury human food fish, whereas it was introduced to the USA as a forage food for Rainbow Trout *Oncorhynchus mykiss*.

Natural Distribution

Islands of Hokkaido and possibly Honshu, Japan.

Naturalized Distribution

North America: USA.

North America

United States

In 1959, a shipment of 3 600 000 eyed Wakasagi eggs from Tokyo, Japan, was received by the California Department of Fish and Game, the resulting fry being released as a forage food for Rainbow Trout in the following six Californian waters: Dodge Reservoir in Lassen County; Dwinnel Reservoir, Siskiyou County; Freshwater Lagoon, Humboldt County; Spalding Reservoir, Nevada County; Sly Park Reservoir (Jenkinson Lake), El Dorado County; and Big Bear Lake in San Bernardino County (Courtenay *et al.*, 1984). Since then there have been subsequent introductions in other reservoirs.

Although Wales (1962) said that the Wakasagi was only known to survive in Freshwater Lagoon, Moyle (1976a) suggested that it could be expected in the lower Klamath and Sacramento river systems and perhaps also in other drainages. This has indeed come about, and the species has now become established in several reservoirs in the Sacramento–San Joaquin drainage, including Folsom Reservoir on the American River (Stanley *et al.*, 1995).

H. nipponensis was introduced into California under the mistaken belief that it was a subspecies, *H. transpacificus nipponensis*, of the native Delta Smelt (Moyle, 1976a). Although Welcomme (1988) treated it as such, since 1970 it has been suggested that the two forms are in fact different species, and this was confirmed by Stanley *et al.* (1995).

Ecological Impact Although Stanley *et al.* (1995) could find no evidence of hybridization between *H. nipponensis* and the native *H. transpacificus*, which in 1993 was listed as a threatened species by both state and federal governments, the two species are now known to be hybridizing in the Sacramento–San Joaquin estuary (P.B. Moyle, pers. comm. 1995), thus jeopardizing the genetic integrity of the native species.

ESOCIDAE

Northern Pike
Esox lucius

The preferred habitat of Pike is still or sluggish canals, ponds and lakes with plenty of aquatic weeds and marginal vegetation. Largely piscivorous, they will also eat frogs, newts, water birds and aquatic mammals. In much of its range the species is a popular sport fish and as such has been widely translocated.

Natural Distribution

Circumpolar: Britain, northern Europe, the USSR, Alaska, Canada and the northern USA.

Naturalized Distribution

Europe: Ireland; Spain. ***Africa:*** Algeria; ? Ethiopia; Morocco; Tunisia; ? Uganda.

Europe

Ireland

Went (1957), from whom much of the following account is derived, has traced the introduction of the Pike into Ireland.

The earliest mention, albeit a negative one, to the presence of Pike in Ireland, occurs in *Topographia Hibernica* by Giraldus Cambrensis, probably compiled between 1185 and 1187, in which he describes 'the sea, river and lake fish, and those that are missing . . . some fine fish are wanting. I mean pike, perch, roach, gardon and gudgeon'.

Giraldus's statement seems confirmed by the fact that there was no name for the Pike in the ancient Celtic language; the later name for the species is *gailliasc*, which means literally a strange or foreign fish, strongly implying an introduction.

What may be the earliest positive reference to Pike in Ireland occurs in Edmund Spenser's poem *Epithalamion*, published in 1595, in which, referring to the River Awbeg, a tributary of the Blackwater in Co. Cork, he wrote:

> The silver scaly trouts do tend full well
> The greedy pikes which use therein to feed.

This suggests that Pike were first introduced to Ireland, probably to stock fish stews, at some time between 1185–1187 and 1595.

The species' spread throughout the country appears, however, to have been slow. In *A Chorographical Description of West or hIar Connaught*, published in 1682, Roderic O'Flaherty wrote that 'there was never a pike or bream as yet engendered in all this countrey nor in the adjacent parts of Mayo or Galway counteys', showing that nearly a century

after Spenser's publication Pike in Ireland were still of only very limited distribution.

In other parts of Ireland, however, it is known that Pike were established before 1682, since it is on record that in the early years of the century they were actually being exported to some places in southern England. It is known also, from the *Civil Survey* of 1654–1656, that Pike occurred in the River Camoge and its tributaries in Co. Limerick. Other waters in Ireland known to have held Pike by the second half of the 17th century include Loughs Iron, Gur, Baneen-Annagh, and Direvreagh, and the Rivers Gaine, Barrow, Nore, Suir and Burren. Thereafter, during the 18th century, references to Pike in Ireland grow apace, and by the 19th century the species had probably achieved its present widespread distribution in alkaline waters throughout the country. As elsewhere, the Pike in Ireland is valued as a sport fish, but is a predator of Brown Trout *Salmo trutta* and young Atlantic Salmon *S. salar.* (See also Fitter, 1959; Kennedy, 1969; Vooren, 1972.)

Spain

According to Calderon Andreu (1955), juvenile Pike (pickerel) imported from France since 1949 have been used on numerous occasions to stock Spanish rivers, where although Vooren (1972) says that 'it is not stated whether the species has established wild populations there, and effects on indigenous species are not mentioned', Welcomme (1988) states that Pike are 'regarded as an important gain as no native predator exists'. Elvira (1995a) lists the species as occurring throughout the country except in the north and Galicia.

Ecological Impact

Pace Welcomme (1988), Rincón *et al.* (1990) found that in the Douro basin Pike had a negative impact on fish assemblages. Introduced to the National Park of Daimiel in central Spain in the 1950s, Pike eradicated the native fish community and had themselves died out by 1986 (Elvira, 1995b). In the Mediterranean basins, Pike (and introduced Mosquitofish *Gambusia affinis* (q.v.) and Largemouth Bass *Micropterus salmoides* (q.v.) pose a threat to the local endemic *Valencia hispanica* (Elvira, 1995a).

Africa

Algeria

Moreau *et al.* (1988) list Pike as introduced from France to Algeria in 1956, where they became established in impoundments on Oued Fodac.

Ethiopia

Tedla and Meskel (1981) record that in 1981 250 000 Pike were stocked by the Italian Stabilimento Ittiogenico di Roma in Lake Tana, Ethiopia, where they apparently became established but where their current status is uncertain.

Morocco

Northern Pike from France and Poland were introduced in 1934 (and in 1960 from Israel) for angling purposes to waters of the Atlas Mountains of Morocco, where they established self-maintaining populations (Welcomme, 1981, 1988).

Tunisia

In 1966, Northern Pike from France were imported to Tunisia where according to Welcomme (1981, 1988) they became established 'in high altitude areas', presumably in the mountains near the Algerian border southwest of Tunis.

Uganda

In the 1960s, Uganda imported some Northern Pike from Israel which became established in aquaculture centres; their present status in the wild, if they occur there, is unknown (Welcomme, 1981, 1988).

Unsuccessful Introductions

In 1951 and 1958, unsuccessful attempts were made to introduce Northern Pike from France to the island of Madagascar (Kiener, 1963; Moreau, 1979; Reinthal & Stiassny, 1991), and in 1979–1981 to the Azores (Goubier *et al.*, 1983).

Translocations

As previously mentioned, Northern Pike have been widely and successfully transplanted for sporting purposes within their natural range. In Europe, for example, they have been planted in high-altitude lakes in the Vosges Mountains of France (Vooren, 1972) and reintroduced to Lake Annecy in 1939 (Hubault, 1955). In North America, Smith (1896) refers to a planting of Pike in Idaho from a hatchery in Illinois.

Ecological Impact

Some of these translocations have been both illegal and detrimental to native species; thus Courtenay and Moyle (1992) say that 'California has spent hundreds of thousands of dollars in recent years attempting to eradicate populations of white bass (*Morone chrysops*) and northern pike (*Esox lucius*) established through illegal introductions. Such introductions, made in reservoirs, have tremendous potential to reduce salmonid populations downstream'. The release of Pike in the Vosges, referred to above, has been condemned on the grounds that the Pike is 'an enemy of the common trout' (Vooren, 1972).

UMBRIDAE

Eastern Mudminnow (Striped Mudminnow)
Umbra pygmea

Both *Umbra pygmea* and *U. krameri* (see below) are tolerant of deoxygenated waters.

Natural Distribution

Lowland streams and swamps on the eastern seaboard of the USA.

Naturalized Distribution

Europe: Belgium; France; Germany; The Netherlands.

Europe

Belgium; The Netherlands

According to De Groot (1985), who points out that at the time of writing there was a Dutch fish farm more or less in the middle of the Eastern Mudminnow's range in Belgium and The Netherlands, it is uncertain in which of these two countries the species first appeared in the wild. In the latter, it is restricted to a limited area in the Peel near Eindhoven in the province of Noord Brabant (Vooren, 1972; De Groot, 1985), where according to the former it has occurred since 1920 and from where Kleijn (1968) says it spread naturally over the border into neighbouring Belgium, perhaps via the Heidemaatschappij fish farm near Valkenswaard. An alternative suggestion postulated by Kleijn is that the fish were deliberately released by private aquarists. Poll (1949) only states that Mudminnows entered Belgium from The Netherlands. De Groot (1985), on the other hand, considered it more likely that the fish had been imported from the USA as a forage species for piscivorous salmonids introduced from the same source. De Groot points out a precedent, quoted by Mulier (1900), who said that in 1891 Max Von dem Borne had imported 80 'American minnows' (the White Shiner *Notropis albedus*) as a forage fish to his trout farm. Although Poll (1949) says the Eastern Mudminnow is an undesirable species, Kleijn (1968) points out that it is found in waters unoccupied by any native species.

In Belgium and The Netherlands, the Eastern Mudminnow occurs in pools, ditches and similar stagnant or sluggish waters in areas of peat bog where no indigenous species can survive, and where it is found it is usually abundant. It is a polyphagous predator, and in spring feeds extensively on amphibian larvae. Its ecological impact, if any, is unknown (Parent, 1950).

France

The Eastern Mudminnow was introduced around 1913 from the USA into a small isolated lake in the Bourbonnais district of northeastern France, and is known to have still been there in 1958 (Spillman, 1959). There seems no reason to suppose that it has since disappeared, but whether it has succeeded in spreading to other nearby waters is unknown. D'Aubenton *et al.* (1983) mention a recently discovered location for the species in Argonne, east of Paris, while Welcomme (1988) says it is also found in 'one small area of the Loire basin' in the southeast.

Germany

Muus and Dahlstrøm (1968) record that the Eastern Mudminnow has been successfully introduced to parts of northern Germany, especially Schleswig-Holstein and Niedersachen, where it is established. Welcomme (1988) says that the species spread naturally to Germany from France, but a deliberate introduction seems more plausible.

Allied Species

Escaped aquarium fish are probably responsible for the naturalization of the Mudminnow *Umbra krameri*, a native of the basin of the River Danube, in Polish waters, where it is 'present in a few localities' (Welcomme, 1988).

Wheeler and Maitland (1973) say that Mudminnows, presumably having escaped from aquaria, appeared in the wild in Britain in 1925, but were seen no more after 1934.

CHARACIDAE

Black Tetra (Blackamoor)
Gymnocorymbus ternetzi

A small (5.5 cm) tropical fish much favoured by aquarists. Although fairly hardy, it prefers soft acidic waters and relatively high temperatures (Welcomme, 1988).

Natural Distribution

The Mato Grosso region of Brazil.

Naturalized Distribution

Asia: Thailand. *South America:* ? Colombia.

Asia

Thailand

Black Tetras were imported from Hong Kong and Japan into Thailand in the 1950s, where Piyakarnchana (1989) says they 'breed successfully in natural habitats'.

South America

Colombia

Welcomme (1988) says that Black Tetras are reproducing successfully in Colombia, but whether in captivity or the wild is unclear.

Piranha species
Serrasalmus sp.

Piranhas are small carnivorous fish that tend to form large shoals which make formidable hunting groups. They have strong jaws and sharp teeth which can tear flesh from their prey with great efficiency, but also feed on seeds and fruits.

Natural Distribution

Rivers of South America.

Naturalized Distribution

Asia: Sri Lanka. *Oceania*: ? Hawaiian Islands.

Asia
Sri Lanka

Welcomme (1988) says that 'a piranha of undetermined species has invaded many Sri Lankan rivers', where it is reproducing, but gives no details regarding the date of introduction or the source.

Oceania
Hawaiian Islands

Sakuda (1993), quoted by Eldredge (1994), recorded the discovery in 1992 of a *Serrasalmus* species in the wild on the island of Oahu, where its current status is undetermined.

Unsuccessful Introductions

Shafland and Foote (1979) describe the establishment in 1963 or 1964 and subsequent history of a breeding colony of *S. humeralis* in Miami, Florida, USA, which Welcomme (1988) says was deliberately eradicated in 1981. The same author states that from time to time other piranha species (including *S. rhombeus* and *S. nattereri*) have escaped from aquarists in the USA, but have never established breeding populations in the wild.

CYPRINIDAE

Bleak
Alburnus alburnus

Silver Bream (White Bream)
Blicca bjoerkna

The Bleak and Silver Bream are small fish of temperate lakes and sluggish streams; they are of only limited commercial and sporting importance.

Natural Distribution

Western Europe east to the Ural Mountains in the former USSR, apart from Italy, Spain, Portugal, the former Yugoslavia and Greece. In Britain, the Silver Bream is confined to southeastern England.

Naturalized Distribution

Europe: Cyprus; Spain.

Europe

Cyprus

In 1972, Bleak and Silver Bream were accidentally introduced with other fish species from Britain to Cyprus, where both are established and breeding. Welcomme (1988) says that Bleak exist only in 'stunted populations', and that although they may be of value as a forage fish for native species, their fecundity gives rise to large numbers which can cause ecological problems. Silver Bream are unpopular locally because of their small size.

Spain

In June 1992, Bleak were collected for the first time in the River Noguera Ribagorzana in the basin of the River Ebro, where they have become established (Elvira, 1995a).

Bighead Carp
Aristichthys nobilis

The Bighead Carp, one of the so-called 'Chinese carps', has been widely translocated and introduced for aquaculture both within and without its native range. However, because its breeding requirements are very rigid, in most countries to which it has been introduced it is only maintained by captive reproduction or repeated restocking. Only where environmental conditions mirror those in its native range, such as in the Danube basin, have naturalized populations developed. Welcomme (1988) suggests that the species may well also be established in other Danubian riverine countries such as Austria, Czechoslovakia and Romania, from which it has not been formally recorded.

Natural Distribution

Eastern Siberia and southern China.

Naturalized Distribution

Europe: Riparian countries of the Danube River basin; ? Austria; Bulgaria; ? Czechoslovakia; Germany; Hungary; ? Romania; former USSR. *North America:* USA.

Europe

Riparian countries of the Danube River basin; ? Austria; Bulgaria; ? Czechoslovakia; Germany; Hungary; ? Romania; former USSR

Welcomme (1981, 1988) says that the Bighead Carp is 'now self breeding and widespread in the Danube Basin, where it forms the basis for angling catches'. Although the source of these naturalized populations is not given, they presumably derive from escapes from aquaculture.

Borisova (1972) lists the Bighead Carp as occurring in inlet and outlet ditches and other natural waters adjacent to the Akkurgan Fish Combine in Tashkent, Uzbekistan, in the former USSR, to which it was accidentally introduced with other Chinese carps from the Far East in the 1960s.

North America

United States

Adult Bighead Carp have been taken in the wild in Alabama, Arkansas, Florida, Illinois, Indiana, Kansas, Kentucky and Missouri (Courtenay *et al.*, 1991), having escaped or been deliberately released from aquaculture centres (Courtenay, 1993), but have established a breeding population only in the Missouri River (Pflieger, 1989).

Aquaculture

In Europe, Bighead Carp are reproducing successfully in captivity in France, Germany, Hungary, Italy, Poland and the former Yugoslavia (Welcomme, 1981, 1988), The Netherlands (De Groot, 1985), the former USSR (Negonovskaya, 1981) and possibly elsewhere. In Asia, they are successfully bred in Indonesia (Muhammad Eidman, 1989), Israel (Ben-Tuvia, 1981; Davidoff & Chervinski, 1984), Japan (Chiba *et al.*, 1989), Korea (Welcomme, 1981, 1988), Malaysia (Ang *et al.*, 1989), the Philippines (Juliano *et al.*, 1989), Singapore (Chou & Lam, 1989), Sri Lanka (De Silva, 1987), Thailand (Piyakarnchana, 1989; De Iongh & Van Zon, 1993), Taiwan (Liao & Liu, 1989) and Vietnam (Welcomme, 1981, 1988). In Africa, they were introduced to Egypt for experimental aquaculture in 1975 (Moreau *et al.*, 1988). In South and Central America, they are reared in captivity in Brazil, Costa Rica, Cuba, the Dominican Republic, Mexico, Panama and Peru (Welcomme, 1981, 1988). In Oceania, Bighead Carp were introduced to Fiji from Malaysia in 1968 'for pond culture and the control of vegetation' (Anon., 1971) and as 'a pituitary donor' (Welcomme, 1988). (See also Shelton & Smitherman, 1984.)

Rosy Barb
Barbus conchonius

The Rosy Barb, a small (14 cm) tropical fish much favoured by aquarists, has been widely introduced outside its natural range, and it may well be naturalized more extensively than is believed (Welcomme, 1988).

Natural Distribution

Northern India and Assam.

Naturalized Distribution

North America: ? Mexico. *South and Central America:* ? Colombia; Puerto Rico. *Australasia:* ? Australia.

North America

Mexico

Although Contreras and Escalante (1984) say that in 1967 Rosy Barbs were unsuccessfully released in the Río Santa Catarina, Monterrey, to mark the opening of the municipal aquarium there, Welcomme (1988) indicates that they have become established.

Unsuccessful Introductions

In North America, unsuccessful attempts have been made to naturalize Rosy Barbs in Florida, USA (Courtenay *et al.*, 1984).

South and Central America

Colombia

Welcomme (1988) says the Rosy Barb, introduced for ornamental purposes, is breeding successfully in Colombia; no further details are given.

Puerto Rico

Sometime prior to 1971 Rosy Barbs imported from the USA were released, presumably from aquaria, in Río Arroyata, a tributary of the Río La Plata, between Cidra and Naranjito near Route 172 in Puerto Rico, where cock fish were later observed in nuptial colouring. A number were collected for culture purposes in the Maricao hatchery in which spawning took place; from here they were stocked in the Loiza reservoir, where they became established (Erdman, 1984; Wetherbee, 1989).

Australasia

Australia

Arthington *et al.* (1983) recorded the escape in 1970 of Rosy Barbs from an aquarium near Brisbane, Queensland, where McKay (1984) reported the species to have been found 'in one suburban creek' when a survey of the area was conducted by the state museum in 1977. Five years later, Allen (1989) said that the Rosy Barb 'has become common in several streams around Brisbane, particularly in the Seven Hills area'. Brumley (1991) however, said the population may not be self-maintaining.

Tawes (Javanese Carp; Punten Carp)
Barbus javanicus/gonionotus

The Tawes is a large migratory barb which is an important species in the commercial fisheries of rivers in southern Thailand and Malaysia. It has been introduced, mainly in South East Asia, as a pituitary donor, for aquacultural purposes, to control aquatic weeds, and to establish commercial fisheries. Fish that have escaped from captivity have become naturalized in various rivers, where they are commercially fished (Welcomme, 1988).

Natural Distribution

Thailand, Malaysia, Laos, Khmer Republic, Vietnam and Java.

Naturalized Distribution

Asia: Celebes (Sulawesi); India; Philippines.
Oceania: Fiji.

Asia

Celebes (Sulawesi)

Originally introduced from Java to Celebes for aquaculture in 1930, the Tawes has become successfully established in the wild and is fished for commercially (Welcomme, 1981, 1988).

India

First imported to India from Indonesia in 1972 for the control of aquatic vegetation, the Tawes occurs locally in the wild in Kalyani in West Bengal (Shetty *et al.*, 1989; Yadav, 1993). In 1986, the Tawes was introduced for aquaculture to the Chetpet fish farm in Chengalpattu in Tamil Nadu (Sreenivasan, 1989).

Philippines

In 1956, the Tawes was first introduced as a pituitary donor from Java to the Philippines, where it is currently established in the wild in some rivers (Juliano *et al.*, 1989).

Oceania

Fiji

Introduced to Fiji from Malaysia in 1968 and/or 1969 for aquaculture and as a pituitary donor, the Tawes is well established in the Rewa River in the southeast of Viti Levu (Anon., 1971; Maciolek, 1984; Andrews, 1985).

Other Introductions

The status of the Tawes in Sri Lanka, to which it was introduced from Java in 1968 (De Silva, 1987), and in Papua New Guinea, where it was imported from Malaysia in 1967 or 1970 (Glucksman *et al.*, 1976; West & Glucksman, 1976; Clements, 1988) is uncertain. In China, where it was introduced from Thailand in 1986, the Tawes is cultured in Guandong Province (Yo-Jun & He-Yi, 1989).

Translocation

The reported introduction of the Tawes to Ranu Lamongan on Java (Payne, 1987), where it provided a thriving fishing industry, represented a translocation rather than an introduction.

Half-banded Barb (Chinese Barb; Green Barb; Half-striped Barb)
Barbus semifasciolatus

This small tropical fish is a favourite of aquarists, by whom it has been widely transported around the world.

Natural Distribution

Southeastern China.

Naturalized Distribution

Asia: Singapore. *Oceania:* Hawaiian Islands.

Asia

Singapore

Johnson (1964, 1973) and Chou and Lam (1989) record the accidental introduction of Half-banded Barbs from China to Singapore where, being tolerant of adverse water conditions, the species has become naturalized in the catchment area.

Oceania

Hawaiian Islands

In 1940, the Half-banded Barb was planted in Nuuanu Reservoir 3 on the island of Oahu where it became somewhat precariously established (Brock, 1952, 1960; Kanayama, 1968; Maciolek, 1984; Courtenay *et al.*, 1991; Courtenay, 1993).

Unsuccessful Introduction

The Half-banded Barb has been introduced without success in Papua New Guinea (Glucksman *et al.*, 1976).

Allied Species

In addition to the foregoing, the following *Barbus* spp. are or may be reproducing in the wild outside their natural range. *Barbus amphigramma* in rivers near Lake Naivasha, Kenya, where it was introduced from Tanzania in 1982 (Muchiri & Hickley, 1991); the European Barbel *B. barbus* in Morocco, to which it was introduced from France around 1920 (Moreau, 1979; Moreau *et al.*, 1988); the Golden Barb *B. gelius* (of central India) in Colombia (Welcomme, 1988); the Blackspot Barb *B. filamentosus* (of southwestern India) on Oahu in the Hawaiian Islands since 1984 (Eldredge, 1994); the Smallmouth Yellowfish *B. holubi* (South Africa) in Lake Kyle, Zimbabwe (Toots, 1970; Ludbrook, 1974; Moreau *et al.*, 1988); the Island Barb *B. oligolepis* (Sumatra) in Colombia (Welcomme, 1988); the Natal Yellowfish *B. natalensis* (South Africa) in rivers and dams in Zimbabwe (Moreau *et al.*, 1988); the Cherry Barb *B. titteya* (Sri Lanka) in Mexico and Colombia (Welcomme, 1988); the Tiger or Sumatra Barb *B. tetrazona* (Borneo and Sumatra) in Enoggera Creek, Brisbane, Australia (McKay, 1989) and Colombia (Welcomme, 1988); and *B. sealei* in Palau (Babelthuap) (Bright & June, 1981).

Zebra Danio
Brachydanio rerio

A small tropical fish with a broad range of environmental requirements, the Zebra Danio is a popular aquarium species.

Natural Distribution

Eastern India.

Naturalized Distribution

North America: USA. *South America:* ? Colombia.

North America

United States

Courtenay *et al.* (1991) say that the Zebra Danio has been collected, but was not then known to be established in the wild, in California, Florida and New Mexico. Courtenay (1993), however, indicates that since 1984 the species has become naturalized in Wyoming.

South America

Colombia

According to Welcomme (1988), the Zebra Danio has appeared in Colombian waters, presumably having been released or escaped from an aquaculture centre; whether it is established in the wild is uncertain.

Goldfish (Golden Carp)
Carassius auratus

Two subspecies of the Goldfish have been recognized, the nominate form *C. a. auratus* (which in China has been kept in domestication at least since the early years of the Sung Dynasty (AD 960–1279)) and *C. a. gibelio*, the Prussian Carp, which is the form that occurs naturally in eastern Europe.

The Goldfish is arguably the most popular of all ornamental fish species, and occurs widely in captivity, in a multitude of 'fancy' morphs, in warmwater aquaria and in coldwater ponds. In the tropics, the distribution of the Goldfish is usually restricted to between 200 and 1000 m asl (Welcomme, 1988). It requires summer temperatures in excess of 20°C in order to breed successfully, and has thus failed to establish naturalized populations in northern Canada or northern Europe (Welcomme, 1981). The species' ecological impact in most of its naturalized range is negligible, although in some countries it tends to produce stunted populations.

C. a. gibelio has been less widely introduced than the nominate race, but has expanded its range naturally into most of the rest of Europe.

Natural Distribution

From the Lena river system of eastern Europe eastward to southern Manchuria, the Amur basin, the Tym and Poronai Rivers of Sakhalin Island, and China.

Naturalized Distribution

Europe: most countries of western and central Europe, including Britain, southern Scandinavia, and Spain. *Asia:* Japan; Korea; Taiwan and Hainan; ? Thailand; ? Vietnam. *Africa:* Madagascar; South Africa; Zimbabwe. *North America:* Canada; Mexico; USA. *South and Central America:* Argentina; Bolivia; Brazil; Chile; Colombia; Peru; Uruguay. *Australasia:* Australia; New Zealand;

? Papua New Guinea. **Oceania:** Hawaiian Islands; Mauritius; Western Samoa; Azores; Fiji.

Europe

As mentioned under 'Naturalized Distribution', Goldfish are now established in the wild in most of the countries of western and central Europe (Vooren, 1972), including Britain, southern Scandinavia, and Spain. In many countries the history of their establishment is unknown; the following are some examples of what is recorded (e.g. by Hervey & Hems, 1968, from whom much of the following is derived) of the species' naturalization in central and western Europe.

Goldfish may first have arrived in *Portugal* as early as 1611, 'after the people of that country had discovered the route to the East Indies by the Cape of Good Hope' (*Loudon's Magazine of Natural History* 3:478), from where they are likely to have spread to *Spain* shortly thereafter; in the latter, they now occur in all drainages except Galicia (Elvira, 1995a).

What may have been the earliest reference to Goldfish in *Britain* occurs in the *Diary* of Samuel Pepys, who on 28 May 1665 wrote: '. . . to see my Lady Pen, where my wife and I were shown a fine rarity: of fishes kept in a glass of water, and will live so for ever; and finely marked they are being foreign'. These are generally assumed to have been Goldfish, though Hervey and Hems (1968) cast doubt on this attribu-

tion. M.E. Bloch in his *Oeconomische Naturgeschichte Der Fische Deutchlands* (Berlin, 1784) suggests an even earlier date for the arrival of the Goldfish in England, during the reign (1603–1625) of James I.

Thomas Gray appears to have been the first English author definitely to mention the species when in 1742 he wrote his *Ode on the Death of a Favourite Cat, Drowned in a Tub of Gold Fishes.*

There is circumstantial evidence that Goldfish were included in the cargo of an East India Company vessel that arrived in London from Macao early in 1692. George Edwards, in his *Natural History of Birds* (1745) wrote: 'The first account of these fishes being brought to England may be seen in Petiver's works, published about *anno* 1691'. However, James Petiver's *Gazophylacium Naturae et Artis*, which mentions two Goldfish imported direct from China, was published not in 1691 but in 1711. Tantalizingly, Petiver does not give a date for this importation, but a drawing of the fish in his book is believed to have been made in 1705 or 1706.

During the 18th century increasing numbers of Goldfish were brought into Britain, where self-maintaining populations are now widely established, mainly in England, and to a lesser extent in Jersey in the Channel Islands, southern Scotland, Wales, and Ireland. (For further details see Lever, 1977.)

The first Goldfish in *France* were to be seen in Paris around 1750, having been landed at the port of Lorient by a ship belonging to the French East India Company.

The earliest reference to the introduction of Goldfish to *Italy* seems to be that in a letter of 6 May 1775 to Richard Bentley from Horace Walpole, who kept large numbers in a pond at his home, Strawberry Hill outside London: 'I have lately given Count Perron some goldfish, which he has carried in his post-chaise to Turin: he has already carried some before. The Russian minister has asked me for some too, but I doubt their succeeding there'.

In 1753–1754 a number of Goldfish were introduced to *Holland* by Count Clifford and the Lord of Rhoon, but over a decade later they had still not managed to breed. Job Baster, who released 16 Goldfish in two ponds in Zeeland in the winter of 1759–1760, was more successful, some of his fish breeding in the latter year.

Goldfish were first introduced to *Scandinavia* sometime before 1740, when Carl Linnaeus received a specimen, which he later described in his *Systema Naturae* (1758), from the Swedish Ambassador to Denmark.

Bloch (see above) states that Goldfish were introduced to *Germany* by at least 1780, when the German Ambassador to Holland brought some back from that country to Berlin.

In *Russia*, Gregory Alesandrovich, Prince Potemkin, the favourite of Catherine the Great, gave a banquet in her honour on 1 April 1791, when Goldfish in silver bowls decorated the tables in the Winter Gardens.

Welcomme (1981, 1988) records that Goldfish have been introduced to the polluted waters of two dams in *Cyprus* (which are suitable for them because of their low oxygen requirements) for ornamental purposes and for mosquito control.

In 1891, *C. a. auratus* was introduced to *Hungary* for ornamental purposes and aquaculture (Weisinger, 1975) where the original stock survived until 1952. In the 1930s more of this form were imported, 'where it has spread throughout natural waters . . . [its] slow growth and dense populations make [it] a pest' (Welcomme, 1981, 1988). In 1954, *C. a. gibelio* was imported to *Hungary* from *Bulgaria* as a bait fish for anglers and for aquaculture, and has become widely naturalized. Again, its slow growth and dense populations have made it a pest (Pinter, 1976).

Asia

Japan

In 1904 the Japanese ichthyologist Mitsukuri wrote: 'There is a record that about four hundred years ago – that is to say about the year 1500 – some common goldfish were brought from China to Sakai, a town near Osaka' (quoted by Hervey & Hems, 1968). (See also Okada, 1966.)

Korea

According to Welcomme (1981, 1988), the Goldfish was first introduced to Korea as recently as 1972, where it is breeding and grows better than local species. Welcomme (1981) says that it is 'susceptible to disease' but (1988) that it is 'more disease free than local species'.

Taiwan and Hainan

According to Okada (1966), the Goldfish is not a native of the islands of Taiwan and Hainan, but occurs there as the result of human introduction 'several centuries ago from mainland China' (Liao & Liu, 1989).

Thailand

According to Amatayakul (1957), the Goldfish was introduced to Thailand (then Siam) in 1692–1697 during the Audhaya Period. Since about 1920, large numbers of aquarium fish species, including Goldfish, have been imported to Thailand, where *C. auratus* is now believed to be established in some waters (Piyakarnchana, 1989).

Vietnam

Welcomme (1981, 1988) says that Goldfish are widely used for stocking lakes and ponds in Vietnam; whether any are established in the wild is uncertain.

Unsuccessful Introductions

In Asia, Goldfish introduced to India (Banerji & Satish, 1989; Natarajan & Ramachandra Menon, 1989; Singh & Kumar, 1989) and to Indonesia (Muhammad Eidman, 1989) failed to become established.

Africa

Madagascar

Goldfish first introduced to Lake Itasy on the island of Madagascar in 1861 and to Lake Alaotra in the 20th century are now widely distributed in suitable high altitude plateaux waters, where they are a valued commercial species (Kiener, 1963; Moreau,

1979; Reinthal & Stiassny, 1991). (See also under Common Carp *Cyprinus carpio*.)

South Africa

Goldfish were recorded in Cape Province as early as 1726 by F. Valentyn, who saw them in a tank in the house of the governor of the Cape (Raidt, 1971; Skelton & Skead, 1984). The first record of Goldfish in the wild in South Africa is by Castelnau (1861), who includes them in his list of freshwater species in the Cape Colony. These fish were probably imported to the Cape by sailors *en route* to Britain from India, and there were doubtless many subsequent introductions from the Far East during the 19th century (De Moor & Bruton, 1988).

In 1941, the Jonkershoek hatchery in Cape Province began distributing Goldfish to various parts of South Africa for ornamental purposes and the control of mosquitoes, and Welcomme (1981) records introductions from Mauritius in 1953 for aquacultural purposes. At around the same time, Goldfish were illegally released in the Liesbeeck dam in the Cape peninsula.

De Moor and Bruton (1988) summarized the distributional records of Goldfish in the wild in South Africa as follows:

DATE	LOCALITY	SOURCE
1938	Near Robertson (Breede River system)	Harrison (1938)
1938	Brandvlei Lake (western Cape)	Harrison (1938)
1949	Liesbeeck River (western Cape)	Harrison & Lewis (1968–1969)
?	Little Princess Vlei (Heathfield)	Harrison (1976)
?	Ponds and dams in Natal	Crass (1964)
?	Umsindusi River (Umgeni system, Pietermaritzburg)	Crass (1964)
?	Upper Tugela River	Crass (1966)
1974	Kowie River (Port Alfred)	Albany Museum
1983	Baakens River (eastern Cape)	King & Bok (1984)
1983	Gansvlei and Klein Riet Rivers (Sak River, Orange system)	Hocutt & Skelton (1983)
?	Aapies River (Pretoria)	
1987	Nthuseleni stream (Umzimkulu system, Underberg)	Albany Museum

Ecological Impact   The few naturalized populations of Goldfish in South Africa appear to have had little if any direct impact on native species; it is, how-ever, likely, that Goldfish have been associated with the introduction of various harmful fish parasites, but the significance, if any, of these has yet to be determined (Bruton & Merron, 1985; De Moor & Bruton, 1988).

Zimbabwe

The Goldfish was one of the earliest, if not the first, alien fish to be introduced to Zimbabwe (then Rhodesia). Its exact date of arrival is unrecorded, but according to Toots (1970) 'it is possible that some early settler had his goldfish bowl securely padded down in his ox-wagon on the way to Rhodesia' (!). Welcomme (1981, 1988) says the species is reproducing in 'widespread and not very numerous populations in some near stagnant pools'.

North America

Canada

Goldfish are currently naturalized in the following Canadian provinces:

New Brunswick. In about 1972 two self-maintaining populations were discovered, one in Killarney Lake, Fredericton, and another in a farm pond just upstream of the Mactaquac dam on the St John River. The latter population was subsequently eradicated, but the former survived in 1981 (Crossman, 1984).

Ontario. Radforth (1944) reported Goldfish in Lakes Erie and Ontario and Dymond (1947) recorded them in the Detroit River. After 1954 they were found in Lake St Clair; Grenadier Pond, Toronto; Musselman's Lake north of Toronto; and Gillies Lake near Timmins (Scott, 1967). Hybridization in Lake Erie with Common Carp *Cyprinus carpio* (q.v.) seems to be adversely affecting the latter's population (Crossman, 1984).

Manitoba. Goldfish were first recorded in Manitoba in 1975, south of Pas, and in Cook's Creek near East Selkiv in the following year (Crossman, 1984).

British Columbia. Goldfish were discovered in a large natural pond at Salmon Arm in 1935 and at around the same time in a small lake in the vicinity of Lac du Bois near Kamloops (Carl & Clemens, 1953; Carl & Guiguet, 1958; Carl *et al.*, 1967).

Goldfish formerly established in the wild in Quebec and Alberta are since believed to have been eradicated. These, and those still naturalized in Canada, are all believed to have originated in escapes or deliberate releases from captivity.

Mexico

Since their introduction from France in the 1890s, Goldfish have become established in the wild in Mexico in the Valley of Mexico (Meek, 1904; Alvarez & Navarro, 1957), in the upper Río Lerma (Romero, 1967) and in Nuevo León (Contreras, 1967). The species is said by Hensley and Courtenay (1980) to be rare in the wild in Mexico. Specific localities given by Contreras (1969) are El Chorro, Coahuila and by Contreras (1978) Potosí, Nuevo León. Welcomme (1981, 1988) says that in Mexico the Goldfish is a popular food fish as well as being valued for ornamental purposes. (See also Contreras & Escalante, 1984.)

United States

According to Courtenay and Hensley (1980b) and Courtenay (1995), the Goldfish was the first alien fish to be introduced to North America, where DeKay (1842) recorded releases into the wild being made in the 1680s.

Courtenay *et al.* (1984, 1986) reported Goldfish to be naturalized 'with apparent broad and stable distributional limits' in California, Delaware, District of Columbia, Georgia, Idaho, Illinois, Indiana, Iowa, Kentucky, Maryland, Massachusetts, Michigan, Nebraska, Nevada, New Hampshire, New York, Ohio, Oklahoma, Pennsylvania, Rhode Island, South Carolina, South Dakota, Tennessee, Texas, Virginia, Washington and Wisconsin. The status of the species in Alabama, Arizona, Arkansas, Colorado, Connecticut, Kansas, Louisiana, Minnesota, Mississippi, Missouri, Montana, New Jersey, New Mexico, North Carolina, North Dakota, Oregon, West Virginia and Wyoming is uncertain. (See also Lachner *et al.*, 1970 and Hubbs, 1982.)

South America

Argentina

According to Navas (1987), the Goldfish is only naturalized in Argentina in some waters in the province of San Juan, where Welcomme (1988) says it was introduced in the 1890s.

Bolivia; Peru; Uruguay

Welcomme (1988) says that Goldfish are established in Bolivia, Peru and Uruguay, but gives no further details.

Brazil

Goldfish introduced to Brazil from Japan in the 1920s are said by Welcomme (1988) to be reproducing successfully.

Chile

Delfin (1901) seems to have been the first authority to refer to the introduction of the Goldfish to Chile. Quijada (1913) confirms Delfin's identification, while Oliver (1949) says that the first introductions took place in 1885 in the region of Peñaflor, Río Andalién, Palomares and Estero Nonguén. Eigenmann (1927) adds to this list Hospital and Llo-Lleo. Welcomme (1988), who gives the date of introduction as 1856, says the species is established and reproducing.

Colombia

First introduced to Colombia from the USA as a forage fish in 1940, the Goldfish is said by Welcomme (1981) to be 'established in Llanos orientales'.

Unsuccessful Introductions

The Goldfish has been unsuccessfully introduced to Puerto Rico (Erdman, 1984; Wetherbee, 1989) and to Hispaniola (Wetherbee, 1989).

Australasia

Australia

Clements (1988) has traced the early history of the Goldfish in Australia, where it was one of the earliest, if not the first, fish introduction.

It seems probable that the first importations of Goldfish arrived in Melbourne in the 1850s, when they were on display in Cremorne Gardens and where they are known to have bred successfully before being washed away in a flood, probably into the Yarra River where they may have become established.

On 23 January 1860, the *Maidstone* left London bound for Melbourne carrying 50 Goldfish for Edward Wilson, editor of the *Melbourne Argus*, but all perished on the voyage. In 1861 and 1862, the *Argus* lists further shipments of Goldfish to Australia, some of which arrived safely, and later in the 1860s the Governor of New South Wales, Sir John Young, is known to have bred 'carp' (according to Arentz (1966) and Rolls (1969) probably Goldfish) in Government House and to have released them in waters between Sydney and Botany Bay. Goldfish are known to have been established in the Sydney area in the mid-1860s, from where early settlers of the outback were encouraged to stock new waters with 'carp', no doubt including Goldfish.

The establishment and spread of Goldfish in Australia has been helped by continued releases of aquarium and surplus bait fish (which is illegal only in Victoria south of the Murray River), and the species is now the most widespread of all naturalized fish species, being found throughout much of the Murray/Darling basin, the inland Cooper's Creek, most eastern coastal streams, some streams in Western Australia, and in numerous lentic waters of southern Australia (New South Wales, Victoria, South Australia, southeastern Queensland, and Tasmania). Although the species' distribution is somewhat patchy in upland areas, where in Victoria and New South Wales it is absent from fast-flowing waters, Goldfish are frequently found in slowly-flowing tableland streams and in dams, shallow lakes, and lagoons. In Tasmania they occur only in a few scattered waters in the north and southeast (Whitley, 1951; Lake, 1959; Weatherley & Lake, 1967; Llewellyn, 1983; McKay, 1984, 1989; Merrick & Schmida, 1984; Allen 1989; Brumley, 1991; Twyford, 1991).

New Zealand

Under the scientific name *Cyprinus carassius* Thomson (1922) recorded the first introduction of 'Golden Carp' to New Zealand in 1864, when A. M. Johnson, then curator of the Canterbury Acclimatization Society, brought some out from England on the *British Empire* which berthed in Lyttleton (Druett, 1983; Lever, 1992; McDowall, 1994b). The Goldfish were, according to Thomson, 'the only survivors of a large and varied assortment of fish', the remainder of which died when a careless deckhand dropped a lump of white-lead putty into their tank.

In 1867, the Nelson Society obtained some 'Carp'

(most likely Goldfish) from their *confrères* in Sydney, Australia, and in the following year the Canterbury Society received a further consignment of Goldfish from Melbourne. In 1867 and 1868, the Auckland Society acquired some 'Russian Carp' (again probably Goldfish) which they released successfully in Lake Pupuke, St John's Lake and elsewhere in the Auckland/Waikato area (McDowall, 1994b).

Frequent introductions of Goldfish to New Zealand continued to be made until the 1960s, when in order to prevent the importation of alien fish diseases the introduction of coldwater fishes was banned (McDowall, 1994b).

There seem to be few records of how Goldfish became established in the numerous waters in New Zealand where they are now to be found, but McDowall (1994b) records how Gilbert Mair, in the *Rotorua Chronicle* of 25 October 1919, described how the species was first released in Lake Taupo. In 1873, a consignment of Goldfish was transported by ship from Auckland to Napier, and thence by wagon and horses to Taupo – the Maori referring to the fish as *morihana* after Constable H.C. Morrison who was in charge of the shipment. From Taupo, Goldfish were subsequently transferred to Lake Rotorua and other lakes in the Rotorua district in 1880 (Burstall, 1980).

Goldfish have been planted in New Zealand waters with a variety of motives: by aquarists wanting to increase the diversity of species; by bored pet-owners; by pastoralists who hoped that they would help to control phytoplankton bloom in nutrient-enriched farm ponds; as forage for introduced Brown Trout *Salmo trutta* (q.v.) and Rainbow Trout *Oncorhynchus mykiss* (q.v.) in some of the lakes of central North Island such as Rotoiti, Rotorua and Taupo; and (according to Haas & Pal, 1984) to try to control mosquitoes. So far as is known at present, their role in controlling algal blooms and as food for trout is negligible (McDowall, 1990b).

Today, Goldfish occur in widely scattered localities throughout North Island, and in South Island especially in waters near Christchurch, Canterbury.

Papua New Guinea

Allen (1991) records that the Goldfish has been introduced to the Lake Sentani region of Irian Jaya, but that the date of introduction and the species' present status there is unknown. This introduction is not mentioned by West (1973), Glucksman *et al.*, (1976), West and Glucksman (1976), nor by Welcomme (1981, 1988).

Oceania

Hawaiian Islands

Goldfish, probably introduced from China by Asian immigrants, are known to have been established in the Hawaiian Islands by 1901. They are now believed to occur in standing waters – mainly reservoirs – on all the main islands (Brock, 1952, 1960; Hida & Thomson, 1962; Kanayama, 1968; Lachner *et al.*, 1970; Maciolek, 1984).

Mauritius

In 1953, Goldfish were imported from Madagascar to Mauritius, where they are now reproducing and widespread (Welcomme, 1981, 1988).

Western Samoa

Maciolek (1984) records the establishment of Goldfish on the island of Savaii in Western Samoa, but gives no further details.

Azores

Goubier *et al.* (1983) say that reports dating from 1792, 1847 and 1885 indicate the presence of Goldfish and Crucian Carp *C. carassius* (q.v.) in Lacs de Furnas and Sete Cidades on the island of São Miguel. Nothing is known about the origin or present status of these populations, but it has been surmised (by Moreira Da Silva, 1977) that eggs may have been transported to São Miguel on the webbed feet of water birds.

Fiji

Eldredge (1994) lists the Goldfish as a successful introduction to Fiji.

Crucian Carp (Prussian Carp)
Carassius carassius

Although a long-standing member of the European ichthyofauna, the Crucian Carp may have spread naturally into much of western Europe or have been introduced together with the Common Carp in the Middle Ages.

Natural Distribution

Widely distributed throughout eastern and central Europe apart from Switzerland, southern Italy and Finland. Regarded as native in Belgium, France, Spain and southern England.

Naturalized Distribution

Europe: Cyprus. *Asia:* ? India; Philippines; ? Sri Lanka. *Oceania:* ? Azores.

Europe

Cyprus

Welcomme (1981, 1988) records the establishment of the Crucian Carp in two dams in Cyprus, to which it was introduced for ornamental purposes and to control mosquitoes, in the knowledge that it has a high tolerance of polluted and deoxygenated waters.

Asia

India

Crucian Carp were successfully introduced from England to the Nilgiri Hills in Tamil Nadu in 1874 or 1878 where they became established in Ooty Lake and some other waters. Their present status there is

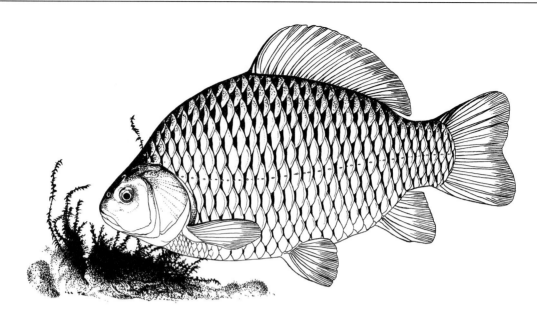

in some doubt; Jones and Sarojini (1952), Welcomme (1981, 1988), Sehgal (1989), Shetty *et al.* (1989) and Yadav (1993) say that they still survive, albeit in reduced numbers since the introduction of the Common Carp; Sreenivasan (1991), however, says that Crucian Carp died out in the 1970s.

Philippines

Crucian Carp introduced from Japan to the Philippines in 1964 have established self-maintaining populations in some ponds (Juliano *et al.*, 1989).

Sri Lanka

The present status of Crucian Carp introduced into streams and reservoirs in Ceylon (Sri Lanka) in 1915 (Fernando, 1971) is uncertain (De Silva, 1987).

Oceania

Azores

See under Goldfish *C. auratus.*

Other Countries

Although De Buen (1959) refers to the naturalization in Chile of '*Carassius carassius* = *Carassius*

auratus', it is clear from his description that he is referring to the Goldfish (q.v.) rather than to the Crucian Carp.

Crucian Carp established in Chicago, USA, in the early 1900s have since died out (Welcomme, 1981, 1988).

Crucian Carp are said to have been unsuccessfully introduced to Kenya (Welcomme, 1981, 1988).

According to Lake (1959), Crucian Carp were introduced to Australia in about 1876, while Whitley (1951) says they were planted in the Queanbeyan River in New South Wales in 1880. Although Weatherley and Lake (1967) and McKay (1984) say they are established in Australia in much the same range as the Goldfish, which was introduced at around the same time, Brumley (1991) says the Crucian Carp 'has never been confirmed in Australia', and it seems likely that early reports referred to incorrectly identified Goldfish (Fletcher, 1986). No other authors consulted confirm the presence of Crucian Carp in Australia.

Quillback
Carpoides cyprinus

The Quillback is a large sucker with broad habitat requirements which is found in muddy rivers as well as in clear lakes. It is a useful commercial and sporting species (Welcomme, 1988).

Natural Distribution

Mississippi River system of the USA.

Naturalized Distribution

North America: Mexico.

North America

Mexico

The introduction of the Quillback from the USA to Mexico at an unrecorded date has resulted in the establishment of commercial fisheries in the northeast of the country (Welcomme, 1988).

Catla (Indian Carp)
Catla catla

The Catla is one of the so-called 'major carps' of India, where it is extensively cultured and frequently used to stock reservoirs. It has been introduced to several countries for aquacultural purposes, but with one exception unsuccessfully.

Natural Distribution

India and Burma.

Naturalized Distribution

Asia: Philippines. *Oceania:* Mauritius.

Asia

Philippines

Introduced to the Philippines from India in 1967, Catla that escaped from captivity have succeeded in establishing self-maintaining populations in a number of rivers (Juliano *et al.*, 1989).

Oceania

Mauritius

Introduced to the island of Mauritius between 1960 and 1973, Catla are established in several reservoirs (Moreau *et al.*, 1988).

Unsuccessful Introductions

Attempts to naturalize Catla in Mauritius, Israel, Japan, Malaysia and the former USSR (Welcomme, 1981, 1988); Zimbabwe (Toots, 1970); Sri Lanka (De Silva, 1987); Thailand (De Iongh & Van Zon, 1993); and China (Yo-Jun & He-Yi, 1989) have proved unsuccessful.

Translocations

Natarajan and Ramachandra Menon (1989) refer to a particularly successful translocation of Catla to Mettur and Sathanoor reservoirs in Tamil Nadu, India.

Chinese Grass Carp (White Amur)
Ctenopharyngodon idella

The Chinese Grass Carp has been widely translocated within its natural range and introduced to countries in South East Asia for aquacultural purposes, and has been introduced elsewhere primarily for the control of aquatic vegetation (Shireman & Smith, 1983). Although it is one of the world's most widely introduced fish species, its strict breeding requirements (Jhingran & Pullin, 1985) – especially the necessity for a water temperature of between 27 and 29°C and a current flow of between about 60 and 150 cm per second – preclude its breeding in most of the waters into which it has been introduced, where populations are maintained by continued restocking with imported or captive-bred stock.

Natural Distribution

Lowland rivers of China and the Amur River basin of the former USSR in far eastern Siberia.

Naturalized Distribution

Europe: countries of the Danube River basin (Czechoslovakia; Hungary; Romania; former Yugoslavia). **Africa:** Ethiopia; Ivory Coast; Sudan. **North America:** Mexico; USA.

Europe

Countries of the Danube River basin (Czechoslovakia; Hungary; Romania; former Yugoslavia).

In Europe the Grass Carp is spreading rapidly throughout the Danube River system where, espe-

cially in *Czechoslovakia*, it has become a valued sporting species. Introduced from China to *Hungary* in 1963 it escaped from captivity and by 1974 had become established in the Tisza River, from where it spread to the Danube. Since 1963, Grass Carp from *Romania* and the former USSR, where the species was translocated in the early 1960s (Dukravets, 1972; Nezdolic & Mitrofanov, 1975; Negonovskaya, 1981), and *Hungary*, have been imported into the former *Yugoslavia*, where they have become naturalized in the Danube (Vooren, 1972; Welcomme, 1981, 1988).

Africa
Ethiopia

In 1975, Grass Carp from Japan were introduced to control aquatic vegetation in Ethiopia, where they became established in Lake Fincha (Moreau *et al.*, 1988). (See also Tedla & Meskel, 1981.)

Ivory Coast

In 1979, Grass Carp from France were introduced to control aquatic weeds in the Ivory Coast, where they became established in Lake Bouaké (Moreau *et al.*, 1988).

Sudan

In 1975, Grass Carp from India were imported by Pisciculture de Shagra to Sudan, where they became established in irrigation canals (Moreau *et al.*, 1988).

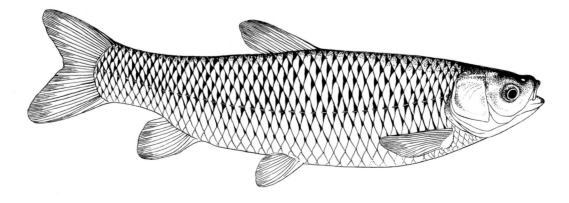

North America

Mexico

In 1964 or 1965, Grass Carp were released in Mexico in Río Cupatitzio, Michoacán; Temazcal, Oaxaca; and Presa de la Boca, Nuevo León. Since 1974–1975 they have been naturalized in the Cupatitzio (Rosas, 1976a, b; Contreras & Escalante, 1984).

United States

Grass Carp were first imported from Malaysia to the USA by the Fish and Wildlife Service for their Fish Farming Experimental Station at Stuttgart, Arkansas, and from Taiwan by Auburn University, Alabama, in 1963 (Guillory & Gasaway, 1978). The species was subsequently distributed, for the biological control of aquatic vegetation, to 11 other states. The earliest escapes and deliberate releases were recorded in Arkansas (Sneed, 1972).

Since 1977, Grass Carp have been caught in the wild in Alabama, Arkansas, Florida, Georgia, Michigan, Missouri, New York, Pennsylvania, and West Virginia, and unconfirmed reports suggest their presence in the Mississippi River in Wisconsin and Minnesota and in the Missouri River in Kansas and Nebraska, upstream to Gavins Point dam near Yankton in South Dakota (Pflieger, 1978; Pierce, 1983; Courtenay *et al.*, 1984, 1986).

In the USA, Grass Carp are presently considered to be naturalized at three places in the lower Mississippi River; near Eudora, Chicot County, Arkansas; near Simmesport, Avoyelles Parish, Louisiana; and near St Francisville, West Feliciana Parish, Louisiana (Conner *et al.*, 1980). In addition, Grass Carp larvae have been collected in the Atchafalaya River near Butte la Rose, St Martin's Parish, Louisiana (Courtenay *et al.*, 1984), and juveniles in a Mississippi backwater near Memphis, Shelby County, Tennessee (Conner *et al.*, 1980). (See also Provine, 1975.)

Unsuccessful Introductions

Grass Carp are being reared successfully in aquacultural centres and/or are acclimatized (see Preface) in natural waters, but have failed to become established in the wild, as follows:

Europe. Cyprus, Italy, Poland (Welcomme, 1981, 1988); Denmark (Jensen, 1987); Belgium (Timmermans, 1978); France, Germany, Greece (Vooren, 1972); The Netherlands (Vooren, 1972; De Groot, 1985); Britain (Lever, 1977); Sweden (Anon., 1971; Jernelov, 1985).

Asia. Afghanistan, Bangladesh, Korea, Pakistan, Vietnam (Welcomme, 1981, 1988); Burma (Chaudhuri & Hla Tin, 1971); India (Yadav, 1993); Indonesia (Muhammad Eidman, 1989); Israel (Ben-Tuvia, 1981); Malaysia (Ang *et al.*, 1989; Chou & Lam, 1989); Philippines (Juliano *et al.*, 1989); Sri Lanka (De Silva, 1987); Thailand (Piyakarnchana, 1989; De Iongh & Van Zon, 1993).

Africa. Egypt, Kenya, Rwanda, (Welcomme, 1981, 1988); Mozambique (Bartley, 1993); South Africa (De Moor & Bruton, 1988); Malawi, Lesotho, Uganda, Tanzania (Moreau *et al.*, 1988).

South and *Central America.* Argentina (Anon., 1971; Navas, 1987); Bolivia, Brazil, Colombia, Costa Rica, Cuba, Honduras, Panama and Peru (Welcomme, 1981, 1988); Puerto Rico (Erdman, 1972, 1984; Ortiz-Carrasquilla, 1980, 1981).

Australasia. New Zealand (Clements, 1988); Papua New Guinea (Glucksman *et al.*, 1976; West & Glucksman, 1976).

Oceania. Hawaiian Islands, Fiji (Maciolek, 1984); Mauritius (Welcomme, 1981, 1988); Guam (FitzGerald & Nelson, 1979).

Control of Aquatic Weeds and Ecological Impact

The control of aquatic weeds by Grass Carp, and their ecological impact, particularly in the USA, have been discussed in detail by Shireman (1984) and Taylor *et al.* (1984), from whom much of the following is derived, and summarized by Lever (1994).

The impact of Grass Carp on the populations of other animals is not well understood, and usually depends on the stocking rate and the amount of vegetation removed. Several studies have shown increases in populations of invertebrates after the removal of Grass Carp, while others have shown zooplankton and benthic communities to be unaffected. The impact of Grass Carp on other fishes is usually a result of the removal of vegetation, which inhibits indigenous species from spawning. However, the spawning of some species, for example cyprinids and centrarchids, may actually benefit from the introduction of Grass Carp. Although these results may appear contradictory, the impact is directly related to the dependence of the fish communities on aquatic macrophytes and the degree of control exerted by Grass Carp, which is usually itself related to the stocking rate.

Since the eradication of macrophytes could cause serious problems for native fishes and other aquatic organisms (through loss of shelter, food resources, and spawning grounds), the selection of appropriate stocking levels of correctly sized Grass Carp is important in order that elimination of weeds is avoided but control is achieved. In man-made waters, which often have monocultures of exotic plants, this is quite easily managed. In natural conditions, however, where polyculture usually prevails, the selective feeding habits of Grass Carp are of primary consideration. At low stocking densities, grazing on preferred, but non-target, species may actually allow target species to increase in biomass, while stocking at densities required to control target species may result in the eradication of some non-target species before foraging on target species commences.

In addition to their direct impact on aquatic animals and plants, Grass Carp may have an indirect impact on other fish species through changes in water quality caused by their feeding habits. The destruction of aquatic macrophytes can cause increased phytoplankton production, and eutrophication usually results in a considerable increase in such undesirable species as blue-green algae (Cyanobacteria). This increase in primary productivity, and concomitant effects on the formation of phytoplankton communities, can cause associated changes in aquatic organisms as a result of alterations in water-quality characteristics.

An interesting example of the indirect impact that Grass Carp can have on local fish communities occurs in the USA, to which the species introduced a parasitic Asiatic tapeworm. The Red Shiner *Notropis lutrensis*, an abundant and widespread fish in eastern North America, has apparently been able to adapt to the effects of the tapeworm, of which it has become a vector. The introduction of the Red Shiner as a bait fish into the Virgin River, Utah, however, has coincided with the decline of the threatened Woundfin *Plagopterus argentissimus*. It is believed that Red Shiners have become competitively superior to Woundfins weakened by tapeworm infestations (Deacon, 1988; Moyle & Leidy, 1992).

Despite the potential problems that can arise from the use of Grass Carp in the management of aquatic weeds, when stocked at appropriate levels they have been shown to be a highly efficient controlling agent.

Common Carp (European Carp)
Cyprinus carpio

The Common Carp is probably the first species of fish to have been widely translocated within its natural range and introduced outside it (Welcomme, 1988). Balon (1974) records that Common Carp were taken from the River Danube by the Romans, by whom they were widely distributed throughout Europe to stock monastic fish stews. During the past century Common Carp have been so widely introduced around the world that they are now almost cosmopolitan wherever suitable conditions prevail. Although a valuable source of human food and in many countries highly regarded as a sporting asset, the species' roiling activities have in some places had a serious ecological impact. Alikunhi (1966) and Sarig (1966) have compiled synopses of biological data on the Common Carp.

Natural Distribution

Originally confined to central Asia east of the Caspian Sea, from where it spread naturally east during the later glaciations to Manchuria and west to the rivers of the Danube basin and the Black and Aral Seas.

Naturalized Distribution

Europe: British Isles; Cyprus; Denmark; France; Germany; Italy; Spain; Sweden. ***Asia:*** Afghanistan; India; ? Indonesia; Israel; Malaysia; Philippines; ? Sri Lanka; ? Taiwan; Thailand. ***Africa:*** Ethiopia; Kenya; Madagascar; Morocco; Namibia; South Africa; Tunisia; Uganda; Zimbabwe. ***North America:*** Canada;

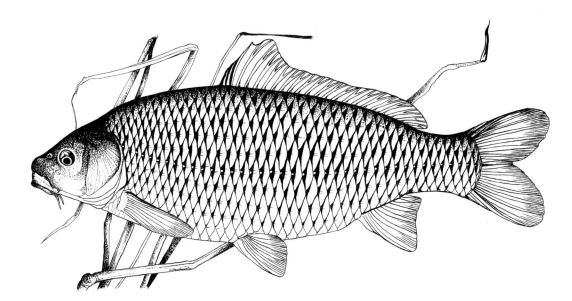

Mexico; USA. ***South and Central America:*** Argentina; Bolivia; Brazil; Chile; Colombia; ? Costa Rica; Cuba; Dominican Republic; Panama; Peru; Uruguay; Venezuela. ***Australasia:*** Australia; New Zealand; Papua New Guinea. ***Oceania:*** ? Azores; ? Fiji; Guam; Hawaiian Islands; ? Mauritius.

Europe

Balon (1974) has summarized the probable steps of the spread and domestication of the Common Carp in Europe.

Except for the *Danube* basin it seems unlikely that Common Carp occurred naturally in Europe as early as the beginning of the Christian era. The Roman poet, Decius Magnus Ausonius (*c.* AD 309–392), made no mention of Common Carp in the Rhine or Moselle in his descriptive poem *Mosella.* The most westerly known natural occurrence of rheophilic Common Carp in the Danube is at its confluence with the Morava River near Devin in Bulgaria. From here, the fish were probably transported to other European countries by Roman traders. Evidence for the natural occurrence of Common Carp in the Danube is provided by the differing provincial names given to the fish by local people; from the use of the same Celtic name outside the western limit of its natural range; from mediaeval manuscripts; and from recent fieldwork.

The Roman author and monk, Flavius Cassiodorus (*c.* AD 485–580) provides evidence in his *Variae* for the Danubian origin of western domesticated Common Carp, when he refers to the *Carpa* of the Danube as a delicacy served to King Theodorus at Ravenna.

In *Italy,* to which they were transferred from the Danube, Common Carp were kept in captivity in specially designed *piscinae,* from which they no doubt from time to time escaped into nearby rivers. After the collapse of the Roman Empire and the advent of Christianity, Common Carp began to be reared for fast-day food in monastic stews. Albertus Magnus of Cologne (*c.* 1206–1280) makes what is probably the earliest mention of the breeding of domesticated Common Carp in his commentary on the *Books of Sentences* written by Peter Lombard, Bishop of Paris (*c.* 1100–1164).

Balon (1974) suggests the following steps in the spread and domestication of Danubian Common Carp.

(1) Transportation west of the Danube during the 1st to 4th centuries AD.
(2) Sporadic western introduction and rearing in captivity in the 5th to 6th centuries.
(3) The beginning of rearing on a mass scale with repeated introductions from the 7th to 13th centuries (to *Germany* and *France* by 1258).
(4) Continued rearing and mass culture in the 14th to 16th centuries (in *Sweden* by 1560).
(5) Intensification of selective breeding and further introductions (e.g. to *Denmark* and *Spain* by 1660) and outside Europe (e.g. to *North America* in 1831; to *Australia* (1860) and to *South Africa* (1896), from the 17th century onwards.

In *China*, the domestication of Common Carp began independently some five centuries earlier than in Europe.

British Isles

In his delightful treatise on fishing, *The Compleat Angler*, first published in 1653, Isaak Walton wrote:

The Carp . . . was not at first bred, nor hath been long in *England*, but is now naturalized. It is said they were brought hither by one Mr *Mascal* . . . doubtless, there was a time, about a hundred, or a few more years ago, where there were no Carps in *England*, as may seem to be affirmed by Sir *Richard Baker* in whose Chronicle [1643] you may find these verses.

> Hops and Turkies, Carps and Beer,
> Came into England all in a year.

Baker's distich, however, post-dates the arrival of the Common Carp in England by at least a quarter of a century, since it is known that hops first arrived in England around 1520 and Turkeys *Meleagris gallopavo* from North America about a decade later.

In his *History of British Fishes* (1859) William Yarrell wrote:

Leonard Mascall [*c.* 1514–1589; clerk of the Kitchen to the Archbishop of Canterbury] takes credit to himself for having introduced the Carp . . . but notices of the existence of the Carp in England occur prior to Mascall's time. . . . In the celebrated *Boke of St Albans*, by Dame Juliana Barnes, or Berners, the Prioress of Sopwell Nunnery . . . Carp is mentioned as 'a deyntous fisshe, but there ben fewe in Englande, and therefore I wryte the lesse of hym'.

Dame Juliana's book was first published in 1486; the edition in which mention of Common Carp is made in an added *Treatyse on Fysshynge with an Angle* was published in 1496.

Went (1950) has traced in some detail the introduction of the Common Carp and Tench *Tinca tinca* (q.v.) into Ireland.

The Journal Book entry of the Royal Society for 29 April 1663 records that 'Mr Boyle was desired to communicate his papers, concerning the manner how my Lord of Corke, his ffather, had Carpes transported into Ireland, where they were not before'. In the course of his address to the Royal Society, Robert Boyle read the following extracts from the Earl of Corke's diary:

3 Sept. 1634. I wrott to the Lord President of Mounster, and as he desired, gave order to Sir John Leek to deliver his Lop. 20 yonge carpes, 10 Tenches . . .
26 Sept. 1640. I wrott by Badnedg to Sir John Leek to furnish Sir Phillip Percival with 40 young carpes, Mr Henry Warren second Remembrancer of Exchequer with 20 Carpes and my daughter the countess of Kildare, with 20 carpes to stoar the new pond withall.
7 Oct. 1640. Letters to Sir John Leek to deliver to Sir Philip percival 40 yonge carpes and some Tenches towards the storing of his ffishe pond.

Since it is known that Richard Boyle, 1st Earl of Cork, lived in Ireland after 1588, the 'carpes' referred to in the above entries in his diary were almost certainly to be translocated from his Irish estate rather than introduced from England; it is thus necessary to seek further back in time for the species' introduction from England. Charles Smith, in his *History of Cork* (1750), commenting on Edmund Spenser's *Faërie Queen* (1590–1596), claims that Common Carp and Tench were to be found in the River Awbeg in Co. Cork during the reign (1603–1625) of James I, and that both species were commonly kept in fish ponds, although Tench were less plentiful. This date has come to be generally accepted as that when Common Carp and Tench were first introduced to Ireland.

The Common Carp is today fairly widely distributed in the British Isles in slowly flowing or still waters in much of central and southern England (including the Channel Islands), in scattered parts of Ireland, and in some parts of Wales and Scotland. Even in southern England, summer water temperatures are often barely high enough to ensure successful spawning, and in many places Common Carp populations only survive by artificial stocking. Although in the British Isles the Common Carp is not highly regarded as a human food fish, it is much prized as a sporting asset. (For further information on the Common Carp in the British Isles see Lever, 1977.)

Cyprus

According to Welcomme (1981, 1988), Common Carp from Israel were introduced to Cyprus in 1966, where they have become 'very popular for angling and eliminate algae from drainage ditches'. They are said to be established in the Mia Milia dam, a fairly shallow reservoir with an abundance of aquatic vegetation.

Asia

Afghanistan

Welcomme (1981, 1988) says that in the 1970s Common Carp from China were introduced to Afghanistan, where they have become established in some reservoirs for which there were no native species available as suitable colonists.

India

Three varieties of the Common Carp have been introduced to India, *C. c.* var. *specularis*, the Mirror Carp; *C. c.* var. *communis*, the 'Bangkok' or Scale Carp; and *C. c.* var. *nudus*, the Leather Carp (Jhingran, 1989).

In January 1939, all three varieties were imported from Sri Lanka to Tamil Nadu, where in Ooty Lake, Sandynulla and Wellington reservoirs and some municipal waters the Mirror variety became established. In 1949, it was also stocked successfully in Kodaikanal, and 5 years later in Yercaud Lake. In all the hill stations Mirror Carp were soon reproducing successfully, and Ooty Lake alone was supplying tens of thousands of fingerlings annually for transplantation elsewhere. Subsequent breeding failure in Ooty and Kodai Lakes was traced to predation of ova by Mosquitofish *Gambusia affinis* (q.v.), which were later eradicated (Jhingran, 1989; Sreenivasan, 1989).

In 1946, 500 Mirror Carp fingerlings were released in Ulsoor Lake, Bangalore, in the state of Karnataka, and in 1947 others from Ootacamund in the Nilgiris were planted in waters at Bhowali in the Kumaon Hills in Uttar Pradesh, from where some were later transplanted to Nohan in the Sirmur district of Himachal Pradesh. In 1948, more were introduced from the Fishery Research Station at Ootacamund, and stocked in Amritsarovara in the Nandi Hills in Kolar and in the Markonahalli Fish Farm in Tumkur, and a year later others from Ootacamund were placed in a lake at Lonavia near Bombay. None of these plantings was very successful. In 1952, yet another shipment of 300 fingerlings was placed in fish farms at Hessarghatta, Bangalore and Markandeya Kolar. In the following year the Mirror Carp were found to have bred at Markandeya, from which fish were subsequently

stocked in the temple pond at Belur, Hassan, where breeding was later reported (Jones & Sarojini, 1952; Chandrasekharaiah, 1989; Sehgal, 1989; Singh & Kumar, 1989).

Encouraged by these results, fingerlings and adults in breeding condition were placed in 38 temple ponds in Karnataka, and in 1956 also in ponds at Krishnarajasagar and Shanthisagar at the former of which successful spawning took place in 1957.

In 1956, the Scale Carp (*communis*) was introduced to the farm at Hessarghatta, and subsequently throughout Karnataka, where because of its hardiness, fast growth and ability to breed in captivity, the culture of Common Carp became widespread. In the wild Common Carp also flourished, especially in Lake Hessarghatta, where in 1960 a fish weighing 30 kg was caught; as the population of Common Carp in the lake increased, the previously clear water became permanently turbid through the fishes' roiling and grazing activities (Chandrasekharaiah, 1989).

In Karnataka, Common Carp have also thrived in reservoirs such as that at Krishnarajasagar, where they have replaced *Puntilus dubius*, *P. karnaticus*, *Cirrihna cirrosa*, *Labeo kontius* and *L. fimbriatus* as the dominant species (Chandrasekharaiah, 1989).

The 'Bangkok' variety (*communis*) was first imported from Cuttack to the state of Bihar in 1958, where after early breeding success in captivity it became established in small reservoirs at Hatia and Neterahat near Ranchi.

In 1959, both Mirror and Scale Carp were transplanted by the Department of Fisheries from the Nilgiris, Bhimtal and Nahan in Himachal Pradesh into various natural waters in Jammu and Kashmir. The stock from Nahan became further established in the Chotanagpur Hills region of Bihar in 1961, in Ding, Dhan, and Shabdol (1960), Galta, Amber, Mount Abu, and Ranthanphore in Rajasthan (1961), Sikkim (1959) and Manipur (1960) (Sehgal, 1989). In the early 1970s, some individuals of the Leather Carp (*nudus*) somehow gained access to ponds near the Nepalese border, where they interbred with *communis* (Banerji & Satish, 1989).

Today, the Mirror Carp in particular is well established in Kashmir, Himachal Pradesh, Bihar, Manipur, Sikkim, Delhi and elsewhere, and the Leather Carp in Ootacamund lakes (Yadav, 1993). Welcomme (1981, 1988) adds Madhya Pradesh, Punjab, Tamil Nadu and Rajasthan to this list.

Ecological Impact In Kashmir, Common Carp are said to be responsible for a decline in numbers of native snow trout *Schizothorax* spp. and in Manipur of *Osteobrama belangeri* (Shetty *et al.*, 1989; Sreenivasan, 1989).

Indonesia

The Common Carp and the Goldfish *Carassius auratus* (q.v.) are believed to be the earliest exotic fishes introduced to Indonesia from China (certainly before 1900). In Malang in East Java and in Bogor in West Java several fish farms were constructed, from where both species doubtless escaped into the wild. Common Carp have been frequently imported during the past century, and some naturalized populations are believed to occur (Muhammad Eidman, 1989).

Israel

The Common Carp was introduced from Europe to Israel between 1931 and 1934 for aquacultural purposes, and is now the country's main cultured fish. It has also been introduced, both accidentally and deliberately, into the wild in lakes, rivers and reservoirs. In the Hulah Nature Reserve it is one of the most abundant fish species, but in Lake Tiberias (Yam Kinneret/Sea of Galilee) in spite of stockings made in 1948 and 1949, it has done less well and no large-scale fishery has developed (Reich, 1978).

Although spontaneous spawning of Common Carp in Israeli waters has not been observed, it has been deduced by the presence of large populations of adults and juveniles in Lakes Tiberias and Hulah, and in some reservoirs such as Tsalmon, west of Lake Tiberias (Ben-Tuvia, 1981). Gophen *et al.* (1983) expressed concern for the future of native species in Lake Tiberias after the introduction of exotics such as Common Carp, but Davidoff and Chervinski (1984) pointed out that despite constant infiltration of Common Carp from nearby fish ponds their numbers in the lake have remained small.

Malaysia

Introduced to Malaysia from China in the early 19th century (Welcomme, 1981), the Common Carp 'has not established itself to any major extent' (Ang *et al.*, 1989), but there are said (by Welcomme, 1981, 1988) to be 'self-breeding populations in some ponds and tin mining pools'.

Philippines

Common Carp were introduced to the Philippines from Hong Kong in 1915; from Formosa (Taiwan) in 1925–1926; the 'Sinjonja variety' from Bogor, Indonesia (1956); the 'Punten variety' and the 'Red variety' from the same source in 1957; and the 'Majalaya Strain' from Indonesia in 1988. They are currently established in the wild in lakes, reservoirs, rivers, paddyfields and ponds, where the naturally breeding population forms a commercial fishery and is a source of ova for captive rearing (Juliano *et al.*, 1989).

Sri Lanka

Mirror Carp fingerlings (var. *specularis*) from Prussia (northern and western Germany) were introduced into Ceylon (Sri Lanka) in 1914 by the Ceylon Fishing Club. Although they grew well in the club's hatchery in Nuwara Eliya they failed to breed, and fared no better when removed to the Abbotsford estate at a lower elevation. They were eventually returned to Nuwara Eliya where they were placed in a large pond in the park; here they at last bred successfully, due, it was believed, to the presence of 'a plentiful supply of water lilies which provide breeding facilities'. Mirror Carp were also introduced into Hatton (1280 m asl) where they bred successfully (Chacko, 1945). Although Welcomme (1981, 1988) says that Common Carp are established in Sri Lanka, De Silva (1987), quoting Fernando (1971), says 'breeding not confirmed'.

Taiwan

'*Cyprinus carpio* or common carp was introduced long ago from Japan and is also being commercially monocultured' (Liao & Liu, 1989).

Thailand

Common Carp were first introduced to Thailand from China in 1912 for aquaculture, and 'is also known to breed in natural waters' (Piyakarnchana, 1989). Welcomme (1981, 1988) says the Common Carp has been introduced from China, Japan, Israel and Germany 'from 1913 onwards', and is 'well established in ponds and the wild'. De Iongh and Van Zon (1993) recorded importations in 1972 (1000), 1973 (15 000) and 1981 (101 700).

Unsuccessful Introductions

In Asia, Common Carp are breeding successfully in

aquaculture but not in the wild in Korea, Nepal and Pakistan (Welcomme, 1981, 1988); in Singapore (Chou & Lam, 1989); in Saudi Arabia and Syria (El Bolock & Labib, 1967); and in China (Yo-Jun & He-Yi, 1989).

Africa

Ethiopia

In 1936 and 1940, Common Carp (and Rainbow Trout *Oncorhynchus mykiss* (q.v.)), both probably from Italy, were planted in Lake Aba Samuel, an artificial impoundment some 25 km south of Addis Ababa which was formed by damming the River Akaki. The Common Carp soon became established, and are now one of the most abundant fish species in the Awash drainage system, where they are a valued sporting asset (Tedla & Meskel, 1981), especially in Lakes Akaki and Koka (Moreau *et al.*, 1988).

Kenya

In 1969, Common Carp from Uganda were introduced to Kenya, where they have replaced indigenous *Tilapia* spp. in the Maxina reservoir on the Tana River, and where their roiling habits have caused turbidity (Welcomme, 1981, 1988).

Madagascar

Common Carp were first introduced from France to Madagascar in 1914, and in 1925 and 1926 respectively to Lakes Itasy and Alaotra, and again in 1959. The subspecies established in the wild in Malagasy waters is the Mirror Carp *C. c. specularis* (Moreau *et al.*, 1988).

Ecological Impact The arrival in Madagascar of several exotic fishes coincided with a corresponding decline in the populations of some native species. For example in Lake Itasy, a large waterbody on the central plateau, although Kiener (1963) cites widespread deforestation and concomitant siltation as a major factor in the decline of the endemic cichlid *Ptychochromoides betsileanus*, the progressive introduction of the Goldfish *Carassius auratus* (q.v.) around the turn of the century, the Common Carp in 1925, the Redbreast Tilapia *Tilapia rendalli* (q.v.) in 1955, the Nile Tilapia *Oreochromis niloticus* (q.v.) in 1962 and the Largemouth Bass *Micropterus salmoides* (q.v.) in 1963, also played an important role in the eventual eradication of *Ptychochromoides* from the lake (Moreau, 1979; Reinthal & Stiassny, 1991).

Morocco

Introduced from France around 1925, Common Carp have 'adapted to local conditions' and are established in the wild in Morocco (Welcomme, 1981).

Namibia

In Namibia (South West Africa) Common Carp have been reported from the following localities: the Kuiseb canyon (Dixon & Blom, 1974); the Von Bach dam in the Swakop River system (Skelton & Merron, 1984); and the Hardap dam on the Fish River (Gaigher, 1975b); they are also said to be widely distributed in farm dams and in other freshwater systems (Skelton & Merron, 1984). In the catchment area of the Omuramba Omatako drainage basin Common Carp are also widespread, and thus threaten invasion of the Okavango delta in northeastern Botswana (Schrader, 1985; Bruton & Van As, 1986; De Moor & Bruton, 1988).

South Africa

De Moor and Bruton (1988) have traced the history of the introduction of the Common Carp into South Africa.

The Cape Argus of 15 September 1859 reported the importation of half-a-dozen Common Carp from England by Mr C. A. Fairbridge, a member of the Cape Legislative Assembly, in order to 'make our . . . barren rivers . . . a source of food'. These fish were placed in the reservoir in the Botanical Gardens in Cape Town. In March 1866, the *South African Advertiser and Mail* recorded the arrival of three Common Carp from England which were released in a pond on the estate of a Mr Ekstein in Rondebosch. Some time later he removed 391 Common Carp from his pond, and offered to present 100 to 'any gentleman having a place fit for their reception' (Harrison, 1966b). There were doubtless many other introductions of Common Carp in the second half of the 19th century, and in 1895 they were reported to be 'in fair abundance' in the dam at Beaufort West (Harrison, 1956). Skelton and Skead (1984) report an early record of 'carp' being seen in the Eerste River in about 1800, but this remains unconfirmed. The earliest introduction of Common Carp from England to the Jonkershoek Hatchery at Stellenbosch was not made until 1896 (Anon., 1944);

in the same year 109 'carp and tench' from Dumfries, Scotland, were placed in the Pirie Hatchery in Kingwilliamstown. The stocking of farm dams began in 1900 (Anon., 1944) and continued until 1921, when the breeding stock was eradicated and efforts were made to prevent the further spread of Common Carp by legislation (Harrison, 1959).

Common Carp are currently established in South Africa as follows:

Natal. In impoundments at Chelmsford, Spioenkop, Wagendrift, Midmar, Albert Falls, Nagle, Klipfontein, Verdruk, Tom Worthington, McHardy, Bloemveld, Tweediedale, Kandaspunt prison, Durnacol mine, Mount Edgecombe experimental station, Paulpietersburg Municipality and Hlobane mine dams, and in Pongolapoort Dam, from where they may have spread upstream into the Phongolo River. In rivers, Common Carp are naturalized in the Umgeni, Bushmans, Umkomaas, Umzimkulu, Tugela, Umzimvubu, Umlaas and Umzimhlava; their degree of success varies from river to river – they are, for example, comparatively uncommon in the main Umgeni and Tugela Rivers but abundant in many of the dams on those rivers and in the main Umkomaas River (De Moor & Bruton, 1988).

Cape. In the Bakens River in the eastern Cape (Heard & King, 1982; King & Bok, 1984); the Breede River in the Bontebok National Park (Braack, 1981); Rondevlei (Harrison, 1948); the lower reaches of the Rivers Berg, Breede, Gouritz, Kromme, Gamtoos, Swartkops, Sundays, Fish, Keiskamma, Buffalo, Nahoon, Kei and Umtata (Jubb, 1965); the Sak River in the Orange drainage (Hocutt & Skelton, 1983); Zeekoeivlei and the Liesbeeck dam near Cape Town and Liesbeeck Lake; Grassridge dam on the Fish River (Cambray *et al.*, 1977); Sandvlei near Muizenberg in the western Cape (Begg, 1976); Paardevlei, Princessvlei, Voelvlei and Misverstand (Picketberg), Brandvlei and Hospital dams (Worcester), and in numerous farm dams in the Worcester, Robertson, Ashton, Bonnievale and Swellendam localities, and in Lake Mentz on the Sundays River (De Moor & Bruton, 1988). Also in the estuary of the Kei River (Plumstead *et al.*, 1985); the Taung region (Vaal-Hartz) (Anon., 1960); the 'Oog' fountain at Kuruman; and in rivers of the Transkei and southeastern Cape below the 800 m contour (Jubb, 1973). (All quoted by De Moor & Bruton, 1988.)

Natal and *Orange Free State.* In the Hans Strijdom dam on the Mogol River (Kleynhans, 1983), the Olifants River in the Kruger National Park (Pienaar, 1978), the Loskop dam (Kruger, 1971), the Hartbeespoort dam on the Crocodile River (Cochrane, 1983), in the Rietvlei dam southeast of Pretoria (Smith, 1983), and in the Doorndraai dam on the Sterk River near Potgietersrus (Batchelor, 1974) – all in the Limpopo River system. (All quoted by De Moor & Bruton, 1988).

In the Orange/Vaal system, Common Carp have been reported from the Vaal dam (Du Plessis & Le Roux, 1965); Barberspan in western Transvaal (Goldner, 1967); the Boskop dam in western Transvaal (Koch & Schoonbee, 1980); the lower Vaal River at Warrenton and the Vaal Hartz Weir (De Moor & Bruton, 1988); the Caledon River near its confluence with the Orange River (Marshall, 1972); Lake Le Roux (Jackson *et al.*, 1983); the Verwoerd dam (Hamman, 1980); and in the Upper Orange and Caledon, the Vaal River system, the middle Orange River system (rarely), and the southern tributaries of the Orange and the upper Fish River (Namibia) (Skelton, 1986). (All quoted by De Moor & Bruton, 1988.)

Common Carp are thus present in the following major catchments in South Africa: the south coast rivers from the Berg in the western Cape to the Umtamvuna in Natal; the Orange/Vaal and Natal coastal catchment from the Umtamvuna to the Tugela, and the Phongolo, Limpopo and Incomati (Bruton & Van As, 1986). Within these catchments, Common Carp seem to have done best in the southern Cape and in the Orange/Vaal systems, and have been least successful in the Incomati and the lowveld areas of the Limpopo basin (De Moor & Bruton, 1988).

De Moor and Bruton (1988: appendix III: 271–272) give details of dates of initial introductions and earliest records (with references) of Common Carp in various localities in South Africa.

Ecological Impact   In South African waters, the coexistence of Common Carp and Moggels or Mud Mullets *Labeo umbratus* in the Sundays and Buffalo Rivers in the eastern Cape has increased the turbidness of the already heavily silted water to such an extent that few other fish apart from the eel *Anguilla mossambica* can survive (Jubb, 1959a). In the Umzimvubu and Umzimhlava Rivers in East Griqualand, Common Carp are said to have had a negative impact on the trout fishery (Harrison, 1948). In the Rondevlei, Common Carp had all but eradicated 'Nitella' plants by 1948, and the increased turbidness of the water had made it unsuitable for stocking with bass (Anon., 1948). In the Vaal River, Mulder (1973) suggested that increased turbidity had hampered such piscivorous species as the Largemouth Yellowfish *Barbus kimberleyensis* in pursuit of its prey, which had contributed to its decline. Roiling of the substrate by feeding Common Carp may inhibit nest-building of cichlids and other species wherever they occur sympatrically (all quoted by De Moor & Bruton, 1988).

In those vegetated vleis (especially Rondevlei in the Cape) in which they have become established, Common Carp have significantly affected the aquatic vegetation on which waterfowl depend, and this together with competition for aquatic food organisms has led to a decline in the birds' populations (Harrison, 1948, 1956).

Although Jubb (1973) suggested that the presence of Tigerfish *Hydrocynus forskahlii* would inhibit the spread of Common Carp, Pienaar (1978) indicated Common Carp might evict the native species which requires clear and well-oxygenated waters in which to hunt its prey.

Not all authorities are agreed about the negative impact of naturalized Common Carp in South Africa; Ashton *et al.* (1986), for example, consider that the Common Carp's involvement in habitat destruction and in muddying water, and its impact on native fish species have been exaggerated, and suggest that increased eutrophication has simply favoured Common Carp and other detritivores.

The introduction of Common Carp outside their natural range has resulted in the spread of fish parasites, including various species of Protozoa, Monogenea, Branchiura, Cestoda and Nematoda (Hoffman & Schubert, 1984); many of these are also parasites of the Goldfish *Carassius auratus* (q.v.), and it seems likely that their spread can be associated with both species. According to Bruton and Van As (1986), the Common Carp has been implicated in the introduction to South Africa of at least seven alien, or suspected alien, fish parasites.

Tunisia

In 1965, Common Carp from France were introduced for aquacultural purposes and to develop a fishery in Tunisia, where they became established in the wild (Moreau, *et al.*, 1988).

Uganda

Moreau *et al.* (1988) say that the Common Carp was introduced from Israel to Uganda in 1957 and 1962, where Welcomme (1981) says it was 'introduced into some highland lakes where it was overfished'. Moreau *et al.* (1988) indicate that the species is still established in the wild in Uganda.

Zimbabwe

Toots (1970) says that the Common Carp was probably first imported to Southern Rhodesia (now Zimbabwe) prior to 1920, although the earliest recorded introduction, by the Rhodesian Angling Society (RAS) was not until 1925, when fingerlings were released in Matopos dam. In the following year 1500 fingerlings from East Griqualand were acquired by the RAS and planted in their waters around Bulawayo. Further importations were made until 1938, after which Common Carp were declared undesirable and further introductions were forbidden. Maar (1960), who conducted an investigation into Common Carp in Zimbabwe, concluded that there was no evidence of their invasion of natural waters or of any negative ecological impact. Toots (1970) said that in RAS waters such as the Umgusa, Matopos and Hillside dams, Common Carp had done exceptionally well.

In 1958, some Mirror Carp (*specularis*) were acquired from Israel by the Fisheries Research Centre at Mazoe, from where some were successfully planted in the Savory (Henderson) and Mazoe dams. Toots (1970) recorded Common Carp as established in the Tuli, Bulawayo, Midlands, Shamva, Mazoe, Ruwa, Macheke and Umtali areas, while Welcomme (1981) says they occur in 'Lake McIlwaine and other dams but not rivers'.

Unsuccessful Introductions

In Africa, Koura and El Bolock (1960) and El Bolock and Labib (1967) record the success in aquaculture but not in the wild of Common Carp introduced to Egypt. Welcomme (1981, 1988) reported the same for Cameroun, the Central African Republic, Ghana, the Ivory Coast and Rwanda, and that the species' status in Nigeria, Sudan, Swaziland and Togo was unknown. Moreau *et al.* (1988) listed unsuccessful introductions to Lesotho, Malawi, Zaire and Zambia.

North America

Canada

MacCrimmon (1968), from whom much of the following account is derived, has traced the introduction of the Common Carp into Canada.

Ontario

The earliest recorded importations of Common Carp, from the United States Fish Commission in Washington, DC, were made in 1880 (Dymond, 1955) when Samuel and B. F. Reasor released 10 fish in a pond on their property at Cedar Grove, York County, Ontario (MacKay, 1963), and others were sent to W. H. Barber of Andover, New Brunswick.

In the following year, Samuel Wilmot, Superintendent of Fish Culture for Canada, acquired 100 Common Carp from Rudolph Hessel in Washington, which were placed, with only moderate success, in ponds at the government-run Newcastle hatchery on Wilmot Creek, Ontario, in the hope of obtaining 'fish that will be welcomed to the poor man's table'.

In 1883, a further 100 Common Carp were sent by the United States Fish Commission to Ontario, and although no further shipments are on record until 1888, when 70 fish were despatched from Detroit, Michigan, it is stated in the Commission's annual report for 1881 that 'the numbers of carp [exported] were greater than recorded', and it is known that Common Carp from private sources (which were sold for $2–$5 per breeding pair), which are unlikely to have been officially recorded, were imported to Ontario during this period.

In the counties of York, Ontario and Durham several mill ponds north and east of Toronto, on inflowing streams of Lakes Simcoe, Ontario and Scugog drainages, were reputedly stocked with Common Carp in the 1880s or early 1890s. The escape of Common Carp from Dyke's Pond, York County in 1896 (MacCrimmon, 1956) is believed to have led to the colonization of Lake Simcoe (see below), while their appearance in Lake Scugog is attributed to the washing out in 1916 of a dam on Brown's Lake on East Creek in Durham County, in which Common Carp had been placed before the turn of the century.

In 1891, the Ontario Game and Fish Commission received several hundred young Common Carp from Washington, DC some of which were planted in a pond on the Grand River near Dunneville, the remainder being released directly into the river. A year later a further shipment was received and placed in the Upper Grand River, 'where it is believed they will increase immensely, furnishing in the future a cheap article of food . . .'. The Annual Report of the Ontario Game and Fish Commission for 1893 stated that 'several barrels of carp were taken this spring . . . they are a very desirable food fish, growing and multiplying rapidly . . .'. In the same year some 'German' Carp were received from the United States Fish Commission, and released in the Upper Grand River. These are the last recorded plantings of Common Carp in the waters of Ontario where, because of the extent of the population, MacCrimmon (1968) describes their distribution regionally.

Lakes Erie and St Clair

It is believed that Common Carp became established in these two lakes by at least 1883, possibly having escaped from rivers in which they had been planted on the USA side of these waters, where a large commercial Common Carp fishery had been developed by 1884 (Smiley, 1886). Smith (1904) claimed that 'with the probable exception of the Illinois River, no body of water appears to be so well stocked as Lake Erie'.

Those Common Carp established in Lake Erie at the mouth of the Grand River by 1892 would seem likely to have originated in the plantings in that river mentioned above, as would those reported at Long Point Bay in the following year, and along the Ontario shoreline at Port Maitland, Port Colborne and in the upper Welland Canal before the turn of the century.

In 1897, Common Carp were first caught in Rondeau Bay, where they were said to be common by around 1900, prior to which considerable populations of Common Carp were established along the Canadian shore of Lake Erie from the upper Niagra River to Lake St Clair and the St Clair and Detroit Rivers, and up the Lake Huron shore to near Southampton.

Lake Simcoe and the Trent Canal system

The escape of Common Carp from Dyke's Pond in York County in 1896, referred to above, resulted in the establishment of a population in Holland Marsh and Cook's Bay on Lake Simcoe (MacCrimmon, 1956), where large numbers were soon being sold as food to local residents. By 1902, Common Carp had appeared along the north shore of Lake Simcoe and in Lake Couchiching further north. From the former, probably via the Trent Canal, they spread northward to Georgian Bay and eastward to the Kawartha lakes. By 1910, Common Carp had become abundant in the Trent Canal as far as Bolsover and Kirkfield, and within a further 5 years they had spread from the canal to most of the Kawartha lakes. The appearance of Common Carp in Lake Scugog (see above) coincides with the spread of the fish through the Trent Canal system, of which it is a part.

Lake Ontario
Common Carp first appeared in the Canadian waters of Lake Ontario probably from those of New York State, where Smith (1892) recorded their presence as early as 1890 following deliberate releases in the lake as early as 1879 and the accidental escape of fish into inflowing streams. By 1901–1902, Common Carp were being reported from near Toronto, probably having escaped from ponds in the Humber and Don River watersheds, and by 1905 they were abundant in the western end of the lake from Toronto to the lower Niagra River, where considerable habitat damage was reported.

It was not until 1907 that Common Carp were observed at the eastern end of Lake Ontario, off Barriefield at the entrance of the St Lawrence River. Within 2 years they were widespread from Brighton to the Bay of Quinte, and by 1909 had reached the Upper St Lawrence, where 2 years later they were said to have reached 'nuisance proportions'. MacCrimmon (1968) suggested three possible origins for these populations: first, an easterly movement along the lake shoreline from Toronto; second, via the Trent Canal system from Lake Simcoe; and third, from the releases and escapes in the waters of New York State mentioned by Smith (1892), or from Cayuga Lake (which flows into Lake Ontario) where Carp were introduced in 1889 (Reed & Wright, 1909).

It is thus apparent that Common Carp were flourishing along almost the entire Canadian shoreline of Lake Ontario between the Niagra and St Lawrence Rivers by the first decade of the 20th century. They now occur in the Ottawa River at least as far upstream as Ottawa.

Lake Huron, Georgian Bay and North Channel
Common Carp in Georgian Bay, where they first appeared in 1905, are traditionally believed to have originated from those at the mouth of the Trent Canal and around Green Island, Waubaushene, and Sturgeon Bay. Within 2 years, Common Carp – mainly *nudus*, but also *specularis* and *communis* – had become common and were being caught commercially in Matchedash Bay.

From the entrance of the Trent Canal, Common Carp seem to have dispersed in both directions along the shore of Georgian Bay. To the west they appeared in Meaford Harbour in 1909, but it was not for another 6 years that they reached Owen Sound Bay. Northward and westward from the entrance to the canal, Common Carp arrived in Honey Harbour by 1908 or 1909, Parry Sound by 1912, and the North Channel by 1914.

Common Carp are presently limited in Georgian Bay to the many fairly small and isolated bays with aquatic vegetation along the inside of the Bruce Peninsula and in the vicinity of the islands fronting Parry Sound and Sudbury, which are believed to support large populations. The principal centres of concentration are in Inner Georgian Bay fronting Simcoe County.

In the North Channel, Common Carp are rare or absent along the shoreline of Manitoubin Island. Discrete populations occur along the north mainland shore of the North Channel, especially near Spanish.

In Lake Huron, Common Carp are known to have been present at the mouth of the St Clair River by 1900, and by 1915 were reported as far north as Southampton. They are currently found along the shoreline fronting Lambton, Huron and Bruce counties, and northward to the apex of the Bruce Peninsula; principal concentrations occur off Lambton County and in the entrances of the main rivers, such as the Sauble at Grand Bend, the Maitland at Goderich, the Penetangore at Kincardine, and the Saugeen at Southampton.

The original Common Carp in the Canadian waters of Lake Huron probably came from the St Clair River, and these fish are likely to have been descended from those planted in the inland waters of Michigan State between 1881 and 1897 (Peterson & Drews, 1957) which became established in the St Clair River and Lake St Clair before 1900. Common Carp could also have reached Lake Huron and the North Channel from Lake Michigan, where they are known to have occurred off the shores of Michigan, Wisconsin, Illinois, and Indiana by 1899 (Baldwin & Saalfield, 1962).

Lake Superior

The first apparent records of Common Carp in Lake Superior appear to date from 1915 at Rossport and 1922 in Thunder and Terrace Bays. Since, however, species other than *C. carpio* were referred to in northwestern Ontario as 'carp', these records have not been confirmed. It was not until 1948 that the first authenticated Common Carp was caught in Lake Superior, at Batchawana Bay, west of Sault Ste Marie.

On the northwestern shore of the lake a Common Carp was taken in 1953 at the mouth of the Naki River (Lake Helen), and others were observed in a lagoon at Nipigon off the Nipigon River in 1956. Identified individuals were caught in Thunder Bay, fronting Port Arthur and Fort William, in 1954 and 1955 respectively (Ryder, 1956), and Common Carp were said to be common at the entrances to the Kaministikwia, McIntyre, and McVicar's creeks. By the mid-1960s, Common Carp are believed to have been widespread in the parts of Thunder, Black and Nipigon Bays between Simpson Island and Fort William.

Traditional belief is that Common Carp in the Canadian waters of Lake Superior are the descendants of discarded anglers' bait fish. Since, however, Common Carp planted at Duluth, Minnesota, USA, in 1889 are believed to have failed to become established, it is possible that some of them or their offspring could have dispersed into Lake Superior. Similarly, Common Carp introduced to Wisconsin and Michigan from 1879 could have spread from those states into the Lake Superior watershed.

Other Ontario Waters

Elsewhere in Ontario, Common Carp are largely restricted to small lakes, impoundments, reservoirs and slowly flowing streams in agricultural areas of the southwest south of the rocks of the Precambrian Shield. Although Common Carp are apparently absent from the northern tributaries of Georgian Bay, they are present in the lower French River which flows out of the lake, and isolated populations are found in Manitouwabing Lake (McKellar township) and Mill Lake (McDougall township) in the Parry Sound district. In the former, Common Carp are said to have been introduced by an early settler, perhaps in the late 19th century. North of the Trent Canal, Common Carp occur in such isolated waters as Elsie Lake (Minden township), Wilbermere Lake (Monmouth township), and Horseshoe Lake in Glamorgan township.

Common Carp in the waters of Ontario are of some commercial value near urban areas, but have a detrimental impact on aquatic vegetation and on native fish species through competition for food, space, and spawning grounds (Scott & Crossman, 1973; Crossman, 1984).

Quebec

The earliest reference in the literature to the presence of Common Carp in the waters of Quebec appears to be by Melancan (1936), who gives no information on the species' introduction or distribution. Vladykov and McAllister (1961), however, reported Common Carp in the St Lawrence River downstream as far as Rivière-du-Loup, 175 km below Quebec City. In the Montreal stretch of the St Lawrence, upriver from Quebec City, Common Carp have occurred since at least 1910, having doubtless dispersed from adjacent waters in Ontario. MacCrimmon (1968) considered that they were then probably also established in Lakes Saint-François, Saint-Louis and Deux-Montagnes, formed by the St Lawrence at Montreal, where their presence is confirmed by Cuerrier *et al.* (1946).

In the Ottawa River, which flows into the St Lawrence upstream of Montreal, Common Carp were first recorded at Pointe-au-Chênes, 80 km above Montreal, in 1941. Subsequently, Common Carp were noted further upriver at Papineauville, and in 1944 at Pointe Gatineau, south of Hull. Construction of the Carillon dam near Pointe Fortune in Ontario formed a major impoundment with favourable Common Carp habitat in both Ontario and Quebec (MacCrimmon, 1968).

In the St Lawrence River, between Montreal and Quebec City, Common Carp were not recorded downstream of Lac Saint-Pierre before 1940.

MacCrimmon (1968) suggests three possible origins of Common Carp in the waters of Quebec: first, possible deliberate or accidental introductions; second, dispersal from the upper St Lawrence River in Ontario and New York State; and/or, third, from Lake Champlain via the Richelieu River and the state of Vermont, where large plantings of Scaled and Mirror Carp took place between 1883 and 1886.

British Columbia

Common Carp now naturalized in several watersheds on the mainland of British Columbia are believed to be descended from those introduced to the

Columbian River system in Washington State in the 1880s (Carl & Guiguet, 1958), which entered the province via the Okanagan Valley around 1912. Whitehouse (1946) suggested their probable dispersal northward up the Columbia River into the Okanagan River and Lakes Osoyoos, Skaha and Okanagan north of Penticton, where Common Carp were first recorded in 1917 (Clemens *et al.*, 1939) and where Dymond (1936) reported them to be fairly common in shallower parts of the lake.

In the watershed of the Fraser River, Common Carp were first noted in 1928 in the Shuswap Lake region of South Thompson River near Armstrong by Carl *et al.* (1967), who also recorded them in Lakes Kalamalka, Woods and Shuswap. It is traditionally believed that Common Carp gained access to the headwaters of the Fraser River from those of the Columbia by way of irrigation ditches or creek courses near Armstrong, where the divide between them is of low elevation (Carl & Guiguet, 1958). From Lake Shuswap, Common Carp dispersed downriver to Popkum Creek near Chilliwack by 1939, to Lake Hatzic by 1944, and to brackish waters at Point Grey by 1946 (Carl *et al.*, 1967).

East of the headwaters of the Fraser River, Common Carp have been observed in Kootenay Lake north of Kaslo and in Shannon Lake west of Kelowna (MacCrimmon, 1968). In the mid-1960s they appeared in the Squamish River on Home Sound, which they probably reached, according to MacCrimmon (1968), by travelling north through the brackish waters of the Pacific Ocean off Vancouver from the Fraser River estuary.

A population of Common Carp formerly established for many years in Glen Lake on Vancouver Island (Carl *et al.*, 1967), believed to have originated from fish that escaped from a nearby pond to which Common Carp from Oregon may have been introduced, are thought to have been poisoned by rotenone in 1961 (MacCrimmon, 1968). More recently, Common Carp have been illegally released in farm ponds in the Courtenay and Oyster Rivers area of Vancouver Island (Crossman, 1984).

In British Columbia, Common Carp are today naturalized in most of the lakes of the Okanagan Valley; in Arrow, Twin, Horne, Christina, and other lakes of the Columbia system; and in lakes in the Fraser system.

Saskatchewan and Manitoba

Common Carp are present in parts of the Nelson River catchment within the borders of both Saskatchewan and Manitoba.

The earliest definite reference to the presence of Common Carp in Saskatchewan is by Atton (1959), who recorded them in the Frenchman River, a tributary of the Milk River, in the Missouri watershed in the extreme southwest in 1921. MacCrimmon (1968) ascribed the failure of this population to expand its range to the general absence of permanent waters and the existence of only semi-desert land to the north and east.

Despite introductions of Common Carp in Manitoba in 1885–1886 and 1889 there is no record of them becoming established, and Hinks (1943) said they were unknown in the province until 1938.

The arrival of Common Carp in the Nelson River catchment in Manitoba is likely to have been a result of dispersal from North Dakota and Minnesota, USA. Atton (1959) noted two possible routes from the headwaters of the Mississippi River: one between the Minnesota River in the Mississippi drainage and Big Stone Lake in the Nelson drainage, and the other between the headwaters of the James River in the Mississippi catchment and the Red River in North Dakota. The possibility of deliberate or accidental human intervention cannot, however, be discounted. The earliest record of Common Carp in the Canadian waters of the Nelson River system was in the Red River at Lockport, Manitoba in 1938 (Hinks, 1943), and shortly afterwards in Lake Winnipeg.

Common Carp began slowly to expand their range northwards through the Nelson River system (Dymond, 1955), reaching the Winnipeg River by 1943 (Hinks, 1943), the Mukutawa River near the northern end of Lake Winnipeg in 1954 (Keleher, 1956), Cross Lake on the Nelson River, north of Lake Winnipeg, in 1956 (Keleher & Kooyman, 1957), and Split Lake, further downstream on the Nelson River, and north of the railway line to Hudson Bay, in 1963.

In the watershed of the Assiniboine River, Common Carp first appeared at Virden in extreme southwestern Manitoba in 1948. Five years later young fish were noted at Kamsack in Saskatchewan (Atton & Johnson, 1955), and in the following year Common Carp were caught in the Qu'Appelle River. In 1955, a single fish was taken in Pasqua Lake (Atton, 1959), and in 1958 Common Carp became established in Last Mountain Lake in the Qu'Appelle area.

In the Saskatchewan River system, Common Carp

became established after 1958 near Saskatoon in the South Saskatchewan River, in the North Saskatchewan and Shell Rivers near Prince Albert, Saskatchewan, and the Par, Manitoba (MacCrimmon, 1968), possibly from the headwaters of the Qu'Appelle via Aiktow Creek (Atton, 1959). By 1966, Common Carp had appeared in the Swan and Red Deer Rivers and in Namew Lake which flows into Cumberland Lake and the Saskatchewan River (MacCrimmon, 1968).

Among the larger lakes, MacCrimmon (1968) was informed of the presence of Common Carp in Lake Winnipeg soon after 1938; in Lake Manitoba by 1948; in Lake Winnipegosis by 1954; in Lake Dauphin by 1955; in Playgreen Lake by 1958; in Cedar Lake by 1959; and in Split Lake by 1963. He considered that 'the carp population of Manitoba and Saskatchewan continues to increase in both range and abundance'.

MacCrimmon (1968) summarized the then distribution of Common Carp in Canada as extending 'eastward from Vancouver Island at 120° long. to the lower St Lawrence River in Quebec at 70° long. The distribution is not continuous, the species being unrecorded from Alberta and from that area of Manitoba and Ontario lying between Lake Winnipeg and Lake Superior. The species ... most southerly occurrence being in southern Ontario at 42° N lat. The most northerly record is that from Manitoba [Split Lake] at 65° N lat.' Although in Ontario most Common Carp populations occurred south of the 45th parallel, in Saskatchewan and Manitoba they were established at least as far north as the 53rd parallel. MacCrimmon (1968) considered that although natural or artificial barriers would inhibit the further natural spread of Common Carp in Canada, there was an abundance of natural or impounded waters with apparently suitable habitats in all provinces if Common Carp were introduced deliberately or accidentally. A westward extension of the Common Carp population from the Prairie Provinces into Alberta seemed probable if the slow spread of the species through Saskatchewan and the Assiniboine River system continued, and perhaps in southern Ontario where Common Carp were absent from many isolated natural and artificial ponds, and also from the central Rideau Canal system to which they had access from both north and south. Natural colonization of the Maritime Provinces of New Brunswick, Nova Scotia and Prince Edward Island, and of Newfoundland, MacCrimmon considered 'remote'.

MacCrimmon (1968) drew attention to the apparent dispersal of Common Carp through saline water on both the east and west coasts (down the St Lawrence to Rivière-du-Loup and the Kamouraska River, and northwards up the coast of British Columbia to the Squamish River), although pointing out that floodwaters discharged by the St Lawrence and Fraser Rivers respectively considerably diluted the salinity. He considered that this probable dispersion through saline water was not surprising in view of the existence of natural populations of the species in saline waters of Eurasia (Nikolsky, 1963) and the culture of Common Carp in brackish waters in the Far East (Johnson, 1954) and Israel (Soller *et al.* 1965).

Finally, MacCrimmon (1968) considered that the then current policy of federal and provincial agencies to construct major impoundments for hydroelectricity and water conservation would create new potential habitats for the establishment of further Common Carp populations in Canada.

Ecological Impact   According to Crossman (1984), in Canada '... probably only one exotic, common carp, can definitely be said to have had a negative impact on the environment and on indigenous fishes'. (See also pp. 106 and 110–111.)

Mexico

Common Carp from Haiti were released in Mexican waters in 1872–1873 (Escalante & Contreras, 1985), and others from Israel were introduced in 1956 (Obregón, 1960). Contreras and Escalante (1984) list the following records of Common Carp in Mexico: Valley of Mexico lakes (Meek, 1904; Alvarez & Navarro, 1957); Sonora border in the Río Colorado and Chihuahua (Allen, 1980); Río Grande (Treviño-Robinson, 1959); lower Río Casas Grandes (Contreras *et al.*, 1976); Río Santa Maria, Baja California (Follett, 1960); Río Conchos (Contreras *et al.*, 1976; Contreras, 1978); upper Río Mezquital (Contreras *et al.*, 1976); Parras basin (Contreras, 1969, 1978); Río Yaqui (Hendrickson *et al.*, 1980); Michoacán (Alvarez & Cortés, 1962); Nuevo León (Contreras, 1967); San Luis Potosí (Alvarez, 1959); and Aquascalientes (Contreras & Contreras, 1985).

Ecological Impact   Common Carp naturalized in Mexico have been associated with the disappearance of indigenous fishes in Casas Grandes, Bustillos, Camargo in Chihuahua, Peña del Aguila and Tunal in Durango, Parras in Coahuila, and San Juan Del Río in Querétaro (Contreras, 1969, 1975, 1978; Contreras *et al.*, 1976).

United States

MacCrimmon (1968) has traced the early history of the introduction of the Common Carp to the USA.

The earliest record is by DeKay (1842), who quotes directly from a letter received from Captain Henry Robinson describing his importation from France of six or seven dozen Common Carp and their subsequent release in a pond on his property near Newburgh, New York, in 1831 and 1832. Forester (1850) said that Robinson planted his Common Carp in five ponds on the bank of the Hudson River between New Windsor and Newburgh, but that 'in process of time a heavy freshet carried away his dams and flood-gates, and a large proportion of his carp escaped into the Hudson'. (See also Phillips, 1883; Smiley, 1883).

In spite of the definite statements about Robinson's introduction, Cole (1905a) questioned the species' correct identification: 'we shall never know whether the fish . . . were true carp or whether he happened when procuring the fish . . . to get hold of the hybrid form'. Cole's doubt was apparently prompted by Baird (1879) who, because he was unable to find any Common Carp in the Hudson River or in the fish market (where all the fish sold as 'carp' were actually the Crucian Carp *Carassius carassius* (q.v.)), suggested that the fish imported by Robinson may have been Goldfish *C. auratus* (q.v.) or hybrids. Forester (1850), however, said that there were then two cyprinid species that had been recently introduced to the USA and that both had 'become entirely naturalized in some of our waters'; these were the Goldfish, then established in the Schuylkill and in some streams in Massachusetts, and the Common Carp living near Newburgh in the Hudson. Today, the year 1831 is generally accepted as that of the earliest introduction of the Common Carp into North America.

Although S. F. Baird of the United States Fish Commission pressed for the introduction of Common Carp to the USA by the Commission as early as 1872, their importation by private individuals preceded by several years that by the Commission.

In 1872, Julius A. Poppe imported five 15-cm Common Carp (the sole survivors of an original shipment of 83) from Reinfeld, Holstein, Germany, to northern California, which he released in a pond on his property in Sonoma Valley, where they were soon breeding (Poppe, 1880). Subsequently, Common Carp reared by Poppe were widely distributed to waters in California and contiguous states (Evermann & Clark, 1931; Miller, 1943; La Rivers, 1962), and as far away as central America and Hawaii. The success of Poppe's Common Carp-rearing programme spurred the newly-formed California Fish Commission into importing 88 Common Carp from Japan in 1877 (Peterson & Drews, 1957).

In 1876, Rudolf Hessel, on behalf of the United States Fish Commission, imported a shipment of Common Carp from Germany, all of which died on the voyage from Europe. Undeterred, in the following year Hessel acquired a second consignment of 345 Common Carp (227 Mirror and 118 Scaled) from Höchst, near Frankfurt (Hessel, 1878; Baird, 1879) which were placed in ponds in Druid Hill Park, Baltimore, from where in the spring of 1878 65 Mirror Carp and 48 Scaled Carp were transferred to ponds at the Washington Monument, Washington Arsenal, and the Smithsonian Institute in Washington, DC (Hessel, 1878).

Thereafter, Common Carp began to be dispersed to many parts of the USA for aquaculture and as a source of food. In 1879, for example, 12 265 (Laycock, 1966) or 6203 (MacCrimmon, 1968) were distributed among 25 states and territories, to be followed in 1880 by a further 31 443, and in 1883 some 260 000 were divided among 298 of 301 congressional districts. During the 1880s and 1890s, the culture of the Common Carp and its distribution became the most important activity of the United States Fish Commission, and led Hessel (1884) to claim that 'the progeny of the 345 carp brought from Germany in 1877 have been distributed to all parts of the United States, and the carp is almost as familiar to our people as any other kind of domesticated animal'. This programme of introductions ceased, however, in 1897, partly in response to increasing complaints about the damage being caused by the species and partly because it was in any case by then widely established (Cole, 1905b). Today, according to Allen (1980) and Courtenay (1993) Common Carp exist in all the contiguous states of the USA, being most abundant in those of the mid-west (Courtenay *et al.*, 1984).

Ecological Impact   Taylor *et al.* (1984) have summarized the ecological impact of the Common Carp in the USA.

Although not introduced, like the Grass Carp *Ctenopharyngodon idella* (q.v.), for the control of aquatic vegetation, the Common Carp's foraging activities result in the removal of weeds both by direct consumption and by uprooting caused by digging through the substrate in search of food. Additional damage to the growth rate and survival potential of rooted aquatics is created by increased turbidity and a concomitant decline in the penetration of light as a result of roiling. That Common Carp are directly responsible for such effects on water quality and aquatic vegetation has been established experimentally by the use of exclosures and the removal of Common Carp from certain waters. Although a decline in the populations of indigenous fishes has frequently followed the establishment of large populations of Common Carp, and cause and effect is almost certain, the evidence remains largely circumstantial. It is, however, strengthened by the fact that after the removal of Common Carp, native fish populations frequently recover. The reasons for the decline of native species is likely to be loss of macrophytes, due to Common Carps' activities, which reduce shelter, forage and spawning sites.

A similar situation exists with regard to the impact of Common Carp on aquatic vegetation and a concomitant deterioration in habitats for waterbirds, the populations of which then all too frequently decline. The evidence for cause and effect is strong, albeit again as yet largely circumstantial: an exception is provided by Buck (1956), who in a controlled study in Oklahoma proved fairly conclusively the impact of introduced Common Carp on populations of three indigenous fish species.

Although often accused in the early days of its naturalization in the USA of being a predator on the eggs of native fishes, the role of the Common Carp in this activity is now known, by stomach contents analysis, to be insignificant. An exception is again provided, by Jonez and Sumner (1954), who showed that predation by Common Carp on the eggs of the Razorback Sucker *Xyrauchen texanus* in Lakes Mead and Mojave, Nevada, may have been an important factor in the decline of the native species in the basin of the Colorado River. Common Carp may indirectly affect the survival of the eggs and larvae of native fishes by their roiling activities and through causing the desertion of nests in response to disturbance by

feeding or spawning, but the significance, if any, of this has not yet been quantified.

Finally, roiling arising from the spawning activities of Common Carp has been cited as detrimental in shallow waters used for spawning by native species – especially centrarchids – although the evidence for this, apart from an experimental study by Forester and Lawrence (1978) on Bluegills *Lepomis macrochirus* and Largemouth Bass *Micropterus salmoides* (q.v.), is largely anecdotal.

(For a full list of references, see Taylor *et al.*, 1984 and Moyle, 1984, and subsequently e.g. Courtenay 1989, 1993; Courtenay & Moyle, 1992; Lever, 1994.)

South and Central America

Argentina

Common Carp have been spasmodically introduced to Argentina during the 20th century for aquaculture and for sport fishing in various lakes and rivers, such as Río de la Plata; Río Paraná in Misiones; Lago San Ramón in Bragado; Lago del Dique; San Roque in Córdoba; and elsewhere. Since about 1850, there has also been some natural diffusion of Common Carp down the Uruguay River from Brazil.

The species is now widely distributed in many small rivers and lakes of the Río Salado basin in central Argentina (Ringuelet *et al.*, 1967; Baigun & Quiros, 1985; Navas, 1987).

Ecological Impact   Navas (1987) considered that attempts should be made to eradicate Common Carp in Argentina because of the environmental damage they cause, which is said to affect the native Argentine Pejerrey *Odontesthes bonariensis*, and because their boniness makes them a poor food for human consumption.

Bolivia

Welcomme (1981, 1988) says that Common Carp from Mexico were successfully introduced to Bolivia in 1945 for aquaculture to produce a cheap source of food and (the Koi variety) for ornamental purposes; Common Carp are currently naturalized in a number of Bolivian lakes and dams.

Brazil

Common Carp from the USA and Europe have been introduced to Brazil on various occasions since 1882 or aquacultural purposes, and are still being reared in some central and southern regions where they are also naturalized in some rivers (Welcomme, 1981, 1988).

Chile

Common Carp from Germany were first introduced to Chile in 1875 as the founder stock of a potential commercial fishery; they are currently established in lagoons, lakes, and ponds on private estates in various parts of the country (De Buen, 1959).

Colombia

The Common Carp was first introduced to Colombia, from an apparently unrecorded source, in 1940 for aquacultural purposes, and is presently naturalized in the Cauca River, Cundinamarca, in the northwest, and in Santander in the south (Campos Cereda, 1970).

Costa Rica

Common Carp introduced to Costa Rica from Taiwan in 1876 as a pituitary donor are well established in polyculture centres, and possibly in the wild (Campos Cereda, 1970).

Cuba

Welcomme (1981, 1988) says that Common Carp introduced to Cuba from the USA and former USSR in 1927 and 1983 for aquaculture and game fishing are now 'established throughout the country'.

Dominican Republic

The Common Carp was introduced to the Dominican Republic from Mexico in 1953 for aquacultural purposes, and now 'breeds in rivers and lagoons throughout the year'; especially in the former it 'reaches a good size' (Welcomme, 1981, 1988).

Panama

According to Welcomme (1981, 1988), Common Carp from the USA, Israel and Colombia, introduced to Panama for aquaculture in 1976, 1979 and 1981, are established and breeding in some lakes, reservoirs and ponds, particularly in Lago Gatún.

Peru

Common Carp introduced to Peru from Japan and China in 1946 and 1960 for aquaculture are established and reproducing in the Ate Valley, Vitarte (Welcomme, 1981, 1988).

Uruguay

Since about 1850, Common Carp have diffused down the Uruguay River from Brazil into Uruguay, where they are now naturalized in the Salto Grande reservoir and in the lower reaches of the Uruguay River (Welcomme, 1981, 1988).

Venezuela

Introduced in 1940 from an apparently unrecorded source to establish commercial fisheries, Common Carp established in Venezuela are said to have 'eliminated some cohabiting native species' (Welcomme, 1981, 1988).

Unsuccessful Introductions

In South and Central America, the Common Carp is recorded as established in aquaculture but not, so far as is known, in the wild in Ecuador, El Salvador, Guatemala, Haiti, Honduras, Nicaragua and Surinam (Welcomme, 1981, 1988), and in Puerto Rico (Erdman, 1972, 1984; Ortiz-Carrasquillo, 1980, 1981).

Australasia

Australia

Although Clements (1988) has claimed that the first Common Carp to be imported to Australia arrived in Hobart, Tasmania, from England on 22 February 1858 on board the *Heather Belle*, and that other shipments followed in the early 1860s, it is generally accepted by other authorities that the earliest introduction was in 1872 by the Geelong and Western District Acclimatization Society of Victoria (Wharton, 1971, 1979). McKay (1984) records the introduction of Common Carp to Queensland in 1888.

Three strains of Common Carp are at present established in Australia; the so-called 'Prospect' strain introduced into Prospect Reservoir, near Sydney, New South Wales; the 'Singapore' or 'Yanco' strain, introduced into canals in the Murrumbidgee irrigation area of New South Wales; and a 'Boolara' strain, introduced to a fish farm at Boolara in Victoria.

Shearer and Mulley (1978) have traced the origins of these three strains of Common Carp in Australia. The Boolara strain appears to be the most recently introduced, and has been responsible for the increase in numbers and expansion of range that has taken place since about 1964. The 'Yanco' and 'Prospect' strains are believed to be those referred to as 'Singapore' and 'Prussian' carp respectively by Whitley (1951). It is believed that the Boolara and Prospect populations are of European origin while the Yanco one came from Asia.

Stead (1929) said that the 'Prospect' strain originated in nine small Scaled Carp which in 1907 he purchased from a dealer in Sydney, by whom they had been imported from Europe, which he released in one of the Bloxsome Ponds adjacent to Prospect reservoir.

In 1908, Stead obtained from the same source half-a-dozen small Mirror Carp which, after a period of acclimatization in a trout hatchery, were also placed in Prospect reservoir. Here the Mirror Carp thrived and increased in numbers, and many were translocated to other waters in New South Wales.

Until the late 1950s, Common Carp in Australia seem to have been ecologically insignificant. At that time, however, an immigrant German fish farmer of Boolara in Gippsland, Victoria, applied to the state government for permission to import 'Prussian' Carp from Germany for aquacultural purposes. Clements (1988) believed this to be the first attempt by a private individual to culture Common Carp in Australia. The request was refused, but the German, undeterred, went ahead with his plans, and in August 1960 almost certainly illegally imported Mirror Carp from Germany, though he claimed they were fish legally translocated from Prospect reservoir (Clements, 1988). By 1961, Common Carp from the Boolara Fish Farm were being sold to other aquaculturists in Victoria.

Reports from overseas revealed how environmentally destructive large concentrations of Common Carp could be. In 1962, the government of Victoria officially declared the species a noxious pest, effectively banning its culture in the state. An eradication programme of Common Carp in Victoria appeared to be successful, until in 1963–1964 it became apparent that Common Carp had been illegally released in many open waters, in at least two of which, the Yallourn Storage Dam in Gippsland and Lake Hawthorn in northwestern Victoria, which overflows

into the Murray River, they had survived. From these two locations, Common Carp spread rapidly throughout the Murray/Darling system of Victoria and New South Wales and into southern Queensland, and through most of the waters of Gippsland; in many of these areas Carp and Goldfish *Carassius auratus* (q.v. have become the dominant fish species (Brumley 1991), and hybrids between them are common (McDowall, 1980).

Evidence for the presence of Common Carp in Western Australia and prior to 1995 in Tasmania i contradictory. In 1971, 120 Common Carp of the ornamental Koi variety were planted in a lake at the Narrows Interchange in Perth, where McKay (1977) quoted by Clements (1988), said they were all believed to have been taken by cormorant *Phalacrocorax* sp. McKay (1984) and Twyford (1991) however, indicate their survival in Western Australia Common Carp of the Boolara strain were reported from several farm dams in Tasmania between 1967 and 1975 (Clements, 1988), when according to Shearer and Mulley (1978) they were eradicated.

In January 1995 a single Common Carp, apparently of the 'Boolara' strain, was discovered in Lake Crescent in the Central Highlands of Tasmania Within the next 3 months a further 300 were found in this lake and four others in the nearby Lake Sorell The size and age of the fish suggested they had been in Lake Cresent for at least 4 years, probably as a result of the release of surplus bait fish. An eradication programme is being undertaken by the Tasmanian Inland Fisheries Commission (Anderson 1995).

The rate at which fish of the 'Boolara' strain extended their range in Victoria between about 1964 and 1984 was rapid, and contrasted with the sedentary nature of those of the 'Prospect' and 'Singapore strains. A commercial fishery for Common Carp became established in Victoria (McKay, 1984). (See also Lake, 1959; Butcher, 1962; Weatherley & Lake 1967; McDowall, 1980; Cadwallader & Backhouse 1983; Merrick & Schmida, 1984; Arthington, 1986 McKay, 1986–1987, 1989; Allen, 1989; Burchmore & Battaglene, 1990; Morison & Hume, 1990.)

Ecological Impact   Although McKay (1984) said that the Common Carp 'is now a threat to Australian fresh waters' and Cadwallader (1978a) claimed that although in rivers and deep lakes Common Carp may have little effect, in shallow slow-flowing waters they an increase water turbidity and disrupt plants, Arthington (1991) could find little evidence to substantiate these claims. Fletcher *et al.* (1985) concluded that Common Carp have not been responsible for increased turbidness in the Lower Goulburn River basin in Victoria, where such increase was attributed largely to hydrologic alterations. The impact, if any, by Common Carp on the turbidity of Australian waters is obscured by the natural turbidity of many of the country's inland waters (Kirk, 1977), and the interaction of hydrological factors with the type of soil and land degradation (Arthington, 1991), although in very high concentrations in shallow, drying waters, Common Carp do on occasion increase turbidity (Fletcher *et al.*, 1985). By disturbing sediments, Common Carp have been accused of increasing nutrients which encourage the growth of toxic cyanobacteria (blue-green algae) in rivers. Were they to become widely established in Tasmania, it is believed that they might adversely affect the island's A$30 million a year trout-fishing industry (Anderson, 1995). (See also Smith & Pribble, 1978; Wharton, 1979; Arthington, 1989; McDowall, 1990b.)

New Zealand

Although Thomson (1922) records that stocks of *Cyprinus carpio* were introduced to New Zealand on several occasions between 1864 and 1881, doubt has been cast on the correct identity of these fish. There are records of this species having been abundant in the Waikato, released in Lakes Mahinapua and Taupo, and being common in Lake Rotorua, but McDowall (1979) was of the opinion that in most of these waters the fish were actually Goldfish *Carassius auratus* (q.v.).

In more recent times, Common Carp have been periodically introduced to New Zealand, sometimes accidentally or deliberately among shipments of Goldfish, but whenever they have been discovered they have been regarded as a potential threat to New Zealand's ecosystem and destroyed. 'There is little if any evidence' wrote McDowall (1990b), 'to support a long-term presence [in New Zealand] prior to the 1960s'.

Koi Carp, an ornamental cultivar of the Common Carp, first appeared in New Zealand in the late 1960s, when they were widely cultured. Some escaped or were deliberately released into the wild,

where they were first reported to be established and breeding in the Waikato River south of Auckland in 1983 (McDowall, 1990b). McDowall's map indicates populations also in tributaries of the Wairoa River north of Auckland and in Taranaki. Although 'designated a noxious species' (McDowall, 1984), they are less likely than wild strains of the Common Carp would be to expand their range (McDowall, 1979).

Papua New Guinea

Between 1959 and 1972 three varieties of the Common Carp – the 'Golden', 'Cantonese' and Mirror (*communis*) – were introduced on numerous occasions to Papua New Guinea, many from Sydney, Australia, and from Singapore. West & Glucksman (1976: table 8) give full details of these introductions.)

Of the three varieties introduced, only the 'Golden' and 'Cantonese' survived after 1969; they found conditions in upland areas to their liking, and bred freely in ponds at Dobel and later at the Highlands Agricultural Experimental Station at Aiyura, from which fingerlings were distributed in 1961 and 1963 respectively. Thereafter, artificial ponds, rivers and lakes were extensively planted, and by the mid-1970s 'nearly every body of water in the highlands even marginally suitable for carp has been stocked' (West & Glucksman, 1976). The wisdom of some of these plantings, especially that in Lake Kutubu in 1965, has subsequently been questioned. Berra *et al.* (1975) recorded the presence of Common Carp in the Laloki, Brown and Goldie Rivers, while Glucksman *et al.* (1976) reported them to be 'widespread' in the following Districts: West Sepik, East Sepik, Southern Highlands, Morobe, Madang, Western, Northern, Central, Bougainville and East New Britain. Allen (1991) said that Common Carp are 'common in a few isolated localities such as the upper Baliem River (Irian Jaya), Lake Kopiago, and the Lower and Middle Sepik and Ramu river systems'. In the Sepik drainage, Common Carp are spreading both upstream and downstream along the lower flood plains (Ulaiwi, 1990), where according to Eldredge (1994), 'significant fisheries have developed.'

Ecological Impact According to Allen (1991), Common Carp in Papua New Guinea have 'an adverse influence on the environment because of their prolific breeding habits and their sucking-mode of feeding which greatly disturbs the substratum causing turbidity. They also compete for available food resources in the form of crustaceans, insects, molluscs, worms, and algae'.

Oceania

Azores

Goubier *et al.* (1983) record the release of 'Prussian' Carp from Germany in Lac Sete Cidades on Île de Saõ Miguel in 1890, and between 1895 and 1913 transfers of Common Carp from Sete Cidades to Lac Furnas; the current status of Common Carp on Saõ Miguel, if indeed they survive, is unknown.

Fiji

Welcomme (1988) says that Common Carp from New Zealand (but see under New Zealand) were introduced for ornamental purposes to Fiji in 1936, where it was 'doubtful' if they were reproducing. According to Anon. (1971), in 1968–1969 Common Carp were imported from Malaysia to Viti Levu for aquacultural purposes and as a pituitary donor. The present status of Common Carp on Viti Levu is unknown.

Guam

Best and Davidson (1981) and Maciolek (1984) record the naturalization of the Common Carp on the island of Guam, but say that no data are available.

Hawaiian Islands

According to Maciolek (1984), no records exist about the introduction of Common Carp (probably by Asian immigrants) to the Hawaiian Islands, where they were established before 1901 and where they are now believed to occur in reservoirs on all the main islands. (See also Brock, 1952, 1960; Hida & Thomson, 1962; Kanayama, 1968; Lachner *et al.*, 1970; Maciolek & Timbol, 1980.)

Mauritius

Moreau *et al.* (1988) say that Common Carp introduced to the island of Mauritius from India in 1976 for aquacultural purposes are reproducing successfully in Ferme Reservoir.

Unsuccessful Introductions   In Oceania, Maciolek (1984) records unsuccessful attempts to introduce the Common Carp to New Caledonia and to Tahiti in French Polynesia.

Ecological Impact Worldwide

De Moor and Bruton (1988) have summarized the general impact that naturalized Common Carp have had in the various countries to which they have been introduced and the overall significance of increased sediment loading of freshwater ecosystems.

Although in many countries throughout their introduced range Common Carp are regarded as a valuable additional source of protein and as a sporting asset, they have had a destructive environmental impact – especially on aquatic habitats and other (usually indigenous) species of fish. In the case of the latter this is due to interspecific resource exploitation; competition for food; predation on spawn; and because of habitat alterations. Their habit of roiling has resulted in the mass destruction of aquatic vegetation and the disturbance of bottom sediments, which increases turbidity. The feeding activities of Common Carp also increase suspensoid levels indirectly through their ingestion of microorganisms from the phosphate-rich substrate, and the subsequent excretion of the phosphate in a soluble form that is readily available to algae. The consequence is a phytoplanktonic bloom that further increases water turbidness and decreases light penetration, adversely affecting submerged aquatic vegetation which then dies and decays, releasing suspensoids into the water.

Heavy suspensoid loading has many deleterious effects on aquatic environments; the denial of light decreases photosynthesis and results in reduced primary productivity, reduced visibility of pelagic food, reduced availability of benthic food because of smothering, and the clogging of fishes' gillrakers and gill filaments. This overall decrease in the availability of food results in slower growth rates, reduced fecundity, and thus impaired population recruitment. Eventually, this causes a decrease in the diversity of habitat niches and in the productivity and populations of fish communities (Bruton, 1985). Increases in sediment loading can also have an impact on benthic invertebrates – detrimental to some but favourable to others (Wiederholm, 1984). Although the presence of Common Carp in a river will not lead to an increase in the overall loading of suspensoids, the disturbance of bottom sediments probably worsens the adverse conditions resulting from increased suspensoid loading as a consequence of excessive erosion. (See also e.g. Contreras & Escalante, 1984; Crossman, 1984; McKay, 1984; Taylor *et al.*, 1984; Welcomme, 1984; Lever, 1994.)

Allied Species

In the late 19th century Gudgeon *Gobio gobio* from France were introduced for aquacultural purposes to Spain, where they are currently established in the wild in the following river drainages: North, Douro, Tagus, Guadiana, Guadalquivir, South, Levant, Ebro and Eastern Pyrenees (Elvira, 1995 a,b).

Giant Danio
Danio malabaricus

The Giant Danio is a popular tropical aquarium fish which has been widely transported around the world.

Natural Distribution

West coast waters of India, and Sri Lanka.

Naturalized Distribution

South America: Colombia.

South America

Colombia

In recording the establishment in the wild of the Giant Danio in Colombia, probably following release from an aquarium, Welcomme (1988) suggested that this was unlikely to be a unique case, and that further records from other tropical countries could be expected.

Spotted Steed
Hemibarbus maculatus

Korean Sharpbelly
Hemiculter eigenmanni

The Spotted Steed is a medium-sized (30 cm) fish with a preference for lakes and slowly flowing rivers. Although it is basically a temperate species, Welcomme (1988) points out that its wide latitudinal distribution suggests a considerable degree of thermal tolerance. The Korean Sharpbelly is a small (15 cm) fish which shares similar habitat requirements with the Spotted Steed.

Natural Distribution

China, Korea, Japan, and the basin of the Amur River.

Naturalized Distribution

Asia: the former USSR.

Asia

USSR

Borisova (1972) says that *Hemibarbus maculatus* and *Hemiculter eigenmanni* were accidentally introduced in 1961 from the Yangtze River in China with shipments of Chinese Grass Carp *Ctenopharyngodon idella* (q.v.) and Silver Carp *Hypophthalmichthys molitrix* (q.v.) to ponds of the Akkurgan Fish Combine, near Tashkent in Uzbekistan. In about 1966 they escaped from the ponds, and became established in some neighbouring waters. Later in the same year they spread along the feeder canals into the Syrdar'ya, Chirchik and Akhangaran Rivers, and in the following year into the Tuyabuguz River reservoir. They are now abundant (together with similarly introduced Bitterling *Rhodeus sericeus* (q.v.)) in pools of the Syrdar'ya and in the mouth of the Chirchik and Akhangaran. Because they grow faster and breed more rapidly than in the Far East, *H. maculatus* and *H. eigenmanni* in Uzbekistan have tended to displace local species and have formed new biocoenoses.

Allied Species

The closely related Common Sharpbelly *H. leucisculus*, which is rheophilous, was established in small numbers in the same waters until 1966 when, apparently unable to adapt to living in stillwaters, it disappeared (Aliyev *et al.*, 1963; Borisova, 1972).

Silver Carp
Hypophthalmichthys molitrix

The Silver Carp is, like the Grass Carp *Ctenopharyngodon idella* (q.v.), one of the group of Chinese carp which has been widely distributed for aquacultural purposes. Being principally planktonophagous, it is extensively used in polyculture, and also in monoculture for the biological control of algae. Like other Chinese carp, Silver Carp do not normally breed outside their natural range, and, except in the Danube basin and the USA, stocks are only maintained by frequent artificial replenishment.

Natural Distribution

China and eastern Siberia.

Naturalized Distribution

Europe: countries bordering the Danube River, including Czechoslovakia, Hungary, Romania, the former USSR, and the former Yugoslavia. **Asia:** India. **Africa:** Ethiopia. **North America:** USA. **Oceania:** Mauritius.

Europe

Countries bordering the Danube River (Czechoslovakia, Hungary, Romania, the former USSR and the former Yugoslavia)

Silver Carp from China and the USSR were introduced for aquacultural purposes to *Hungary* and *Romania* in 1963 and 1964 respectively; they first spawned in the wild in the Tisza River in 1973.

In the former *Yugoslavia* and in *Czechoslovakia*, Silver Carp have been imported on several occasions since 1963 from Hungary, Romania and the former USSR, and are known to be breeding in the Danube and other low-lying waters (Welcomme, 1981, 1988).

Borisova (1972) lists the Silver Carp as among those species introduced to Uzbekistan in the former *USSR*, where it is established in inlet and outlet ditches and other natural waters adjacent to the Akkurgan Fish Combine in Tashkent.

Asia

India

Most artificial impoundments in India have flourishing growths of phytoplankton (in particular the blue-green alga *Microcystis aeruginosa*) that are not grazed by native fishes. In an attempt to control these blooms, Silver Carp from Japan were experimentally introduced in 1959 to the Central Inland Fisheries substation at Cuttack, from where they were later deliberately transplanted to two reservoirs, Kulgarhu in 1969 and Getalsud in 1974; in 1971–1972 they also

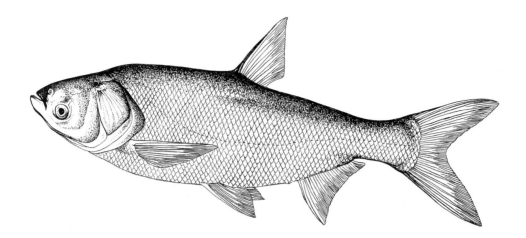

gained accidental access to a third, at Govindsagar, as a result of escaping during a flash flood from the nearby Deoli fish farm. In all three waters the fish increased in size impressively, and by 1974 they were known to be reproducing in Govindsagar (Banerji & Satish, 1989; Chandrasekharaiah, 1989; Jhingran, 1989; Natarajan & Ramachandra Menon, 1989; Sehgal, 1989; Shetty *et al.*, 1989; Singh & Kumar, 1989; Sreenivasan, 1991; Yadav, 1993).

Ecological Impact   'The silver carp has become one of the major controversial species in Indian aquaculture' (Shetty *et al.*, 1989). 'Govindsagar [10 500 ha] is the classical example of silver carp making inroads into the trophic niches of Indian fishes' (Jhingran, 1989).

Silver Carp possess a specialized structure of gill-rakers well adapted to grazing on plankton, on which they feed during their ontogeny, favouring especially phytoplankton, of which they consume great quantities. Silver Carp, which in the Far East mature at between 5 and 6 years, have undergone extreme physiological changes in the lower and milder latitudes of India, where they reach breeding condition at the age of only 1 year. This physiological adaptation is ecologically important; Silver Carp consume three times as much food as does the native Catla *Catla catla*, (q.v.) with little improvement in food conversion efficiency. Catla, which are renowned for their high rate of growth and heavy weight, share with Silver Carp much the same ecological niche. The breeding and feeding vigour of Silver Carp in India could lead to its proliferation at the expense of the resident Gangetic major carps the Rohu *Labeo*

rohita (q.v.), the Mrigal *Cirrhinus mrigala* and, in particular, the Catla.

Since its establishment in Govindsagar, the percentage of Silver Carp in the annual catch has been steadily increasing, with a corresponding fall in the yield of Catla. The catch of Silver Carp rose from 0.06 t (0.01%) in 1977–1978 to 334 t (61.53%) in 1987–1988. The corresponding figures for Catla were 209.87 t (28.57%) and 36.71 t (6.76%). Natarajan (1989) indicates a similar situation in Kulgarhi reservoir.

The possibility of Silver Carp becoming established in other impoundments, or in such rivers as the Ganga and Brahmaputra, could be disastrous for the Catla which, being monotypic, is of special concern to conservationists (Jhingran, 1989).

Africa

Ethiopia

In 1975, Silver Carp from Japan were introduced to Ethiopia, where they became established in the Pincha barrage (Moreau *et al.*, 1988). (See also Tedla & Meskel, 1981.)

North America

United States

According to Courtenay *et al.* (1991), the Silver Carp 'is so widespread in the United States that its establishment seems assured; only confirmation of a feral breeding population is needed before this species is added to the list of established exotics'. Welcomme (1981, 1988) says the Silver Carp 'has escaped into open waters in Arkansas, where it may compete with

Ictiobus bubalus [the Smallmouth Buffalo] if it reproduces'.

Oceania

Mauritius

In 1976, Silver Carp from India were introduced to the island of Mauritius, where they have become established in some reservoirs (Moreau *et al.*, 1988).

Unsuccessful Introductions

Silver Carp are being reared successfully in aquacultural centres but have failed to establish wild populations as follows:

Europe. Belgium, Cyprus, France, Germany, Greece, Poland, (Welcomme, 1981, 1988); Denmark (Jensen, 1987); The Netherlands (Vooren, 1972; De Groot, 1985).

Asia. Bangladesh, Korea, Vietnam (Welcomme, 1981, 1988); Indonesia (Muhammad Eidman, 1989); Israel (Ben-Tuvia, 1981; Gophen *et al.*, 1983; Davidoff & Chervinski, 1984); Japan (Chiba *et al.*, 1989); Malaysia (Ang *et al.*, 1989; Chou & Lam, 1989); Philippines (Juliano *et al.*, 1989); Sri Lanka (De Silva, 1987); Taiwan (Liao & Liu, 1989); Thailand (De Iongh & Van Zon, 1993; Piyakarnchana, 1989).

Australasia. Papua New Guinea (West & Glucksman, 1976).

Rohu
Labeo rohita

One of the Indian 'major carps', the Rohu has been widely translocated outside its natural range in the subcontinent to stock reservoirs and for aquacultural purposes, where its bottom-feeding habits make it well suited for rearing in artificial ponds and for polyculture.

Natural Distribution

Ganges River basin, India.

Naturalized Distribution

Asia: Philippines. *Oceania:* ? Mauritius.

Asia

Philippines

Evidence on the status of the Rohu in the Philippines, to which it was introduced from India in 1964, is contradictory. Welcomme (1981) says that there are 'some self-breeding populations in rivers', but that although 'better adapted to lakes than local species it does not breed'. Juliano *et al.* (1989), however, says that Rohu have 'become a part of natural populations of lakes and reservoirs and contribute to natural fish production'.

Unsuccessful Introductions   In Asia, the Rohu is breeding in captivity but not in the wild in Indonesia (Muhammad Eidman, 1989); Japan, Malaysia and the former USSR (Jhingran & Pullin, 1985); Sri Lanka (De Silva, 1987); Thailand (Piyakarnchana, 1989; De Iongh & Van Zon, 1993); and China (Yo-Jun & He-Yi, 1989).

Africa

Unsuccessful Introductions

In Africa, the Rohu is successfully breeding in aquaculture centres but not in the wild in Madagascar (Reinthal & Stiassny, 1991) and Zimbabwe (Toots, 1970).

Oceania

Mauritius

Welcomme (1988) says that the Rohu is 'widespread throughout the island' of Mauritius, to which it was introduced from India between 1961 and 1975 and where it is naturalized in artificial impoundments (Moreau *et al.*, 1988).

Allied Species

Moreau *et al.* (1988) list the introduction in 1972 of *L. altivelis* to Zimbabwe, where it became established in dams and reservoirs. The same authors also list the introduction to Zimbabwe in 1976 of *Neobola brevianalis*, which became established in cold man-made waters.

Sunbleak
Leucaspius delineatus

The Sunbleak is a tiny fish, seldom reaching a length of more than 8 cm. It is one of the few cyprinids known to show parental care. It favours still or slowly flowing waters.

Natural Distribution

Widely distributed in Europe from the Caspian Sea westward to the rivers of Belgium and The Netherlands.

Sunbleak may have entered parts of northern France by natural dispersal through the canal system of southern Belgium.

Naturalized Distribution

Europe: British Isles (England).

Europe

British Isles (England)

The following account of the introduction to, and status of, the Sunbleak in England is derived from Farr-Cox *et al.* (in press).

Since 1990, the Sunbleak (colloquially known as the 'Motherless Minnow' from its German name 'Moderlieschen') has been caught in a number of localities in southern England, following introductions as an ornamental species and perhaps through accidental importations in shipments of other European fishes. Since then it has dispersed from the sites of introduction both naturally and anthropogenically.

Farr-Cox *et al.* (in press) give the following details of the species' occurrence in the wild in England to mid-1995:

DATE	COUNTY	LOCALITY
1990	Somerset	King's Sedgemoor Drain
1994	Somerset	Bridgwater and Taunton Canal; River Parrett
1995	Somerset	Wych Lodge Lake (Blackdown Hills east of Pitminster); Combe Lake (a recently excavated angling water near Langport, filled by floodwater from the River Parrett in the winter of 1994–1995); River Brue; Whites River (a feeder of the Brue)
c. 1987 and 1994–1995	Hampshire	Broadlands Lake (near Romsey)
1995	Hampshire	Two Lakes Fishery (near Romsey); Stoneham Lakes (near Eastleigh)

In both Somerset and Hampshire the Sunbleak has expanded its range fairly rapidly and widely, and was said by Farr-Cox *et al.* (in press) to be 'present in enormous numbers'. The species' success can be attributed to a number of adaptive advantages, among which are the care it takes of its young, its sexual precociousness, and because its diminutive size renders it relatively immune from predation. The Sunbleak's small size has probably also assisted its dispersal because it is difficult to prevent such tiny fish from escaping through outflow screens from fishery ponds; because they are difficult to sort from other cyprinid juveniles when fry are being netted for transfer elsewhere; and because they are easily swept away when rivers are in spate, and quickly become established in floodwaters.

The Sunbleak's naturalization in Hampshire is

believed to have resulted from two of these factors, which may well have also played a part in the establishment of the Somerset populations. In the former county the original importation took place in 1986–1987, when the then proprietor of Crampmoor Fishery is believed to have brought in some Sunbleak from Germany in a consignment that included Pumpkinseeds *Lepomis gibbosus* (q.v.). At the time Two Lakes Fishery was in the same ownership, and there was an interchange of fish between the two fisheries. Both waters drain into streams that join and enter the River Test below Romsey, and Broadlands Lake – through which the Test flows – is situated some 2 km downriver of the confluence. Although the fast-flowing waters of the Test are not ideal for Sunbleak, the river has clearly acted as a means of the species' dispersal into the lake, where it was first recorded – and misidentified as the Bleak *Alburnus alburnus* – around 1987.

Until 1990, fry netted in Broadlands were sold (as Roach *Rutilus rutilus* and Bream *Abramis brama*) for stocking other waters, fry of different species, if noticed, being thought to be Bleak. The lakes at Stoneham, which support large populations of Sunbleak, was one fishery thus stocked, and it is safe to assume that any stillwater fishery supplied with fry from Broadlands Lake between around 1986 and 1990 also holds Sunbleak.

The species' dispersal in Somerset is more difficult to document. The Somerset levels are low-lying, with interconnecting artificial drains and semi-natural rivers. The King's Sedgemoor Drain receives water from the River Parrett in the latter's lower freshwater reaches, and it seems possible that Sunbleak have spread between these two systems during periods of high water. The Bridgwater and Taunton Canal is fed by the River Tone, which flows into the tidal stretch of the River Parrett. In winter, and at other periods of high water, these tidal reaches are less saline than normal and at this time Sunbleak could perhaps have travelled between the Canal and the Parrett by this route. The River Brue flows near the Parrett, but there are no interconnecting waterways; the estuaries of both rivers converge at Burnham-on-Sea, where of course the water is permanently saline.

Farr-Cox *et al.* (in press) believed that Sunbleak might also have been spread accidentally by man. In Somerset several angling clubs control fishing rights in two separate waters holding Sunbleak, and anthropogenic transportation seems the most likely means of accounting for the species' occurrence where natural physical connections are absent.

Ecological Impact   The ecological impact, if any, of the Sunbleak in England has yet to be determined. A preliminary study in Somerset revealed that water fleas (Cladocera) and small aquatic insects formed the species' staple diet; although no evidence of fish remains was found, the widespread occurrence of the alien in enormous numbers seems likely to result in interspecific competition with some native cyprinids. The Sunbleak's small size renders its further spread almost impossible to control, while its degree of parental care gives it an in-built advantage over indigenous cyprinids (Farr-Cox *et al.*, in press). Although in the Bridgwater and Taunton Canal native Roach, Rudd *Scardinius erythrophthalmus*, and Bream are said by local anglers to be already declining, the National Rivers Authority, which is monitoring the Sunbleak's spread, points out that in continental Europe it coexists satisfactorily with native species.

Orfe (Ide)
Leuciscus idus

The Golden Orfe, an ornamental variety, is widely cultivated, especially in western Europe, by aquaculturists, but has rarely become established in the wild.

Natural Distribution

Eastern and central Europe east of the River Rhine, from eastern Scandinavia eastward to the basins of the Arctic Ocean, White Sea and the River Volga, and rivers of the northern coast of the Black Sea in the former USSR.

Naturalized Distribution

Europe: British Isles. **North America:** United States.

Europe
British Isles

The earliest known introduction of Orfes to England appears to be one reported to Frank Buckland by Lord Arthur Russell (son of the Duke of Bedford) in March 1874:

> Ever since I first saw these splendid fish in the ponds of the Imperial Palace, Laxenburg, near Vienna I determined to introduce them if possible into England. . . . My first attempt a year ago was unsuccessful, two gold orfes only survived the batch my brother [Ambassador at Berlin] had obtained in Berlin. My second and successful attempt has been accomplished with the assistance of Mr Kirsch, Director of the Association for Pisciculture at Wiesbaden; he sent me one hundred and fourteen golden orfes of last year's breed, about two inches [5 cm] long each, and two large specimens. . . . they were all deposited, without a single loss, in one of the Duke of Bedford's ponds at Woburn Abbey.

Numerous other attempts have been made over the years to naturalize Orfes in the British Isles, but few have been successful. At one time or another Orfes have been established in the following British waters (English except where stated) (Lever, 1977):

COUNTY	LOCALITY
Avon	Winford Brook (Bristol)
Bedfordshire	Woburn Abbey
Berkshire/ Buckinghamshire	River Thames (Henley)
Cambridgeshire	Oakington
Cheshire	Lymon Vale
Cornwall	Hayle Kimbro and Stithians
Denbighshire (Wales)	Bodnant Gardens
Devon	Lundy Island (Bristol Channel)
Dorset	Moigne Combe Pond (Moreton) and Owermoigne (Dorchester)
Dumfriesshire (Scotland)	?
Essex	Wivenhoe (Colchester) and Loughton
Gloucestershire	Tortworth Lake (Falfield)
Hampshire	Cemetery Lake (Southampton Common); River Kennet (Cookham); River Test
Huntingdonshire	Hartford
Lincolnshire	River Ouse (Ely) and Bullwants Pond, Mablethorpe
Middlesex	River Ember (Hampton Court)
Perthshire (Scotland)	Killiecrankie
Surrey	Shottermill (Haslemere); Westcott (Dorking); River Thames (Chertsey)
Worcestershire	Spetchley Lake

According to the Duke of Bedford, the original stock of Orfes at Woburn died out as a result of competition with other species, and this may well explain the species' failure to become naturalized in many other British waters.

Unsuccessful Introductions In the late 19th/early 20th century Golden Orfe, imported from Germany and France, were extensively cultivated as ornamental fish in The Netherlands where, however, they have never become naturalized (Vooren, 1972; De Groot, 1985).

North America
United States

The Orfe was first imported by the United States Fish Commission in 1877 (Baird, 1879). Evidence of the species' present status in the USA is fragmentary and contradictory. Lachner *et al.* (1970), quoting Whitworth *et al.* (1968), say 'it has definitely established itself [in the Golden form] in a Connecticut pond [in East Lyme, New London County] since 1962 or 1963'. Courtenay *et al.* (1984, 1986), however, say that this population, plus others previously established in York and Delaware Counties, Pennsylvania; on Long Island, New York; in the Chanango River, New York; in the Potomac River, Virginia; and in Penobscot County, Maine, have all been eradicated. The same authors say that Orfe may be established in some private waters in Pennsylvania, and that 'Although we presently are unaware of any established population of the ide in the United States, the likelihood of finding such a population is great because of its past history and because sources in New Jersey and Florida are distributing the fish. Moreover, the species is often misidentified as goldfish' (q.v.). Courtenay and Stauffer (1990), Courtenay *et al.* (1991) and Courtenay (1993) say the Orfe is established in Maine, and that it has been collected but is not established in Connecticut, New York, Pennsylvania and Tennessee.

Dace
Leuciscus leuciscus

Although normally rheophilic, Dace also occur in lakes. They are a popular anglers' bait-fish.

Natural Distribution

Northern Europe and Asia, from England eastward to Siberia and from Sweden south to southern France. Absent from the Iberian peninsula, Italy and Greece.

Naturalized Distribution

Europe: British Isles (Ireland).

Europe

British Isles (Ireland)

Went (1950) has traced the origin of the Dace (and Roach *Rutilus rutilus* (q.v.)) in Ireland.

In about 1889, Mr J. C. Truss travelled from England to Ireland to fish for Pike *Esox lucius* (q.v.) in the River Blackwater in County Cork, bringing with him as bait some Dace and Roach. The cans in which these fish were kept were swept away in a flood, and some 2 years later both species were found to have become established in the river. The Dace have done better than the Roach, and are considered to be a pest when they intrude into Atlantic Salmon *Salmo salar* and Brown Trout *S. trutta* fisheries. According to Vooren (1972), the Dace 'has apparently not spread much into other Irish waters'.

Opsariichthys uncirostris

A small rheophilic fish which favours the lower reaches of sluggish rivers.

Natural Distribution

China, Japan, Korea, and the basin of the Amur River.

Naturalized Distribution

Asia: the former USSR.

Asia

USSR

This species, known locally as the 'Troyegub', is one of those which have become established in ponds of the Akkurgan Fish Combine near Tashkent in Uzbekistan, to which it was accidentally introduced, together with some other small cyprinids such as *Hemiculter eigenmanni* (q.v.), in consignments of Chinese Grass Carp *Ctenopharyngodon idella* (q.v.) and Silver Carp *Hypophthalmichthys molitrix* (q.v.) from the Yangtze River in China in 1961. According to Borisova (1972), the subspecies involved 'is most similar to *O. uncirostris amurensis*'. In about 1966 it escaped from the ponds, and became established in some neighbouring waters. Later in the same year it travelled along the interconnecting feeder canals into the Syrdar'ya, Chirchik and Akhangaran Rivers, and in the following year into the Tuyabuguz River reservoir. It is now found in large numbers in pools on the Syrdar'ya and in the mouth of the Chirchik and Akhangaran (Borisova, 1972).

Ecological Impact   Because it grows more rapidly and is more fecund than in its native range, *O. uncirostris* in the USSR has partially displaced, through predation and competition for food, some local species, such as gobies (Gobiidae), and formed new biocoenoses (Aliyev *et al.*, 1963).

Fathead Minnow
Pimephales promelas

The Fathead Minnow is a popular bait-fish in the USA, where it has been widely transplanted through the release or escape of surplus bait.

Natural Distribution

From the Great Slave Lake in the North West Territories of Canada and Alberta eastward to Nebraska in the USA, south to Chihuahua in northern Mexico.

Naturalized Distribution

Europe: Belgium; France; Germany.

Europe

Belgium; France; Germany

Introduction and Ecological Impact   The introduction of the Fathead Minnow as a bait-fish to Europe in 1983–1984 and its subsequent release into the wild by Belgian, French and German anglers, resulted in the introduction at the same time of the pathogen *Yersinia ruckeri*, which is the causal agent of enteric redmouth disease (Michel, 1986); this is now infecting wild and cultured trout and eels and is spreading throughout northern Europe (Welcomme, 1988).

Unsuccessful Introduction

In 1957, 150 Fathead Minnows from Florida were introduced to four reservoirs and several farm ponds in Puerto Rico as a potential forage for Largemouth Bass *Micropterus salmoides* (q.v.), by whom they were quickly eliminated (Erdman, 1984; Wetherbee, 1989).

Stone Moroko
Pseudorasbora parva

The Stone Moroko is a small (10 cm) fish accidentally introduced with shipments of Chinese carps and Grass Carp from the Far East.

Natural Distribution

China, Japan and Korea and the basin of the Amur River.

Naturalized Distribution

Europe: Austria; Czechoslovakia; Germany; Hungary; The Netherlands; Romania; former Yugoslavia. ***Asia:*** former USSR. ***Oceania:*** Fiji.

Europe

Austria; Czechoslovakia; Germany; Hungary; The Netherlands; Romania; former Yugoslavia

The date of introduction and the exact source of Stone Morokos now established, by diffusion, in waters throughout the Danube River system and elsewhere in Europe are unrecorded. The species' high rate of reproduction, and consequent dense populations, make it a potential competitor with the fry of native fishes (Wohlgemuth & Sebela, 1987). A.C. Wheeler (pers. comm. 1995) reports its presence in some Austrian lakes, and suspects that it is widely naturalized in eastern Europe. S.J. de Groot (pers. comm. 1995) says that in The Netherlands the first Stone Moroko was caught in the province of South Limburg around 1994–1995; it is believed that the species came from German waters, to which it had gained access from the Romanian Danube.

Asia

USSR

In 1961, Stone Morokos were accidentally introduced, together with Chinese Grass Carp *Ctenopharyngodon idella* (q.v.) and Silver Carp *Hypophthalmichthys molitrix* (q.v.) from the Yangtze River in China to the Akkurgan Fish Combine near Tashkent in Uzbekistan, where the species is known as the 'Chinese chebachok'. In about 1966 it escaped from the ponds and became established in adjacent waters. Later in the same year it penetrated along the interconnecting feeder canals into the Syrdar'ya, Chirchik and Akhangaran Rivers, and by the following year into the Tuyabuguz River reservoir. In 1968, the Stone Moroko was found in the Saryksu River in Fergana Province and in ponds at the Kokand Fish Farm, and 2 years later it ascended to the middle reaches of the Chirchik River, when it was also reported in the Koshkadar'ya River basin and in the Chimkurgan and Pochkamara reservoirs. It now occurs in large numbers in pools of the Syrdar'ya River and in the mouth of the Chirchik and Akhangaran Rivers (Borisova, 1972).

Because it grows and breeds more rapidly than in its native range and is of high ecological valency, the Stone Moroko in Uzbekistan has partially displaced some valuable local native food species and has formed new biocoenoses (Aliyev *et al.*, 1963; Borisova, 1972). It is extensively preyed on, as are the other small exotics in the area, by such species as the Zander or Pike–Perch *Stizostedion lucioperca* and Asp *Aspius aspius*, which have greatly increased in numbers since the establishment of the aliens.

Oceania

Fiji

According to Welcomme (1988), in 1984 the Stone Moroko was accidentally introduced from Japan to Fiji, where it is breeding and established.

Three-lined Rasbora
Rasbora trilineata

The Three-lined Rasbora is one of the many species in this genus to have been widely dispersed around the world by tropical aquarists.

Natural Distribution

Borneo, Sumatra and Thailand.

Naturalized Distribution

South America: Colombia.

South America

Colombia

The Three-lined Rasbora has been introduced, at an apparently unrecorded date and from an unknown source, to Colombia, where it has escaped and has become established in the wild; 'it is surprising that escapees of this and other *Rasbora* species have not been reported from tropical countries other than Colombia' (Welcomme, 1988).

Rosy Bitterling
Rhodeus ocellatus

According to Welcomme (1988), the Rosy Bitterling is so widely established in China, Japan and Korea 'that the species was for a long time considered native'.

Natural Distribution

Taiwan.

Naturalized Distribution

Asia: China; Japan; Korea. *Oceania:* Fiji.

Asia

China; Japan; Korea

Welcomme (1988) says that the Rosy Bitterling was introduced for ornamental purposes from the island of Taiwan to the Chinese mainland, and thence to Japan and Korea, in 'ancient' times. Nakamura (1955), however, who says that the subspecies involved is the nominate one, states that it was accidentally introduced to Japan in a consignment of Chinese Grass Carp *Ctenopharyngodon idella* (q.v.) fry from China, and was first found in the wild in Japan in 1942. Since then it has been expanding its range in Japan, especially during the 1980s, and through competition and hybridization is threatening the survival of the native subspecies *R. o. smithi* (Chiba *et al.*, 1989). It seems probable that the Rosy Bitterling was introduced to Korea at around the same time and in the same manner.

Oceania

Fiji

Welcomme (1988) says that in 1984 the Rosy Bitterling was accidentally introduced to Fiji, where it is reproducing in the wild.

Bitterling
Rhodeus sericeus

The Bitterling is a small (7–8 cm) popular coldwater aquarium and anglers' bait-fish well known for laying its eggs, through a pendulous ovipositor, into the mantle of various species of freshwater mussels (Anodonta).

Natural Distribution

Much of continental Europe eastward from the basins of the Rivers Seine and Loire in France, rivers debouching into the southern Baltic Sea, the Neva basin, Transcaucasia, Asia Minor, rivers flowing into the Aegean Sea, the basins of the Amur, Sungari, Ussuri and Uda Rivers, the Tym and Poronai Rivers on Sakhalin Island, south into northern China.

Naturalized Distribution

Europe: British Isles (England). *Asia:* former USSR. *North America:* USA.

Europe

British Isles (England)

The earliest evidence for the possible presence of Bitterling in England dates from the early 1900s, when Hardy (1954) was told that, under the name of 'Prussian Carp', they were being caught in a pond in St Helens, Lancashire. Wheeler and Maitland (1973), however, cast doubt on the correct identification of these fish, and it may well be that the species did not become established in England until the 1920s.

Before the Second World War, Bitterling were discovered in the disused arm of a canal at Blackbrook, near St Helens, and others are known to have occurred in the neighbouring Leg of Mutton dam. After the war, the canal was found still to contain large numbers of Bitterlings, and following an exhibition of some specimens by the Merseyside Naturalists' Association it was revealed that they were also present in 10–12 other waters in south Lancashire, but that their numbers had declined in Leg of Mutton dam. It is assumed that their original appearance in the wild was the result of aquarists disposing of unwanted stock, and that their spread was due to the release by anglers of surplus bait.

Hardy (1954) recorded Bitterling at the following sites in Lancashire: Collins Green Flashes; Duckery Flashes, Derbyshire Hill; part of the Southport Sluice; the rock hole at Bold; some ponds in the Haydock area; near the Black Horse Inn at Rainhill; and in the Knowsley area. Bitterling also became established locally in Cheshire (where they were identified by A.C. Wheeler as apparently belonging to the European race *R. s. amarus*) and in the Praes Branch of the Shropshire Union Canal.

In 1925, Bitterling were released in a pond in Fife, Scotland, where within a decade they had all disappeared. In 1953, large numbers of Bitterling were transported as bait by anglers to Esthwaite Water, Lancashire, and to Rydal Water and Grasmere in Westmorland (now Cumbria); what happened to them is unknown.

The current status of the Bitterling in England is uncertain, but it seems probable that overall the population may have declined since the 1960s. (For further details see Lever, 1977.) They are common in the River Ouse near Ely, and probably occur elsewhere in the river (A. C. Wheeler, pers. comm. 1995).

Asia

USSR

For details see under *Hemibarbus maculatus*.

North America

United States

Dence (1925) and Myers (1925) were the first to report the discovery of Bitterling in the wild in the USA, in Saw-Mill Creek, Tarrytown, Westchester County, New York. The former author, who refers to them as 'Bitter Carp', says they were first found on 16 September 1923, and that they were said to be 'common and breeding'. The latter author states that Bitterling from Europe were 'often imported to New York City as an aquarium fish', so their appearance in the wild in the USA is likely to have had the same origin as that in England. Bade (1926) gives the year of their discovery as 1924. No Bitterlings have been seen n Saw-Mill Creek since 1951, and it is assumed that none survive there (Schmidt *et al.*, 1981).

Greeley (1937) recorded two small Bitterlings in the Bronx River, east of the Saw-Mill, and Schmidt *et al.* (1981) collected some there in 1979, but considered that their distribution in the warmer months is corre-lated with and is probably dependent on the very local distribution of the mussel *Anodonta cataracta* in which they lay their eggs, whose numbers have sharply declined as a result of industrial pollution (Courtenay & Stauffer, 1990). (See also Lachner *et al.*, 1970; Courtenay *et al.*, 1984, 1986; Courtenay, 1992.)

Roach
Rutilus rutilus

The Roach is an adaptable fish that can survive in poorly oxygenated, slightly polluted, eutrophic or even brackish water. It favours lakes or sluggish rivers with plenty of vegetation. Its abundance, and wide distribution in otherwise sparsely populated waters, make it an important prey species for piscivorous ani-mals. In Europe it is also highly valued by both sport anglers and commercial food fisheries.

Natural Distribution

Europe and western Asia, from England to the cen-tral former USSR and from northern Scandinavia to the Black and Aral Seas.

Naturalized Distribution

Europe: British Isles (Ireland); Cyprus; Spain. *Africa:* Morocco. *Australasia:* Australia.

Europe

British Isles (Ireland)

For details of the introduction of the Roach into Ireland see under *Leuciscus leuciscus.* In Ireland the Roach has spread enormously, and is hybridizing with both the Common Bream *Abramis brama* and the Rudd *Scardinius erythrophthalmus* (q.v.); concern has been expressed that it could affect the abundance of the latter species (A.C. Wheeler, pers. comm. 1995).

Cyprus

Welcomme (1981, 1988) says that in 1972 Roach from England were successfully introduced for angling purposes to Cyprus where, however, for some reason they are said not to be popular.

Spain

Between 1910 and 1913, Roach from France were introduced to Spain for stocking purposes by the Catalonia Regional Fisheries Service (Elvira, 1995b), where they are now naturalized only in the Ebro and East Pyrenees drainages (Elvira, 1995a).

Africa

Morocco

Moreau *et al.* (1988) list the Roach as established in the wild in natural waters of Morocco, where it is spreading in small lakes.

Australasia

Australia

Victoria seems to have been the only Australian state to have imported the Roach directly from abroad. The first shipment arrived in Melbourne from England on the *Lincolnshire* on 10 June 1861. In June 1863, the Acclimatization Society of Victoria (Lever, 1992) released 20 into a reservoir at Middle Gully, and more were later placed in a number of other waters, including the Yarra and Werribee Rivers, where they became established (Clements, 1988). Other importations of Roach from England are believed to have been made until around 1880, and some were unsuccessfully translocated from Victoria to New South Wales and possibly also to Western Australia.

In September 1975, juvenile Roach were first recorded in Burrumbeet Creek near Ballarat, from where they shortly afterwards colonized Lake Burrumbeet (2000 ha), in the headwaters of the

Mount Emu/Hopkins river system, where according to Clements (1988) they now occur in huge numbers. More recently, they have been found in Lake Purrumbete (where they compete with introduced Brown Trout *Salmo trutta* (q.v.) and Rainbow Trout *Oncorhynchus mykiss* (q.v.)) and in Eildon Weir (Clements, 1988).

Roach in Australia are restricted to some rivers and streams in southern Victoria, especially the lower reaches of the Yarra, Werribee and Maribrynong, and to some still waters such as the Melton reservoir; in some of the waters in which they are found Roach reach very high numbers (Weatherley & Lake, 1967; Cadwallader & Backhouse, 1983; McKay, 1984, 1989; Merrick & Schmida, 1984; Arthington, 1986, 1989, 1991; Clements, 1988; Allen, 1989; Brumley, 1991). According to Allen (1989), Roach also occur in certain coastal drainages of southern New South Wales.

Ecological Impact   In Australia Roach are generally regarded as pests, and being mostly bottom-feeders, their roiling activities cause turbidity in some waters. They are also said to compete for food with more economically valuable introduced trout and with some native species.

Unsuccessful Introductions

Unsuccessful attempts have been made to introduce Roach to Madagascar (Reinthal & Stiassny, 1991) and to New Zealand (McDowall, 1984, 1994b; Clements, 1988).

Allied Species

Goubier *et al.* (1983) say that a species introduced to the Azores in 1879 and known there universally as the Bleak *Alburnus alburnus* is actually *Rutilus macrolepidotus* from the Río Mondego in Portugal. It may in fact be a *Leuciscus* species, and is possibly a synonym for another species (A. C. Wheeler, pers. comm. 1995).

Rudd
Scardinius erythrophthalmus

The Rudd is a fish of slowly flowing rivers and canals and weedy still waters. It is not highly regarded by anglers as a sport fish, but it is sometimes used by them as bait.

Natural Distribution

Most of Europe apart from the Iberian peninsula, southern Greece and northern and central Scandinavia, eastward through the former USSR to Siberia and central Asia.

Naturalized Distribution

Europe: Spain. ***Africa:*** Morocco; Tunisia. ***North America:*** ? Canada; USA. ***Australasia:*** New Zealand.

Europe
Spain

Between 1910 and 1913, Rudd from France were introduced to Spain for stocking purposes by the Catalonia Regional Fisheries Service (Elvira, 1995b) where they are now naturalized only in the Ebro and East Pyrenees drainages (Elvira, 1995a).

Africa
Morocco

Welcomme (1981, 1988) records the introduction of the Rudd to Morocco in 1935, where he says it is established and breeding.

Tunisia

In 1965, Rudd were successfully introduced from France to Tunisia, where they became established in the wild (Moreau *et al.*, 1988).

Unsuccessful Introductions

Reinthal and Stiassny (1991) list Rudd as unsuccessfully introduced to the island of Madagascar.

North America

Canada

On 24 October 1990, an adult Rudd was caught in a recently constructed channel leading into Thompson's Bay on the Canadian side of the St Lawrence River, in Leeds County, Ontario (Crossman *et al.*, 1992). The nearest previous records were of fish taken on the USA side of the St Lawrence between Grindstone Island and Jacques Cartier State Park, New York (Klindt, 1990–1991; quoted by Crossman *et al.*, 1992). 'As a result', suggest Crossman *et al.* (1992), 'this species should be watched for in Ontario as far downstream as Lake St Lawrence'.

United States

Myers (1925) recorded the discovery of Rudd 'long ago' in Central Park Lake, New York, from where they were subsequently transferred to several other waters, and that 'a number of years ago' Rudd were released in the lake in Hudson County Park, Jersey City, where they apparently thrived, 'for specimens are often caught by urchins angling for *Fundulus*'. From both of these sites Rudd have since disappeared. Greeley (1937) recorded Rudd in the Roeliff–Jansen Kill in the lower Hudson watershed in Columbia County, New York, and in the early 1950s a number were taken in Cascadilla Creek near Ithaca in Tompkins County, New York; Rudd are believed to have since disappeared from the latter location (Courtenay *et al.*, 1984).

Cahn (1927) and Greene (1935) say that Rudd were introduced to Wisconsin in 1917 by the state's Conservation Department, probably as a food and sport fish, where they became established in Oconomowoc Lake in Waukesha, where they are believed no longer to occur.

Evidence in the literature on the status of the Rudd in the USA is somewhat contradictory. Lachner *et al.* (1970) say it 'is caught by anglers in New York State'. Courtenay *et al.* (1984, 1986) say that Rudd were then naturalized only in Cobbosseeconte Lake near Augusta, in Kennebec County, Maine; in Copake Lake in the Taghkanic Creek drainage; and below the dam at the outlet of Robinson Pond in the Roeliff–Jansen Kill. (Elsewhere, Courtenay *et al.* (1984) cast doubt on the existence of the Maine population.) Welcomme

(1981, 1988) says the Rudd is 'possibly established in New York, New Jersey and Wisconsin'. Courtenay and Stauffer (1990) and Courtenay *et al.* (1991) say it is naturalized in Maine and New York, and has been collected in Arkansas, Illinois, Kansas, Missouri, New Jersey, Oklahoma, Texas, Virginia and Wisconsin. Crossman *et al.* (1992) say that Rudd have been taken by commercial fishermen in the New York waters of Lake Ontario. Courtenay (1993) says that Rudd were first released in the USA in the 1890s, and are presently established in Kansas, Maine, Nebraska and New York.

Klindt (1990–1991; quoted by Burkhead & Williams, 1991) says that translocation of live bait from the southern USA is believed to be responsible for the occurrence of Rudd in the waters of western New York State. Released bait–fish also account for the establishment of Rudd by 1989 in the Great Lakes basin (Mills *et al.*, 1994).

Ecological Impact

Because of its ability to hybridize with the native Golden Shiner *Notemigonus crysoleucas*, Burkhead and Williams (1991) have suggested that Rudd in Canada and the USA may pose a threat to the genetic integrity of the indigenous species.

Australasia

New Zealand

Thomson (1922) records that Rudd were among the fish shipped to New Zealand from England on board the *British Empire* in 1864, but that none survived the voyage. Clements (1988) quotes from the *Auckland Star* of 5 June 1971 that a Mr Smith of Mussey near Auckland 'acquired 10 rudd four years ago'. McDowall (1994b) says that fertilized Rudd ova were 'privately, illegally, and irresponsibly' smuggled into New Zealand by air, and that the resulting fish were subsequently released into many lakes, ponds and farm dams in Auckland and Northland, in some of which they became established. The *Auckland Star* of 27 June 1973 reported that 'rudd are now caught from the Bay of Islands to the Kaimais', so it must be presumed that the fish were released for the benefit of local anglers. Coates and Turner (1977) quote from the *New Zealand Herald* of 25 August 1973 that Rudd were first imported in 1964. According to McDowall (1990b), the ova came from a commercial fish farm in Essex, England, and were selected from golden–coloured stock.

From their release sites Rudd spread to many other New Zealand localities, and are now widely established in still and slowly flowing waters in, and north of, the lower Waikato River valley in North Island, where they also occur in Lake Rotomanu near New Plymouth in Taranaki (McDowall, 1990b). In South Island, Rudd have been reported from near Christchurch (having presumably been illegally translocated from North Island), but McDowall (1994b) says they no longer occur there. (See also Cadwallader, 1977, 1978b; McDowall, 1979; Brumley, 1991.)

Ecological Impact   McDowall (1984) says that the illegal importation of Rudd to New Zealand and the translocation of Tench *Tinca tinca* (q.v.) led to the adoption in 1980 of the Noxious Fish Regulations whereby the possession of certain named fish species including Rudd, was declared illegal, and the confiscation of any fish held and the prosecution of their possessor was authorized.

White Cloud Mountain Minnow
Tanichthys albonubes

The White Cloud Mountain Minnow is a species of temperate waters that is popular with aquarists.

Natural Distribution

River gorges in the White Cloud Mountain range, China.

Naturalized Distribution

Africa: ? Madagascar. ***South and Central America:*** Colombia.

Africa

Madagascar

Kiener (1963), Moreau (1979) and Reinthal and Stiassny (1991) list the White Cloud Mountain Minnow as having been introduced to Madagascar, but give no details of its present status. It was not collected by Reinthal and Stiassny during their survey of Malagasy waters in 1988.

South and Central America

Colombia

Welcomme (1988) says that White Cloud Mountain Minnows, introduced for ornamental purposes to Colombia at an unrecorded date and from an unknown source, have escaped into the wild where they are established and breeding.

Tench
Tinca tinca

The Tench is a species of cool and sluggish waters with an abundance of submerged aquatic vegetation, and is able to flourish in poorly oxygenated conditions. In some parts of its natural range it is fished commercially for food, and attempts have been made to rear it artificially (Welcomme, 1988).

Natural Distribution

Europe, from Britain, southern Sweden and Denmark to countries bordering the Mediterranean, eastward to central Asia.

Naturalized Distribution

Europe: British Isles (Ireland); Finland; Norway; ? Spain. **Asia:** India; Indonesia. **Africa:** Morocco; South Africa; Tunisia; Zimbabwe. **North America:** Canada; USA. **South and Central America:** Chile. **Australasia:** Australia; New Zealand.

Europe

British Isles (Ireland)

For the history of the introduction of the Tench into Ireland see under Common Carp *Cyprinus carpio.* The Tench is less widely distributed in Ireland than the Common Carp, though since 1956 it has expanded its range through transplantations into many small lakes (Went, 1950; Kennedy & Fitzmaurice, 1970; Vooren, 1972; Lever, 1977).

Finland

According to Welcomme (1981, 1988), since 1936 'self-breeding stocks [of Tench] are found in many waters in Finland where the species is considered useful for … stocking into eutropic waters'.

Norway

Welcomme (1981, 1988) records that in 1820 Tench from The Netherlands were introduced, for an unrecorded purpose, into Norway, where they are currently 'found in a few lakes in the southeast of the country'.

Spain

Elvira (1995b) considered that Tench were 'presumably introduced' to Spain in the Middle Ages.

Asia

India

Francis Day (1876) recorded that in 1866 he had unsuccessfully introduced ova, obtained for him by

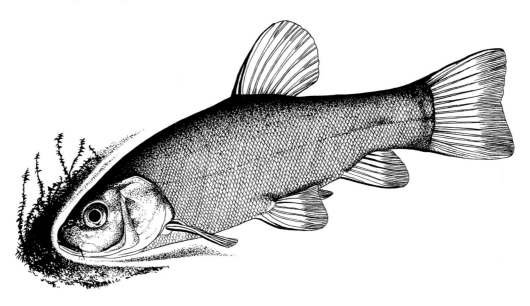

Frank Buckland of the Acclimatization Society of the United Kingdom (Lever, 1992), of both Brown Trout *Salmo trutta* (q.v.) and Tench to Ootacamund, Madras, where the majority died.

A few years later, Mr M'Ivor, superintendent of the government gardens at Ootacamund, was more successful in importing young Brown Trout, Common Carp *Cyprinus carpio* (q.v.) and Tench, some of which arrived safely. Day (1876) records that in 1873 in the 'Neilgherry' (Nilgiri) hills he saw Tench in several waters, which 'were doubtless bred from those Mr M'Ivor took out to the hills'. Fingerlings were also transplanted to ponds and later in the Shevaroy Hills (Molesworth & Bryant, 1921; Jones & Sarojini, 1952). Natarajan and Ramachandra Menon (1989) and Sreenivasan (1989) say that Tench from England were introduced to still and weedy lakes in Tamil Nadu in 1874, where they bred in Ooty and Yercaud Lakes and in the Sandynulla stream. Tench transplanted from Ooty to Kodaikanal Lake eventually died out.

Evidence in the literature of the status of Tench in India is fragmentary and sometimes contradictory. Natarajan and Ramachandra Menon (1989) say that by 1978 Tench no longer occurred in the Nilgiri Hills, due to their inability to compete with introduced Mirror Carp *C. c.* var. *specularis* (q.v.). Welcomme (1981, 1988), who gives the date of introduction to India as 1870 (as does Sehgal, 1989), says that the Tench is 'very limited in its distribution, at present confined to Ooty Lake, Tamil Nadu'. Yadav (1993), however, who says the date of introduction was 1874, states that Tench 'have now fully acclimatised into cold waters of Nilgiris', and that 'attempts are being made to transplant them in plains also'.

Indonesia

Muhammad Eidman (1989) says that Tench from The Netherlands were first introduced to Indonesia by the Inland Fisheries Department in 1927, and that they are now well established and breeding.

Unsuccessful Introduction

In Asia, unsuccessful attempts have been made to naturalize the Tench in Lake Kinneret (Lake Tiberias/Sea of Galilee), in Israel (Reich, 1978; Ben-Tuvia, 1981; Gophen *et al.*, 1983; Davidoff & Chervinski, 1984).

Africa

Morocco; Tunisia

Moreau *et al.* (1988) record the introduction of Tench from France in 1935 to Morocco, where they are now established in the wild and reproducing in several lakes. From Morocco, Tench were successfully introduced to Tunisia in 1965 (Moreau *et al.*, 1988).

South Africa

De Moor and Bruton (1988) have summarized what is known of the introduction of Tench to South Africa.

In October 1896 109 'carp and tench' were unsuccessfully imported from Dumfries, Scotland, to the Pirie hatchery, Kingwilliamstown. The latter species was first successfully introduced to South Africa from Surrey, England (probably from D. F. Leney's Fish Farm) in 1910 and 1911, supposedly as a forage fish for Largemouth Bass *Micropterus salmoides* (q.v.) (Anon., 1944). This, however, seems improbable, as the latter species was not introduced to South Africa from England until 1927, and a more likely reason seems to be as a sport fish for anglers.

Tench were stocked in numerous dams and rivers in the Cape, but in most of them unsuccessfully. Today they are said to occur only in Paardevlei (where individuals up to nearly 6 kg have been caught) and Helderview dams in the Cape; in the Breede River in the Bontebok National Park, Bredasdorp (Jubb, 1978; Braack, 1981); possibly in a farm dam near Grahamstown in the Nuewjaars River (Bushman's system); and in a few farm dams in the Zwartberg/Franklin districts. De Moor and Bruton (1988: appendix IV: 272) give details of dates of initial introductions and earliest records (with references) of Tench in various localities in South Africa.

Ecological Impact   Tench in South Africa are regarded as passive invaders that have displayed little if any tendency to colonize new waters or have a major influence on those ecosystems in which they occur. Why this should be is as yet unknown. Their habit of bottom feeding, however, does create a certain amount of turbidity, though nothing like as much as that of Common Carp *Cyprinus carpio* (q.v.). In Paardevlei there has been an overall deterioration in water quality and a decline in underwater aquatic vegetation since the introduction of Tench, but this alteration has been partially a consequence of a rise in the water level of the dam (Anon., 1944). Welcomme (1988) says the Tench in South Africa is 'regarded as a useful species within the rather narrow climatic range in which it survives'.

Zimbabwe

Hey and Van Wyck (1950) and Toots (1970), quoting from the *Rhodesia Agricultural Journal* (Vol. 20 (5): 1920), say that Tench are likely to have been imported to Zimbabwe (then Rhodesia) from South Africa sometime before 1920. In 1938, the Rhodesia Angling Society imported 300 Tench which they planted in waters under their control. In 1958, the Fisheries Research Centre imported 22 Tench fingerlings from the Jonkershoek hatchery in the Cape, and placed them in the Mazoe dam where they bred successfully.

Ecological Impact Toots (1970) says that Tench in Zimbabwe do not have any negative impact in those waters in which they occur, and suggests that they 'could well be used in impounded waters together with snail-eating indigenous species as an effective snail consumer'.

Unsuccessful Introductions

In Africa, Tench have been unsuccessfully introduced to Madagascar (Kiener, 1961; Moreau, 1979; Reinthal & Stiassny, 1991), Namibia (South West Africa) (De Moor & Bruton, 1988), and Tunisia (Welcomme, 1981, 1988).

North America
Canada

Tench imported from Lakes Christina (according to Dymond (1936) since about 1915), Osoyoos (1941), and Tugulnuit (after 1958) in the Columbia River catchment of British Columbia have generally been assumed, since no records of actual plantings are known, to be the result of dispersal up river in the Columbia system of the progeny of fish introduced to some small lakes in Spokane County, Washington, and Washington County, Oregon, USA, in 1895 (Carl *et al.*, 1967; Scott & Crossman, 1973). Tench have also been reported from Vancouver Island, but Crossman (1984) was unable to verify this claim. They are also said to have been collected in Alberta.

United States

The Tench was first imported by the United States Fish Commission from Germany in 1877 for food and as a sporting asset (Baird, 1879), and in 1890–1891 was distributed to Colorado, Indian Territory, Indiana, Kansas, Maryland, Michigan and Missouri (Baird, 1893). In 1889, when federal fish ponds in Washington, DC, were flooded, Tench escaped into the Potomac River (Baird, 1893).

Smith (1896), who said that the Tench 'has been somewhat extensively planted in the United States by the national fish commission', recorded the introduction in 1885 of 20 Tench to a private water in Virginia City, Nevada, and of yearling Tench in 1895 to various Pacific states, as follows: Older Springs, Washington County, Oregon (50); Fourth of July Lake, Fetz Lake, and a pond in Spokane County, Washington (400); and Diamond Lake, a lake and a pond in Kootenai County, and a pond in Latah County, Idaho (758, of which 500 were placed in Diamond Lake). Dymond (1936), on the other hand, says that the origin of Tench in Washington is some brought to Seattle for the World Fair in 1909 (he does not say for what purpose), 'and afterwards dumped into a large goldfish pond on the campus of the University of Washington. From here some were taken to Lake Union'.

In 1922, between 12 and 24 10–15 cm Tench, the survivors of a much larger shipment from Italy, were released in a reservoir on their ranch near Lobitas Creek in San Mateo County, California, by Pietro and Joseph Balanesi, where they became established, and from where Tench were later successfully translocated to other reservoirs in much of San Mateo and Santa Cruz Counties (Shapovalov, 1944).

In the USA, Tench are naturalized in California (Moyle, 1976a,b); Colorado (Beckman, 1974); in a small lake in Litchfield County, Connecticut (Webster, 1941); in Idaho (Simpson & Wallace, 1978); New Mexico (Koster, 1957); and Washington (Wydoski & Whitney, 1979); and possibly also in Delaware, Maryland and New York (Schwartz, 1963). Tench are believed by Bond (1973) not to have occurred in Oregon since the mid-1960s. The species has been taken but is not known to be established in Arizona (Minckley, 1973), Missouri (Baughman, 1947), and Virginia (Courtenay *et al.*, 1991). (See also Chapman, 1942; Lachner *et al.*, 1970; Courtenay & Hensley, 1980b; Courtenay *et al.*, 1984, 1986.)

South America
Chile

Introduced to Chile for aquacultural purposes in 1908, Tench became established in Lautaro

(Eigenmann, 1927), central Chile (Mann, 1954) and Valparaíso (De Buen, 1959), where they are believed still to survive.

Australasia

Australia

Tench were one of the first species of fish to be introduced to Australia. In 1858 two shipments, including Tench, arrived from England in Tasmania, one on the *Heather Belle* on behalf of Morton Allport of Hobart (Clements, 1988) and another on behalf of the Royal Society of Tasmania, referred to in the Report of the Royal Commission on the Fisheries of Tasmania (1882), which stated that Tench 'were introduced by the late Captain Langdon, and had permanently established themselves in the still waters of such rivers as the Clyde, South Esk and the Jordan, especially where weeds are plentiful' (quoted by Weatherley, 1959). (This latter shipment is also mentioned in the *Melbourne Argus* of 7 January 1862.) These fish were propagated by the Royal Society of Tasmania, which by early 1862 was releasing young Tench in various Tasmanian waters.

The first Tench to reach the Australian mainland were probably those included in a mixed consignment of fish imported from England by Edward Wilson, editor of the *Melbourne Argus*, on the *Lincolnshire*, which berthed in Melbourne on 14 October 1860. Reports in the *Argus* in 1861 indicate that a further shipment of Tench arrived in Melbourne on the same vessel on 10 June of that year.

In 1864, the Acclimatization Society of Victoria (Lever, 1992) acquired 36 Tench from Tasmania, which were released in the Big Hill reservoir at Sandhurst, Bendigo. In the following year Gilbert Duncan freed 11 small Tench in Lake Wendouree at Ballarat, and in 1866 Alfred T. Seal received a further 24 which he placed in ponds in the Botanical Gardens at Ballarat, where they became established. In March 1871, a Mr Cathcart imported a number of Tench from Tasmania to Victoria, where they were later freed in the Barwan River, Geelong (*Geelong Advertiser*, 23 March 1871; quoted by Clements, 1988). Five years later the Geelong and Western District Fish Acclimatization Society received some 400 'English perch and tench' from Tasmania, and in association with the Colac Fish Acclimatization Society planted them in various waters in the

Geelong/Colac area. The Geelong and Western District Society continued to distribute Tench to waters under its control at least until the 1880s (Clements, 1988).

The *Australasian* of 21 July 1866 records an introduction of Tench, probably from Tasmania, to Sydney, New South Wales, and on 2 May 1871, 24 from the same source were sent to the Revd Canon Sharpe of Bathurst (Clements, 1988). Merrick and Schmida (1984), Allen (1989) and Brumley (1991) all say that Tench were first introduced to Australia in 1876, the first named authors specifying Victoria. Weatherley and Lake (1967) give the date of the species' introduction to New South Wales as 1886.

Weatherley (1959, 1962) has summarized the distribution of Tench in Tasmania. They occur in the South Esk River system – the South Esk, Meander, Macquarie and Lake Rivers – in northern Tasmania, and in the Clyde, Jordan and Coal Rivers in the south. They are also found in Lake Tiberias (440 m asl), a plant-filled lagoon some 65 km north of Hobart which, when its waters rise, overflows into the Jordan. According to McDowall (1980), Tench occur in Tasmania 'primarily in the Derwent River'. The waters in which Tench occur in Tasmania are still and well-vegetated as in their native range.

The distribution of Tench in Tasmania is very similar to that of the Perch *Perca fluviatilis* (q.v.), except that the latter occurs in Lake Echo (from which the Tench is absent, although it is found in the lower reaches of the River Dee which flows out of the lake) but not in Lake Tiberias, where the only other fish apart from the Tench is the Eel *Anguilla australis*. It is noteworthy, as Weatherley points out, that the only two species of fish to be found in Lake Tiberias are physiologically extremely tenacious.

The distribution of Tench in Tasmania is thus limited principally to slowly flowing rivers in the north and the south, and they have been unable to colonize highland lakes by dispersal into them via outflowing rivers. Their absence from other apparently suitable rivers is almost certainly because they have not been introduced to them by human intervention. It seems highly unlikely that they could increase their range into new river systems by dispersal through the sea, as Weatherley (1959) found they were unable to tolerate salinities in excess of around 14.58‰ (less than half that of sea water) for any length of time.

On the Australian mainland the Tench is well established and locally abundant in the Murray/Darling Rivers system west of the Great Dividing Range in New South Wales, Victoria, and South Australia, where it is fished for commercially, mainly for shark and lobster bait (Clements, 1988), and in a few waters elsewhere. Its distribution is, however, discontinuous, and it does not occur in fast-flowing upland waters, preferring instead sluggish and weedy sections of lower river reaches, their tributaries and associated lakes. In western New South Wales it is said to have declined since the increase of Common Carp *Cyprinus carpio* (q.v.). In Victoria, Tench are also common south of the Great Dividing Range, especially in some western lakes, and in the Lachlan River just above its confluence with the Murrumbidgee (Whitley, 1951; Weatherley & Lake, 1967; Cadwallader & Backhouse, 1983; McKay, 1984, 1989; Merrick & Schmida, 1984; Allen, 1989; Brumley, 1991).

New Zealand

Tench were included in the first consignment of fish shipped from England to New Zealand by A.M. Johnson on board the *British Empire* in 1864, but none survived the voyage. In 1867, the Canterbury Acclimatization Society acquired some Tench from the Royal Society of Tasmania, which in 1871 were said to 'have successfully multiplied'. In 1868, the Nelson and Southland Societies also received some Tench, from the Royal Society of Tasmania and Morton Allport respectively. In the same year, Frank Buckland of the Acclimatization Society of the United Kingdom (Lever, 1992) sent out a number from England on the *Celestial Queen*, but 'one day one of the ship's boys who was changing the water for the fish, got them into a bucket of fresh-water and emptied it over the ship's side, instead of so doing with a bucket containing the stale water that had been drawn off' (Thomson, 1922).

The earliest record of the release of Tench in New Zealand recorded by Thomson (1922) was in 1869, when the Otago Acclimatization Society liberated 18 in the Ross Creek reservoir, Dunedin. In 1880, the same society despatched 30 to Otekaike and a like number to Elderslie, both in Oamaru; in 1887, when the latter ponds were drained, large numbers of Tench were distributed throughout Oamaru (Thomson, 1922).

Clements (1988) considered it probable that further shipments of Tench arrived in New Zealand from Tasmania until the mid-1870s, by which time he claimed the species 'was widespread and completely acclimatised'. There appears, however, to be no evidence to support this claim, and Thomson (1922) said that Tench were established only in a few waters in South Canterbury, but in none south of Oamaru, and near Hokitika and a few other places on the west coast.

For a century after their introduction the fate of Tench in New Zealand was unknown. In 1940, the Otago Acclimatization Society confirmed the species' survival near Oamaru, and by the 1960s it was known to exist only in a few waters around Oamaru (McDowall (1979) mentions the Waiareka and Kakanui Rivers) and Timaru, having disappeared from its former localities on the west coast. In the late 1950s, the Marine Department transferred some Tench from near Oamaru to a few small lakes in Wellington, in one of which – Lake Kopureherehe near Otaki – they became established (McDowall, 1990b, 1994b).

In the late 1960s and 1970s, Tench were illegally released in waters in and to the north of the Waikako River, in many of which they became established. In New Zealand, Tench occupy the same types of water as in their native range and Australia – well-vegetated stillwaters and slowly flowing lower reaches of rivers. Although Clements (1988) claimed that 'both in Australia and New Zealand [the Tench] is generally regarded as an unfortunate introduction and something of a pest', in the latter country at least it has become a highly esteemed and trophy-sized sport fish (McDowall, 1994b), and appears to be ecologically insignificant (McDowall, 1990b).

Vimba vimba

Vimba vimba (in German, Zährte; in Swedish, Vimma) is a small cyprinid with a markedly projecting and rounded snout and without barbels, resembling *Chondrostoma nasus*.

Natural Distribution

Rivers of eastern Europe and western Asia, including the lower reaches of the Elbe and Weser flowing into the North Sea, those rivers flowing into the Baltic Sea in southern Sweden, Finland, Germany, Poland and the former USSR, and those debouching into the Black and Caspian Seas.

Naturalized Distribution

Europe: The Netherlands.

Europe

The Netherlands

Cazemier and Heesen (1989) reported that on 24 January of that year a single specimen of *Vimba vimba* was caught in a 20-m deep gravel pit which is near and is joined to the River Neder-Rijn, close to the village of Heteren; it is believed that the species arrived in The Netherlands by natural diffusion from the Rivers Lippe and/or Dhünn, both tributaries of the Rhine. S. J. de Groot (pers. comm. 1995) says that since then the species has increased its range and is now being caught in many Dutch waters.

COBITIDAE

Oriental Weatherfish (Japanese Weatherfish; Dojo; Canton Loach; Oriental Weatherloach)
Misgurnus anguillicaudatus

The Oriental Weatherfish is a small eel-like coolwater loach well adapted to life in muddy and deoxygenated waters. It is used by man as an aquarium fish, in aquaculture, and in capture fisheries (Welcomme, 1988). It can move across land, and obtains oxygen from the air by gulping at the water surface, absorbing it through the vascular hind gut and expelling it through the vent. It is reported to aestivate (Allen, 1984).

Natural Distribution

Eastern Asia, from the basins of the Tugur and Amur Rivers and Sakhalin Island in the north, south through Korea, Japan and Taiwan to Hainan and North Vietnam, west to the headwaters of the Irrawaddy River in Burma.

Naturalized Distribution

Asia: Philippines. *North America:* ? Mexico; USA. *Australasia:* Australia. *Oceania:* Hawaiian Islands; Palau Islands (Babelthuap).

Asia
Philippines

The Oriental Weatherfish was introduced to the Philippines from Japan sometime before 1937. It has become naturalized in some highland waters in Trinidad Valley and in rice terraces in Bontoc Province, where so far as is known it has had no harmful effects (Juliano *et al.*, 1989). Welcomme (1981, 1988) says it was introduced in 1942, and that 'it forms the basis for a fairly important fishery'.

North America
Mexico

The Oriental Weatherfish was imported at an unrecorded date from the Far East to rear as a food fish in the Chapingo Fish Hatchery in Mexico, from which it was released into surrounding canals when the hatchery closed before 1961; its present status, if

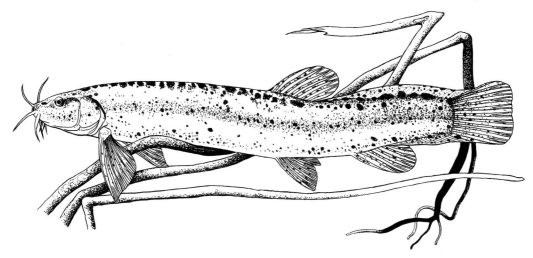

indeed it survives, is unknown (Contreras & Escalante, 1984).

United States

In 1958, Oriental Weatherfish were discovered in a private pond on Hy Meadow Farm near Holly, Oakland County, Michigan, and in the Shiawassee River upstream and downstream of Hy Meadow. They are believed to have been the progeny of some that escaped from the Sunset Water Gardens in Holly (to which they had been imported from Kobe, Japan) in or about 1939 (Schultz, 1960).

In 1968, Oriental Weatherfish were found in a portion of the Westminster flood control channel in Orange County, California. It is thought they were descended from some that escaped from the neighbouring Pacific Goldfish Farm in the 1930s. St Amant and Hoover (1969) reported them to be restricted to a 5 km stretch of the canal by tidal action to the west and intermittent water flow to the east.

Courtenay *et al.* (1987) recorded another population, probably released by or escaped from aquarists, in the Boise River system, Ada County, Idaho. Courtenay *et al.* (1991) said that Oriental Weatherfish have been taken but are not established in Florida and Oregon.

In 1939, an aquaculture centre that was breeding Oriental Weatherfish for sale to aquarists released some in the waters of Lake Michigan, where according to Mills *et al.* (1994) they still occur. (See also Lampman, 1949; Lachner *et al.*, 1970; Moyle, 1976a,b; Courtenay *et al.*, 1984, 1986; Courtenay & Stauffer, 1990; Courtenay, 1993.)

Australasia

Australia

The first Weatherfish to be found in the wild in Australia was a gravid hen fish taken in Lake Burley Griffin in the Australian Capital Territory on 24 September 1980; it was followed by the collection in January 1984 of a juvenile in Ginninderra Creek which flows out of Lake Ginninderra, also in the ACT (Allen, 1984). In the same year, Allen found further juveniles in the Yarra River, Victoria, about 1 km

downstream of the Warrandyte bridge, and Weatherfish have since been found in several other localities in Victoria and the ACT (Lintermans *et al.,* 1990).

Burchmore *et al.* (1990) reported the discovery in January 1989 of a population of Weatherfish in the Wingecarribee River, New South Wales. Treatment of the river with rotenone ichthyocide revealed that the population, which may have been established for many years, was fairly widespread and abundant, and fears have been expressed that it could gradually disperse downstream into the Hawkesbury Nepean River system, which is the second largest coastal drainage catchment in New South Wales, and thence to similarly degraded muddy waters throughout the catchment. There is also an unconfirmed report of a breeding population of Weatherfish in Western Australia.

Oceania

Hawaiian Islands

Brock (1960) says the Oriental Weatherfish was established on the islands of Kauai, Oahu and Maui before 1900. It is widely distributed in both standing waters and streams and is currently expanding its range. It was probably imported by Asian immigrants, and is a popular bait fish (Maciolek, 1984). (See also Brock, 1952; Hida & Thomson, 1962; Kanayama, 1968; Lachner *et al.*, 1970; Maciolek & Timbol, 1980; Courtenay, 1993).

Palau Islands (Babelthuap)

According to Bright (1979) and Bright and June (1981), Oriental Weatherfish were successfully introduced to the Palau Islands as a source of food by the Japanese at some time between 1914 and 1944.

Other Records

In about 1978 and 1995 respectively, single specimens of the Oriental Weatherfish were caught in the River Roding, Essex, below the effluent from a fish importer's holding tanks, and in a pond in Hertfordshire, both in England (A. C. Wheeler, pers. comm. 1995).

CATASTOMIDAE

Bigmouth Buffalo
Ictiobus cyprinellus

Black Buffalo
Ictiobus niger

The Bigmouth Buffalo is the largest of the American suckers, reaching up to 1 m in length, and in suitable conditions can become extremely abundant at the expense of other species. It favours large, slowly flowing rivers, lakes and oxbows. The smaller Black Buffalo prefers faster-flowing water than its congener.

Natural Distribution

The Mississippi River drainage from southern Canada and North Dakota eastwards to Pennsylvania and south to the coast of the Gulf of Mexico. The Black Buffalo extends its range into northeastern Mexico.

Naturalized Distribution

Europe: ? Romania; ? former USSR. *Central America:* Cuba.

Europe

Romania; USSR

Both the Bigmouth Buffalo and the Black Buffalo have been introduced, at an unrecorded date and from an unknown source, to the former USSR, where their status is undetermined. Welcomme (1988), however, points out that the Soviet origin of the suc-cessful Cuban stocking (see below) suggests the establishment, if only in aquaculture, of both species in the former USSR.

In the 1980s, both buffalo were introduced for aquaculture purposes to the Danube River delta in Romania, where their status is uncertain.

Central America

Cuba

In 1981, Bigmouth Buffalo and Black Buffalo were imported for aquaculture purposes from the former USSR to Cuba, where Welcomme (1988) says they are established and breeding.

Unsuccessful Introduction

The Bigmouth Buffalo has been unsuccessfully introduced to Lake Kinneret (Lake Tiberias/Sea of Galilee) in Israel (Ben-Tuvia, 1981; Davidoff & Chervinski, 1984).

Translocations

The Bigmouth Buffalo has been successfully translocated outside its natural range in North America to Arizona and California, and the Black Buffalo to Arizona (Welcomme, 1988).

ICTALURIDAE

White Catfish
Ictalurus catus

The White Catfish is a medium-sized (60 cm) catfish that lives in both lentic and lotic waters.

Natural Distribution

Rivers of the east coast of the USA from New York in the north south to Florida.

Naturalized Distribution

Central America: Puerto Rico.

Central America

Puerto Rico

How White Catfish entered Puerto Rico is, according to Erdman (1984), a mystery, but they may have been included in a shipment of Channel Catfish *I. punctatus* (q.v.) from Baltimore, Maryland, USA. Welcomme (1981, 1988) gives their date of entry to Puerto Rico as 1938. They are now naturalized in Guajataca, Caonillas, Dos Bocos, Guayabal and Garzas reservoirs (Erdman, 1972), and in Matrullas reservoir (Ortiz-Carrasquillo, 1980, 1981). Welcomme (1981, 1988) says they were introduced for recreational and subsistence fishing, and that they are only 'marginally successful and of limited value for sport'. (See also Wetherbee, 1989.)

Unsuccessful Introduction

Juliano *et al.* (1989) records the unsuccessful introduction of White Catfish to the Philippines.

Black Bullhead (Black Catfish; Small Catfish)
Ictalurus melas

Brown Bullhead (Brown Catfish; Dwarf Catfish; Horned Pout)
Ictalurus nebulosus

The Black Bullhead is a small (40 cm) catfish that favours slowly flowing rivers, lakes, ponds and swamps. Since no direct ecological homologue exists in many parts of Europe it has colonized the marshy backswamps of some rivers, where it tends to form dense and stunted populations (Welcomme, 1988).

The slightly larger (45 cm) Brown Bullhead prefers deeper and less turbid water than *melas*, being found mainly in well-vegetated and muddy-bottomed ponds and rivers. Both species are widely kept in coldwater aquaria and ponds and for biological laboratory study.

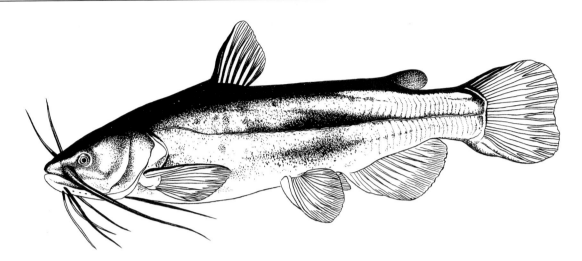

Because the correct identity of these two species, particularly in Europe, has been confused, they are here treated together.

Natural Distribution

Central and eastern North America, from southern Ontario, the Great Lakes and the St Lawrence River south through the basin of the Mississippi and Ohio Rivers to the Gulf of Mexico, and from Montana eastward to the Appalachian Mountains.

Naturalized Distribution

Europe: Austria; Belgium; British Isles (England); Czechoslovakia; Denmark; France; Germany; Hungary; Italy; The Netherlands; Norway; Poland; Romania; Spain; Switzerland; the former USSR; the former Yugoslavia. **North America:** Mexico. **Central and South America:** Chile; Puerto Rico. **Australasia:** New Zealand.

Europe

Austria; Belgium; British Isles (England); Czechoslovakia; Denmark; France; Germany; Hungary; Italy; The Netherlands; Norway; Poland; Romania; Spain; Switzerland; the former Yugoslavia; the former USSR

Wheeler (1978), from whom much of the following account is taken, has unravelled the history and identity of North American bullheads introduced to Europe.

During the 19th century there were numerous importations of ictalurid bullheads to Europe from North America. In *France* they were introduced in 1871 (Vivier, 1951a), while Von dem Borne (1890) and Schindler (1957) recorded them as occurring in *Germany* in 1885, where Tortonese (1967) dated the introduction in the 1880s. Some of those introduced to *France* were washed away in the sewer system into the River Seine, where in 1879 and 1894 individuals were caught near the Pont d'Austerlitz (De Groot, 1985). Smiley (1885) and Kendall (1910) list the following shipments of bullheads to Europe: 1884, 100 to Ghent, *Belgium*, of which 95 arrived alive (see also Lefebvre, 1883); 1885, 50/49 to *Germany*; 1885, 50/48 to *England* (the National Fish Culture Association in South Kensington, London (Lever, 1992)); 1885, 30 (or 80) to Amsterdam, *Netherlands*; 1885, 100/81 to *France* 1892, 502 to *Belgium*, where some were liberated in ponds at Lommel in Limburg (De Groot, 1985); 1892, 76 to *Germany*; 1903, 400 to *Belgium*. Most of these fish were said to be *I. nebulosus*, although *punctatus* (q.v.) and possibly *natalis*, were also included. They apparently came from sloughs in the basin of the Mississippi where young fish were trapped after seasonal floods subsided.

Between 1910 and 1913, *I. melas* was introduced from France to *Spain* for stocking purposes by the Catalonia Fish Service (Elvira, 1995b), where it is now established in the Douro, Tagus, Ebro and Eastern Pyrenees drainages (Elvira 1995a).

Welcomme (1981, 1988) gives the following additional information on the movement of bullheads to and within Europe: from Germany to *Hungary* in 1902; from the USA to *Norway* (1890); from Germany to

Poland (1905 (Pennekamp, 1905)); from Italy to *England*; from the USA to the former *Yugoslavia* (1905–1907); and specifically *I. nebulosus* to *France* between 1930 and 1950. Welcomme says the species imported into *France* from the USA in 1885 was *melas*. In *The Netherlands*, Welcomme says the species introduced from the USA was *melas*, and that *nebulosus* occurs there possibly through 'migration' (dispersal) from Belgium – probably via the Meuse canal (De Groot, 1985). In *Austria*, Welcomme implies the presence of bullheads is by natural diffusion through the Danube.

According to Anon. (1971), *I. nebulosus* was one of a number of North American fishes introduced to *Czechoslovakia* in the late 19th century, where it 'increased rapidly and occupied tributaries and pools of the River Labe (Elbe)'.

Thus, by the end of the 19th century North American bullheads were widely naturalized in many places in Europe. By the mid-20th century they occurred in numerous rivers in *France* (Vivier, 1951a), in *The Netherlands* (Nijssen & De Groot, 1974), *Belgium* (Wheeler & Maitland, 1973), and *Germany* (Ladiges & Vogt, 1965). From Germany, bullheads were introduced into the former *USSR* where they became established in White Russia and the western Ukraine (Berg, 1964–1965). They became widely naturalized in the basin of the River Danube, where Bǎnǎrescu (1964) reported their presence in *Romanian* waters and indicated their occurrence in those of the former *Yugoslavia* and *Hungary*, a distribution confirmed by the map in Blanc *et al.* (1971). In *Denmark* they were reported by Brunn and Pfaff (1950) to be 'very rare' (see also Jensen, 1987). In *Switzerland* they are found in Lac des Brenets and the Rhône Genevois (Muus and Dahlstrøm, 1968; Vooren, 1972). They occur discontinuously in *England*, probably as a result of isolated plantings (Wheeler & Maitland, 1973), although one population is known to have become established near North Weald, Essex (Lever, 1977), and a population estimated to number around 300 occurs in a pond in Berkshire (Lever, 1977).

For many years it was generally accepted that the species of bullhead introduced to and now widely naturalized in Europe was *Ictalurus (Ameiurus) nebulosus*. Study of the literature (e.g. Redeke, 1941) recording critical determinations and of specimens from both Europe and England, however, (for full details see Wheeler, 1978) reveals the presence of both *I. nebulosus* and *I. melas*, and leaves no doubt that the latter is widely distributed. Evidence for the widespread occurrence of *I. nebulosus*, on the other hand, is less persuasive, but it is known to be established in some waters in western Europe, particularly in The Netherlands where it is the dominant ictalurid.

Wheeler (1978) found that the majority of bullheads imported into *England* come from Italy and are *I. melas*. When, however, they originate in Germany, Belgium or The Netherlands, it is possible that both *melas* and *nebulosus* are included.

Status and Ecological Impact   Welcomme (1981, 1988) summarizes the current status and ecological impact of bullheads in Europe as follows:

COUNTRY	STATUS	IMPACT
Belgium	?	A voracious and undesirable species viewed as a pest
British Isles (England)	In ponds to which introduced . . . some escape to rivers. *I. nebulosus* less abundant	Seemingly no problems
France	Very common in ponds and canals all over the country	?
Germany	In parts of the country	Regarded as a nuisance
Hungary	Rapid spread to all natural waters. Stocks declined after 1950s in most waterways	Originally caused considerable damage but since decline is merely a nuisance
Italy	Widespread. Well established in certain water bodies (see also Mazzola, 1992)	Regarded as a nuisance
The Netherlands	Very rare (*melas*) locally common (*nebulosus*)	? No particular problems
Norway	In two or three small lakes	Unimportant
Poland	Spread in some waters but not now abundant	Regarded as a nuisance
Former USSR	In Ukraine	Forms basis for commercial fishery
Former Yugoslavia	In both natural waters and ponds	Undesirable

North America

Mexico

Introduced to Mexico at an apparently unrecorded date for rearing in captivity for food, *I. melas* has become established in the wild in Río Yaqui near Ciudad Obregón (Branson *et al.*, 1960) and in the Río Casas Grandes (Contreras *et al.*, 1976; Hendrickson *et al.*, 1980), and Ríos Pánuco, Lerma-Santiago and Balsas (Contreras & Escalante, 1984).

Translocations

Bullheads have been widely transplanted west of the Rocky Mountains outside their natural range in North America, particularly *I. nebulosus* to Oregon and Washington, USA (Smith, 1896; Chapman, 1942), and both *nebulosus* and *melas* to British Columbia, Canada, where their presence may be the result of natural dispersal from the Columbia River (Carl & Clemens, 1953; Carl & Guiguet, 1958; Carl *et al.*, 1967).

South and Central South America

Chile

Although Welcomme (1988) indicates that both *I. melas* (1907) and *I. nebulosus* (1908) have been introduced to Chile for commercial fishing and are established there, De Buen (1958, 1959) specifies that the characteristics of the species naturalized in Lagunas Lo Prado and Pudahuel, in the estuary of the Angostura River, approximate to the nominate subspecies *I. m. melas*.

Puerto Rico

The United States Bureau of Fisheries introduced *I. n. nebulosus* into Carite reservoir in 1915–1916: it remains restricted to reservoirs in east–central and southeastern Puerto Rico, such as Carite, Comerío, Melanía, Patillas and probably La Plata, and in the middle Río La Plata (Erdman, 1984). The Marbled Bullhead *I. n. marmoratus* was introduced to western Puerto Rico in 1946, where it now occurs in Caonillas, Dos Bocas, Guajataca, Loco and Luchetti (Yauco) reservoirs; drainage canals in the Lajas Valley; the Río Hondo in the Guanajibo drainage and Caño Tiburones (Barceloneta), and probably elsewhere (Erdman, 1984). (See also Wetherbee, 1989.)

Hildebrand (1935a) reported the introduction of *I. melas* to Puerto Rico in 1914; this appears to be misidentification of *I. nebulosus* released in the preceding year.

Australasia

New Zealand

Arthur (1881) refers to the introduction to New Zealand in 1877 of 'the Catfish (*Pimelodus cattus*)' (*Ictalurus nebulosus*) by the Auckland Acclimatization Society, which received the species from Thomas Russell, a New Zealander then living in San Francisco, California. Hunter (1915) said that from two consignments, 225 fish were released into St John's Lake in Auckland, but Sherrin (1886) and Thomson (1922) put the number at 140. Thomson (quoting Sherrin) noted that 'they were lost sight of for a time, but reappeared in considerable numbers in 1884 ... in 1885 they were caught in hundreds, and were sent to many parts of the provincial district'. By 1916, they were said to be abundant in St John's and Takapuna Lakes.

In 1885, 30 Brown Bullheads were transferred from Auckland to Wellington where they were freed in a pond at Petone; what became of them is unknown. At about the same time, others were sent to Hokitika in Westland, where they were successfully released in Lake Mahinapua, and in ponds near Ashburton (Thomson, 1922) from which they have since disappeared (McDowall, 1990b). Bullheads were clearly not appreciated in New Zealand, since Hunter (1915) reports requests to the USA that no more be sent.

From the 1950s to the late 1960s, Bullheads were known in New Zealand only from the lower Waikato River in Auckland (Patchell, 1977) and from Lake Mahinapua but, in common with several other naturalized fishes, have become much more widely distributed in the 1970s–1990s, appearing in such places as Lake Wairarapa, Masterton, Paraparaumu, and Tauranga (McDowall, 1979). Why this should be so is unclear, but McDowall (1994b) considers that the Bullhead's wide distribution north of the Waikato and eastward into the Hauraki plains results from the activities of commercial eel fishers. Bullheads and eels live sympatrically in New Zealand, and McDowall believed that the former were accidentally transferred between rivers hidden in eel-catchers' fyke nets. However, McDowall considered that the appearance of Bullheads in more distant waters, such as the Wairarapa and the lower Wanganui River, was not caused by eel catchers, while the release of Bullheads in Lake Taupo in the late 1970s 'defies any sensible explanation, and must have been the action of some irresponsible biological hooligan'. Today, large numbers of Bullheads continue to be caught by eel fishers, by whom they are sold for pet food and fish meal (McDowall, 1994b).

Unsuccessful Introduction

Maciolek (1984) records the unsuccessful introduction of *I. nebulosus* to the Hawaiian Islands. In 1984 the species was introduced from the USA to China where it is cultured in Hubei Province and Beijing (Yo-Jun & He-Yi, 1989).

Channel Catfish
Ictalurus punctatus

The Channel Catfish is a large (125 cm) species of warm temperate waters which is of great importance to aquaculture and of considerable value in sports fisheries. In Central and South America it forms the basis of a growing commercial aquaculture industry for both overseas and home consumption. Given the high demand for catfish, Welcomme (1988) believed that the species will be introduced to further countries for rearing in fish farms.

Natural Distribution

Central drainages of North America from the prairie provinces of Canada and the Hudson Bay region to the Great Lakes and the St Lawrence basin, south to Florida and northern Mexico.

Naturalized Distribution

Europe: ? British Isles (England); Cyprus; Italy. **South and Central America:** ? Brazil; ? Cuba; Dominican Republic; Puerto Rico. **Oceania:** Hawaiian Islands.

Europe

British Isles (England)

The earliest documented introduction of 'North American catfish' to Britain seems to have occurred before 1892, when they are recorded as having been introduced to 'Loch Uisg: Lochbuie, island of Mull', Scotland. Whether these fish were *punctatus* is unknown.

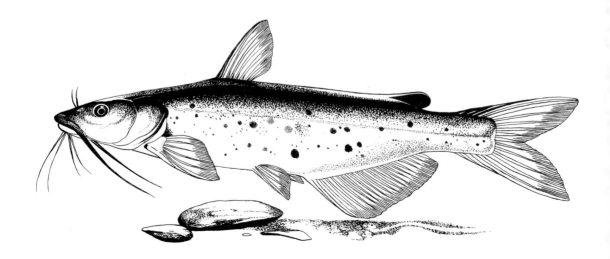

No further importations are on record until, in 1969, 900 10-cm young Channel Catfish, imported from the USA ostensibly to stock aquaria, were released for sporting purposes in the following waters: the conjoined North and South Lakes (200 and 250 fish respectively) at Wraysbury near Staines, Buckinghamshire; the Fleet Lake (250) at Thorpe near Chertsey, Surrey; the Sailing Club Lake (100) at Paper Court Farm near Ripley, Surrey; and the Car Park Lake (100) at Yateley near Sandhurst, Surrey (all in southeastern England). As the stocking density was too low (the recommended figure is 3750–5000 per hectare), and since Channel Catfish require warm and moving water with a summer temperature of 27–32°C for successful spawning, their long-term survival must be in some doubt (Wheeler & Maitland, 1973; Wheeler, 1974; Lever, 1977). Channel Catfish are currently being imported into Britain under ornamental licences; several are said to have been caught in the River Trent in Nottinghamshire, and they have been stocked in various fisheries (A. C. Wheeler, pers. comm. 1995).

Cyprus

Welcomme (1981, 1988) records the introduction in 1975 of Channel Catfish from the USA to Cyprus for angling and aquacultural purposes, and the establishment of 'self-breeding populations in the wild'.

Italy

Welcomme (1988) records the introduction in 1976 of Channel Catfish from the USA to Italy, where he indicates they are established and breeding. Mazzola (1992) includes *punctatus* among those species 'having characterized a type of lake aquaculture and above all the development of recreational activities like sports fishing' in Italy.

Unsuccessful Introductions

In Europe, Channel Catfish have failed to become established in Belgium (Lefebvre, 1883), the former USSR (Kudersky, 1982), and the former Yugoslavia (Welcomme, 1981, 1988).

Asia

Unsuccessful Introductions

In Asia the Channel Catfish has been unsuccessfully introduced to China (Yo-Jun & He-Yi, 1989),

Indonesia (Muhammad Eidman, 1989), Japan (Chiba *et al.*, 1989), Korea (Welcomme, 1981, 1988), the Philippines (Juliano *et al.*, 1989) and Taiwan (Liao & Liu, 1989).

Africa

Unsuccessful Introduction

Welcomme (1981, 1988) records the unsuccessful introduction of the Channel Catfish to Nigeria.

North America

Translocations

The Channel Catfish has been widely and successfully translocated outside its natural range in North America, specifically to the Pacific states of the USA (Smith, 1896) and Mexico (Vergara de los, Rios, 1976; Contreras & Escalante, 1984).

Ecological Impact

In Mexico, transplantation of the Channel Catfish appears to have been associated with the replacement of native fishes in the Río Colorado (Miller, 1963; Contreras & Escalante, 1984).

South and Central America

Brazil

Channel Catfish introduced from the USA in 1972 are said by Welcomme (1981), somewhat ambiguously, to be established 'not in north-east Brazil … unsatisfactory, further tests planned in south'.

Cuba

Welcomme (1981) says that Channel Catfish from Mexico were introduced to Cuba in 1979, where they are established near Havana.

Dominican Republic

Welcomme (1981, 1988) says that Channel Catfish introduced to the Dominican Republic from the USA in 1970 have escaped from aquaculture centres and have become established in 'some rivers and ponds'.

Puerto Rico

Channel Catfish were first introduced to Puerto Rico by the United States Bureau of Fisheries in 1938, and subsequently to Cidra, Dos Bocas and Loiza reservoirs, where they have become naturalized and provide a useful fishery, although their rapid increase in numbers has sometimes led to stunting (Erdman, 1984). (See also Wetherbee, 1989.)

Oceania

Hawaiian Islands

In 1953 and 1954, a total of 38 Channel Catfish from California were imported to the Hawaiian Islands, where for 5 years unsuccessful attempts to propagate the species were made in the Nuuanu hatchery, some of the adult fish dying through the use of hormone injections designed to induce spawning. In 1958, 1715 young Channel Catfish were imported from Fort Worth, Texas, and released in Wahiawa reservoir on Oahu. In the following year, a further 4500 from Oklahoma were planted in Wahiawa reservoir and 1300 in the Waialua River on Kauai. Brock (1960) said that a few Channel Catfish had been taken in Wahiawa reservoir 'indicating both survival of the original planting and excellent growth of the fish'. Maciolek (1984) said that since 1953 over 200 000 Channel Catfish had been imported from the USA to Hawaii, Kauai, Maui and Oahu, on the last of which they 'are now widely distributed'. Welcomme (1988), on the other hand, says they are 'established in one reservoir on Oahu Island'. (See also Hida & Thomson, 1962; Kanayama, 1968; Lachner *et al.*, 1970.)

Unsuccessful Introductions   In Oceania, Channel Catfish have failed to become naturalized in Guam and on Tahiti in French Polynesia (Maciolek, 1984).

Margined Madtom
Noturus insignis

Madtoms are small (10 cm) catfishes with venom glands at the base of the pectoral spines. The Margined Madtom favours moderately flowing waters with a muddy bottom and plenty of aquatic vegetation.

Natural Distribution

The USA, from North and South Dakota to Texas, and from New York to Florida.

Naturalized Distribution

North America: Canada.

North America
Canada

In 1971, four Margined Madtoms were collected from the stream tributary to Lac La Pêche in Quebec (Rubec & Coad, 1974), who suggested that this constituted an introduction, possibly through the release by anglers of surplus bait. Subsequent collections indicating other populations and a slightly wider distribution led Crossman (1984) to suggest that this proposed origin may need revaluation.

SILURIDAE

Wels (European Catfish; Danube Catfish; Sheat Fish)
Silurus glanis

A large (up to 3 m) catfish which favours slowly flowing or still waters. Predatory and largely piscivorous, the Wels is chiefly nocturnal, spending the day concealed among vegetation near the bottom. In Europe it is of considerable value for both commercial and sporting fisheries.

Natural Distribution

Central and eastern Europe, mainly east of the River Rhine to the southern former USSR.

Naturalized Distribution

Europe: British Isles (England); Spain.

Europe

British Isles (England)

Francis Day in *The Fishes of Great Britain and Ireland* (1880) describes the capture in Ireland in about 1827 or 1828 of 'a unique example of a fish which some have considered may be the *Silurus glanis*' in a tributary of the River Shannon above Lough Allen.

The earliest attempt to naturalize the Wels in Britain noted by Day appears to be one referred to by Llewellyn Lloyd in his *Scandinavian Adventures* (1854), in which he wrote that 'Through the indefatigable exertions of Mr George D. Berney of Morton, Norfolk, the silurus was last year introduced into England . . .'. No more was heard of this introduction, which it can be concluded was a failure.

Under the heading 'The Arrival of the Silurus Glanis in England' a second attempt, irresponsibly sponsored by *The Field* magazine, was described in the columns of that publication on 17 September 1864 by James Lowe, joint honorary secretary with Frank Buckland of the Acclimatization Society of the United Kingdom (Lever, 1992):

> That much desired fish, the Silurus, has at last been brought alive to this country, after various failures. The success is entirely due to the intelligent enterprise and perseverance of Sir Stephen B. Lakeman, who

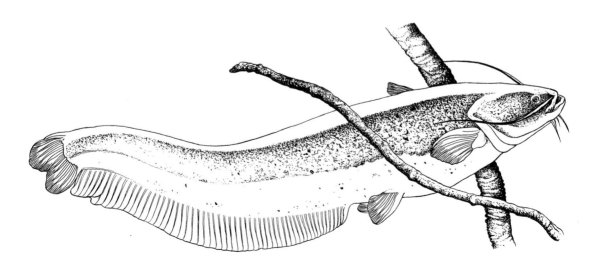

himself accompanied the fish all the way from Bucharest, a distance of 1800 miles; and on Thursday night I had the pleasure of assisting Mr Francis Francis [Piscatorial Director of the Acclimatization Society] in placing fourteen lively little baby-siluri in a pond not far from the fish hatching apparatus belonging to the Acclimatization Society on Mr Francis's grounds at Twickenham [London]. . . . The fourteen little siluri (or silureses) which have arrived are what remain of thirty-six of the same species which started from Kopacheni . . . on the banks of the Argisch, a tributary of the Danube.

At a later date Frank Buckland 'took down ten of them to my friend, Higford Burr, Esq; Aldermaston Park, Reading, and turned them out into a large pond in front of the house. Some three years afterwards this pond was let dry – the silurus had entirely disappeared'. In about 1865, Sir Joshua Rowley released some young Wels into a lake with an outlet into the River Stour in Suffolk, where they survived until at least 1894 (*The Field*, 8 September). Fitter (1959) records that in 1872 the Marquess of Bath introduced some small Wels to a lake at Frome in Somerset, where they ate so many Brown Trout *Salmo trutta* that in 1875 the lake had to be drained and the Wels removed.

On 27 October 1880, Lord Odo Russell (Ambassador at Berlin, and a son of the Duke of Bedford) brought 70 1-year-old Wels from Germany to his family estate at Woburn Abbey in Bedfordshire. This is by far the largest and most successful introduction of Wels to Britain, although Fitter (1959) suggests there may have been some later importations to Woburn from the Danube. The Wels at Woburn flourished, and in 1951 a few small ones were transferred to the lake at Claydon Park in Buckinghamshire. Wheeler and Maitland (1973) reported the spread of Wels from Woburn to other waters on the Bedfordshire/Buckinghamshire border – two of which are at Little Brickhill and Tiddenfoot Pit. This is the main area in England where Wels survive today, although it is believed that they are currently extending their range. (For further details see Lever, 1977.)

Spain

In 1974, Wels from Germany were introduced for angling purposes to Spain (Elvira, 1995b), where they are now locally common in the lower reaches of the River Ebro (Elvira, 1995a).

Unsuccessful Introductions

In Europe, the Wels has been unsuccessfully introduced to Belgium, Cyprus and Italy (Welcomme 1981, 1988), Denmark (Jensen, 1987) and The Netherlands (De Groot, 1985).

Allied Species

Devick (1991) and Eldredge (1994) list the Bronze Corydoras *Corydoras aenus* (Callichthyidae), a native of Venezuela and Trinidad south to La Plata, as occurring in the wild on Oahu in the Hawaiian Islands in 1984.

CLARIIDAE

Walking Catfish (Thai Catfish)
Clarias batrachus

The Walking Catfish is, as its name implies, able to move on land and, when it does so, it breathes atmospheric air. It possesses auxiliary specialized breathing organs – sac-like structures containing multibranched extensions well-supplied with blood vessels for respiration – which open off the gill arches. Walking Catfish live in ponds or seasonal pools that may dry out in prolonged droughts. On these occasions they are able to travel overland to new water bodies by making snake-like movements and by using their pectoral fins as 'legs'. Alternatively, they can submerge themselves in mud at the bottom of their pool and remain in a state of suspended animation until the rains return. They do not appear to have a preference for any particular quality of water, living in both clear, well-oxygenated waters and in muddy swamps and ditches with little dissolved oxygen, and are tolerant of high salinities.

Within its native range the Walking Catfish is a popular species for aquaculture, where because of its high tolerance of anoxic conditions, its fecundity and rapid growth, dense populations and thus very high yields are obtained.

Natural Distribution

South East Asia from eastern India, Bangladesh, Burma and Sri Lanka to the Malay archipelago.

Naturalized Distribution

Asia: Celebes; ? China; Philippines; Taiwan. ***North America:*** USA. ***Australasia:*** Papua New Guinea. ***Oceania:*** Guam.

Europe
Unsuccessful Introduction

A. C. Wheeler (pers. comm. 1995) says that following release from aquaria, a population of Walking Catfish became established in the 1970s in the heated effluent of the Pilkington glass factory in the St Helen's Canal, Lancashire, England. This population, which never reproduced, has since died out.

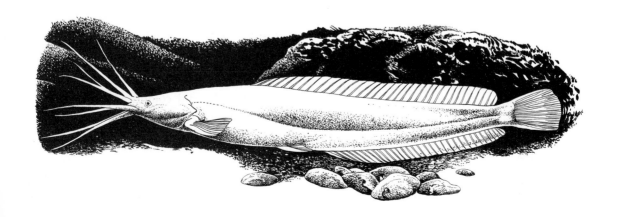

Asia

Celebes

In 1939, Walking Catfish from Java were introduced for aquaculture and commercial fishing to Celebes, where they have become widely naturalized; they are generally regarded as an undesirable addition to the island's ichthyofauna, and attempts have been made to eradicate them (Vooren, 1968).

China

In 1978, Walking Catfish from Thailand were imported for experimental aquaculture to China, where the species has become 'established in ponds in Kwantung Province, where it has proved useful for small-scale pond culture' (Welcomme, 1981, 1988). It is also distributed in southern Yunnan Province (Yo-Jun & He-Yi, 1989). Whether it is also established in the wild is unknown.

Philippines

In 1972, (1974 according to Welcomme (1981, 1988)) Walking Catfish from Thailand were introduced for aquacultural purposes to the Philippines, where they are now established in lakes and rivers in which they completely dominate, and indeed have nearly eliminated, the indigenous *C. macrocephalus*, to which they are inferior as a food fish (Juliano *et al.*, 1989). In Lake Lanao on Mindanao, Walking Catfish prey extensively on native cyprinids (Payne, 1987).

Taiwan

Also in 1972 and also from Thailand, Walking Catfish were introduced to Taiwan (Chen, 1976), where their speedy growth rate has helped them to become a popular food fish. They have hybridized with the native Chinese Catfish *C. fuscus*, which has now almost disappeared, whereas the hybrid has spread all over the island (Liao & Liu, 1989).

Unsuccessful Introduction

In Asia, the Walking Catfish has been unsuccessfully introduced to Hong Kong, where it is less popular than the native *C. fuscus* (Welcomme, 1981, 1988).

North America

United States

Courtenay and Miley (1975) and Courtenay (1978), from whom much of the following account is

derived, have traced the introduction, spread, and ecological impact of the Walking Catfish in the USA.

So far as is known, all stocks of the species imported to the USA comprised young 3–4 cm albinos shipped from Bangkok, Thailand. They were introduced in 1964 or 1965 to Penagra Aquariums west of Deerfield Beach, Broward County in southeastern Florida, where they were placed in outdoor ponds dug in drainage land which, in times of flood, overflow into nearby drainage canals. Between 1965 and 1967, Walking Catfish escaped into these waterways either from the ponds or from a vehicle transporting them between Miami and Parkland and quickly became established. These drainage canals form an extensive interconnected network that drains Lake Okeechobee and the inland Water Conservation areas into the Atlantic Ocean via the Intracoastal Waterway. The only barriers to the dispersal of fish within this system are flood-control and salinity structures comprising dams with flood gates. These are ineffective in preventing the spread of fish that can travel overland.

In 1967, the Florida Game and Fresh Water Fish Commission placed the Walking Catfish on its restricted list of exotic species, thus banning both further importations and possession of the fish. This caused the dumping of the species by fish farmers in the Tampa Bay area, Hillsborough County, into nearby streams, where it became established.

It is interesting to note that because albinos are easily seen by predators, within 2–3 years of their escape (when they were confined to three counties in Florida) natural selection had ensured reversion to the normal dark-brown to grey colour phase (Courtenay & Stauffer, 1990).

Within a decade of their escape, Walking Catfish were naturalized over some 8750 km² in 10 counties of southern Florida, or 18% of the State. By the late 1970s, the known range of the species on the lower east coast of Florida extended from just north of Fort Pierce, St Lucie County, to near Homestead, Dade County, a distance of around 215 km, achieved in about 12 years. This extension of range covered over 25% of Florida, and comprised Brevard, Broward, Charlotte, Collier, Dade, DeSoto, Glades, Hendry, Highlands, Hillsborough, Lee, Manatee, Martin, Monroe, Okeechobee, Palm Beach, Polk, Sarasota and St Lucie Counties, and possibly also Hardee and Indian River Counties, and by the early 1980s perhaps also Orange and Osceola Counties (Courtenay *et al.*, 1984, 1986). Thus, by the late 1970s, Walking Catfish occurred in most fresh waters of the southern half of peninsular Florida (Courtenay, 1978, 1979a). 'That probably represents a record for range expansion by a newly introduced fish' (Courtenay & Stauffer, 1990). Individual Walking Catfish have also been taken, but are not known to be established, in California, Georgia, Nevada and Massachusetts (Courtenay *et al.*, 1984, 1986).

A number of factors have combined to militate in favour of the colonization of Florida by the Walking Catfish; among these are its fecundity; its rapid growth rate; its ability to survive in dense populations; its resistance to desiccation; its tolerance of high degrees of salinity, which enables it, should the need arise, to colonize new fresh waters via the sea; and its capacity to travel overland. Although in the winter of 1970, low temperatures killed thousands of Walking Catfish in parts of southern Florida, subsequent cold winters have had a much reduced impact on the population. Courtenay (1978) believed that 'cold-resistant segments of the established populations have become selected and acclimatized' to occasional harsh winters. These factors have collectively led many authorities to predict the colonization of new areas of Florida, and perhaps even adjoining states, by the Walking Catfish.

Ecological Impact   Predation on indigenous fishes by Walking Catfish in Florida occurs principally in the dry season (November to April/May), when low water conditions force native species and the alien into deeper, often isolated, refugia (Courtenay, 1970; Courtenay & Ogilvie, 1971; Courtenay *et al.*, 1974). Under these conditions, Walking Catfish both compete for food (Lachner *et al.*, 1970) and prey heavily on native fishes, and will frequently kill large bass *Micropterus* spp. without eating them (Courtenay & Miley, 1975). It is ironic, as

Courtenay and Stauffer (1990) point out, that the greatest negative impact exerted by Walking Catfish in Florida has been on fish farmers, who have been forced to surround their ponds with protective fencing to prevent the exotic preying on cultured fish, which are far more favoured for human consumption. 'This species', Lachner *et al.*, (1970) conclude, 'is a severely harmful competitor, for it apparently reduces the entire freshwater community that it invades to one common denominator, more walking catfishes'. 'The present threat to fish farming interests, and the potential threat to native fishes, are thought to be serious' (Courtenay & Miley, 1975). (See also Buckow, 1969; Idyll, 1969; Courtenay & Robins, 1973; Minckley, 1973; Silverstein & Silverstein, 1974; Courtenay, 1975; Gore & Doubilet, 1976; De Man, 1983; Courtenay, 1989; Courtenay *et al.*, 1991; Courtenay, 1992, 1993.)

Australasia

Papua New Guinea

'First introduced to the Lake Sentani region (date unknown), but now also found in the Vogelkop Peninsula on the Prafi Plain near Manokwari and in the Sorong district' (Allen, 1991). As the presence of the Walking Catfish in Papua New Guinea is not mentioned by Welcomme (1988), it may be assumed that its introduction, or at least naturalization, occurred between about 1987 and 1990.

Oceania

Mariana Islands (Guam)

In 1910, Walking Catfish from Manila in the Philippines were introduced to the island of Guam in the Marianas, where they have become naturalized (Fowler, 1925; Herald, 1961; Kami *et al.*, 1968; Best & Davidson, 1981; Maciolek, 1984).

Chinese Catfish (Oopu Kui; Asiatic Catfish; White-spotted Catfish; White-spotted Clarias)
Clarias fuscus

The Chinese Catfish is widely cultured in parts of mainland China and on the island of Taiwan, where its tolerance of poorly oxygenated waters and its supplementary breathing organs make it ideal for culture in paddy-fields ('rice-fish culture'). The Chinese Catfish is also widely transported to Chinese communities throughout South East Asia, where it is valued for its medicinal properties and also for human consumption (Welcomme, 1988).

Natural Distribution

Southern China.

Naturalized Distribution

Oceania: Hawaiian Islands.

Oceania

Hawaiian Islands

The Chinese Catfish was introduced to the Hawaiian Islands before 1901, probably by Asian immigrants. It is now widely naturalized in streams on Hawaii, Kauai, Maui and Oahu, and is increasing and spreading (Brock, 1952, 1960; Herald, 1961; Kanayama,

1968; Maciolek & Timbol, 1980; Maciolek, 1984; Courtenay *et al.*, 1991; Courtenay, 1993).

Allied Species

The African Catfish *C. gariepinus (lazera)*, a native of the Rivers Niger and Nile, was introduced in 1972–1973 from the Central African Republic to Cameroun, Congo, Gabon, Ivory Coast and Zaire for experimental aquaculture and to kill stunted populations of *Tilapia*. It is known to be breeding successfully in captivity in the Ivory Coast, and 'there is uncertainty in most areas as to whether the species has become established in natural waters' (Welcomme, 1988).

The African Catfish has also been imported for aquacultural purposes to Cyprus and The Netherlands, where its status is uncertain (Welcomme, 1981, 1988); in 1985 to the Philippines, again with unknown results (Juliano *et al.*, 1989); and in 1981 to China (Yo-Jun & He-Yi, 1989).

The catfish *C. macrocephalus* was introduced from Thailand to Malaysia in 1950 (Tweedie, 1952), where it has bred naturally in swampy ditches, paddy-fields and pools, but competes unsuccessfully with the native Catfish *C. batrachus* (Ang *et al.*, 1989), and to China from the same source in 1982 (Yo-Jun & He-Yi, 1989).

LORICARIIDAE

Suckermouth Catfishes (Armoured Catfishes)
Hypostomus spp.

Suckermouth catfishes are resistant to desiccation and adapted to both poorly oxygenated, slowly flowing waters and well-oxygenated torrents, and are thus widespread in Latin American waters (Welcomme, 1988).

Natural Distribution

The genus *Hypostomus* occurs from Panama and Costa Rica in Central America southward to the basin of the Río de la Plata in South America, apart from Pacific drainages in Peru and Chile.

Naturalized Distribution

North America: USA. *Oceania:* Hawaiian Islands.

North America
United States

Breeding populations of at least three morphologically distinct, but so far unidentified, species of suckermouth catfishes, believed to be derived from escapes from aquaria, are naturalized in the USA in Florida, Nevada and Texas.

In Florida, the largest population occurs in Six Mile Creek, near Eureka Springs, east of Tampa, Hillsborough County, to which suckermouth catfishes escaped from nearby fish farms (Burgess, 1958), and from which they have spread into the Hillsborough River, where they are now abundant (Courtenay & Stauffer, 1990). A second *Hypostomus* species has been established in Florida in the Snapper Creek drainage in Dade County since 1959,

probably as a result of the dumping of unwanted fish from private aquaria (Courtenay *et al.*, 1974). Rivas (1965) reported *H. plecostomus* to be established in a rockpit in west Miami.

Minckley (1973) recorded suckermouth catfishes as naturalized 'in a warm spring in southern Nevada' (identified by Courtenay and Deacon (1982) as Indian Spring, 62 km northwest of Las Vegas, Lark County), since at least 1966. Although the population may have declined somewhat due to collection of adults for sale to local pet shops, predation on eggs by an introduced snail *Melanoides tuberculata*, and competition from other introduced fishes, it nevertheless remains abundant and is breeding successfully (Courtenay & Deacon, 1982). It is believed to originate in fish released by aquarists.

Barron (1964) was the first to report the establishment of suckermouth catfishes in the headwaters of the San Antonio River, Texas, where Hubbs *et al.* (1978) recorded them as a major component of the fish fauna in the spring run. Hubbs (1982) suggested that the species present may be *H. plecostomus*. The population is derived from fish that escaped from the San Antonio Zoological Gardens (Barron, 1964; Hubbs *et al.*, 1978). (See also Courtenay & Robins, 1973; Courtenay *et al.*, 1984, 1986, 1991; Courtenay, 1993.)

Oceania
Hawaiian Islands

Eldredge (1994) lists suckermouth catfishes *Hypostomus* spp. as occurring since 1984 on Oahu in the Hawaiian Islands.

Radiated Ptero (Sailfin Catfish)
Pterygoplichthys multiradiatus

The Radiated Ptero or Sailfin Catfish is a popular warmwater aquarium species.

Natural Distribution

The Río Magdalena, Colombia; Guyana; throughout the Amazon basin and the Río San Francisco, Brazil; tributaries of the upper reaches of the Amazon in Bolivia, Venezuela and Peru; and the Río de la Plata in Argentina and Paraguay.

Naturalized Distribution

North America: USA. ***Oceania:*** Hawaiian Islands.

North America

United States

Individual specimens of the Radiated Ptero have been collected in the wild in waters of southeastern Florida since 1971 (Courtenay *et al.*, 1984, 1986). In June 1983, breeding and establishment in the upper Río Vista Canal system near Miami in Dade County was confirmed by the Florida Game and Fresh Water Fish Commission, followed later in the same year by proof of reproduction in Broward County and in Lake Osborne in Palm Beach County to the north (Courtenay *et al.*, 1986; Courtenay & Stauffer, 1990). (See also Courtenay *et al.*, 1991; Courtenay, 1993.)

Oceania

Hawaiian Islands

In 1986, the Radiated Ptero was found to be naturalized on Oahu in the Hawaiian Islands (Courtenay *et al.*, 1991; Courtenay, 1993; Eldredge, 1994).

Allied Species

Courtenay *et al.* (1991), Devick (1991), Courtenay (1993) and Eldredge (1994) list an ***Ancistrus*** sp. (bristlenosed catfish) as introduced from South America to Oahu in the Hawaiian Islands, where it has been established since 1985.

BELONIDAE

Courtenay *et al.* (1991) and Courtenay (1993) list the Freshwater Long Tom *Strongylura kreffti* of Australia and New Guinea as established in fresh and marine waters in the Hawaiian Islands since 1988. Devick (1991) and Eldredge (1994) lists the Asian Needlefish or Stickfish *Xenentodon cancila* as occurring in the wild on Oahu in the Hawaiian Islands in 1988.

CYPRINODONTIDAE

Plains Killifish
Fundulus zebrinus

A small cyprinodont able to tolerate a fairly wide range of water temperature; the Plains Killifish is used as a bait-fish by anglers.

Natural Distribution

Southeastern Montana to Missouri and Texas, USA.

Naturalized Distribution

North America: Mexico.

North America

Mexico

The Plains Killifish is established in three separate localities of the Río Grande in Mexico between El Paso and the confluence of the Pecos River (Hubbs *et al.*, 1977; quoted by Contreras & Escalante, 1984), probably as a result of the release of excess bait by fishermen (Contreras & Escalante, 1984).

Allied Species

Between 1970 and 1973, the Common Killifish or Mummichog *F. heteroclitus* was introduced from North America to Spain (Elvira, 1995b), where it is currently naturalized in the drainages of the Rivers Guadiana and Guadalquivir (Elvira, 1995a). Its successful establishment poses a potential threat, through interspecific competition, to the endemic tooth carps *Aphanius iberus* and *Valencia hispanica*.

Giant Rivulus (Hart's Rivulus; Trinidad Rivulus)
Rivulus harti

The 10 cm Giant Rivulus is the largest species in its genus. It is a popular ornamental aquarium fish and in its native range is also used for human consumption (Welcomme, 1988).

Natural Distribution

Northern Brazil, eastern Colombia, Venezuela, and some neighbouring offshore islands, including Trinidad.

Naturalized Distribution

North America: ? USA.

North America

United States

Evidence on the status of the Giant Rivulus in the USA is fragmentary and contradictory. According to St Amant (1970), surplus aquarium fish from

Venezuela were released in California around 1967, where Moyle (1976a,b) said that 'evidence for permanent breeding populations [is] uncertain'. Courtenay *et al.* (1984) say that the Giant Rivulus was formerly established in Riverside County, California, from which it had since disappeared (see also Shapovalov *et al.*, 1981). Welcomme (1988) says that 'some individuals escaped from aquaria in Florida [presumably an error for California] but the population that was established was later eradicated'. Courtenay *et al.* (1991) and Courtenay (1993), however, indicate that the Giant Rivulus is still established in California.

Allied Species

Although the Mangrove Rivulus ***R. marmoratus*** may be a native of Puerto Rico, where it lives in mangrove lagoons and reproduces in fresh and brackish waters, Erdman (1972) believed that it might have been introduced at an unrecorded date.

ORYZIATIDAE

Japanese Rice Fish
Oryzias latipes

The Japanese Rice Fish is a popular aquarium species, and is also used in experimental laboratory research and for the control of mosquitoes.

Natural Distribution

Japan, Korea, eastern China and eastern Khazakstan.

Naturalized Distribution

Central America: Cuba.

Central America

Cuba

In 1976, Japanese Rice Fish from the former USSR were introduced to control mosquitoes in Cuba, where they are established and breeding (Welcomme, 1988).

Unsuccessful Introductions

Japanese Rice Fish formerly established in Suffolk County, New York (Welcomme, 1981) and California (Welcomme, 1988), have since disappeared (Courtenay & Hensley, 1980b). An introduction to the Hawaiian Islands was also unsuccessful (Welcomme, 1988).

POECILIIDAE

Pike Killifish (Pike Top-minnow)
Belonesox belizanus

The 10–20 cm Pike Killifish is so called because it both resembles a small Northern Pike *Esox lucius* and hunts its prey in much the same manner. In its native range it is believed to have suffered considerably from the use of insecticides sprayed to kill mosquito larvae in the muddy backwaters and marshes in which it lives.

Natural Distribution

From southern Mexico through Guatemala. Honduras and Nicaragua to Costa Rica.

Naturalized Distribution

North America: USA

North America

United States

In November 1957, Pike Killifish imported from Yucatan, Mexico, were dumped in a canal bordering SW Eighty-seventh Avenue in southeastern Dade County, Florida, after a research grant funding their use in laboratory experiments ceased (Belshe, 1961; Miley, 1978). Whether the fish were released by laboratory technicians or by importers is unknown (Courtenay *et al.*, 1984). Pike Killifish are currently established and abundant in canal systems and saline cooling canals (40‰) in southeastern Dade County (Belshe, 1961; Rivas, 1965; Lachner *et al.*, 1970; Courtenay & Robins, 1973; Courtenay *et al.*, 1974; Miley, 1978; Courtenay *et al.*, 1984; 1986; 1991; Courtenay, 1993).

Pike Killifish were formerly also established near San Antonio, Texas, from where they have since disappeared (Hubbs *et al.*, 1978).

Ecological Impact   In times of drought, when water remains only in the main canals, Pike Killifish, voracious predators, have been known to eradicate all small native forage fishes where they occur; their predation on the Mosquitofish *Gambusia affinis* (q.v.) has had a marked effect on that species' ability to control mosquito larvae (Belshe, 1961; Lachner *et al.*, 1970; Courtenay & Robins, 1973; Miley, 1978; Turner, 1981).

In experimental studies of predation by Pike Killifish on four native cyprinodontids in Dade County, Miley (1978) found that the Mosquitofish and the Sailfin Molly *Poecilia latipinna* (q.v.) were much more heavily predated than the Least Killifish *Heterandria formosa* and the Blue Killifish *Lucania goodei* because the two former species live near the surface. Where aquatic vegetation, in the form of *Hydrilla verticillata*, provided cover, predation on all four species was significantly reduced. In a particularly weedy canal, native poeciliids and cyprinodontids comprised 98% of the number and almost 60% of the biomass of all native fishes: after the vegetation was removed, these figures fell to 15 and 23% respectively, most dramatically in the numbers of Mosquitofish which fell from 210 (2.4% of the biomass) to nil. Furthermore, where Pike Killifish occurred, native poeciliids and cyprinodontids constituted 54% of the number and 34% of the biomass, compared with 97% and 58% respectively where Pike Killifish were absent. (See also Taylor *et al.*, 1984.)

Allied Species

According to Welcomme (1988), the Ten-spotted Livebearer *Cnesterodon decemmaculatus*, a small ornamental aquarium fish, has been introduced from Argentina to Chile, where it is reproducing. No further details are given, and the species is not mentioned by De Buen (1959).

Mosquitofish (Topminnow; Gambusia)
Gambusia affinis

The Mosquitofish, a small livebearer (females reach up to 6 cm and males 3.5 cm in length) has been introduced to every continent (except Antarctica) to control malarial mosquitoes, and 'is now possibly the most widely distributed fresh-water fish in the world' (Krumholz, 1948). It prefers still or slowly flowing waters, and its tolerance of high and low temperatures (from 38 to 4.5°C) (Otto, 1973); of high salinity (Ahuja, 1964); of oxygen depleted waters and pollutants; and its fecundity, have enabled it to become naturalized over a wide range of climatic and environmental conditions (Lloyd *et al.*, 1986).

Past opinions on its effectiveness in the control of mosquitoes have differed (see e.g. Gerberich & Laird, 1968), but it is now generally accepted that the Mosquitofish is little or no more efficient than many native larvivorous fish species (Haas & Pal, 1984). The use of biological controlling agents declined with the introduction of pesticides, but concern about the effects of the latter on the environment have led to a resurgence of interest in the use of biological controls (Welcomme, 1988).

Records of most introductions do not specify which subspecies was used, but Lloyd *et al.* (1986) considered that in the majority of cases both were probably involved. (But see under Natural Distribution.)

Natural Distribution

North America, from southern Indiana and Illinois south to Florida and northwards along the Atlantic slope to New Jersey in the USA, and in Mexico south to the Río Cazones, Veracruz.

The Mosquitofish has been divided into two subspecies, the nominate *G. a. affinis* from the west of its range (from southern Indiana to Alabama and the mouth of the Río Grande) and *G. a. holbrooki* from the east (from eastern Alabama and Florida to New Jersey). The two races occur allopatrically except in southern Alabama and northwestern Florida (Krumholz, 1948). Recent research, however, indicates clear taxonomic separation at the species level (Arthington & Lloyd, 1989). To avoid confusion with the literature, the earlier division into subspecies is here retained.

Naturalized Distribution

Europe: ? Austria; Bulgaria; Corfu; Corsica; Cyprus; France; ? Germany; Greece; Hungary; Italy; Monaco; Portugal; Romania; Spain; former USSR; former Yugoslavia. ***Asia:*** ? Afghanistan; ? Brunei; China; Hong Kong; India; Iran; Iraq; Israel; Japan

Philippines; ? Saudi Arabia; Singapore; Syria; Taiwan; Thailand; Turkey; Yemen. **Africa:** Algeria; ? Central African Republic; Egypt; Ethiopia; Kenya; Madagascar; Morocco; South Africa; Sudan; Zimbabwe. **North America:** Canada. **South and Central America:** Argentina; Bolivia; Chile; Hispaniola (Haiti & Dominican Republic); Peru; Puerto Rico. **Australasia:** Australia; New Zealand; Papua New Guinea. **Oceania:** Bermuda; Cook Islands; Fiji; French Polynesia; Guam; Hawaiian Islands; Kiribati; Maldive Islands; Mariana Islands; Micronesia; New Hebrides; Samoa; Solomon Islands.

Europe

According to Elton (1958) and Muus and Dahlstrøm (1968), Mosquitofish have become naturalized in all European Mediterranean countries and in those around the Black Sea.

In *Bulgaria,* to which Mosquitofish were introduced in the mid-1950s, they are widely distributed in a variety of aquatic habitats, but are said to be only effective in the control of mosquito larvae when used in conjunction with insecticides. In cold winters they have a poor rate of survival (Haas & Pal, 1984).

Mosquitofish were introduced to the island of *Corsica* in the western Mediterranean in 1926 (Gerberich & Laird, 1968), where they are said to be 'very efficient in ditches and wells which are well looked after' (Haas & Pal, 1984), and to the island of *Cyprus* in the eastern Mediterranean in 1929 and again about 1954 (Gerberich & Laird, 1968). Shortly after the Second World War they were imported to south and south-western *France,* where although they are often abundant they seem unable to control mosquito larvae (Haas & Pal, 1984). On the mainland of *Greece,* on the other hand, where they were introduced in 1928–1929, they appear to be very effective where they occur at high densities (Motobar, 1978; Haas & Pal, 1984); Stephanidis (1964) describes their negative impact on the native fauna of a small lake on the Greek island of *Corfu,* where they have probably eradicated *Valencia letourneuxi,* which has also become extremely rare where *Gambusia* occurs on the western coast of the Greek mainland (A.C. Wheeler, pers. comm. 1995). Mosquitofish are widely distributed and successful in *Italy,* (where the subspecies is said by Artom (1924) to be *G. a. holbrooki*), to which they were introduced between 1919 and 1922 and again in 1927 (Krumholz, 1948; Welcomme, 1981, 1988,

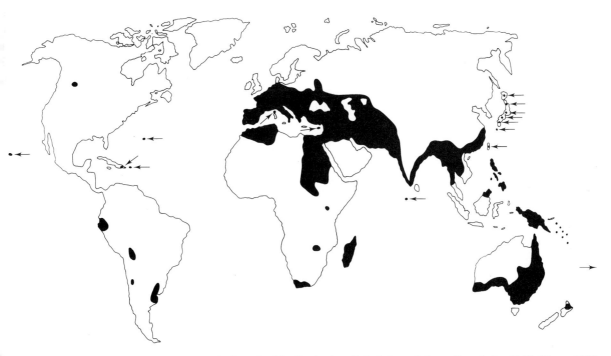

Naturalized distribution (discontinuous) of Mosquitofish *Gambusia affinis* (adapted from Krumholz, 1948; Elton, 1958; Lloyd *et al.,* 1986).

1991; Mazzola, 1992). Mosquitofish were introduced to *Monaco* in the mid-1960s and soon after the Second World War to *Portugal* where they are now disseminated widely, especially in the River Sado (Haas & Pal, 1984), and to *Spain* in 1921 (Meynell, 1973), where they are widely distributed except in the north and Galicia (Elvira, 1995a). Introduced to *Romania* in 1950–1953, Mosquitofish survive only in 'one warm and sulfurous lake' (Haas & Pal, 1984). In 1925, Mosquitofish were translocated within the former *USSR* (Motobar, 1978), where Vooren (1972) says they are established only 'near industrial warm water discharges in the vicinity of Moscow'. Borisova (1972) lists them as among those species accidentally introduced in 1961 in consignments of Grass Carp *Ctenopharyngodon idella* (q.v.) and Silver Carp *Hypophthalmichthys molitrix* (q.v.) from the Far East to the Akkurgan Fish Combine near Tashkent in Uzbekistan, where they became established in nearby inlet and outlet ditches and other natural waters. In 1924 and 1932, Mosquitofish were introduced to Croatia in the former *Yugoslavia*, and between 1947 and 1954 to Macedonia (Gerberich & Laird, 1968), where their 'presence in marshes is thought to be responsible for the absence of *Anopheles* [mosquito] larvae therein' (Haas & Pal, 1984).

In northern Europe, Schroder (1928) records the introduction of Mosquitofish to *Germany* in 1921, and Gerberich and Laird (1968) mention their introduction to *Austria* at an apparently unrecorded date. In both countries their present status is unknown. In *Hungary*, where they were introduced in 1939, 'populations have persisted in two thermally heated ponds' (Welcomme, 1988).

Asia

Afghanistan

Motobar (1978) records the introduction of Mosquitofish to Afghanistan in 1970. Haas and Pal (1984) say they were introduced around 1980, and that they are 'alone not epidemiologically satisfactory' in the ditches, rice paddies, pools, lakes and marshes where they occur.

Brunei

Haas and Pal (1984) indicate the introduction of Mosquitofish to Brunei in about 1935, but give no further details.

China

Yo-Jun and He-Yi (1989) record the introduction of Mosquitofish to China several decades ago, and that they are currently naturalized in many localities in Shanghai and Jiansu Provinces.

Hong Kong

Haas and Pal (1984) record the introduction of *Gambusia affinis* to Hong Kong, but give no further details.

India

Mosquitofish from Italy were first imported to Mysore in 1928; in 1937 some were sent by the Department of Public Health to Travancore, but due to a sudden change in temperature and inadequate packing only around 150 survived the journey. A second consignment of between 600 and 800 in the following year was more successful, the fish being placed in a hatchery at Perungadavila near Neyyattinkara (Gopinath, 1942), from where some were transferred to several other states as far north as the Punjab. In 1929, more were imported from Ceylon (Sri Lanka) by the Madras Fisheries Department, which stocked them in waters on Krusadai Island, and others were transferred from Bangalore to the city of Madras in 1930 (Chacko, 1948; Jones & Sarojini, 1952). In 1929 and 1930, Mosquitofish were unsuccessfully introduced to Tamil Nadu (Natarajan & Ramachandra Menon, 1989; Sreenivasan, 1989), where in Ooty Lake in the Nilgiri Hills they adversely affected the Mirror Carp *Cyprinus carpio* var. *specularis* (q.v.) fishery by eating their eggs; only after Mosquitofish were eradicated could the Mirror Carp fishery be re-established (Shetty *et al.*, 1989). Yadav (1993) says that the Mosquitofish 'is now widely distributed in parts of India'.

Iran

Motobar (1978) indicates the introduction of Mosquitofish to Iran in 1928. Haas and Pal (1984) say that they were introduced in 1976, and that they have achieved 'a certain reduction' in *Anopheles* larvae.

Iraq

Mosquitofish were first introduced to Iraq in 1928 (Al-Daham *et al.*, 1977). Haas and Pal (1984) indicate an introduction around 1960, and that they became established in 'numerous ditches, streams, pools, a few ponds, lakes, and marshes'.

Israel

The Mosquitofish is the only non-commercial alien fish intentionally planted in inland waters of Israel, to which it was introduced in the 1920s and where it is now widespread and abundant and may be competing with the native *Aphanius mento* (Ben-Tuvia, 1981; Davidoff & Chervinski, 1984).

Japan

Sawara (1974) says that Mosquitofish were first introduced to Japan 'in the 1910's, and are now widely distributed mainly in the Kanto Plain, Ryukyu Islands and others'. In 1969, Mosquitofish from Kasai, Edogawa-ku, Tokyo, were released in a pond in the Botanical Gardens of the University of Tokyo, Koishikawa, Bunkyo-ku, where they established a self-maintaining population (Sawara, 1974). Chiba *et al.* (1989) say that Mosquitofish were originally introduced to Japan from Taiwan in 1916.

Philippines

'When returning to the Philippine Islands from the United States in 1913 I secured two dozen mosquito fish at Honolulu [Hawaii] . . . and brought them to Manila. The offspring of these fish now number many thousands and are being widely distributed throughout the Philippine Islands . . .' (Seale, 1917). Juliano *et al.* (1989) record them as established in brackish ponds.

Singapore

Chou and Lam (1989) say that Mosquitofish have become 'established permanently in low salinity waters' in Singapore.

Saudi Arabia

In 1983, *G. a. holbrooki* was discovered in the Hofuf Oasis in the Eastern Province of Saudi Arabia (Ross, 1984). Whether it is established and breeding is unknown.

Syria

Introduced to Syria before 1968 (Gerberich & Laird, 1968) and again in 1979, Mosquitofish became established in some still and flowing waters (Haas & Pal, 1984).

Taiwan

Introduced from the USA in the early 1920s (and possibly previously in 1911), the Mosquitofish 'spread all over Taiwan as a wild fish and also came to be considered as a good experimental fish' (Liao & Liu, 1989).

Thailand

The Mosquitofish was introduced to Thailand between 1919 and 1929 jointly by the Fisheries Department, the Ministry of Agriculture and Cooperative, and the Ministry of Public Health. 'The species is now widely distributed in natural waters, especially in the urban region of the big cities' (Piyakarnchana, 1989).

Turkey

In Turkey, Mosquitofish introduced at an apparently unrecorded date have successfully controlled *Anopheles*, *Culex*, and *Aëdes* mosquito larvae (Haas & Pal, 1984).

Yemen

Introduced at an apparently unknown date to Yemen, Mosquitofish became established in streams, pools, and reservoirs (Haas & Pal, 1984).

Unsuccessful Introductions

Mosquitofish were introduced into Indonesia in 1929 when they escaped or were released into open waters from aquaria; they have since disappeared (Muhammad Eidman, 1989). They have also apparently been unsuccessful in Malaysia (Haas & Pal, 1984).

Africa

Algeria

G. a. holbrooki was introduced from the east of the species' North American range to Algeria in 1930, where it became widely established. At Cuelma it has markedly reduced the larvae of *Anopheles labranchiae*, and at Quargla has eliminated them (Haas & Pal, 1984). Gerberich and Laird (1968) say the date of introduction was 1926.

Central African Republic

The status of the Mosquitofish in the CAR, to which it was introduced in 1958, is uncertain. Welcomme (1981) said it 'disappeared after several months', but later (Welcomme, 1988) that it was reproducing.

Egypt

Introduced to Egypt in 1929 (Motobar, 1978), the Mosquitofish is having 'no evident effect on mosquitoes' (Welcomme, 1981, 1988).

Ethiopia

In 1938, the Establishmento Ittiogenico di Roma introduced 1000 Mosquitofish of the form *G. a. holbrooki* from Italy into Lake Tana, where they proved effective in controlling mosquitoes (Tedla & Meskel, 1981).

Kenya

Muchiri and Hickley (1991) record the successful establishment of a *Gambusia* sp. (presumably *affinis*) in Lake Naivasha.

Madagascar

G. a. holbrooki was first introduced to the island of Madagascar in 1929, and specifically to Lake Alaotra in 1940 and to Lake Itasy 10 years later (Kiener, 1963; Moreau, 1979). It is now widely distributed, but apparently prefers to feed on fish larvae rather than mosquito larvae (Reinthal & Stiassny, 1991). Welcomme (1981) says it 'developed in waters of the plateaux'. In 1988, Reinthal and Stiassny (1991) collected Mosquitofish in the following Malagasy waters; the River Sampazina and a small stream flowing out of Lac Vert in the northeast central rainforest; and in the Rivers Namorona and Botely in the Ranomafana National Park.

Morocco

First introduced to Morocco in 1929 (Gerberich & Laird, 1968), Mosquitofish have been found to be uncharacteristically intolerant to variations in temperature but tolerant of polluted waters. Since their introduction, the density of mosquito larvae in rice paddies has declined by between 90 and 95%, although the larvae can be difficult for the fish to swallow and their predation can be hindered by aquatic vegetation (Haas & Pal, 1984).

South Africa

De Moor and Bruton (1988), from whom much of the following account is derived, have traced the history of the introduction of Mosquitofish to South Africa.

Mosquitofish were first imported to South Africa in 1936 to control mosquitoes and as a forage fish for bass *Micropterus* spp. (Jubb, 1976, 1977). There were no official importations of Mosquitofish to the Jonkershoek hatchery, but stocks of both *G. a. affinis* and *G. a. holbrooki* were obtained from private individuals in 1936 and 1944 respectively, and were distributed to various localities (Anon., 1944). De Moor and

Bruton (1988: appendix V: 272–273) give dates of initial introductions (with sources) of Mosquitofish in South Africa.

Recent records of Mosquitofish in the Transvaal and southern Cape are from Groenvlei (near Knysna) (Van der Merwe, 1970); Victoria Lake, Germiston, in the Vaal catchment; Maryvale reed swamp, Blesbokspruit, in the Vaal system; the Knysna and Hoekraal River systems; and the Platkloof River in the Berg River system. Pre-1975 records occur for the Limpopo River, Messina; Princess Vlei; and Karatara River, Sedgefield, in all of which Mosquitofish probably still survive (De Moor & Bruton, 1988).

Ecological Impact Mosquitofish in South Africa seem to have had little negative impact on native fishes. In Groenvlei, for example, where a population has been naturalized for many years, they do not seem to have affected either *Atherina breviceps* or *Gilchristella aestuaria*, two small indigenous species which still occur in large numbers. In an artificial outdoor fish pond in Grahamstown, however, *G. affinis* was observed preying on young *Tilapia sparrmani*, and their known predation on young bass reduces their value as a forage fish for adult bass. Similarly, their inability to penetrate thickly matted aquatic vegetation (a favourite habitat for mosquito larvae) lessens their usefulness as a biological controlling agent (Myers, 1965; De Moor & Bruton, 1988).

Sudan

Introduced to Sudan from Italy in 1929 (Motobar, 1978), the Mosquitofish is established in irrigation canals in Gézira cotton fields, where it is susceptible to predation by *Clarias gariepinus* (Moreau *et al.*, 1988).

Zimbabwe

G. a. affinis was first imported to Salisbury, Rhodesia (now Harare, Zimbabwe) 'in a bucket from New York' in 1925. The progeny of these fish, and later importations, are still present in some waters (mainly artificial garden pools and aquaria), but have not been very successful in controlling mosquitoes (Toots, 1970).

Unsuccessful Introductions

Mosquitofish have been introduced in Africa apparently unsuccessfully to Ghana and the Ivory Coast (Welcomme, 1988), and to Senegal and Uganda (Haas & Pal, 1984).

North America

Canada

In 1924, some 200 Mosquitofish from California were imported to Alberta by the Canadian Entomological Branch, around 25 of which were in the same year 'planted in the warm waters overflowing from the Cave and Basin swimming pool at Banff' (Mail, 1954). These waters overflowed into an extensive swampy area below the pool, in which mosquitoes bred prolifically, and where the Mosquitofish became established (Nelson, 1984). The hotsprings are situated about 1.5 km southwest of Banff on the southwest side of the Bow River in the Banff National Park.

In 1929 'top minnows (presumably *Gambusia affinis* or *Lebistes reticulatus*)' were introduced to a pond near Kelowna, British Columbia, where they failed to survive the winter (Carl & Guiguet, 1958; Carl *et al.*, 1967).

In 1958, 50 Mosquitofish were released in a pond near the University of Manitoba, Winnipeg, where they increased to around 3000 before in 1961 the pond dried out and all the fish died (Smith, 1960).

The identity of some fish discovered in 1968 in the sulphur hotspring area of Third Vermillion Lake, just across the Bow River from the Cave and Basin site, is somewhat confused. McAllister (1969) refers to them as both '*Gambusia affinis*' and as 'guppies' [*Poecilia reticulata*]. Crossman (1984) has assumed them to be Mosquitofish.

Translocations

Mosquitofish have been widely translocated in North America outside their natural range, for example to Arizona (Courtenay, 1992), Nevada (Deacon *et al.*, 1964; Hubbs & Deacon, 1964; Courtenay & Deacon, 1982); Montana (Brown & Fox, 1966); and the Great Lakes (Mills *et al.*, 1994).

South and Central America

Argentina

Introduced from Texas, USA, to Argentina at an apparently unrecorded date, Mosquitofish are well established in waters of the Río Tercero, Córdoba; in Rosario de Santa Fé; and in public park lakes in Buenos Aires (Navas, 1987).

Bolivia

According to Welcomme (1981), the Mosquitofish is established in 'most subtropical regions' of Bolivia, where it 'eats fry of other fishes' (Welcomme, 1988).

Chile

Welcomme (1981, 1988) says Mosquitofish were introduced to Chile in 1937, where Mann (1954) states that they occurred in northern and central Chile. De Buen (1959) records the subspecies introduced as *G. a. holbrooki*.

Hispaniola (Haiti and Dominican Republic)

The Mosquitofish established on the island of Hispaniola, to which it was introduced at an apparently unrecorded date and where it is widely distributed in valleys and lowland streams of the interior, is believed to be *G. a. holbrooki*. Where the two species occupy the same waters, it seems unable to compete with the larger endemic *G. hispaniolae/dominicensis* which is found in the Valle Neiba/Cul de Sac region (Wetherbee, 1989).

Peru

Introduced to Peru in 1940, Mosquitofish are said by Welcomme (1988) to be established and breeding.

Puerto Rico

According to Welcomme (1988), Mosquitofish were first introduced to Puerto Rico in 1914. Hildebrand (1934), however, says they were first introduced around 1923, when he found them in Patillas reservoir. Erdman (1984) reported them also in Melanía reservoir near Guayama, but said the species seems to have spread little in the past 50 years. He added that it is often confused with *Poecilia vivipara*, which could account for some inconsistencies in distributional records.

Australasia

Australia

Kailola (1981) records the introduction of Mosquitofish 'into Sydney in 1925 and to Brisbane in 1929. From 1935 introductions were made to parts of New South Wales, and in 1940 to Darwin. They were liberated in irrigation ditches at Nedlands, Western Australia, in 1934, and in 1940 introduced by health authorities to Broome . . .'. Rolls (1969) reports the importation of a small number from the Middle East in 1926 by a Sydney health inspector, William Vogwell, who bred them in ponds at Redfern. During the Second World War, Mosquitofish were widely liberated by both military personnel in camps and by health authorities in public areas (McDowall, 1980; McKay, 1986–1987; Clements, 1988).

Merrick and Schmida (1984) recorded the Mosquitofish as 'locally abundant with a range extending to 8 drainage divisions. In the North-east Coast division it is present in the coastal swamps of the Maroochy River as well as the North Pine and Brisbane River systems and the Gold Coast swamps. The distribution extends throughout the South-east Coast, Murray–Darling, South Australian Gulf and Southwest Coast divisions. There are also restricted ranges in the Timor Sea, Lake Eyre and Western Plateau divisions; although *G. affinis* appears to have increased its range in central Australia since the 1974 floods'.

Arthington and Lloyd (1989) updated and summarized the distribution of Mosquitofish in Australia, where they are now 'locally abundant in inland and coastal streams and rivers within 10 major drainage divisions on the Australian mainland'. They are most frequently found in warm, slowly flowing or lentic waters with plenty of aquatic vegetation, and in backwaters and billabongs of lotic systems (McDowall, 1980; Arthington *et al.*, 1983; McKay, 1984), and also occur in freshwater ponds, swamps, marshes, lagoons and lakes, and in saline lakes, geothermally heated springs, and such artificial habitats as the cooling waters of power stations. Although Mosquitofish are able to survive in saline waters, they have not so far colonized river estuaries (Arthington & Lloyd, 1989).

In southeastern Queensland, Mosquitofish are found in particularly high densities in modified and/or polluted urban and suburban creeks overgrown with alien grasses, water hyacinth and *Azolla*, and in more degraded waters may constitute over 90% of the fish biomass (Arthington *et al.*, 1983). Although found principally in disturbed habitats in Queensland, Mosquitofish also occur in near-pristine environments (Arthington & Lloyd, 1989).

In central and southeastern Australia, Mosquitofish are established in nearly every type of suitable aquatic habitat, ranging from such comparatively unspoiled waters as Cooper Creek and the Finke River system in central Australia to waterholes in the arid centre and south of the country, and permanent streams and other urban and suburban wetlands around Adelaide, Melbourne and Sydney (Lloyd *et al.*, 1986), and polluted drains, ditches, lakes, pools, and saline evaporation basins (Lloyd, 1987). In the extreme southeast of South Australia, the distribution of Mosquitofish is limited, probably due to the scarcity of many permanent water bodies (Arthington & Lloyd, 1989).

In southwestern Western Australia, Mosquitofish are abundant in both still and flowing waters, where they range along the west coast from the Hay River to the Greenough River system north of Perth, where numerous shallow lakes provide ideal habitats. Although in lentic waters in the Northern Swan Coastal Plain, where Mosquitofish are virtually universal, indigenous fishes are uncommon and irregularly distributed, in some lotic waters such as Moore River, Ellen Brook, Canning River and North Dandalup River, native fishes seem relatively unaffected by the presence of Mosquitofish (Arthington & Lloyd, 1989).

The map in Twyford (1991) incorrectly indicates Mosquitofish to be widely distributed in the Northern Territory and in the eastern half of Tasmania.

All Mosquitofish identified in Australian waters belong to the eastern subspecies *G. a. holbrooki* (but see under Natural Distribution).

Mosquito Control   Myers (1965), and many other researchers, have concluded that the Mosquitofish is little or no better at controlling mosquitoes than many other fish species, and that mosquito larvae are frequently a minor constituent of the Mosquitofish's diet. In many instances, Mosquitofish have suffered heavy mortality when fed exclusively on mosquito larvae, and those that survive show poor growth rate and delayed maturation (Ravichandra Reddy & Pandian, 1972). At low densities, Mosquitofish may actually be of benefit to mosquito larvae by preying on their invertebrate predators, such as Notonectidae (aquatic bugs) and Odonata (dragonflies and damselflies) (Stephanidis, 1964; Hurlbert *et al.*, 1972).

In Australia, mosquitoes pose problems in three distinct habitats – domestic, freshwater and saltwater – in each of which a different species is responsible (Lloyd *et al.*, 1986). It has been found that some of the most important biting pests and disease vectors are relatively unaffected by fish predation, due to the temporal and spatial patchiness of their breeding ecosystems; these include *Culex sitiens*, *C. fatigans*, *C. annulirostris*, *Aëdes vigilax*, *A. normanensis*, *A. notoscriptus* and *A. aegypti* (Lloyd *et al.*, 1986). Although it has been claimed that at high densities Mosquitofish are effective at controlling mosquito larvae in permanent swamps, marshes and pools, this has yet to be confirmed.

Many authorities (e.g. Myers, 1965; McDowall, 1980) consider that such native fishes as hardyheads *Craterocephalus*, smelts *Retropinna*, rainbowfishes *Melanotaenia*, gudgeons *Hypseleotris* and *Mogurnda*, and *Galaxias* are as good, or better, at controlling mosquitoes as *Gambusia*, although this has not so far been proven.

Ecological Impact   According to McKay (1984), the Mosquitofish in Australia 'may be the most harmful of our introduced fishes'. Myers (1965) considered that almost everywhere that it has become established it has eventually eradicated most or all of the smaller indigenous fish species, and has seriously affected the young of commercially important game or food fishes.

Interference competition and resource exploitation competition (see Lever, 1994) are both exhibited by introduced Mosquitofish in Australian waters. Taylor *et al.* (1984) have pointed out that naturalized fishes normally have generalist feeding requirements and show trophic opportunism, and there is thus frequently a considerable overlap in the diet of naturalized and native species. The foods of *Gambusia affinis* and native Australian melanotaeniids, atherinids, galaxiids and a retropinnid are very similar (Lloyd, 1990). Rainbowfish *Melanotaenia duboulayi* appear to alter their feeding habits when in the presence of high densities of Mosquitofish, becoming more herbivorous and less dependent on terrestrial insects that drop into the water. When, in this situation, they do prey on other animals, they tend to select larger species, whereas Mosquitofish choose smaller prey. These dietary shifts reduce the overlap in food selection between the two species, and may contribute to the division of food resources (Arthington & Lloyd, 1989; Arthington, 1991). Predation on eggs and fry, aggression, and habitat disruption resulting in loss of macrophytes required by the native species for nesting sites may also be responsible for its decline in urban waters inhabited by Mosquitofish (Arthington *et al.*, 1983).

Other indigenous Australian fishes show niche shifts in the presence of Mosquitofish. Lloyd (1987), quoted by Arthington (1991), reported an expansion of niche width rather than a more predictable reduction. As a corollary to dietary niche shifts where Mosquitofish occur, some native fishes show shifts in both abundance and distribution. In the lower River Murray there is a mutually excluding pattern of distribution between Mosquitofish and the Pygmy Perch *Nannoperca australis* (Lloyd, 1987). Other species, such as melanotaeniids, atherinids, galaxiids and a retropinnid occur in greatly reduced numbers where they exist sympatrically with exotic Mosquitofish and Green Swordtails *Xiphophorus helleri* (q.v.) than in

waters without them (Arthington *et al.*, 1983; Lloyd, 1987; Arthington, 1991).

The ecological impact on native fishes of interference competition resulting from aggression and finnipping by Mosquitofish is believed to be considerable (Arthington, 1991). Lloyd (1987, 1990) found that Mosquitofish will attack indigenous species up to double their size, and can cause such damage to scales and caudal fins that disease may result. McKay (1978, 1984) suggested that the densities of Mosquitofish and Swordtails in some creeks could overwhelm native fishes.

The decline of autochthonous fishes in Australia and elsewhere has most frequently occurred in disturbed and polluted environments to which many exotic species, because of their broad ecological tolerance, flexible habitat requirements, and trophic opportunism, are well adapted (Arthington *et al.*, 1983; Courtenay & Stauffer, 1990; Arthington, 1989, 1991; Arthington & Lloyd, 1989). However, Mosquitofish and other poeciliids have also become established in undisturbed Australian waters (Lloyd, 1987), and may be responsible for the decline of a native rainbowfish *Rhadinocentrus ornatus* in the almost pristine environment of the dystrophic sedge Eighteen-Mile-Swamp on North Stradbroke Island, Queensland (Arthington & Lloyd, 1989; Arthington, 1991). *G. affinis* and *R. ornatus* both feed mainly on terrestrial insects on the surface of water and on small aquatic invertebrates, and there may thus be some competitive displacement, although direct predation by Mosquitofish on *ornatus* eggs and fry cannot be ruled out (Arthington, 1991). Myers (1965) names the Mosquitofish 'the fish destroyer', and there is some evidence that it may kill fish beyond the fry stage.

In coastal Queensland streams around Brisbane with many *Gambusia*, *Poecilia* and *Xiphophorus* fishes, nine indigenous surface-feeding or mosquito-larvae-eating fishes (*Melanotaenia*, *Pseudomugil*, *Craterocephalus* and *Retropinna*) are often rare or absent; wherever the introduced species do not occur, however, indigenes are normally abundant (Arthington *et al.*, 1981). In such streams, Mosquitofish prey on small terrestrial insects and on immature instars of aquatic insects, deliberately choosing very small prey items (Arthington, 1989). Such size selectivity could well affect the structure of invertebrate communities by changing recruitment of various taxa (Arthington, 1991), including beetles, backswimmers, rotifers, crustaceans and molluscs (Hurlbert *et al.*, 1972; McKay, 1977).

In some streams and lakes in the vicinity of Perth, Western Australia, Mosquitofish are almost the only species present, native fishes having nearly or completely disappeared (Merrick & Schmida, 1984).

The Mosquitofish is a remarkably successful colonist in Australia, especially in those waters from which predatory fish are absent and those not too severely changed by human activities, in many of which it has become the dominant species, sometimes to the total exclusion of native fishes. Its widespread naturalization is due to a combination of its tolerance of a broad spectrum of environmental factors and high fecundity, which have together enhanced its ability to compete and survive (McKay, 1984).

Mosquitofish do not appear to perform any useful roles in Australian waters, and although they may form part of the diet of water birds and other indigenous piscivores, they are avoided by other fish whenever there is a choice of food (Lloyd, 1987). They are also considered to be less useful in controlling mosquito larvae than some native fishes (see above under Mosquito Control), and they are believed to be partially responsible for the decline of various small indigenous fishes, including some that play an important role in controlling mosquitoes (Arthington *et al.*, 1983; Lloyd *et al.*, 1986; McKay, 1978, 1984; Arthington, 1989, 1991). In contrast to the claim by Roots (1976) that Mosquitofish 'proved ideal for introducing into tropical and subtropical regions as a biological control to combat the mosquito', 'the general view is that *G. holbrooki* is a pest or ecological "weed"' (Arthington, 1991).

New Zealand

According to McDowall (1990b), the first shipment of Mosquitofish to New Zealand arrived from Hawaii in 1928, but none survived the voyage; a second, from Sydney, Australia, 2 years later, suffered the same fate. A third consignment from Hawaii arrived on the *Aorangi* in the same year (1930), the survivors being placed in a small pond in the Auckland Botanical Gardens. Early in 1931 they were said to be 'breeding fast', and some were sent to the Cawthron Institute in Nelson and to the Wellington Acclimatization Society's hatchery at Masterton for experimental purposes and to try to determine their possible effects on such native species as whitebait, bullies and mudfish; the results of this research are unrecorded.

The earliest release of Mosquitofish into the wild noted by McDowall (1990b) was in Lake Ngatu, in Northland, in 1933, and in 1936–1937 others were sent from Auckland to New Plymouth. On the outbreak of the Second World War in 1939, when work ceased on the new Broadcasting House in Wellington, some Mosquitofish were released in the flooded basement, and others were placed in swamps near a newly constructed army camp at McKay's Crossing.

Although the Mosquitofish was introduced to New Zealand, as elsewhere, to control mosquitoes, interest in the fish was soon abandoned. It remains unclear in how many other localities Mosquitofish were planted, but only those in Northland persisted for any length of time, and they are presumed to be the source of more recently established populations elsewhere. According to McDowall (1990b), 'In recent times there has been a significant expansion of [the species'] occurrence. Mosquitofish can now be expected almost anywhere there are suitable habitats, in and north of the Waikato and Bay of Plenty. Mosquitofish were released into Lake Tarawera, in the Rotorua district, and the species is now present there in several areas, often where warm streams and springs flow into the lake. It is also found in some places in the King Country, Hawke's Bay and elsewhere. It is now so abundant in some areas that it has been described as an "infestation" . . . e.g. [it is] the most abundant fish species in the Whangamarino Swamp'.

The spread of Mosquitofish in New Zealand is likely to be limited by the species' preference for still or slowly flowing waters, especially those with extensive weedy margins (McDowall, 1990b).

The subspecies present in New Zealand has been identified as the nominate western from *G. a. affinis* (but see above under Natural Distribution).

Papua New Guinea

Mosquitofish were first introduced to Papua New Guinea in March 1930 – West and Glucksman (1976) say to Rabaul on East New Britain and Allen (1991) says to Irian Jaya, and later to Papua New Guinea. The original stock came from Sydney, Australia, and were placed in a pond in the Rabaul Botanic Gardens, from where their progeny were subsequently distributed to numerous outstations.

In the early stages of their naturalization Mosquitofish were believed to have considerably reduced the mosquito population in the Lakunai

Swamp near Rabaul, where the fish continued to be reared in captivity and distributed to all parts of the then Trust Territory. They were first imported to the territory of Papua in 1933, when a shipment from Rabaul was received in Port Moresby by the public health authorities, by whom they were released in some rainwater catchment tanks and a billabong near the Laloki River, 20 km from Port Moresby; from there, subsequent floods enabled the fish to gain access to the Laloki River and neighbouring swamps (West & Glucksman, 1976).

Contemporary reports (e.g. Giblin, 1936) suggest that Mosquitofish were quite effective in controlling mosquito larvae, and their distribution continued into the Second World War when, with the advent of DDT, demand for them declined, although some introductions and/or translocations continued to be made until at least 1972.

Glucksman and West (1976: table 2) give full details of all known introductions of *Gambusia affinis* to Papua New Guinea, where they are today widely distributed in all districts (Glucksman *et al.*, 1976), especially in the Laloki, Brown and Goldie Rivers (Berra *et al.*, 1975), and in the Sepik and Ramu systems, in Lake Kutubu, and around such major cen-

tres of human population as Port Moresby, Lae and Madang (Allen, 1991). Allen (1991) adds that the species' present status in Irian Jaya is unknown.

Oceania

Bermuda

In 1928, the Bermuda Government Health Department imported Mosquitofish from the eastern USA, following a programme of extensive ditching of the islands' peat marshes to create better drainage and to provide permanent waterways in which the fish could survive. The introduction proved extremely successful, and resulted in the eradication in the salt marshes of the Yellow Fever Mosquito *Aëdes aëgyptii*, and provided a valuable additional source of food for herons, kingfishers and other piscivorous birds (D.B. Wingate, pers. comm, 1995). Mosquitofish remain extremely abundant in numerous ditches, pools, marshes and reservoirs throughout Bermuda (Haas & Pal, 1984).

Because data on the introduction of *Gambusia affinis* to other oceanic islands are somewhat sparse, they are presented here in tabular form (adapted from Maciolek (1984) and Haas & Pal (1984)).

ISLANDS (OF SPECIAL CONCERN)	FIRST INTRODUCED	SOURCES
Cook Is. (Rarotonga, Mitiaro)	?	Krumholz, 1948; Welcomme, 1988
Fiji (Viti Levu)	1930s	Krumholz, 1948; Ryan, 1980; Andrews, 1985; Welcomme, 1988
French Polynesia (Tahiti)	?	Krumholz, 1948; Welcomme, 1988
Guam	?	Brock & Yamaguchi, 1955; Brock & Takata, 1956; Best & Davidson, 1981; Shepard & Myers, 1981; Haas & Pal, 1984; Welcomme, 1988
Hawaiian Islands	15 September 1905 (less than 420 from Seabrook, Texas, USA). 'Now among the more abundant and widely distributed exotics . . . useful in controlling mosquito larvae in certain, artificial habitats' (Maciolek, 1984).	Van Dine, 1907; Seale, 1917; Krumholz, 1948; Brock, 1952, 1960; Hida & Thomson, 1962; Kanayama, 1968; Maciolek, 1984; Welcomme, 1988
Kiribati		
1. Line Is. (Fanning, Washington)		Guinther, 1971; Maciolek, 1984; Welcomme, 1988
2. Gilbert Is., Marshall Is. (Jaluit), Palau (Babelthuap)	*c.* 1935	Gressit, 1961; Maciolek, 1984; Haas & Pal, 1984; Welcomme, 1988
Maldive Is.	?	Haas & Pal, 1984
Mariana Is. (Pagan, Saipan, Tinian)	?	Krumholz, 1948; Best & Davidson, 1981; Welcomme, 1988
Micronesia (Ponape, Pulusuk)	?	Krumholz, 1948; Welcomme, 1988
New Hebrides	?	Haas & Pal, 1984
Samoa (Aunuu, Tutuila, Swains, Savaii)	*c.* 1935	Krumholz, 1948; Haas & Pal, 1984; Welcomme, 1988
Solomon Is.	*c.* 1930	Glucksman *et al.*, 1976; Clements, 1988

Allied Species

It was formerly believed that the Domingo Mosquitofish *Gambusia hispaniolae/dominicensis*, a native of Haiti, occurred near Alice Springs in the Lake Eyre division of the Northern Territory of Australia (e.g. Merrick & Schmida, 1984; McKay, 1986–1987; Welcomme, 1988; Allen, 1989). Lloyd and Tomasov (1985), however, have shown that these fish are *G. affinis holbrooki* with atypical scale counts.

One-spot Livebearer (Dusky Millions Fish)
Phalloceros caudimaculatus

Although the very small (females *c.* 6 cm) One-spot Livebearer has been widely transported around the world by aquaculturists, the two recorded international introductions were made to control mosquitoes (Welcomme, 1988).

Natural Distribution

From Río de Janeiro in Brazil southward to Uruguay and Paraguay.

Naturalized Distribution

Africa: Malawi. *Australasia:* Australia.

Africa

Malawi

In 1956, One-spot Livebearers from Brazil were introduced to Nyasaland (Malawi), where naturalized populations became established in the Bwumbwe Dam and in the Ruo River (Jubb, 1977).

Australasia

Australia

Nothing is known about the date of the introduction of the One-spot Livebearer to Australia, where it is believed to be established only in some creeks and drains around Perth in Western Australia (Trendall & Johnson, 1981), and where it was originally misidentified as the Mosquitofish *Gambusia affinis* (q.v.). Its present status there is uncertain (Merrick & Schmida, 1984; Arthington, 1986, 1989, 1991; Allen, 1989; Arthington & Lloyd, 1989; McKay, 1989).

Sailfin Molly (Black Molly)
Poecilia latipinna

One of the most popular of all aquarium fishes, the colourful Sailfin Molly exists in numerous 'fancy' varieties. It has been recorded in the wild in many places, either as the result of escapes or releases from aquaria or through introductions for the control of mosquitoes (Welcomme, 1988). It is well able to survive in oxygen depleted, almost anaerobic, waters (McKay, 1984).

Natural Distribution

Fresh, brackish and salt waters from the Gulf of

Mexico and the Atlantic slope through North Carolina, USA, to central Veracruz, Mexico.

Naturalized Distribution

Asia: Philippines; ? Singapore. **North America:** Canada. **South America:** Colombia. **Australasia:** Australia; New Zealand. **Oceania:** Guam; Hawaiian Islands.

Asia

Philippines

The Sailfin Molly was introduced from Mexico to the Philippines to control malarial mosquitoes (Vooren, 1968), where it has become 'a pest in brackish ponds' (Juliano *et al.*, 1989).

Singapore

The presence of this species in Singapore is deduced from the fact that Australian introduction material came from this source' (Welcomme, 1988). There is, however, no indication as to whether Sailfin Mollies are established in the wild in Singapore, and the species is not mentioned by Chou and Lam (1989).

North America

Canada

In 1968, Sailfin Mollies were discovered breeding in the same hotspring near Banff, Alberta, as Mosquitofish *Gambusia affinis* (q.v.). It is believed that they were originally released from local aquaria around 1960 (McAllister, 1969), and they are known to have still been there in 1981 (Crossman, 1984; Nelson, 1984).

Translocations

Outside their natural range in North America, Sailfin Mollies have been widely translocated within the USA, for example to California (Moyle, 1976a,b; Shapovalov *et al.*, 1981; Courtenay, 1992); Florida (Courtenay & Robins, 1973; Courtenay *et al.*, 1974); Nevada (Deacon *et al.*, 1964; Hubbs & Deacon, 1964; Courtenay & Deacon, 1982); and to Mexico (Contreras & Escalante, 1984).

South America

Colombia

Welcomme (1988) records the establishment of the Sailfin Molly in Colombia, but gives no further details.

Australasia

Australia

As in the case of the One-spot Livebearer *Phalloceros caudimaculatus* (q.v.) little is known about the introduction of the Sailfin Molly to Australia, where since around 1968 or 1978 (accounts differ) it has been established in a few creeks and drains at Hervey Bay and in Sandgate Lagoon north of Brisbane, Queensland (McKay, 1978, 1984, 1986–1987, 1989; Arthington, 1986, 1989, 1991; Allen, 1989; Arthington & Lloyd, 1989). Clements (1988), who refers to the species as the Black Molly, names the site at Hervey Bay as Pialba-Urangan. McKay (1986) says that in 1985 Sailfin Mollies were recorded in a water trough at Monto, Queensland. The origin of these populations is believed to be releases from aquaria.

New Zealand

The Sailfin Molly was introduced to New Zealand many years ago, where it was first reported in the wild by Winterbourn and Brown (1967) in the Waipahihi stream, a thermal water which flows into the north-eastern shore of Lake Taupo on North Island. This population died out following a rise in water temperature associated with a change in the management of the stream and its catchment, and it now occurs, probably as the result of a second release by aquarists, in very large numbers in geothermally heated swamps between Tokaanu and Waihi, at the south-western corner of Lake Taupo (McDowall, 1990b). Natural spread of the Sailfin Molly in New Zealand is considered highly unlikely because of colder neighbouring waters (McDowall, 1979, 1984).

Oceania

Guam

The Sailfin Molly is established and breeding in the wild in Guam (Best & Davidson, 1981; Shepard & Myers, 1981; Maciolek, 1984). The source and date of introduction of the original stock are apparently unrecorded.

Hawaiian Islands

The Sailfin Molly was imported with the Mosquitofish *Gambusia affinis* (q.v.) from Seabrook, Texas, USA (Welcomme, 1981, 1988 incorrectly says from Central America) to the island of Oahu on 15 September 1905 (Van Dine, 1907), where it became established at Moanalua. In 1940, a second introduction, mainly of the black variety, was made to the Nuuanu reservoir on the same island. Both of these introductions, which numbered together less than 420, were made to control mosquito larvae. In 1960, 10.5 kg of Sailfin Mollies were unsuccessfully translocated from the Kapalama Canal on Oahu to Port Allen and Nawiliwili on Kauai. Brock (1960) recorded Sailfin Mollies as 'well established in brackish water areas of Oahu and Molokai'. Maciolek (1984), however, says that the species was introduced to all the main islands (Hawaii, Kauai, Maui, Molokai and Oahu), and that it is currently established in estuaries on all of them, where with the Mosquitofish it is 'among the more abundant and widely distributed exotics', and where these two species and other topminnows 'have been useful in controlling mosquito larvae in certain artificial habitats such as urban ponds and ditches, water tanks, effluent ponds, and reservoirs'. (See also Brock, 1952; Kanayama, 1968; Maciolek & Timbol, 1980.)

Allied Species

Courtenay and Robins (1973) and Courtenay *et al.* (1974) record the establishment of the Swordtail Molly *P. petenensis* in Hillsborough and Palm Beach Counties, Florida, USA, and of the Broadspotted Molly *P. latipunctata* in Palm Beach County.

Shortfin Molly (Mexican Molly)
Poecilia mexicana

The taxonomic status of *P. mexicana* is not entirely clear, and it is sometimes regarded as a component of the *P. sphenops* complex (Stauffer, 1984). The identification of *P. mexicana* in the USA is, however, now fairly certain (Courtenay *et al.*, 1984). Although not a popular aquarium species, the Shortfin Molly has been widely used to control mosquitoes in Pacific Islands.

Natural Distribution

The Atlantic slope of Central America from the Río San Juan in the Río Grande basin, Nuevo León, Mexico, and the Pacific slope from the Río del Fuerte basin, Sonora, Mexico, south to the Caribbean slope of Colombia, the Río Tuira in the Pacific slope of eastern Panama, and the Netherland Antilles.

Naturalized Distribution

North America: USA. **Oceania:** Fiji; Hawaiian Islands; American Samoa; Western Samoa; Society Islands (Tahiti).

North America

United States

The Shortfin Molly is currently naturalized in six or seven states in the USA. In California, fish that escaped from an aquarium became established in a drainage canal system south of Mecca, Riverside County (St Amant, 1966, 1970; St Amant & Sharp, 1971; Mearns, 1975; Hubbs *et al.*, 1979; Courtenay *et al.*, 1984; Courtenay & Stauffer, 1990). In Idaho, Shortfin Mollies are established in thermal waters in the Bruneau River below the Blackstone Grasmere Road Bridge, Bruneau Hot Springs, Owyhee County (Courtenay *et al.*, 1984, 1986; Courtenay & Stauffer, 1990); in Montana in Trudeau Pond, Madison County (Brown, 1971; Courtenay *et al.*, 1984; Courtenay & Stauffer, 1990); and in Nevada, in Rogers Spring near Overton, in Blue Point Spring, in several other springs and in most of the Moapa River, Clark County, and also in Lincoln County (Deacon *et al.*, 1964; Hubbs & Deacon, 1964; Lachner *et al.*, 1970; Courtenay & Deacon, 1982; Courtenay *et al.*, 1984

Courtenay & Stauffer, 1990); all these populations seem to have resulted from escapes or releases from aquaria. Hahn (1966) reported the discovery in the previous year of a large population of Shortfin Mollies in hot springs and associated drainages, including a small tributary of Rock Creek, at the deserted town of Valley View, Saguache County, Colorado; according to Courtenay *et al.* (1984) and Courtenay & Stauffer (1990), the present status of this population is uncertain. Most of these populations are believed to date from the 1960s (Courtenay, 1993). In the late 1980s, two established populations of this species were found near Brownsville and San Benito, Cameron County, Texas (Courtenay & Stauffer, 1990).

Fishes belonging to the species complex of which *P. mexicana* is a member have been collected in Dade County, Florida (Courtenay & Robins, 1973; Courtenay *et al.*, 1974, 1984, 1986; Courtenay & Stauffer, 1990), but although it is believed that they were formerly established, their current status is unknown. In Arizona, Shortfin Mollies have been caught on one occasion in a canal near Phoenix, Maricopa County (Minckley, 1973).

Oceania

Fiji

Ryan (1980) and Andrews (1985) record the presence of Shortfin Mollies on the island of Viti Levu, Fiji. The source and date of the introduction of the original stock appear to be unknown.

Hawaiian Islands

Since about 1960, Shortfin Mollies have been well established in streams on the island of Oahu (Maciolek, 1984). The source and date of the introduction of the original stock are apparently unrecorded.

Samoa

Maciolek (1984) received information that Shortfin Mollies were established on Savaii in Western Samoa and also in American Samoa. Attempts in both Western and American Samoa to breed Shortfin Mollies as live bait for fishing have proved uneconomic (Eldredge, 1994). Once again, the date of introduction of the original stock and its source are apparently unknown.

Society Islands (Tahiti)

Maciolek (1984) was informed of the presence of Shortfin Mollies on the island of Tahiti in the Society Islands, French Polynesia. The date of introduction of the original stock and its source are apparently unrecorded.

Guppy (Millions Fish)
Poecilia reticulata

The Guppy (named after the 19th century geologist and naturalist Robert J.L. Guppy, who settled on the island of Trinidad) is a small (up to 6 cm) and extremely abundant and resilient tropical fish, well able to survive temperatures as low as 15°C for short periods and to live in brackish estuarine and poorly oxygenated fresh water. Its ability to mature rapidly and its high fecundity have earned the species its alternative name, and have enabled it to spread rapidly to new localities. It is an extremely popular aquarium fish, occurring in many brilliantly coloured varieties, and has additionally been widely used for the control of mosquito larvae. It also eats the eggs of other fish, and has been blamed for the decline of several native species.

Natural Distribution

Northeastern South America from western Venezuela to Guyana, the Netherlands Antilles, islands off the coast of Venezuela, Trinidad, Barbados, St Thomas and Antigua, and possibly Margarita and Tobago (but see below under Lesser Antilles).

Naturalized Distribution

Europe: ? British Isles (England); The Netherlands. ***Asia:*** India; Singapore; Sri Lanka. ***Africa:*** Kenya; Madagascar; ? Namibia; ? South Africa; Uganda. ***North America:*** Mexico; USA. ***South and Central America:*** Colombia; Cuba; Hispaniola (Haiti and Dominican Republic); Jamaica; ? Lesser Antilles; Peru; Puerto Rico. ***Australasia:*** Australia; New Zealand; Papua New Guinea. ***Oceania:*** Canary Islands; Cook Islands; Fiji; French Polynesia; Guam; Hawaiian Islands; Mauritius; New Caledonia; Palau Islands; Western Samoa; Vanuatu.

Europe

British Isles (England)

In the mid-1960s, two colonies of Guppies became naturalized in England. One was established in the heated cooling water discharged from the Central Electricity Generating Board's power station in the River Lee (or Lea) in Hackney, Essex, in northeast London. The poor quality of this stretch of water provided an effective barrier to many potentially competing and/or predatory native fish species apart from the Three-spined Stickleback *Gasterosteus aculeatus,* and this may have assisted the Guppies in becoming established (Meadows, 1968; Lever, 1977).

In the early 1970s, pollution in the Lee decreased, and other native fishes were able to become re-established. This, together with the closure of the power station with a consequent lowering of the water temperature and a further reduction of pollution, favouring native species, caused the extinction of the Guppy population (Wheeler & Maitland, 1973; Wheeler, 1974; Lever, 1977).

The second colony of Guppies, believed to originate from the dumped stock of a failed pet shop, became established in 1963 in the 350 m Church Street stretch of the St Helens Canal in Lancashire, where the water was artificially heated by discharge from the nearby glass factory of Pilkington Brothers (Wheeler & Maitland, 1973; Wheeler, 1974; Lever, 1977). This population too has died out. (See also under Redbelly Tilapia *Tilapia zillii.*)

The Netherlands

For many years Guppies have been naturalized in the heated cooling water discharge of an iron works at Ijmuiden (Hoogovens) in The Netherlands (Nijssen & De Groot, 1974; De Groot, 1985). The source and date of introduction of the original stock appear to be unknown.

Asia

India

Guppies were first imported to India, to control mosquitoes in Madras, by a Major Selley in 1908 or 1909, but were all reported to have died (Prashad & Hora, 1936). In 1946, however, Guppies were found to be thriving in the Rameshwaram temple water tank, from where they were later successfully transferred to other localities in Madras (Chacko, 1948; Jones & Sarojini, 1952). Today, Guppies are abundant in a few municipal ponds of Ranchi, Bihar (Banerji & Satish, 1989), and are 'well distributed' in Rameshwaram, Thanjuvar, Madras city, Malabar,

Cuddapah, Kurnool, and elsewhere in southern India, and have also occurred in the Kulbhor River near Loni in Maharashtra (Yadav, 1993). (See also Natarajan & Ramachandra Menon, 1989; Shetty *et al.*, 1989.)

Singapore

According to Chou and Lam (1989) the Guppy is 'firmly established in Singapore's inland waterways'. When and from where it was introduced is apparently unrecorded.

Sri Lanka

Welcomme (1981, 1988) says that Guppies were introduced to Ceylon (Sri Lanka) in the 1930s, and are 'abundant in canals around Colombo'. The species is not mentioned by Fernando (1971) nor by De Silva (1987).

Unsuccessful Introductions

In Asia, Guppies are recorded as having been introduced unsuccessfully to Indonesia (Muhammad Eidman, 1989), the Philippines (Juliano *et al.*, 1989), and Bangladesh, Burma, Hong Kong, Thailand, Malaysia and Singapore (Haas & Pal, 1984).

Africa

Kenya

Guppies from Uganda were introduced to control mosquitoes in Kenya in 1956, and became established in the central Tana River, where they may be the cause of a decline in native cyprinodonts (Moreau *et al.*, 1988), and in Lake Naivasha (Muchiri & Hickley, 1991).

Madagascar

In 1988, Reinthal and Stiassny (1991) found locally caught Guppies, which had been introduced to Madagascar at an apparently unrecorded date, for sale in a number of fish markets, but did not collect any in the wild.

Namibia

In May 1988, Guppies were found at the Kuruman Dog and Lake Otjikoto sinkholes, some 20 km west of Tsumeb, in South West Africa (Namibia) (De Moor & Bruton, 1988). Whether any survive there today is unknown, but if they do invasion of the Kafue and Okavango swamps and Lake Liambezi has been predicted (De Moor & Bruton, 1988).

South Africa

Guppies from Barbados were imported to the Jonkershoek fish hatchery in 1912 to be reared for mosquito control, but failed to survive the Cape winter (Siegfried, 1962). Since then, the species has probably been frequently released into the wild by aquarists or has escaped from captivity, but Guppies have only been recorded in natural waters locally in northern Natal (specifically in the Kranzkloof Nature Reserve in 1986 and in the Kenneth Stainbank Nature Reserve), and from Empangeni, Zululand, where they are 'reported to be plentiful in numerous small streams' (De Moor & Bruton, 1988). It has been predicted that they could successfully penetrate the lower reaches and floodplains of the Umfolozi, Mkuze, Phongolo, Maputo, Incomati and Limpopo Rivers, where 'their disruption of these valuable natural systems would be irreversible' (De Moor & Bruton, 1988).

Uganda

In 1948 Guppies from the USA were introduced to control mosquitoes in Uganda, where they became established in many small streams and ditches where they posed a threat to various endemic cyprinodonts (Moreau *et al.*, 1988). Haas and Pal (1984) say that Guppies were introduced to one lake and many large ponds in Uganda, where they proved ineffective at controlling mosquito larvae and were eaten by native fishes.

Unsuccessful Introductions

Haas and Pal (1984) list an unsuccessful attempt to establish Guppies in Senegal. Moreau *et al.* (1988) do likewise for Nigeria.

North America

Mexico

The release of aquarium stock has resulted in the naturalization in Mexico of Guppies in Parras (Contreras, 1969) and Monterrey (Contreras, 1978), and also before 1961 in several places in the Río Balsas (Morelos), and in the upper Río Pánuco, around Lago de Chapala, Laguna Cortés and Todos Santos, Cabo San Lucas, and Baja California Sur (Contreras & Escalante, 1984).

United States

Deacon *et al.* (1964) recorded the discovery in December 1960 of Guppies in Blue Point Spring in the Lake Mead National Recreational Area above the Overton arm of Lake Mead, 68 km northeast of Las Vegas in Clark County, Nevada, in which they may have been released in 1957. In August 1961, others were found in the Preston Town Spring, White Pine County. In October 1980, Courtenay and Deacon (1982) found large numbers of Guppies in Indian Spring in Clark County, but none surviving in Blue Point Spring.

Hubbs *et al.* (1978) reported Guppies to be abundant in 1977 in a stenothermal ditch draining the San Antonio Zoo, Texas, into the San Antonio River. (See also Hubbs, 1982.) Guppies are also naturalized in Arizona (Minckley, 1973), Idaho (Simpson & Wallace, 1978), near High Springs, Alachua County, Florida (Courtenay *et al.*, 1984) and Wyoming (Baxter & Simon, 1970). Courtenay *et al.* (1974) listed the Guppy as established in Hillsborough and Palm Beach counties, Florida, but Courtenay *et al.* (1984) suggested that populations in the former are probably not self-sustaining, while the one in Boca Raton, Palm Beach, died out when its habitat dried in the late 1970s. St Amant and Hoover (1969) reported the introduction of Guppies to California, where they may be established locally in sewage treatment ponds (Moyle, 1976a,b). All these populations originated in releases from aquaria.

Ecological Impact   In some waters of the south-western USA, introductions of Guppies have been particularly damaging to endemic autochthonous fishes (Courtenay & Stauffer, 1990). Specifically, the population of Guppies at Preston, Nevada, has adversely affected that of a localized native cyprinodont *Crenichthys baileyi* (Deacon *et al.*, 1964). (See also Lachner *et al.*, 1970; Courtenay *et al.*, 1991; Courtenay, 1993.)

Unsuccessful Introduction

Although Welcomme (1988) lists the Guppy as breeding in the wild in Alberta, Canada, the large population formerly established in the Cave and

Basin hot springs in the Banff National Park in th 1960s (McAllister, 1969) died out around 197 (Crossman, 1984; Nelson, 1984).

South and Central America

Colombia

Introduced to Colombia to control mosquitoes i 1940, Guppies are currently 'widespread throughou warmer zones of the country' (Welcomme, 1981).

Cuba

Garcia Avila *et al.* (1991) record the successful intro duction of Guppies, from an unknown source and a an unknown date, as a biological controlling agent c mosquito larvae in deoxygenated lakes and pollute ditches and drains on Isla de la Juventud, Cuba.

Hispaniola (Haiti and Dominican Republic); Jamaica

Wetherbee (1989) says that Guppies have been intro duced to Hispaniola and Jamaica in the Greate Antilles, where they are 'common in lowlan streams'. He gives no source or date of introductio of the original stock.

Lesser Antilles

Rosen and Bailey (1963) suggested that the presenc of Guppies in the Lesser Antilles may be the result c human intervention.

Peru

Welcomme (1988) says that in 1940 Guppies wer introduced to control mosquitoes in Peru, wher they are reproducing.

Puerto Rico

Guppies were first recorded in Puerto Rico b Erdman (1947), who found them to be abundant i Adjuntas, Aibonito, Comerío and Cayey, and b Oliver-González (1946) who reported their preda tion on *Schistosoma mansoni*. Today, Guppies ar found in streams throughout the island from se level to 700 m asl (Erdman, 1984); Ortiz-Carrasquill (1981) recorded them in Río Matrullas up t 736 m asl. Welcomme (1981, 1988) says the origina stock was accidentally released from aquaria in 1935

Australasia

Australia

The date and source of the introduction of the Guppy to Australia are apparently unrecorded, but the species is currently widely established in urban and suburban drains and creeks around Cairns, Innisfail, Ingham, Mackay, Rockhampton, Gladstone and in Ban Ban springs near Gayndah in eastern Queensland (McKay, 1978, 1986–1987). Further south, the species is more scattered, occurring locally in clear freshwater springs flowing into Barambah Creek, a tributary of the Burnett River (Arthington & Lloyd, 1989). McKay (1984) says that in 1977 it was abundant in Ithaca, Toowong and Seven Hills Creek around Brisbane, but by 1981 had been reduced to a relic population only in Seven Hills Creek.

McKay (1986–1987) reported Guppies in the Gove town lagoon (since about 1976) and in Sadgroves Creek near Darwin in the Northern Territory, while Clements (1988) refers to their presence 'in very large numbers' in the lagoon at Nhulunbuy on the Gove peninsula.

Most populations of Guppies in Queensland have reverted to the normal wild coloration, but occasionally highly coloured 'deltatail' males are collected, suggesting that new stock is frequently being released near human habitation in Queensland, and in the Northern Territory where populations have been repeatedly eradicated (McKay, 1986–1987). The wide distribution of Guppies in ditches draining sugar-cane plantations suggests they are commonly used to control mosquito larvae, and some of the spread of the species in Queensland may be attributable to releases for this purpose (Lloyd, 1986; McKay, 1986–1987). (See also Rosen & Bailey, 1963; Arthington *et al.*, 1983; Allen, 1989; Arthington, 1989; McKay, 1989.)

Ecological Impact Guppies share a similar diet with Mosquitofish, but are less aggressive. The impact they have on Australian endemic fishes has yet to be determined, though where the two exotics are common in disturbed Queensland waters native species tend to be rare (McKay, 1978; Arthington *et al.*, 1983; Arthington, 1989, 1991). 'It is interesting to note', McKay (1986–1987) points out, 'that not only do guppies survive when released into creeks containing [the more aggressive] *Gambusia*, but frequently become the dominant species, selecting the more open flowing sections of the stream, and avoiding the stagnant swampy backwater which is usually the province of *Gambusia*'.

New Zealand

Exactly when the Guppy was first introduced to New Zealand is apparently unrecorded, but it seems probable that it was one of the earliest warmwater fishes to be imported. The date of the earliest releases from aquaria is also unknown, but the population(s) around Reporoa have been established for many years – possibly since as early as the 1920s. Formerly also occurring in the Waipahihi, a stream that flows into the northeastern corner of Lake Taupo in North Island, from which they disappeared when the stream flow was modified, increasing the water temperature, Guppies are today confined to a few small geothermal streams in the Volcanic Plateau on North Island, and to several small warm water streams near Reporoa, such as the Kawaunui, a tributary of the Waiotapu, and portions of the upper Waikato River near its confluence with the Kawaunui (McDowall, 1979, 1990b).

Papua New Guinea

The Guppy was first reported in drains in Boroko, a suburb of Port Moresby, in 1967, and within a few years had spread into waters draining into the nearby Waigani swamp, which forms part of the Laloki River system (West, 1973). Berra *et al.* (1975) recorded Guppies in the Laloki, Brown and Goldie Rivers. The origin of these populations is believed to be escaped or released aquarium fish (Glucksman *et al.*, 1976; West & Glucksman, 1976).

Ecological Impact Because of its rapid breeding the Guppy is regarded as a pest in Papua New Guinea, where in some streams around Port Moresby it has excluded all native fish species (Allen, 1991).

Oceania

Hawaiian Islands

Brock (1952, 1960) says that Guppies were first introduced to the island of Oahu in 1922, where they became naturalized in streams flowing through Honolulu and into Kaneohe Bay, and also in some other streams on the windward side of Oahu. Maciolek (1984) says there were subsequent introductions to Kauai, Maui and Molokai, and that Guppies are now established and increasing in streams on all five major islands. (See also Hida & Thomson, 1962; Kanayama, 1968; Maciolek & Timbol, 1980.)

Because information on Guppies elsewhere in Oceania is sparse and fragmentary, it is given here in tabular form, adapted from Haas and Pal (1984) and Maciolek (1984).

ISLANDS (OF SPECIAL CONCERN)	DATE OF INTRODUCTION	SOURCES
Canary Islands	?	Elvira, 1995b
Cook Islands (Rarotonga, Mitiaro)	?	Maciolek, 1984
Fiji (Viti Levu)	?	Ryan, 1980; Maciolek, 1984
French Polynesia (Tahiti)	?	Fowler, 1932b; Maciolek, 1984
Guam	?	Brock & Takata, 1956; Best & Davidson, 1981; Shepard & Myers, 1981; Maciolek, 1984
Mauritius	*c.* 1920. Established in streams, small pools, large ponds, lakes and marshes. Eaten by *Tilapia*	Haas & Pal, 1984

ISLANDS (OF SPECIAL CONCERN)	DATE OF INTRODUCTION	SOURCES
New Caledonia	?	Laird, 1956; Maciolek, 1984
Palau Islands (Babelthuap)	By Japanese, between 1914 and 1944	Bright, 1979; Bright & June, 1981; Maciolek, 1984
Western Samoa	?	Fowler, 1932a; Maciolek, 1984
Vanuatu (Tanna)	?	Laird, 1956; Maciolek, 1984

Allied Species

Chou and Lam (1989) say that the Liberty Molly *P. sphenops* has become 'established permanently in low salinity waters' in Singapore. Since at least 1950 both *P. sphenops* and the Cuban Limia *P. vittata* have been naturalized in the Hawaiian Islands (Courtenay *et al.*, 1991; Courtenay, 1993), the latter in the lower reaches of such streams as Moanalua, Nuuanu and Manoa, and in the Kapalama Canal, flowing through Honolulu on the island of Oahu, and in some estuaries (Brock, 1960). Courtenay and Robins (1973) list *P. sphenops* as established and breeding in Florida, USA, but Courtenay *et al.* (1974) say that it 'showed total lack of establishment'. Welcomme (1988) lists the Yucatan Sailfin Molly *P. velifera* as breeding in Colombia.

Porthole Livebearer
Poeciliopsis gracilis

The Porthole Livebearer is a small (up to 5 cm) popular warmwater aquarium species that in the wild favours moderately fast-flowing waters.

Natural Distribution

The Pacific and Atlantic slopes of southern Mexico, Guatemala and Honduras.

Naturalized Distribution

North America: USA. ***South America:*** Venezuela.

North America

United States

On 27 July 1974, four Porthole Livebearers, a species previously unrecorded in California, were collected by A.J. Mearns in an irrigation canal at the junction of US Highway 111 and Johnson Avenue, 1.6 km south of Mecca, Riverside County. Further individuals were collected on 17 November 1974 at the same location, where at least a dozen others were observed schooling with young Shortfin Mollies *Poecilia mexicana* (q.v.) and Red Shiners *Notropis lutrensis*. The presence of young fish and the broad spectrum of sizes led Mearns (1975) to conclude that the fish were established in the canal, where he believed they had been released by aquarists or had escaped from a neighbouring tropical fish farm. The species' contin-

ued presence was confirmed by Moyle (1976a,b), Hubbs *et al.* (1979), Welcomme (1981), Courtenay *et al.* (1986, 1991), Courtenay and Stauffer (1990) and Courtenay, (1993).

South America

Venezuela

Welcomme (1988), who does not mention the population in California, says the Porthole Livebearer is reproducing in Venezuela.

Green Swordtail
Xiphophorus helleri

The range of freshwater habitats of the Green Swordtail – springs, streams, rivers, lagoons, rock pools and swamps, from sea level to an altitude of 1600 m – is mirrored by the number of forms, each differing in shape, colour, and tail development, in which the species is found. It is one of the most popular aquarium fishes, and has been bred in a number of varieties. Welcomme (1988) believed that it is probably more widely naturalized than records reveal.

Natural Distribution

The Gulf of Mexico basins from the Río Nautla, Veracruz, Mexico, southward to northwestern Belize.

Naturalized Distribution

Asia: Sri Lanka. *Africa:* Madagascar: Namibia; South Africa. *North America:* USA. *South and Central America:* ? Colombia; Hispaniola (Haiti and Dominican Republic); ? Jamaica; Puerto Rico. *Australasia:* Australia; New Zealand; Papua New Guinea. *Oceania:* Fiji; Guam; Hawaiian Islands.

Asia

Sri Lanka

Welcomme (1981, 1988) says that Green Swordtails, introduced to Sri Lanka in the 1960s, escaped from aquaria and became naturalized in Laxpana (or

Laxapana) dam. This population is mentioned neither by Fernando (1971) nor by De Silva (1987).

Unsuccessful Introduction

Natarajan and Ramachandra Menon (1989) record the unsuccessful introduction in Asia of Green Swordtails to Tamil Nadu, India.

Africa

Madagascar

In 1988, Reinthal and Stiassny (1991) collected Green Swordtails, which had been introduced to Madagascar at an apparently unrecorded date, in the following waters: a small stream flowing out of Lac Vert at Perinet in the northeast central rainforest; and in the Rivers Namorona, Fanolafana, Sakafiana, Vintanona, Botely and Fotobohitra in the Ranomafana National Park.

Namibia

In May 1988, Green Swordtails were collected in the Lake Otjikoto sinkhole, 20 km west of Tsumeb, Namibia (South West Africa) (De Moor & Bruton, 1988).

South Africa

De Moor and Bruton (1988) have summarized what is known about the establishment of the Green Swordtail in South Africa, to which it was imported from Mexico by the aquarium trade, or according to Moreau *et al.* (1988) as forage for the introduced Largemouth Bass *Micropterus salmoides* (q.v.).

The earliest record of Green Swordtails in the wild was in 1974, when a single specimen was taken in a tributary of the Crocodile River (Incomati system) in the Transvaal lowveld (Anon., 1974; Jubb, 1976, 1977). The name of the tributary was not given, but it is assumed to have been the Gladdespruit, where the species was later collected by Appleton (1974). There are also more recent records in Natal, which are probably derived from several other separate releases from aquaria.

The Green Swordtail is presently naturalized in the lower reaches of the Gladdespruit (Appleton, 1974). It also occurs in semi-natural ponds in Congella Park, Durban; in the Umpambinyoni River, Natal (De Moor & Bruton, 1988); and in several lakes and artificial waters in the Transvaal (Moreau *et al.*, 1988).

Ecological Impact   The ecological impact of the Green Swordtail in South Africa has yet to be assessed, but as it is known to eat the fry of other species it could become a nuisance were it to expand its range (Jubb, 1976, 1977). Fortunately, the Gladdespruit population is restricted to a pool which is isolated from the rest of the river by a series of waterfalls (Anon., 1974).

North America

Unites States

The Green Swordtail is currently naturalized in the USA in Florida, Idaho, Montana, Nevada (as a hybrid with the Southern Platy *X. maculatus* (q.v.)), and possibly also in California.

In Florida, Green Swordtails are established in a pond at the Satellite Beach Civic Center, Brevard County, in some canals and roadside ditches near Ruskin, Hillsborough County, and in Palm Beach, Polk and Manatee Counties (Courtenay *et al.*, 1974; Dial & Wainwright, 1983), who also reported the presence in Palm Beach, Brevard and Hillsborough Counties of *X. helleri* × *X. maculatus* hybrids, and in Brevard and Hillsborough of hybrids between *X. helleri* × *X. variatus*. All the Florida populations originate in releases from private aquaria or escapes from aquarium fish farms.

In Montana, Green Swordtails have been released by aquarists into Trudeau Pond, Madison County (Brown, 1971). Courtenay *et al.* (1987) found them to be established in the Kelly Warm Spring in the Grand Teton National Park, Wyoming, and in thermal waters in Warm Springs Creek and Barney Hot Spring, Clark County, in southern Idaho.

La Rivers (1962) recorded the presence of Green Swordtails, released from aquaria, in Rogers Spring, Clark County, Nevada, but Deacon *et al.* (1964) believed the site was actually the neighbouring Blue Point Spring. Green Swordtails were breeding in Indian Spring, Clark County, in 1975, but in 1980 the only *Xiphophorus* present was a hybrid between *X. helleri* and *X. maculatus*, and there were no *Xiphophorus* in either Blue Point or Indian Springs (Courtenay & Deacon, 1982, 1983).

In California, St Amant and Hoover (1969) collected Green Swordtails near an aquarium fish farm in Westminster, Orange County, and they have also been taken in a drainage canal near Oasis, Mono County, and near Mecca, Riverside County

Courtenay *et al.*, 1984). Moyle (1976b), however, doubted if Green Swordtails were fully established in California.

Green Swordtails were formerly breeding in Rock Springs, Maricopa County, Arizona, but the population was wiped out in a flood in 1965 (Minckley, 1973). (See also Courtenay *et al.*, 1986, 1991; Courtenay, 1992, 1993.)

Unsuccessful Introduction

Although Welcomme (1981, 1988) lists the Green Swordtail as established and breeding in Alberta, Canada (McAllister, 1969), it has not been recorded there since well before 1981 (Crossman, 1984; Nelson, 1984).

South and Central America

Colombia

Welcomme (1988) says that the Green Swordtail is breeding in Colombia, but whether in the wild or only in aquaria is unclear.

Hispaniola (Haiti and Dominican Republic); Jamaica

Wetherbee (1989) records the introduction at an unknown date and from an unknown source of the Green Swordtail (which he incorrectly assigns to the genus *Poecilia*) to Hispaniola and Jamaica (and also to Puerto Rico), but says that he has caught it in only one stream in northern Hispaniola.

Puerto Rico

Erdman (1947) recorded Green Swordtails from Quebrada Honda, just above its junction with Río La Plata near Aibonito. More recently, he found them also in the Arecibo drainage near Utuado (Erdman, 1984). When and from where they were imported is apparently unknown.

Australasia

Australia

Green Swordtails from Singapore were first reported in Australia, in Brisbane, Queensland, in about 1966 (Arthington & Lloyd, 1989). McKay (1978) found them in a total of 19 creeks in the environs of Brisbane, including Ennogera and Ithaca, and also in Kedron brook; in all, they were later discovered in 17

of the 36 drainage systems in the Brisbane region. Subsequently, they were also found at Gladstone between Marlborough and Cairns. These populations have been attributed to a combination of the release of surplus aquarium stock; the escape of juveniles from aquarists' fish ponds; the flooding of domestic ornamental ponds during cyclonic rains (McKay, 1978); and movement by school children (Arthington & Lloyd, 1989). In Queensland, the Green Swordtail has been found in waters as far west as Ipswich (Allen, 1989). (See also Rosen & Bailey, 1963; McKay, 1977, 1984, 1989; Arthington *et al.*, 1981, 1983; Merrick & Schmida, 1984; Clements, 1988.)

Ecological Impact   In coastal Queensland streams with large populations of *Gambusia*, *Poecilia* and *Xiphophorus*, indigenous freshwater fishes, especially the surface-feeding or mosquito-eating *Melanotaenia*, *Pseudomugil*, *Craterocephalus* and *Retropinna*, are normally rare or absent. In Enoggera Creek, above the reservoir, where alien fishes are absent, eight native species are common. Immediately below the reservoir where *Gambusia* and *Xiphophorus* are present, only, five native species occur (McKay, 1984).

New Zealand

The Green Swordtail has been a favourite ornamental fish in New Zealand for many years. It is only within the last decade, however, that it has become established in the wild, in the Waipahihi stream, which flows into the northeastern corner of Lake Taupo in North Island, where it is believed to have been released by an aquarist (McDowall, 1990b).

Papua New Guinea

Introduced to waters in the Wau/Bulolo district of Papua New Guinea, the Green Swordtail was abundant in streams near Wau in 1981. The source of these fish is unknown, but they were probably released by an aquarist (Allen, 1991). As the Green Swordtail is not mentioned by West and Glucksman (1976) or Glucksman *et al.* (1976), the introduction presumably took place between 1976 and 1981.

Oceania

Fiji

Ryan (1980) and Andrews (1985) record the successful introduction at an unknown date and from an unknown source of the Green Swordtail to the island of Viti Levu.

Mariana Islands (Guam)

Best and Davidson (1981) and Shepard and Myers (1981) report the successful introduction, again at an unknown date and from an unrecorded source, of Green Swordtails to Guam.

Hawaiian Islands

Green Swordtails were first introduced to streams near Honolulu and Kaneohe Bay on Oahu in 1922; in 1940 Nuuanu reservoir was stocked, and at a later date many of the leeward streams on Kauai were also planted with *helleri* × *maculatus* hybrids (Brock, 1952, 1960). Subsequently, Green Swordtails were success-fully introduced on all the main islands, where they are currently established and spreading (Hida & Thomson, 1962; Kanayama, 1968; Maciolek & Timbol, 1980; Maciolek, 1984).

Translocation/Ecological Impact

Contreras and Escalante (1984) reported 'massive hybridization' between *X. helleri*, *X. maculatus* (q.v.) and *X. variabilis* (q.v.) when these species were translocated within Mexico to Ojo de la Peñita, near Monterrey, and between each of them and the endangered endemic Monterrey Platyfish *X. couchianus*. La Peñita supports one of only three surviving populations of the endemic species, which has suffered 'massive gene swamping'.

Southern Platyfish (Moonfish)
Xiphophorus maculatus

Like the Green Swordtail, the Southern Platyfish (or Platy) is a common and popular aquarium species, which Welcomme (1988) believed is probably more widely established in the wild than reports would suggest.

Natural Distribution

The Gulf of Mexico basins from Ciudad Veracruz, Mexico, south to northwestern Belize.

Naturalized Distribution

Africa: Madagascar; ? Nigeria. **North America:** USA. **South and Central America:** ? Colombia; Hispaniola (Haiti and Dominican Republic); Jamaica; Puerto Rico. **Australasia:** Australia. **Oceania:** Hawaiian Islands; Palau Islands (Babelthuap).

Africa

Madagascar

In 1988, Reinthal and Stiassny (1991) collected Southern Platyfish, which had been introduced to Madagascar at an apparently unrecorded date, in the following waters: a small stream flowing out of Lac Vert at Perinet, in the northeast central rainforest; in the River Namorona, in the Ranomafana National Park; and in a small stream flowing into the River Mangoro, in the central east coast area.

Nigeria

Welcomme (1981) says that the Southern Platyfish is 'possibly established' in Nigeria, but later (Welcomme, 1988) states that the species' status in the country is 'unknown'.

North America

United States

The Southern Platyfish is currently naturalized in Florida and, as a hybrid, in Nevada, and is also listed by Courtenay *et al.* (1991), who give no further details, as established in Colorado.

In Florida, Southern Platyfish are breeding in canals at the Satellite Beach Civic Center, Brevard County (Dial & Wainwright, 1983), and in some roadside ditches near Ruskin and possibly in other localities in Palm Beach and Hillsborough counties

(Courtenay *et al.*, 1974; Dial & Wainwright, 1983). They are common only locally, and are nowhere abundant or the dominant species. The former population is derived from releases from aquaria, while the latter originated in escapes from nearby fish farms (Courtenay *et al.*, 1984). Hybrids between *X. maculatus* and the Green Swordtail *X. helleri* (q.v.) have occurred in Palm Beach, Brevard and Hillsborough counties, and between *X. maculatus* and the Variable Platyfish *X. variatus* (q.v.) in Brevard County (Courtenay *et al.*, 1974).

In Nevada, La Rivers (1962) reported the presence of Southern Platyfish in Rogers Spring, Clark County, but Deacon *et al.* (1964), who give the date of introduction as about 1957, believed the location was actually the nearby Blue Point Spring. An apparent *maculatus* × *helleri* hybrid is presently established in Indian Springs, also in Clark County (Courtenay & Deacon, 1982, 1983; Dial & Wainwright, 1983; Courtenay *et al.*, 1984, 1986; Courtenay & Stauffer, 1990).

St Amant and Hoover (1969) and Hubbs *et al.* (1979) reported Southern Platyfish near a fish farm in Westminster, Orange County, California, where they had apparently disappeared by 1980 (Courtenay *et al.*, 1984). Hubbs (1972) and Courtenay *et al.* (1991) indicate the collection, but not establishment, of the species in Texas.

South and Central America

Colombia

Welcomme (1988) lists the Southern Platyfish as introduced to and breeding in Colombia, but whether in the wild or only in captivity is unclear. No source or date of the introduction is given.

Hispaniola (Haiti and Dominican Republic); Jamaica

Wetherbee (1989) (who incorrectly assigns the species to the genus *Poecilia*) reported the introduction of Southern Platyfish from Mexico to Hispaniola and Jamaica (and also to Puerto Rico), where he says they are established 'in lowland streams'. The source and date of introduction of the original stock are not mentioned.

Puerto Rico

Erdman (1947) reported Southern Platyfish from Quebrada Honda at its junction with Río La Plata at La Plata, 11 km northwest of Aibonito. Subsequently he recorded them in the Loiza drainage near Loiza reservoir and at Río Abajo Forest Station north of Utuado (Erdman, 1984).

Australasia
Australia

According to McKay (1978), the establishment of Southern Platyfish in Queensland may have been due to the deliberate stocking of a deep, isolated pool in the upper Brisbane River for breeding aquarium stock, from which they colonized the river during a flood. Since then, they have dispersed naturally to other reaches of the upper Brisbane River (Arthington *et al.*, 1983; McKay, 1984; Arthington & Lloyd, 1989), and now occur in three of the 36 drainage systems in the Brisbane region (McKay, 1984, 1986–1987). Further north in Queensland, Southern Platyfish occur at Hervey Bay, where they were released from an aquarium fish farm in the late 1960s (McKay, 1978), and at Babinda, Tully, Gordonvale, and in the Barron River (McKay, 1978, 1986; Allen, 1989).

Oceania
Hawaiian Islands

Southern Platyfish were introduced to Oahu in 1922, and in 1940 to Nuuanu reservoir on the same island (Brock, 1952, 1960). They were subsequently released in other reservoirs on Hawaii and Maui, and are currently established on all three islands (Hida & Thomson, 1962; Kanayama, 1968; Maciolek, 1984).

Palau Islands (Babelthuap)

Between 1914 and 1944, Southern Platyfish were introduced by the Japanese to the Palau Islands (Babelthuap), where they are established and breeding (Bright, 1979; Bright & June, 1981; Maciolek, 1984).

Unsuccessful Introductions

In Asia, Natarajan and Ramachandra Menon (1989) record an unsuccessful attempt to introduce the Southern Platyfish to Tamil Nadu, India.

Translocation/Ecological Impact

See under *X. helleri*.

Variable Platyfish (Variegated Platyfish; Variated Platyfish)
Xiphophorus variatus

The Variable Platyfish (or Platy) is another common and popular aquarium species that has been widely transported around the world by aquaculturists.

Natural Distribution

Endemic to Mexico, from southern Tamaulipas to eastern San Luis Potosí and northern Veracruz.

Naturalized Distribution

North America: United States. ***South America:*** ? Colombia. ***Ocean:*** ? Hawaiian Islands.

North America
United States

Variable Platyfish are currently naturalized in Florida and Montana, and have been collected in Arizona and California.

In Florida, they occur in Gainesville, Alachua County (Burgess *et al.*, 1977) probably as a result of deliberate releases by students at the University of Florida (Courtenay *et al.*, 1986; Courtenay & Stauffer, 1990). They are also known to have escaped from aquarium fish farms into canals and roadside ditches in Palm Beach, Brevard, Hillsborough (Tampa Bay), and Manatee counties, in none of which they are believed to survive today. The Shortblade Swordtail *X. v. xiphidium* has failed to establish in Florida (Courtenay & Robins, 1973; Courtenay *et al.*, 1974).

In Montana, Brown (1971) recorded naturalized populations of Variable Platyfish, released by aquarists, in thermal outflows in Beaverhead, Granite and Madison counties, in the first of which they remain extant (Courtenay *et al.*, 1986; Courtenay & Stauffer, 1990).

Variable Platyfish formerly occurring in Orange and Riverside counties, California (St Amant & Hoover, 1969; St Amant & Sharp, 1971; Courtenay *et al.*, 1974; Moyle, 1976b) failed to establish viable populations. A population breeding in the Salt River at Tempe, Maricopa County, Arizona, between 1963 and 1965, probably following releases by Arizona State University students, was destroyed by a flood (Minckley, 1973); Minckley also recorded the presence, but not establishment, of Variable Platyfish in drains near Yuma, in Yuma County, Arizona. (See also Courtenay *et al.*, 1991; Courtenay, 1993.)

South America
Colombia

Welcomme (1988) lists the Variable Platyfish as breeding in Colombia, but whether only in captivity or also in the wild is unknown. The date and source of the introduction are not given.

Oceania
Hawaiian Islands

Maciolek (1984) was informed that sometime after 1960 the Variable Platyfish was introduced to a pond on the island of Oahu, where it may be established.

Translocation/Ecological Impact

See under *X. helleri*.

ATHERINIDAE

Pejerrey
Odontesthes bonariensis

The Pejerrey, a coolwater predator, is highly valued for human consumption. Largely lacustrine, it also thrives in affluent rivers (Welcomme, 1988). All international introductions have been made from Argentina.

Natural Distribution

Uruguay and Argentina. In Uruguay the Pejerrey occurs from Punta del Este in the east westward to Nueva Oalmira at the mouth of the Río Uruguay. In Argentina it is found from Buenos Aires south to Bahía Blanca.

Naturalized Distribution

Europe: Italy. *Asia:* Israel. *South America:* Bolivia; Chile; Peru.

Europe
Italy

Welcomme (1988) says that between 1970 and 1975 the Pejerrey was introduced to Italy for aquacultural purposes and to establish fisheries, and that it has become 'very successful' in Lago di Nemi, 25 km southeast of Rome. This introduction is not mentioned by Mazzola (1992).

Asia
Israel

In 1967, the Pejerrey was introduced into the Baruch reservoir east of Haifa, where although it has become established its rate of growth is reported to be slow (Ben-Tuvia, 1981).

South America
Bolivia; Peru

According to Welcomme (1988), the Pejerrey dispersed from the site of its (unnamed) original introductions in Bolivia in 1945 and 1962 to the Río Desaguadero and Lago de Poopó, and later to Lago Titicaca on the Bolivian/Peruvian border, where it arrived in Peruvian waters in 1955.

Ecological Impact   In Lago Titicaca the establishment of the Pejerrey and the Rainbow Trout *Oncorhynchus mykiss* (q.v.) has had a negative impact on the populations of some native fishes. Although the majority have managed to survive the arrival of the alien predators, one species, *Orestias cuvieri*, has not been seen since the late 1960s, and the populations of others such as *O. pentlandii*, *Trichomycterus rivulatus* and *T. dispar* have considerably declined. This fall in numbers may not have been due solely to predation by or competition with the introduced species, of which the Rainbow Trout is probably the more important, but also to overfishing and environmental mismanagement (Hanek, 1982; quoted by Welcomme, 1988). The native fishes were unable to support a thriving fishery, whereas since the introduction of the Rainbow Trout and Pejerrey the fishery in Lago Titicaca is now estimated to have a potential of between 75 000 and 100 000 t a year, present catches in Peruvian waters averaging 6000 t annually (Welcomme, 1988).

Chile

In Chile, where it was introduced between 1940 and 1942 for food, sport and aquaculture, the Pejerrey is known to exist only in waters such as Lago Penuales in the vicinity of Valparaíso (De Buen, 1959).

Ecological Impact   Although it is of some commercial value in Chile, the Pejerrey may have contributed to the elimination of some *Orestias* and *Trichomycterus* fish species (Welcomme, 1988).

Unsuccessful Introductions

Although being successfully reared in captivity in Japan, where it has been repeatedly stocked since 1966 to provide a sport fishery in Lakes Ashinoko and Tsuku, the Pejerrey has failed to become established in the wild (Chiba *et al.*, 1989). In Morocco and Brazil, to which it was introduced for aquacultural purposes, the Pejerrey has also apparently failed to survive in natural waters (Welcomme, 1988).

CHANNIDAE

Snakehead (Striped Snakehead; Chevron Snakehead; Snakehead Murrel)
Channa striata

The Snakehead (*Channa striata* = *Ophiocephalus striatus*) usually frequents oxygen-depleted waters, but possesses auxiliary air-breathing organs in its gill chambers which enable it to utilize oxygen from the air. It can survive for prolonged periods out of the water and, so long as its skin remains damp, can endure lengthy droughts by burrowing into the mud. These characteristics make it particularly valuable for use in commercial fisheries and for culture in anoxic tropical waters, and also for long-distance transportation both overland and by seafarers who use it for food on their voyages, and have introduced it to various oceanic islands.

Natural Distribution

India, Sri Lanka, South East Asia, China and the Philippines.

Naturalized Distribution

Asia: Indonesia. *Africa:* Madagascar. *Australasia:* Papua New Guinea. *Oceania:* ? Fiji; Hawaiian Islands; Mauritius; New Caledonia.

Asia

Indonesia

Although Indonesia is usually included in the Snakehead's native range, Welcomme (1988) says it was introduced there (Welcomme (1981) specifically mentions the Sunda Islands) from southern China in 'ancient' times, and is now 'well established', according to Allen (1991) west of Weber's Line. It is not mentioned as an introduced species in Indonesia by Muhammad Eidman (1989).

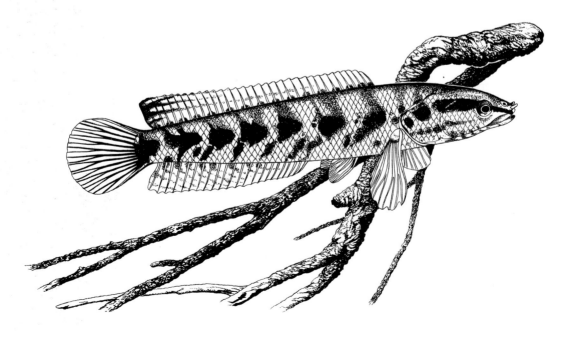

Africa

Madagascar

Although Welcomme (1988) says that the Snakehead, which was introduced to Madagascar in 1978, has since disappeared, Reinthal and Stiassny (1991) collected it in 1988 in the River Mangoro in the northeast central rainforest, and in Lac Itasy on the central plateau.

Ecological Impact   Raminosa (1987), quoted by Reinthal and Stiassny (1991), found that in Madagascar the Snakehead and the introduced Largemouth Bass *Micropterus salmoides* (q.v.) were together displacing the endemic cichlid *Paratilapia polleni* over much of its freshwater range on the island.

Australasia

Papua New Guinea

Allen (1991) says that in 1989 he collected specimens of the Snakehead in streams near Bintuni on the Vogelkop Peninsula, Irian Jaya, and that these were the first examples taken in Papua New Guinea, to which he believed the species had been introduced by immigrants from Indonesia.

Oceania

Fiji

Snakeheads have been introduced to, and may be established on, the island of Viti Levu (Van Pel, 1959; Devambez, 1964; Maciolek, 1984). The date and source of the introduction appear to be unknown.

Hawaiian Islands

Brock (1952, 1960) says that the Snakehead was introduced to the island of Oahu before 1900, where it became established and 'not uncommon'. (See also Jordan & Evermann, 1905; Hida & Thomson, 1962; Maciolek 1984; Courtenay *et al.*, 1991; Courtenay, 1993.)

Mauritius

Welcomme (1988) says the Snakehead was introduced at an apparently unrecorded date for sporting purposes to Mauritius, where it is breeding but 'restricted to two reservoirs'.

New Caledonia

Devambez (1960) and Maciolek (1984) say the Snakehead is established in New Caledonia. The date and source of the introduction appear to be unknown.

Unsuccessful Introduction

Maciolek (1984) records the unsuccessful introduction of the Snakehead to the island of Guam in the Marianas.

General Ecological Impact

Allen (1991) says the Snakehead is 'a voracious predator of fishes, crustaceans, frogs, snakes, and insects'. Its potential for causing harm in ecologically sensitive waters is therefore considerable.

Allied Species

Between 1906 and 1919, *Channa maculata* was imported from Taiwan to Japan, where it is now established in the wild.

PERCICHTHYIDAE

Striped Bass
Morone saxatilis

The Striped Bass is a diadromous species that migrates long distances upriver from coastal waters and estuaries to spawn. It is a valuable sporting and commercial asset.

Natural Distribution

From the St Lawrence River, Canada, southward to the St John's River in Florida, USA, and discontinuously from the Suwannee River, Florida, west to Lake Pontchartrain in Louisiana.

Naturalized Distribution

North America: Mexico. *Africa:* ? South Africa.

North America
Mexico

Burgess (1980) records the presence of Striped Bass in the Río Colorado and near Del Río in the Río Grande, Mexico. Contreras and Escalante (1984) say they were probably introduced by USA state agencies for sporting purposes and to rear in captivity for food, and that they are established in the Río Grande and possibly in the Río Colorado south of Lake Mead.

Translocations and Ecological Impact

Striped Bass have been translocated outside their eastern natural range in North America particularly to California (Albrecht, 1964), where Courtenay and Moyle (1992) say that hundreds of thousands of dollars have been spent in recent years in an attempt to eradicate them and Northern Pike *Esox lucius*, both of which have become established through illegal transplantations and both of which, when planted in reservoirs, have great potential to reduce populations of salmonids downstream.

In Canada, Crossman (1984) reports that Striped Bass occasionally caught in the sea off the coast of British Columbia have probably spread north from fish translocated to the west coast of the USA.

Africa
South Africa

Welcomme (1981, 1988), quoting Moreau (1979), says that Striped Bass were introduced in 1971 to form a sport fishery in South Africa, where they have become established. De Moor and Bruton (1988), however, do not mention this species in South Africa.

Unsuccessful Introductions

Doroshev (1970) refers to the failure to establish Striped Bass in the former USSR. Attempts to naturalize the species on Kauai in the Hawaiian Islands in 1920 or 1922 were also unsuccessful (Brock, 1952, 1960; Kanayama, 1968; Randall & Kanayama, 1972; Maciolek, 1984). Whitley (1951) reports a failed planting of Striped Bass in Sydney Harbour, Australia, around 1930.

Allied Species

Although the natural range of the potamodromous White Bass *M. chrysops* reaches as far south as the upper Río Grande, Contreras and Escalante (1984) do not regard the species as a native of Mexico, where Hubbs *et al.* (1977) record it in the middle and lower Río Grande; Contreras and Escalante (1984) found it also in the Río San Juan to Doctor Coss, Nuevo León.

Erdman (1984) reported a failed attempt to introduce adult White Bass to Loiza reservoir in Puerto Rico, but considered that a possible future introduction of fingerlings might be successful.

CENTROPOMIDAE

Nile Perch
Lates niloticus

The Nile Perch is a large (up to 2 m in length and 200 kg in weight) voracious and opportunistic piscivore, which in some parts of Africa is one of the most important human food-fishes. It is also regarded as a valued sporting asset.

Natural Distribution

The Congo, Volta, Nile and Niger river systems, and Lakes Albert, Chad and Turkana.

Naturalized Distribution

Africa: Kenya; Tanzania; Uganda. (For convenience, intranational translocations are here treated as international introductions.)

Africa

Kenya; Tanzania; Uganda

The proposal to introduce the Nile Perch to Lake Victoria, which is situated between Kenya, Uganda and Tanzania, was first mooted as long ago as the late 1920s, on the grounds that the Perch would prey on the numerous haplochromine cichlids in the lake, converting these generally underutilized fish to a form of protein more readily available to local fishermen, and would provide a new sporting fish for anglers (Harrison *et al.*, 1989; Worthington, 1973, 1989). Those who argued against the introduction (e.g. Graham, 1929; Fryer, 1960) cited the well-known fact that a predatory fish can never produce the same biomass as its prey species, and pointed out that the endemic species flocks, which were of the greatest interest to evolutionists, would suffer irreparable damage, and that the important tilapia fishery would also be affected. (Barlow and Lisle (1987) and Miller (1989) have shown that following the introduction of Nile Perch to Lake Victoria, the lake's fish biomass decreased to such an extent that around 4 kg of native fish were lost for every kilogram of Nile Perch.)

Nevertheless, in the face of expert opinion, in the mid-1950s Nile Perch from Lake Albert, Uganda, were experimentally translocated to Lake Kyoga above the Murchison Falls, also in Uganda. In the late 1950s, Nile Perch were presumably deliberately and secretly translocated or accidentally escaped

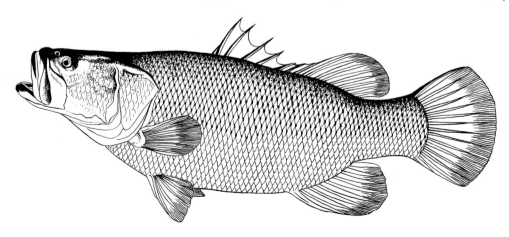

from a nearby fish farm at Kijansi (Payne, 1987) into Ugandan waters of Lake Victoria (Barel *et al.*, 1985; Ribbink, 1987; Worthington, 1989; Craig, 1992), since in 1957 they were first reported from the lake (Payne, 1987). In 1960, fishermen began catching Nile Perch in the northernmost Ugandan waters of Lake Victoria (Harrison *et al.*, 1989).

Once it was evident that a naturalized population had become established in the lake, more Nile Perch were in 1959 translocated into Ugandan waters of Victoria from Lake Albert, and later from Kenya's Lake Turkana into Kenyan waters (Acere, 1988; Craig, 1992). Since then, Nile Perch have gradually colonized the whole lake, spreading clockwise around the 4200 km of shoreline, reaching the. southwestern waters by the early 1980s (Harrison *et al.*, 1989). They first appeared in Kenyan waters in 1966, where their expansion was explosive. In the Nyanza Gulf, in the northeast of Lake Victoria, where the shallow water provides an ideal habitat, by 1981 some 60% of the catch comprised Nile Perch; since then the figure has risen to 80%. In more open waters, the proportion of Perch was around 30%. By 1978, they had invaded Tanzanian waters, where 4 years later they accounted for 6.4% of the total catch, a figure that has since considerably increased (Payne, 1987). Although Nile Perch were first recorded in the south of Lake Victoria as early as 1964 (Ogutu-Ohwayo, 1988), large numbers were not reported until 1983, and the species is now found virtually throughout the lake (Ogutu-Owayo, 1990a,b).

It is, perhaps, worth pointing out that the Nile Perch in Lake Victoria do not, strictly speaking, represent a true 'introduction' as correctly defined, but rather 'translocations' from Uganda and Kenya into their own national waters of the lake. The species' colonization of Tanzanian waters of Lake Victoria is the result of natural dispersal of these translocated stocks.

It is believed that more than one subspecies of *Lates niloticus* may occur in Lake Victoria (see Ribbink, 1987; Craig, 1992); in the following discussion, comments apply to all of these forms.

Ecological and Socio-economic Impact: Historical View

Lake Victoria, the largest tropical lake (around 68 600 km²) and the second largest in the world, supports a naturally rich and diverse fish community that formerly sustained economically important cichlid fisheries. Although for many years over-exploita-

tion of Lake Victoria's fish has been a cause for concern, the introduction of Nile Perch greatly exacerbated the problem.

Most of the fish in Lake Victoria at the time of the Nile Perch's introduction were small haplochromine cichlids (of which there were more than 300 endemic species) (Witte *et al.*, 1992), the majority of which are only a few centimetres in length, and tilapia species (also cichlids) which attain a length of around 30 cm (Ribbink, 1987).

Some, though by no means all, of the serious problems caused by the naturalization of Nile Perch in Lake Victoria were accurately predicted. For example, large numbers of economically important tilapia and haplochromine cichlids were formerly used as human food over a wide area; since the arrival of Nile Perch nearly all the native fish species of commercial value have virtually disappeared – not only herbivores, as might be expected, but also catfish *Bagrus* spp. There are also indications of an increase in phytoplanktonic growth, macrophytes, prawns, benthic organisms, and of the native pelagic cyprinid *Rastrineobola argentea* (Witte *et al.*, 1992). According to Ogutu-Owayo (1990a), Nile Perch now prey 'heavily' on *R. argentea*.

Nile Perch which, having all but exhausted the supply of tilapia and haplochromines, now survive mainly on the prawn *Caridina nilotica* (an important link in the lake's food-chain) or as cannibals (Worthington, 1989), at present comprise some 80% of the fish biomass in the lake, and are caught in large numbers. Because of their large size and oily flesh, Nile Perch cannot be sun-dried like the small-endemic species, but have to be smoked or fried to preserve them, and some islands and lakeside woods are being deforested to meet the demand for firewood and other purposes. Furthermore, local fishermen who were previously able to subsist on small scale tilapia and haplochromine fisheries are now being superseded by others able to afford the larger and stronger gill-nets necessary to capture Nile Perch (Reynolds & Gréboval, 1988). An export market in Nile Perch fillets has recently become established (Achieng, 1990).

The naturalization of Nile Perch in Lake Victoria has not only caused ecological upheaval but has also been responsible for a serious loss to evolutionary science. The lake has supported an extraordinary community of cichlid fishes from which have evolved several hundred endemic species. The loss of these species flocks arising from explosive speciation before they have been scientifically described is inestimable (Barel *et al.*, 1985). It has even been suggested (quoted by Harrison *et al.* (1989) that 'the cichlid fishes of Lake Victoria are considerably more important than the finches of the Galapagos Islands' – a claim that some might regard as rather extravagant.

Nile Perch in Lake Victoria are grossly overexploiting their prey, and this has been responsible for their explosive growth during the past 35 years (Payne, 1987). It is at least possible that starvation, combined with intensive fishing by man, may cause their eventual decline (Ribbink, 1987; Miller, 1989).

This overexploitation of their prey by Nile Perch in Lake Victoria is a classic example of the refutal of the claim made by Thomson (1922) who wrote:

> In regard to any natural enemy it is, of course, absolutely certain that it cannot exterminate, but can only keep in check, the animal it is intended to cope with. If it does more, then its own means of livelihood are imperilled, or it has to find other victims.

Since Thomson's time rather different views have come to prevail. In *The Independent* newspaper of 18 April 1988, Professor John R. Krebs, FRS said:

> An ecosystem may have the appearance of balance and co-evolved harmony; but that's simply a byproduct of each species separate activity. People perhaps used to say that predators do not over-exploit their prey; now one thinks of them doing their best to if they can.

Elaborating on this, Krebs wrote (pers. comm. 1992), that his opinion 'derives from the view that the evolutionary process of natural selection does not act to favour harmonious properties or communities and ecosystems, but rather acts to favour efficient performance (transfer of resources such as food into reproductive output) at the level of the individual)'.

In recent years it has become apparent that Nile Perch may not be exclusively responsible for the decline of local fish stocks in Lake Victoria. Surveys carried out in 1986 and 1988 revealed that although in most areas haplochromines were rare and Nile Perch were ubiquitous and abundant, in many places numerous haplochromine species were to be found in a wide variety of habitats, especially in inshore waters. These localities were all in reserves where fishing is prohibited; in nearshore and pelagic waters, where fishing is permitted, all small native species

were rare or absent (Anon., 1988; Harrison *et al.* 1989). Payne (1987) suggested that the move to large-mesh gill-nets for Perch had removed the pressure on the smaller cichlids, with a consequent resurgence of tilapia – not the native species but the introduced Nile Tilapia *Oreochromis niloticus* (q.v.) which is found naturally in the same waters as Nile Perch, which prey upon them (Ogutu-Ohwayo 1990a) but with which they may have established some kind of ecological equilibrium. This suggests that previous overfishing with fine-meshed nets may have been at least partially responsible for the decline of native fish species in Lake Victoria (Anon. 1988).

The claim by Zaret (1974) that 'the relative absence of [serious ecological impact] . . . following the introduction of the predatory Nile Perch, which some authors predicted would create problems . . .' is in marked contrast to the view expressed by Miller (1989) that 'Lake Victoria provides one of the most dramatic examples of the damage that can be done by introducing exotic fish species into a lake', and the accusation by Barlow and Lisle (1987) that 'never before [has] man in a single ill advised step placed so many vertebrate species simultaneously at risk of extinction and also, in doing so, threatened a food resource and a traditional way of life of riparian dwellers'. Even today, there is controversy among ichthyologists about the precise role that Nile Perch have played in the ecology and socioeconomy of Lake Victoria.

Published material on the Nile Perch in the lake is extensive; recent references include Ligtvoet (1989), Reynolds & Gréboval (1988), Watson (1989), Bruton (1990), Ribbink (1991), Baskin (1992), Halliwell (1992), Laycock (1992) and, most comprehensively, *The Impact of Species Changes in African Lakes* (edited by Tony J. Pitcher and Paul J.B. Hart, 1995), from which much of the following updated analysis is derived.

Impact on Fish Stocks

The traditional fishery of Lake Victoria concentrated on *Oreochromis esculentus*, *Bagrus docmac* and *Labeo victorianus*, which were taken in small numbers with primitive equipment. An increasing market, coupled with improved fishing methods, brought new species into the fishery and increased fish mortality. By the mid-1950s, catches were much lower than at the turn of the century. The serious depletion of important food-fishes was probably a result of a combination of factors, including the introduction in the 1950s of four exotic tilapiine species (*Oreochromis niloticus, O. leucosticus, Tilapia zillii* and *T. melanopleura*); severe alterations in the water level due to heavy precipitation in the early 1960s; and the increasing use of beach and fine-mesh nets. The introduction in the late 1950s of *Lates niloticus* at first made little contribution to the lake's fishery. Since the early 1970s, however, *L. niloticus* has contributed an increasing percentage of the growing total catch and is now the principal species taken, with significant contributions also from introduced *O. niloticus* and the endemic Dagaa or Omena *Rastrineobola argentea*. There is growing evidence that the lake is becoming eutrophic, resulting in an increased incidence of algal blooms and fish mortality caused by anoxic conditions. The Lake Prawn *Caridina nilotica* has become more abundant, and the Water Hyacinth *Eichhornia crassipes* is a recent arrival. Evidence from the south of the lake suggests that the decline in the haplochromine fish community is more probably a consequence of overfishing rather than of heavy predation by *L. niloticus* (Kudhongania & Chitamwebwa, 1995).

Limnological Alterations

The classic symptoms of eutrophication, including decreased water transparency, cyanobacterial blooms, raised nutrient concentrations, and hypolimnetic deoxygenation, all currently occur in Lake Victoria, and fish disperse daily to avoid oxygen-poor layers. Alterations in the phytoplankton community have changed available food sources to primary consumers, but these grazers have been much suppressed by higher trophic levels. The introduction of alien fish species greatly modified the phytoplankton, zooplankton and fish communities in the lake, and had a serious negative effect on water quality, which in turn has adversely affected fish species. Although pollution by man has contributed to this deterioration in water quality, climatic alteration and, in particular, the introduction of *L. niloticus* and the four tilapiine species, are the principal factors (Ochumba, 1995). Stocking the lake with large-bodied cichlids, which would be less vulnerable to predation by *L. niloticus* and could probably reduce algal and detrital densities, is recommended (Ochumba, 1995).

Impact on Native Tilapiines

The impact, if any, of *L. niloticus* on native tilapiines in Lakes Victoria and Kyoga, where it coexists with the successful alien tilapiine *O. niloticus*, is difficult to assess. If it occurs, it is certainly less than that provided, in the form of competition for spawning sites, nurseries, food, habitat, and possibly hybridization, by the four introduced tilapiines (Twongo, 1995).

Diversity and Stability of Native Fish Stocks

Since the naturalization of *L. niloticus* and alien tilapiine fishes in Lakes Victoria, Kyoga and Nabugabo, the total fish catch has increased but species diversity has drastically declined. Prior to the increase in the stocks of introduced species, fish diversity in Lake Victoria was higher than in Lake Albert, the source of the alien species. Now the position is reversed, and there have been similar declines in species diversity in Lakes Kyoga and Nabugabo, where the dominants are the aliens *L. niloticus* and *O. niloticus* and the native *Rastrineobola argentea*. The composition of fish species in these lakes appears to have stabilized, although the total catch has in some places begun to decline. In some lakes haplochromines are staging a recovery, especially where cover, such as that provided by the invasive and spreading Water Hyacinth *Eichhornia crassipes*, supplies refugia. The three dominant species are the most important commercially, and the other two constitute the principal prey of *L. niloticus* (Ogutu-Ohwayo, 1995).

Dynamics of Haplochromine Cichlids in Mwanza Gulf

The disappearance in Lake Victoria of some 200 endemic haplochromine cichlids, around two-thirds of the total haplochromine community, coincided with the increase and spread of *L. niloticus*. The loss of detritivores has resulted in increased microbial decomposition, which in turn has resulted in reduced oxygen concentrations in water near the bottom, while the absence of herbivores has promoted algal blooms. The decline of the haplochromine community appears to be due at least in part to increased predation by *L. niloticus*, and has resulted in eutrophication of the lake and an overall simplification of its ecosystem. The slight increase of some haplochromine species in shallow waters could be a result of a reduction of the *L. niloticus* populations in those areas; alternatively, it could be caused by decreased fishing mortality through the use of larger mesh nets intended for *L. niloticus*. This new ecosystem is more unstable than its predecessor, and it seems probable that the long-term *L. niloticus* fishery is unsustainable (Witte *et al.*, 1995).

Analysis of Species Changes

An analysis, by Bundy and Pitcher (1995), of 20 years of catch data suggests, however, that predation by *L. niloticus* is unlikely to be the sole cause of species changes in Lake Victoria. The decline of the haplochromine community began before *L. niloticus* became widespread and abundant, and is likely to have been triggered by increased fishing pressure. Predation by *L. niloticus* merely accelerated a decline that was already in progress.

Assessment of the *L. niloticus* Fishery

An assessment, by Pitcher and Bundy (1995), of the total Lake Victoria catch data between 1979 and 1990 indicates that the current *L. niloticus* fishery is overexploited and unsustainable, and projections predict a collapse of the stock within a few years if the present expansion of fishing pressure persists and immature fish continue to be harvested. Such a collapse would prove a disaster for local human communities. The fishery is estimated to have produced a net gain amounting to some US$ 280 million during the past 30 years, and has provided direct employment for at least 350 000 people, and possibly in excess of a million when ancillary activities are included. Pitcher and Bundy (1995) recommend that the fishing fleet be reduced to around 14 000 and that a minimum size limit for fish of 50 cm be introduced; this should produce a sustainable *L. niloticus* fishery of some 250 000 t for export annually.

Development of the *L. niloticus* Fishery

Allegations that *L. niloticus* in Lake Victoria is responsible for the loss to riparian people of vital subsistence and commercial resources, and that it is an unpopular food are, suggest Reynolds *et al.* (1995), factually incorrect; indeed, the species' actual impact, they claim, has been precisely the reverse.

Regarding its other alleged impacts in the lake, the situation is less clear. While it is acknowledged that the success of *L. niloticus* as a provider of food and wealth came at a cost of considerable ecological disturbance, the exact degree is difficult to assess. The species' impact on endemic haplochromine flocks has already been discussed.

Although local deforestation caused by heavy demands for wood-fuel for fish-smoking and domestic requirements has resulted in some adverse social and environmental problems, demand for charcoal for cooking purposes and the market for sawn timber in urban localities have been the main contributory factors to deforestation of coastal Lake Victoria.

The success of *L. niloticus* as a valuable subsistence and commercial food fish has, Reynolds *et al.* (1995) concede, resulted in socio-economic displacement of some local fishermen (who are unable to afford the more expensive nets required) and lower-income consumers, who hitherto have depended on cheap fish as a valuable source of protein. The expansion of the *L. niloticus* fishery has resulted in a widening of local and regional markets and the development of industrial processing to satisfy the increasing export demand. This, of course, has placed greater pressure, referred to above, on *L. niloticus* populations.

As small-scale fishermen are gradually squeezed out of business, they are likely to turn their attention to immature *L. niloticus*, to tilapiine species, to the endemic pelagic *Rastrineobola argentea*, and to the prawn *Caridina nilotica*. The potential yield of the endemic *Rastrineobola* has been estimated at around 1 million tonnes a year and, as Reynolds *et al.* (1995) point out, it is uncertain whether the species can survive the combination of increased fishing pressure and predation by *Lates*, which has opportunistically switched to *Rastrineobola* as one of its primary prey species.

Reynolds *et al.* calculate that, if the *L. niloticus* fishery stabilizes at its present high level, the total net benefit could amount by the early 2000s to US$ 1 billion. Even if the fishery were to decline, Reynolds *et al.* (1995) estimate the net benefit resulting from the *L. niloticus* fishery over a period of 30 years to be not less than US$ 560 m.

Although the future of *Lates niloticus* and its fishery in Lake Victoria is uncertain, its past history has, as Reynolds *et al.* point out, been an economic and commercial success. Lake Victoria and its ecosystem can never be restored to its pre-*Lates* state, and planning should therefore concentrate on maximizing the advantages that have accrued and minimizing the ecological disadvantages. Excessive fishing, pollution, and perhaps above all the incursion of *Eichhornia crassipes* are, Reynolds *et al.* suggest, the principal threats to the future of Lake Victoria.

Socio-economic Impacts

Socio-economic impacts on Lake Victoria fisheries result primarily from rapidly rising catches of *L. niloticus* in the 1980s, and the less dramatic increase in the yield from tilapias; they have been summarized, by Harris *et al.* (1995), as individual and household impacts (incomes, ease of entry to the fishery, quality of fishermen's lives, geographic mobility, employment opportunities, household consumption, and household health); community and organizational impacts; and national, international and transnational impacts. The effects of the changes resulting from *Lates* and tilapiine introductions will continue to impinge on the lake, the fishery and the dependent riparian inhabitants for many years to come.

Other Introductions

Welcomme (1988) refers to an introduction of Nile Perch from Ethiopia to Cuba in 1982–1983 for aquacultural and sporting purposes, where they are said to be reproducing only 'artificially'. In 1979, Nile Perch were released in a heated reservoir in Texas, USA, to try to establish a sport fishery (Welcomme, 1981), presumably unsuccessfully as the introduction is not mentioned by Welcomme (1988). McKay (1984) refers to a proposal in 1968 to import Nile Perch for recreational fishing in Queensland, Australia, a plan that was abandoned in 1986.

Allied Species

Fuchs (1987), Preston (1990) and Eldredge (1994) record the successful introduction of the Barramundi *L. calcarifer* for aquacultural purposes to brackish waters of Tahiti in French Polynesia, and its failure to become established on Guam in the Mariana Islands.

CENTRARCHIDAE

Rock Bass
Ambloplites rupestris

As its name implies, this polyphagous predatory sunfish lives mainly among rocks, stones and boulders, in shallow streams and lakes. In some places it is regarded as a sporting asset and as a food fish.

Natural Distribution

Southern Canada, from Lake Winnipeg, Manitoba, eastward to the coast, and in the USA from the Great Lakes east to Vermont, and south to the Gulf coast.

Naturalized Distribution

Europe: ? British Isles (England); France. **North America:** ? Mexico.

Europe

British Isles (England)

Some time before 1937, a population of Rock Bass became established in a flooded 1.2-ha gravelpit within 3 km of the northeastern outskirts of Oxford. The origin of this colony is unrecorded (Fitter, 1959; Vooren, 1972; Wheeler & Maitland, 1973; Wheeler, 1974; Lever, 1977).

France

According to Welcomme (1981, 1988), Rock Bass were accidentally introduced to the Loire valley in southeastern France before 1978, where they are currently established in 'calm ponds and canals'.

Unsuccessful Introduction

In 1887, Rock Bass were introduced to Germany, where although they later spawned successfully they eventually died out (De Groot, 1985).

North America
Mexico

Miller and Chernoff (1980) found Rock Bass in the Río Piedras Verdes at Colonia, Juárez, and in the Casas Grandes basin of Chihuahua, northern Mexico. (Welcomme (1988) incorrectly gives the date of their 'spread' to Chihuahua as 1983.) Contreras and Escalante (1984), who say they were introduced for rearing as a food fish, were uncertain about their status in Mexico.

Translocations

Outside their natural range in the USA, Rock Bass have been translocated to several states as a food and sport fish.

Unsuccessful Introduction

Erdman (1974) records an unsuccessful attempt in 1916 to introduce Rock Bass to Puerto Rico. (See also under *Lepomis gulosus.*)

Redbreast Sunfish (Longeared Sunfish)
Lepomis auritus

Welcomme (1988) says that most of the limited international introductions of the Redbreast Sunfish have resulted in self-maintaining populations.

Natural Distribution

Rivers of the Atlantic slope from New Brunswick, Canada, to Florida, USA.

Naturalized Distribution

Europe: Italy. *North America:* ? Mexico. *Central America:* Puerto Rico.

Europe
Italy

Besana (1910), Thienemann (1950) and Vooren (1972) record the introduction of the Redbreast Sunfish into mountain lakes in northern Italy (Welcomme (1988) incorrectly says in 1957), where in conjunction with the Largemouth Bass *Micropterus salmoides* (q.v.) its dense and stunted population has caused a decline in numbers of the native *Alburnus a. alborella* in Lakes Varano and Manate. In Lake Monterosi, the Sunfish population has been reduced following predation by Bass.

Unsuccessful Introduction

In 1891, Max Von dem Borne introduced Redbreast Sunfish into Germany, where they failed to become established (De Groot, 1985).

North America
Mexico

Lee (1980) records the occurrence of Redbreast Sunfish in areas of Amistad (Coahuila) and Falcon (Tamaulipas) reservoirs, but does not mention specific localities. Contreras and Escalante (1984), who suggest that they were introduced perhaps as a forage species or for rearing as a food fish in aquaculture, say their status in the wild in Mexico is unknown.

Translocations

The Redbreast Sunfish has been successfully transplanted to several states outside its natural range in the USA.

Central America
Puerto Rico

In June 1957, Redbreast Sunfish from the St Johns River were included in a consignment of fishes despatched to Puerto Rico from the US Fish and Wildlife Service hatchery in Welaka, Florida. Almost all the fish survived the journey to Maricao, where they were soon spawning in the hatchery ponds (Erdman, 1984). Redbreast Sunfish were later successfully stocked (Welcomme (1981) says for 'recreational and subsistence fishing') in Coama, Dos Bocas, Garzas, Guajataca, Guineo, Loco and Matrullas reservoirs (Ortiz-Carrasquillo, 1980), and also in Río Jayuya where spawning took place (Erdman, 1984). Redbreast Sunfish in Puerto Rico are also established in the upper reaches of Río Manatí, in the Río Maricao near the fish hatchery, and further downstream in the Río Guanajibo drainage (Erdman, 1984).

Green Sunfish
Lepomis cyanellus

A species of smaller creeks, brooks and ponds, the Green Sunfish has been widely introduced around the world, mainly as a forage fish for black bass *Micropterus* spp. (q.v.).

Natural Distribution

East-central North America west of the Appalachians, from Ontario, Canada, westward to eastern North Dakota, south to Georgia and northeastern Mexico.

Naturalized Distribution

Europe: Germany. ***Asia:*** Korea; Philippines. ***Africa:*** Morocco; ? South Africa; ? Swaziland; ? Zambia. ***South America:*** Brazil. ***Oceania:*** Mauritius.

Europe
Germany

Muus and Dahlstrøm (1968) and Vooren (1972) say the Green Sunfish has been introduced from the USA to Germany, where 'it occurs in the wild in the vicinity of Frankfurt'.

Asia
Korea

Welcomme (1981, 1988) says that in 1969 Green Sunfish from Japan were introduced to Korea, where they are established and breeding.

Philippines

Juliano *et al.* (1989) list the introduction in 1950 of Green Sunfish from California, USA, to the Philippines, where they are established in swamps in highland areas.

Unsuccessful Introduction

Welcomme (1981, 1988) says that the Green Sunfish has been introduced from the USA for aquacultural purposes to Japan. The species is not mentioned by Chiba *et al.* (1989).

Africa
Morocco

Welcomme (1981, 1988), quoting Moreau (1979), says that Green Sunfish have been introduced by private individuals from the USA to Morocco, where they have become established.

South Africa

Welcomme (1981, 1988), quoting Jackson (1976), says that in 1939 the Green Sunfish was introduced from the USA to South Africa as forage for black bass *Micropterus* spp. (q.v.), where it has become 'widespread in slow flowing temperate waters. Usually occupies smallish, stagnant and well-vegetated dams, which become overpopulated with small stunted [fish]; as a result regarded as a pest'. The species is

not mentioned by De Moor and Bruton (1988), so this is likely to be a result of misidentification, possibly for the Bluegill *L. macrochirus* (q.v.).

Swaziland

Welcomme (1981, 1988), quoting Clay (1972), says that in 1939 Green Sunfish from South Africa were introduced as a forage fish for black bass to Swaziland, where they had become 'well established and breeding by 1940'. There is, again, no mention of this introduction by De Moor and Bruton (1988), and another case of misidentification appears probable.

Zambia

According to Welcomme (1988), Green Sunfish have been introduced from the USA to Zambia for aquacultural purposes, where they are reproducing, though whether in the wild or only in captivity is not stated.

Unsuccessful Introductions

Welcomme (1981, 1988), quoting Toots (1970), says that in 1940 Green Sunfish from South Africa were imported as forage for black bass to Southern Rhodesia (Zimbabwe), where 'after initially breeding prolifically in dams and ponds [they] declined in numbers and failed to become established'. *L. cyanellus* is in fact not mentioned by Toots, who does, however, describe the introduction to Zimbabwe of the Bluegill *L. macrochirus* (q.v.). This is clearly another case of mistaken identity by Welcomme, who also refers to the unsuccessful introduction of Green Sunfish to the Congo, Kenya and Madagascar, on the last of which he says they were released in 1954 as forage for Largemouth Bass *Micropterus salmoides* (q.v.), but where they 'disappeared through too much competition with other species and over-cropping by *M. salmoides*'.

South America
Brazil

Welcomme (1981, 1988) says that in the 1930s Green Sunfish from the USA were introduced to Brazil as forage for bass, where they have become established.

Oceania
Mauritius

Welcomme (1981, 1988) says that in 1950 Green Sunfish were successfully introduced from the USA as a sport fish to the island of Mauritius, where stunted populations have become established.

Translocations

Within North America, the Green Sunfish has been extensively transplanted outside its natural range, apart from in the extreme northwest.

Pumpkinseed (Common Sunfish; Sun Bass; Yellow Sunfish)
Lepomis gibbosus

The Pumpkinseed has a rather more northerly distribution that *L. auritus* and *L. cyanellus* and is thus more tolerant of low temperatures; this has enabled it to become widely established throughout much of central and eastern Europe, partially by natural diffusion through canals and rivers. It inhabits small streams, pond and lakes, in both lowlying and upland regions, favouring still and clear shallow waters with plenty of aquatic vegetation. In many countries it is regarded as an ecological pest.

Natural Distribution

Southern Canada and the USA from the Great Lakes and North Dakota east to the Atlantic coast and south to Texas and Florida.

Naturalized Distribution

Europe: Belgium; British Isles (England); Czechoslovakia; France; Germany; Hungary; Italy;

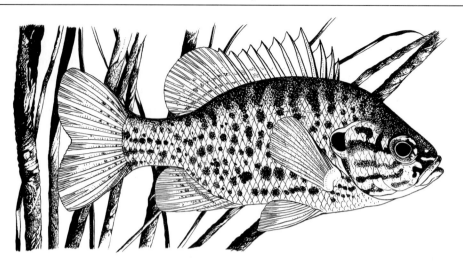

The Netherlands; Poland; Romania; Spain; Switzerland; the former Yugoslavia. **Africa:** Morocco. **South and Central America:** Chile; Guatemala.

Europe

Belgium; British Isles (England); Czechoslovakia; France; Germany; Hungary; Italy; The Netherlands; Poland; Romania; Spain; Switzerland; the former Yugoslavia

The Pumpkinseed was first imported to Europe, to *Germany*, in 1881 by Max Von dem Borne, who acquired 500 from the USA for experimental rearing at his hatchery at Berneuchen. Four years later, Pumpkinseeds first arrived from the USA in *France*, where shortly afterwards they managed to escape or were released into the wild, and from where in 1886 125 were sent to Von dem Borne in *Germany*, who in the same year also acquired a further consignment direct from the USA. Today, Pumpkinseeds in *Germany* occur mainly in Bavaria. Raveret-Wattel (1900) reported Pumpkinseeds as established in natural waters in *France*, where Welcomme (1981, 1988) says more were imported from the USA between 1930 and 1950 and where they are now 'widely spread throughout lowland rivers'.

In the late 19th century, Pumpkinseeds from the USA were introduced to *Czechoslovakia*, where they became established (Anon., 1971). Between about 1880 and 1920, they were imported from the USA to *Italy*, where they have become naturalized in Lago Maggiore (Grimaldi, 1972) and in other lakes in northern and central *Italy* (Welcomme, 1981, 1988). In *Poland*, to which Pumpkinseeds were introduced at an unrecorded date, they are only very locally distributed (Welcomme, 1981, 1988). In *Switzerland*, where the date of their introduction is also unknown, Pumpkinseeds occur in the River Aar and Lac du Neuchâtel (Muus & Dahlstrøm, 1968) and in Lac Lugarno (De Groot, 1985).

Pumpkinseeds are found in large numbers in Lake Balaton in *Hungary* (Thienemann, 1950), and are well established in the lower reaches of the River Danube in *Hungary* (where they were introduced before 1904) and *Romania* (Nikolsky, 1961; Wheeler & Maitland, 1973) and the former *Yugoslavia*, where Welcomme (1981, 1988) describes them as 'widespread'.

In the early years of the 20th century, Pumpkinseeds were introduced to *The Netherlands*, where they have become locally common in clear and well-oxygenated streams and moorland pools in the province of Noord Brabant (Looyen, 1948; Parent, 1950; Vooren, 1972; De Groot, 1985). Brouwer (1925) reported Pumpkinseeds to be established in rivers and canals in *Belgium* just south of Noord Brabant (the Kempen). De Groot (1985) suggests that Pumpkinseeds probably entered *The Netherlands* from *Belgium*, but the dispersal may well have been in the opposite direction.

Between 1910 and 1913, Pumpkinseeds from the USA were introduced to *Spain* for stocking purposes by the Catalonia Regional Fisheries Service (Elvira, 1995b); they are presently established in the wild in the drainages of the Rivers Douro, Tagus and Guadiana, and in the East Pyrenees (Elvira, 1995a).

There appear to have been few attempts to naturalize the Pumpkinseed in the *British Isles*, although it has become established and has bred in several waters – during the First World War at Groombridge in East Sussex (England) and at Newport-on-Tay, Fife (Scotland); since 1938 near Crawley, West Sussex, and in 1953 near Bridgwater, Somerset (both England). The two former populations have since died out. Pumpkinseeds have also been found in a lake on the Broadlands estate in Hampshire (Farr-Cox *et al.*, in press), and since 1973 in several waters in and around London (A.C. Wheeler, pers. comm. 1995).

Ecological Impact   Welcomme (1981, 1988) summarizes the ecological impact of Pumpkinseeds in Europe as follows:

COUNTRY	IMPACT
Belgium	A voracious species locally achieving pest status in some calm waters
British Isles	?
France	Not appreciated due to predation on young of more favoured species
Germany	Competes with other species
Hungary	Regarded as a pest when at peak abundance but has since declined and now accepted as harmless
Italy	Has caused decline of *Alburnus alborellus* and *Perca fluviatilis*
The Netherlands	?
Poland	?
Romania	?
Switzerland	Generally regarded as undesirable
Yugoslavia	Regarded as an undesirable trash fish which competes with local species eating eggs and larvae

Africa
Morocco

In 1937 and 1959, Pumpkinseeds from France were introduced to Morocco, where they became established in Lacs d'Ouezanne and Oued Mellah (Moreau *et al.*, 1988).

South and Central America
Chile

Welcomme (1981, 1988) says the Pumpkinseed is established in the wild in Chile, where according to Vooren (1968) it poses a threat to the local ichthyofauna. This introduction is not mentioned by De Buen (1959), so presumably took place during the 1960s.

Guatemala

Welcomme (1981, 1988) lists the Pumpkinseed as established since 1960 in Lake Atitlán in southern Guatemala.

Unsuccessful Introductions

In Africa, Pumpkinseeds have been unsuccessfully introduced to the Congo, and in South America to Venezuela (Welcomme, 1981, 1988).

Translocations

Outside its natural range in North America, the Pumpkinseed has been translocated to many western states of the USA (Smith, 1896; Chapman, 1942), and to central and western provinces of Canada (Carl & Guiget, 1958; Crossman, 1984).

Warmouth
Lepomis gulosus

As well as the two international introductions described below, the Warmouth has been widely translocated within the USA.

Natural Distribution

The eastern USA from Kansas, Iowa, southern Wisconsin, Minnesota and western Pennsylvania, southward to Florida.

Naturalized Distribution

North America: Mexico. **Central America:** ? Puerto Rico.

North America

Mexico

Some authors include the Río Grande, Mexico, as within the Warmouth's natural distribution, but Contreras and Escalante (1984) do not accept this contention, and say that fish are established in Marte R. Gómez reservoir, Tamaulipas, and in San José de Gracia in Aguascalientes, and also in Mal Paso reservoir, Zacatecas, as a result of human intervention.

Central America

Puerto Rico

According to the US Bureau of Fisheries stocking records (published in *Informe del Comisionado de Agricultura 1934–1935*), 1200 Rock Bass *Ambloplites rupestris* (q.v.) were included in a consignment of Largemouth Bass *Micropterus salmoides* (q.v.), Bluegills *Lepomis macrochirus* (q.v.) and Brown Bullheads *Ictalurus nebulosus* (q.v.) shipped to Puerto Rico in 1916. It is believed that the Rock Bass failed to establish themselves, but in 1971 a Warmouth was netted, together with Bass, Bluegills and Bullheads, under 'poma rosa' bushes in Carite reservoir. Since Warmouth can be confused with Rock Bass, this fish could have been a survivor of a breeding population descended from the 1916 stocking (Erdman, 1984).

Bluegill
Lepomis macrochirus

The Bluegill is probably the most widely distributed of all the sunfish species, having been translocated and introduced for aquacultural purposes; as a forage for bass *Micropterus* spp.; and for recreational sport fishing. It occurs in lakes, ponds, and densely vegetated and slowly flowing streams and rivers.

Natural Distribution

The Mississippi drainage of the USA, from the coasts of Virginia and Florida westward to northern Mexico and Texas, and from western Minnesota to western New York.

Naturalized Distribution

Asia: Japan; Korea; Philippines. **Africa:** Morocco; South Africa; Swaziland; Zimbabwe. **South and Central America:** ? Brazil; Panama; Puerto Rico; Venezuela. **Oceania:** Hawaiian Islands; Mauritius.

Asia

Japan

Introduced in 1960, Bluegills 'have readily established themselves in rivers and lakes and are spreading rapidly over the Japanese main islands' (Chiba *et al.*, 1989).

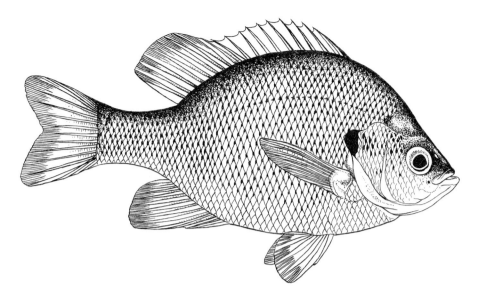

Korea

In 1969, Bluegills from Japan were introduced for aquacultural purposes to Korea, where they have become established (Welcomme, 1981, 1988).

Philippines

In 1950, Bluegills from California, USA, were introduced to the Philippines, where they have become established locally in some highland swamps (Juliano *et al.*, 1989).

Africa

Morocco

In 1960, Bluegills from the USA were introduced privately to Morocco, where they became established in Nfout and Daourat reservoirs (Moreau *et al.*, 1988).

South Africa

In 1938, Bluegills from Maryland, USA, were imported to the Jonkershoek hatchery, Stellenbosch, and to hatcheries in Natal (Hey & Van Wyck, 1950; Harrison, 1952) to rear as a forage fish for Largemouth Bass *Micropterus salmoides* (q.v.) and for recreational angling. From those sent to Natal, 19 were subsequently transferred to the Howick hatchery and eight to a pond on Everdon farm. Within 2 years of their importation, Bluegills were reported to be breeding successfully and their offspring were later distributed to other localities (Anon., 1944). De

Moor and Bruton (1988: appendix VI: 273) give details of the first introductions of Bluegills (with references) to various areas in South Africa.

De Moor and Bruton (1988), from whom much of the following account is derived, have summarized the present distribution of Bluegills in South Africa.

Cape Bluegills are established in the dams of the Orange River catchment, but are absent from the river systems draining into the Verwoerd dam (see below under *Orange Free State*) (Jubb, 1972). They are also found in the Kromme River (Jubb, 1959a), the Breede River (Cambray & Stuart, 1985), Voelvlei, the Berg River, the Riviersonderend system, dams in the Eerste River catchment, Groenvlei (Krysna) (Van der Merwe, 1970), Tyume River (Keiskamma system) (Gaigher, 1975a), and in the Kraai River, a southern tributary of the upper Orange River. In a survey of the Olifants River in 1963–1964, Bluegills were found to be abundant in the Clanwilliam dam and further north in the river, but did not occur in the river at Keerom. Van Rensburg (1966) also recorded Bluegills in the Doorn and Rondegat tributaries of the Olifants. More recently, Bluegills have been found downstream of the Clanwilliam dam, including the lower reaches of the Doring River.

Natal Bluegills are established in midland lentic waters but are usually absent from coastal rivers (Crass, 1964). They have been recorded from the Oribi gorge in the Umzimkulwana River (Bourquin & Mathias, 1984); in the dam on the Umzinto River in the Vernon Crookes Nature Reserve; near Park

Rynie (Bourquin & Van Rensburg, 1984); and in the upper Umgeni River (Crass, 1966). According to McVeigh (1984), Bluegills are also established in the Midmar, Wagendrift and Craigieburn dams.

Transvaal and *Orange Free State*   Bluegills were said by McVeigh (1984) to be only sparsely distributed in the Transvaal, where they are recorded only in some farm dams along the Crocodile River downstream from Nelspruit, in the Amersfoort area, and in the Shiljjalongubo dam near Barberton, on a tributary of the Lomati River in the Incomati system. They have also been reported in the Sterkfontein dam on the Wilge River in the Vaal system; in the Klipplaatdrift dam on the Elands River, a tributary of the Olifants River in the Limpopo system; and in a dam on the Levubu River in the Limpopo system (De Moor & Bruton, 1988).

In the Orange Free State, the Mozambique Tilapia *Oreochromis mossambicus* (q.v.) was introduced as a forage fish for Largemouth Bass, and there are no confirmed records for Bluegills in the province, although McVeigh (1984) considered that they might occur in the Clocolan and Ficksburg districts.

Within the Orange-Vaal system, Bluegills have been reported from the upper Orange and Caledon Rivers (Skelton, 1986). Their presence in the Vaal system has yet to be confirmed, but as they do not appear to occur in the Verwoerd dam (see above under Cape) they are believed not to be present in the main Orange River (Hamman, 1980) or Lake Le Roux (Jackson *et al.*, 1983).

Ecological Impact   Bluegills have a tendency to overpopulate the waters into which they are introduced, resulting in the appearance of stunted populations; they also prey on the young of more desirable fish and compete with them for food (Jubb, 1965).

In North America, Bluegills have coevolved with other native centrarchids such as black bass *Micropterus* spp., and are thus likely to be less susceptible to predation by them than are autochthonous South African fishes. The occurrence of bass may therefore be to the advantage of Bluegills in their competition with native species, since the latter are likely to be preferred prey (De Moor & Bruton, 1988).

In the Kromme River, on both sides of the Churchill dam, populations of indigenous *Barbus senticeps/afer* have declined in the presence of Largemouth Bass and Bluegills, and long stretches of

the Kouga River populated by Bluegills have been found to be virtually devoid of native *Barbus* spp. In rivers containing indigenous predators such as eel but without exotic species, however, *Barbus* species are thriving (De Moor & Bruton, 1988).

In the Clanwilliam Olifants River, many *Barbus* species have markedly declined, mainly as a result of the naturalization of Largemouth Bass, Smallmouth Bass *M. dolomieui* (q.v.) and Bluegills (Gaigher 1981). The impact on native fishes is almost certainly greater when two or more exotic species, rather than just one, are introduced; the native *Barbus* species then suffer predation of young and adults by both species of bass and by Bluegills, and are also forced to compete with the latter for food and other resources. At the same time, as previously mentioned, the presence of bass gives Bluegills a selective advantage over other indigenous prey species (De Moor & Bruton 1988).

In the absence of bass, however, Bluegills do appear to be vulnerable to predation by indigenous predators such as the Sharptooth Catfish *Clarias gariepinus*, and have also been excluded from many artificial waters by the introduction of Mozambique Tilapia, large individuals of which have been reported to prey on small fishes (Bruton & Boltt 1975). The presence of native predators may well be the reason why Bluegills have failed to become established in many coastal rivers of Natal (Crass, 1964) and in most waters of the Transvaal, where such predators as catfish, yellowfish and tigerfish occur (McVeigh, 1984).

De Moor and Bruton (1988) suggest that control of Bluegills in South Africa might be achieved through further introductions of *O. mossambicus* and transplantations of *C. gariepinus* and *B. kimberleyensis* though they warn that the possible impact of the movements of these species would have first to be carefully assessed.

Swaziland

Bluegills were introduced from South Africa to Swaziland as forage for black bass *Micropterus* spp. as follows: to the Black Umbeluzi River above the Hawane Falls in 1940 and 1942; to the Poponyane River (Pigg's Peak) and the Mtitshane River (a tributary of the Motshane River) in 1941; and to dams situated on tributaries of the Komati River at Hereford's farm in 1950 (Clay, 1972). Bluegills in

Swaziland are now well established in many waters (Clay, 1972).

Zimbabwe

Toots (1970) has documented the introduction of Bluegills to Zimbabwe (Southern Rhodesia), where fish from South Africa were in 1940 stocked in dams in and around Bulawayo, Salisbury (Harare), and elsewhere. Thereafter, Bluegills spread rapidly all over the country, and at first were recommended as the preferred forage for black bass *Micropterus* spp. This view later changed in favour of native species, and Bluegill numbers greatly declined. By 1969, Bluegills were said to have disappeared from waters in and around Salisbury, but they survive elsewhere as a sporting fish (Moreau *et al.*, 1988).

Unsuccessful Introductions

In Africa, Bluegills have apparently failed to establish viable populations in the Congo, Kenya and Zambia (Welcome, 1981, 1988). According to Robbins and MacCrimmon (1974), Bluegills have also been introduced to Lesotho, but whether successfully is unclear.

Introduced to Madagascar in 1954 as a forage for Largemouth Bass (Kiener, 1963; Moreau, 1979), Bluegills disappeared due to competition from other species and overcropping by bass (Welcomme, 1981; Reinthal & Stiassny, 1991).

South and Central America

Brazil

Welcomme (1981, 1988) lists the Bluegill as reproducing in Brazil, but whether in the wild or only in aquaculture is unclear. No information on the introduction is given.

Panama

In 1917 and 1925, 800 and 500 juvenile Bluegills respectively were imported from the USA to Panama by the Canal authorities, where they were unsuccessfully released in Lake Gatún (Hildebrand, 1938). According to Dr Abelardo De Gracia, Jr (pers. comm. (1987) to González (1988)), between 1925 and 1935 Bluegills were successfully introduced to Laguna de San Carlos and to Mataahogado in Panama Province. Finally, in 1955 Bluegills were planted in Chiriquí

Province (González, 1988). The species is currently established in Panama in Laguna de San Carlos and in the volcanic lagoons of Chiriquí (González, 1988).

Puerto Rico

The Bluegill is probably the most common member of its genus naturalized in Puerto Rico, where it was first introduced by the US Bureau of Fisheries in Carite and Comerío reservoirs in 1915 for sport and subsistence fishing and as forage (Robbins & MacCrimmon, 1974). It is now established throughout the island in reservoirs, farm ponds, and some rivers where the current is slow, such as La Plata from Aibonito to Comerío. In 1957, Bluegills occurred in large numbers in Cartagena Lagoon, where they have since been replaced by introduced Mozambique Tilapia *Oreochromis mossambicus* (q.v.) (Erdman, 1984).

Venezuela

In the late 1950s, Bluegills were introduced to Venezuela, where they became established only in Laguna Potrerito in which they are subject to stunting and are reported to eat young fish (Welcomme, 1988).

Unsuccessful Introductions

In Central America, Bluegills were successfully introduced to El Salvador in 1955, but 2 years later were eradicated because they had become a pest and were stunting (Welcomme, 1988).

Oceania

Hawaiian Islands

In 1946, 14 Bluegills from California, USA, were introduced to a private lily pond near Kaneohe Bay on Oahu, where they were soon breeding and from where young stock was subsequently transferred to suitable waters on all the main islands (Brock, 1952, 1960). More Bluegills were imported to the Hawaiian Islands at a later date, and they are now naturalized in reservoirs on all the principal islands (Maciolek, 1984). (See also Needham, 1949; Randall, 1960; Hida & Thomson, 1962; Kanayama, 1968; Lachner *et al.*, 1970; Maciolek & Timbol, 1980.)

Maciolek (1984) draws attention to an interesting phenomenon on the island of Kauai, where in at least two reservoirs an intergeneric hybrid between

L. macrochirus and *Micropterus salmoides* appeared spontaneously in 1963. The hybrid, known locally as the 'Blue Bass', is superficially intermediate between the two parent species, and may be fertile; if so, its appearance might possibly lead to the evolution of a new species.

Mauritius

Welcomme (1988) records the introduction in 1950 of Bluegills to the island of Mauritius for sporting purposes, where stunted populations have become established.

Translocations

Outside its natural range in North America, the Bluegill has been widely and successfully transplanted to many states in the USA (Smith, 1896; Chapman, 1942) and to numerous localities in Mexico (Contreras & Escalante, 1984).

Redear Sunfish
Lepomis microlophus

The Redear Sunfish is a species of warm, clear, quiet waters with an abundance of aquatic vegetation. It feeds largely on molluscs, and is thus a potential biological controlling agent of aquatic snail vectors of schistosomiasis (bilharzia); the species is also used as a forage fish for recreational and subsistence fishing (Welcomme, 1988).

Natural Distribution

From southern Indiana, USA, the lower Atlantic slope and the Gulf slope drainages west to Texas, and south to peninsular Florida.

Naturalized Distribution

Africa: Morocco; South Africa. ***North America:*** ? Mexico. ***Central America:*** ? Panama; Puerto Rico. ***Oceania:*** Mauritius.

Africa
Morocco

In 1940, Redear Sunfish from the USA were introduced privately to Morocco, where they became established in Lac Dayet'roumi (Moreau *et al.*, 1988).

South Africa

According to Moreau *et al.*, (1988) Redear Sunfish from the USA were introduced to South Africa in 1937, where they became naturalized in freshwater rivers in the Cape. De Moor and Bruton (1988) do not mention this introduction.

North America
Mexico

Contreras and Escalante (1984) doubt the inclusion of the lower Río Grande River in Mexico in the native range of the Redear Sunfish, which is recorded by Hubbs *et al.*, (1977) in that river and by Hendrickson *et al.* (1980) in the Río Yaqui at Presa Novillo. It was probably introduced to Mexico as a forage species for black bass and for rearing in captivity for human consumption; its status in the wild is uncertain.

Translocations

The Redear Sunfish has been translocated outside its natural range in the USA, in particular as a 'panfish' to the Great Lakes as long ago as the 1920s (Mills *et al.*, 1994).

Central America
Panama

Welcomme (1988) says the Redear Sunfish has been introduced, from an unknown source and at an unknown date, as a forage fish to Panama, where it is reproducing. González (1988), however, makes no mention of this species in Panama.

Puerto Rico

Although stocking records, dates, and sources have not been traced, the Redear Sunfish is known to have been introduced to Loiza reservoir from stock obtained from the US Fish and Wildlife Service hatchery at Welaka, Florida, USA, and to have been established and reproducing in Garzas and Guajataca reservoirs since before 1948 (Erdman, 1984). Elsewhere, Erdman (1984) says the introduction took place in 1957.)

Ecological and Economic Impact   Redear Sunfish have proved a valuable importation to the fresh waters of Puerto Rico, where they have been released in around 50 farm ponds and in five lakes. When stocked in farm ponds with Largemouth Bass *Micropterus salmoides* (q.v.) they have been found to be the best sportfish combination in the island, and are a useful source of food. Redear Sunfish also feed on snails, including the pulmonate *Biomphalaria glabrata*, which are common in farm ponds and act as an intermediate host for the trematode blood fluke *Schistosoma mansoni*. In some lakes, Redear Sunfish populations are said to have declined due to predation by *Tilapia* spp. (Ferguson, 1978; Erdman, 1984).

Oceania

Mauritius

In 1944, Redear Sunfish were experimentally introduced from the USA to the island of Mauritius, where according to Moreau *et al.*, (1988) they are established in one unnamed river.

Allied Species

Welcomme (1988) also records the successful introduction to Cuba of an unidentified *Lepomis* species in 1928, as a forage fish for Largemouth Bass *Micropterus salmoides* (q.v.) introduced in the same year.

According to Dr Abelardo De Gracia, Jr (pers. comm (1987) to González (1988)), between 1925 and 1935 Orangespotted Sunfish *Lepomis humilis* were successfully introduced to Laguna de San Carlos and to Mataahogado in Panama Province, Panama; in the former, where they are well established, in spite of their small size they are eaten by the local people.

Redeye Bass (Southern Red-eyed Black Bass)
Micropterus coosae

The Redeye Bass has a very restricted natural distribution in the USA, where it is almost entirely rheophilic.

Natural Distribution

Coosa River system of Alabama and Georgia, United States.

Naturalized Distribution

Central America: Puerto Rico.

Central America

Puerto Rico

Redeye Bass, which are better adapted to living in rivers than Largemouth Bass *M. salmoides* (q.v.), were introduced from the Flint River near Helen, Georgia, USA, to Puerto Rico in 1958–1959 for initial culture in the fish hatchery at Maricao. Their progeny were subsequently released in farm ponds near Maricao, Utuado and Naranjito, and in the Pitahaya and Jayuya

Rivers (Erdman, 1972). Redeye Bass became established in the Río Maricao near the hatchery and, in 1963, in the Jayuya upstream from the town of Jayuya, where in the following year Erdman (1984) observed Redeye Bass and Redbreast Sunfish *Lepomis auritus* (q.v.) fingerlings in a rocky pool named El Cantil. Redeye Bass have also been reported from Caonillas reservoir, downriver from Jayuya (Erdman, 1984).

According to Robbins and MacCrimmon (1974),

high water temperatures would seem to limit the dissemination of Redeye Bass in Puerto Rico.

Translocations

Robbins and MacCrimmon (1974) found few records of translocations of Redeye Bass within the USA where Welcomme (1988) mentions transfers only to California.

Smallmouth Bass
Micropterus dolomieui

Although in the USA the Smallmouth Bass is an important predatory sport fish, international introductions – apart from to South Africa – have not been very successful, probably because of the species' preference for clear, fast-flowing cool and temperate waters (Welcomme, 1988).

Natural Distribution

From Minnesota in the USA eastward to Quebec, Canada, south to the Tennessee River in Alabama, and west as far as eastern Oklahoma.

Naturalized Distribution

Europe: ? Belgium; ? France; Sweden. *Asia:* ? Hong Kong; Vietnam. *Africa:* South Africa; ? Zimbabwe. *North America:* Mexico. *Central America:* Belize. *Oceania:* ? Fiji; Hawaiian Islands.

Europe

Evidence for the introduction and naturalization of Smallmouth Bass in Europe is fragmentary and often contradictory.

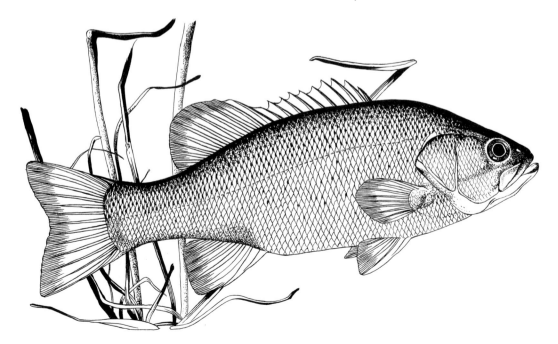

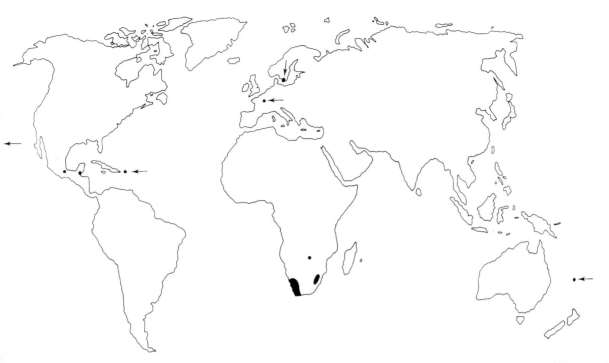

Naturalized distribution of Smallmouth Bass *Micropterus dolomieui* (adapted from Robbins & MacCrimmon, 1974).

Belgium

Smallmouth Bass were first introduced to Belgium from Germany by Max Von Dem Borne between 1885 and 1890 (Von Dem Borne, 1892). Some of these fish were released in the Semois River, and elsewhere, and in the Semois River, according to Robbins and MacCrimmon (1974), a self-maintaining population survives, albeit in decreasing numbers due to overfishing. Although Vooren (1972) said that Smallmouth Bass had been 'recently' released in the Semois, Robbins and MacCrimmon (1974) were unable to trace any releases since those of the late 19th century. Welcomme (1981, 1988) lists Smallmouth Bass in Belgium as 'disappearing and only of very local interest'.

France

'Bass' of unspecified species were first introduced, without success, to France in 1869. Although Robbins and MacCrimmon (1974) believed these to be *M. dolomieui* (on what evidence is unclear), Renault (1951) says that the first introduction of the species to France took place in 1883, while Von Dem Borne (1892) reported sending Smallmouth Bass to aquaculturists in Paris between 1884 and 1890. Raveret-

Wattel (1900) referred to the presence of Smallmouth Bass in natural waters in France, where their continued presence was confirmed by Monod (1949). Although Smallmouth Bass have from time to time been reported from other regions of France (e.g. Charentes, Lake Uzein (Basses-Pyrénées), and between Palavas and Perals (Herault)) they now possibly occur only in that portion of the River Meuse in France near the border with Belgium, and in the Semois, a tributary of the Meuse (Spillman, 1961).

Sweden

Between 1884 and 1890, Max Von Dem Borne sent several consignments of Smallmouth Bass from Germany to aquaculturists in Sweden (Von Dem Borne, 1892). In 1915, 250 were despatched by the US Bureau of Fisheries to Sweden, where they were placed in four lakes. In 1922, 1000 young-of-the-year and 20 adult fish from Von Dem Borne's hatchery at Berneuchen, Germany, were unsuccessfully liberated in a small lake on an island in the Stockholm archipelago, while in the following year an unrecorded number of fingerlings and 3-year-old fish from Germany were planted in ponds at Billesholm, Skåne (Robbins & MacCrimmon, 1974). In 1929, 260 young-of-the-year fish were sent by the US Bureau of Fisheries to the Royal Swedish Agricultural Society, where the 220 survivors were unsuccessfully placed in two ponds at Österbybruk in Uppland (Alm, 1929).

In 1956, 400 Smallmouth Bass fingerlings were acquired from the Ontario Department of Lands and Forests, Canada, and flown to Sweden from Toronto. All except four, which had been left at the Copenhagen aquarium in Denmark *en route*, died during the following winter in the Aneboda hatchery in southern Sweden. The survivors and their offspring were forwarded from Copenhagen to Sweden in 1958, and placed in privately owned ponds at Mörrum near Karlshamn in southeastern Sweden, where they soon died (Robbins & MacCrimmon, 1974).

It was not until 1960 that Smallmouth Bass were successfully introduced to Sweden; in November of that year the Sport Fishing Institute of Washington, DC, USA, in co-operation with the Virginia Commission of Game and Inland Fisheries, shipped 50 fingerlings and 12 yearlings from the Shenandoah River, Virginia to Svängsta, where in 1961 they spawned successfully in the ponds at Mörrum. In 1962, two further shipments of Smallmouth Bass from the USA, comprising 900 fingerlings and 57 mature fish, were released in Lake Vita Vatten (45 ha) near Svängsta, where they spawned successfully and from where fish were later transferred to another lake nearby (Borgstrom, 1966).

As Robbins and MacCrimmon (1974) point out, the Swedish environment (and climate) is marginal for the naturalization of Smallmouth Bass, many of which fail to overwinter successfully. A few waters in southern Sweden do, however, continue to support self-maintaining populations.

Unsuccessful Introductions

Robbins and MacCrimmon (1974) list the following countries in Europe to which Smallmouth Bass have been unsuccessfully introduced (additional references in parentheses): Austria; British Isles (England and Scotland) (Fitter, 1959); Czechoslovakia (Anon., 1971); Denmark (Jensen, 1987); Finland; Germany (Thienemann, 1950; Bauch, 1955; Ladiges & Vogt, 1965; De Groot, 1985); Hungary; Italy (Vooren, 1972, quoting Zaccagnini, 1965); The Netherlands (Mulier, 1900; Vooren, 1972; De Groot, 1985); Norway; Poland; and Switzerland.

Asia

Hong Kong

Around 1935, Smallmouth Bass were introduced by the government to five or six reservoirs in Hong Kong to establish a sport fishery, where Robbins and MacCrimmon (1974) said that the species still occurred in the Jubilee reservoir; fluctuations in water level and the relatively low nutrient quality of the water seem to be the main limiting factors of the species in Hong Kong.

Vietnam

In 1952, Smallmouth Bass fingerlings from the USA were flown to Vietnam as a present for Dao Dai, where they were released for sport fishing in Grand Lake (Xuen-Huong) near Dalat City in the central highlands, at an elevation of 1200 m. On a diet of *Carassius, Tilapia* and cyprinids, the Smallmouth Bass flourished, and are said to provide good angling (Robbins & MacCrimmon, 1974).

Unsuccessful Introductions

Robbins and MacCrimmon (1974) list unsuccessful introductions of Smallmouth Bass in Asia to Guam in the Mariana Islands (1962) and Japan (around 1930).

Africa

South Africa

In the hope that they would 'fill the position left vacant in the present distribution of imported game fish and would improve the fishing waters which lie between the upper trout waters and the sluggish lower reaches held by the largemouth species' (Harrison, 1940), in 1937 Smallmouth Bass from the Lewistown hatchery in Maryland, USA, were imported to the Jonkershoek hatchery in the Cape where 29 arrived safely (Anon., 1944; Harrison 1953), and later spawned successfully. By late 1938 the Jonkershoek hatchery began to distribute Smallmouth Bass fingerlings, some 258 of which were sent to the Howick hatchery for the Natal Administration which, in co-operation with the Cape Piscatorial Society, had been responsible for the original importation; 500 were planted in waters in the Cape; and some were sent to the Rand Piscatorial Association in Johannesburg (and to Swaziland)

Robbins & MacCrimmon, 1974). In the same year, 300 Smallmouth Bass fingerlings were released in the Berg River in the western Cape, where spawning took place in 1940, and where early in the following year fingerlings were said to be abundant. The Breede and other rivers in the western Cape, and even the acidic and peaty water of Steenbras reservoir (which supplies Cape Town's water), were soon supporting thriving Smallmouth Bass populations (Robbins & MacCrimmon, 1974).

In 1952, a hatchery opened at Umgeni in Natal, where Smallmouth Bass were reared for subsequent stocking in various waters in the province (Pike, 1980).

Harrison (1949) reported Smallmouth Bass to occur in the following waters in the Cape: Steenbras reservoir (where, unlike the introduced Brown Trout *Salmo trutta* (q.v.) and Rainbow Trout *Oncorhynchus mykiss* (q.v.), they are reported to be able to breed); the Lower Berg River; the Dwars River (in the Breede River system); the Lower Breede and Olifants Rivers, and the Buffalo River downstream of the stretch occupied by trout; and the upper and lower Kubusie and the lower Umtata Rivers.

Robbins and MacCrimmon summarized the then known establishment of Smallmouth Bass in South African waters. In addition to those mentioned above, they occurred in the Olifants (Clanwilliam) River and in the Umtata River in the Transkei, Cape Province. In the Olifants, they had displaced the redfin *Barbus* as the dominant species (Jubb, 1967). They were also reported to occur in a river in the Knysna area and in numerous ponds in the south-western Cape.

In Natal, Smallmouth Bass were naturalized only in the Umgeni River near Howick; in the Mooi, for some distance downstream of Rosetta; and in the Bushmans River below Estcourt (Crass, 1964), but were stocked and found in the lower reaches of most of the rivers in the region. They occurred only between altitudes of 1000 and 1300 m asl.

In the Transvaal, Robbins and MacCrimmon (1974) reported the establishment of Smallmouth Bass in a number of upland dams and reservoirs, such as Stanford, Ebenezer, Jerico and Witklip reservoirs. In 1970, the Nwanedzi and Luphephe dams, near Messina, were stocked with 5000 *M. dolomieui* and the same number of *M. salmoides* (q.v.).

Robbins and MacCrimmon (1974) described the then distribution of the Smallmouth Bass in South

Africa as 'rather limited', and suggested that hot summers and high water turbidity would prevent the species' further dispersal downstream, and that its only possible expansion of range would be into upland trout waters. Dams and natural barriers, such as waterfalls, would prevent spread upriver.

The Smallmouth Bass has been an extremely successful colonist in South Africa, where it is currently established throughout much of the southern, south-western and eastern Cape, and in the eastern Transvaal. De Moor and Bruton (1988: appendix VII: 273–274) give dates of introductions of Smallmouth Bass (with references) to various localities in South Africa.

De Moor and Bruton (1988), from whom much of the following account is derived, have outlined the present wide distribution and ecological impact of Smallmouth Bass in South Africa.

Cape Smallmouth Bass occur in the Olifants, Berg, and Umtata Rivers (Jubb, 1965), and in the Breede River system, Bredasdorp, in the Bontebok National Park (Braack, 1981). They are also found downstream of the Rooikrantz dam in the Buffalo River, and especially in the Laing dam, but are uncommon elsewhere in the river (Jackson, 1982), and in the Bongolo dam (Swart Kei, Great Kei system) and in the Gamtoos system. Jubb (1965) recorded Smallmouth Bass in numerous dams in the Orange River catchment but not in the main river. They are also found in the Baviaanskloof River and in the Paul Sauer dam, tributaries of the Coega River in the Gamtoos system. They are, or were, believed to occur in the Elands, Swartkops and Kromme Rivers (Jubb, 1971).

By the mid-1980s, Smallmouth Bass were reported to be naturalized in lotic waters throughout almost the whole of the Olifants River system in the western Cape, although in many localities they appeared to be suffering from a shortage of food. Smallmouth Bass seem to coexist well with Banded Tilapia *Tilapia sparrmani* and Bluegills *Lepomis macrochirus* (q.v.). They also assemble below natural barriers, such as waterfalls, and prey on indigenous fish that are washed downstream (A. Smith, pers. comm. to De Moor & Bruton, 1988).

In the Berg and Breede Rivers, Smallmouth Bass seem to be thriving more than in the Olifants system, probably because of the abundance of Common Carp *Cyprinus carpio* (q.v.), Banded Tilapias and

Bluegills in the former rivers. Smallmouth Bass are also found in the Smalblaar and Holsloot Rivers but have failed to become established in the Hex River, as in times of drought it partially dries out, cutting itself off from the Breede River (A. Smith, pers. comm. to De Moor & Bruton, 1988).

Natal   The distribution of Smallmouth Bass in Natal remains substantially as reported above by Crass (1964).

Transvaal   Smallmouth Bass are established in the headwaters of the Mogol River (Limpopo system) (Kleynhans, 1983); Sabie River (Incomati system); Roodeplaat dam (Pretoria, Limpopo system) (Viljoen & Van As, 1985); Rietvlei dam (Pretoria) (Smith, 1983); Blyde and Treur Rivers (Limpopo system) (Pott, 1981); and the Ebenezer dam on the Letaba River. Between 1969 and 1977, some 20 000 Smallmouth Bass were stocked in the Boskop dam near Potchefstroom on the Mooi River in the Vaal system, where they may still occur but where successful reproduction is doubtful (Koch & Schoonbee, 1980).

Ecological Impact   'The introduction of *M. dolomieui* has had a devastating effect on indigenous species in various parts of southern Africa' wrote De Moor and Bruton (1988). Although habitat destruction and siltation are perhaps the most important factors in the decline of indigenous fish populations in South Africa, Skelton (1987) considered that predation by introduced exotics played a significant part in putting at risk nine of the twelve threatened species of freshwater fishes in the Cape, and Gaigher (1978) listed the Smallmouth Bass as the most destructive of all the introduced predators. A description of the impact that the species has had in the various river systems where it is established in South Africa, derived from De Moor and Bruton (1988), follows.

Olifants River system, Clanwilliam   Before the introduction of Smallmouth Bass in 1943 and 1945, *Barbus capensis* and other autochthonous species were abundant, whereas by 1960 *capensis* had drastically declined, although in some stretches of the river where *M. dolomieui* did not occur the native *B. serra* and *Labeo seeberi* were still common (Van Rensburg, 1966). A survey carried out in the following decade revealed that the populations of all three indigenes and some others had fallen dramatically as a result of

predation by *M. dolomieui* and *M. salmoides*, and their distribution had been severely curtailed. The endemics *B. calidus* and *B. phlegethon* had been eliminated in the main river, and the latter was considered to be in serious danger of extinction. Endemic species were able to survive only where natural barriers such as waterfalls prevented colonization by *M. dolomieui* or where higher temperatures hindered invasion by trout (Gaigher, 1973). The presence of *M. dolomieui* and *M. salmoides* also had a negative impact on the distribution of *Austroglanis gilli* and *B. erubescens* – the latter being restricted to two small streams divided by a waterfall from the rest of the system (Gaigher *et al.*, 1980). In total, some eight endemic fishes of the Olifants system are listed as threatened by Skelton (1987), all of which have, to a greater or lesser extent, been affected by *M. dolomieu* (especially) and *M. salmoides*.

It is thus apparent that, as well as habitat degradation and heavy siltation, predation by introduced bass has been a major factor in the decline in the populations of the native species. Both Rainbow Trout (in 1897) and Brown Trout (in 1906) were released in the Olifants River, but the serious decline of indigenous species occurred long after their introduction but shortly after that of bass, providing strong circumstantial evidence that predation by the two trout species had a minimal impact compared with that of the two species of bass.

Berg River   In 1930, Largemouth Bass were introduced to this river, where 4 years later a survey revealed the presence of large numbers of native fishes, including *Barbus andrewi*, *B. burchelli*, *Sandelia capensis* and *Galaxias zebratus*. Smallmouth Bass were introduced in 1938–1939, and by 1943–1944 the populations of these indigenes had declined, *B. andrew* and *S. capensis* only surviving in slowly flowing stretches of the river. It was believed that the former had been a major factor inhibiting the naturalization of trout in the Berg; after their decline, trout began to become established (Harrison, 1952, 1953). Since then, numbers of *B. burgi* have also fallen in the river, due largely to pollution and water extraction for irrigation purposes, but predation by bass and trout is believed to have been a contributory factor (Skelton, 1987).

Breede River   Smallmouth Bass were introduced in 1939–1940. In a survey of this river carried out in 1977–1978, it was found there had been a severe decline in the populations of native species

(Cambray & Stuart, 1985). Although a variety of other factors were involved – including the extraction of water for irrigation, siltation, pesticide pollution, increased salinity, and canalization of the banks – the presence of *M. dolomieui* and other introduced predators contributed to the decline. No specimens of *B. burchelli* have been collected at the same location as either bass or Rainbow Trout, and no indigenous species were found in the main river (Cambray & Stuart, 1985).

Buffalo River After the introduction of Smallmouth Bass before 1949 downstream of Rooikrantz dam, *Barbus trevelyani* disappeared from the lower reaches of the river (Jubb, 1979).

Mogol River (Limpopo system) The date of the introduction of Smallmouth Bass to this river has not been recorded, but the species' presence in the upper reaches may explain why these waters are devoid of other fishes. It has been anticipated that in the main river the alien may threaten the survival of local populations of *Barbus brevipinnus* (Kleynhans, 1983).

Treur River (Limpopo system) In 1957, a new endemic, *Barbus treurensis*, was discovered in the Treur and Blyde Rivers. Between that year and 1981, Smallmouth Bass were, with extraordinary irresponsibility given their known impact on native species, introduced to the Treur, where the combined effect of predation by trout and Smallmouth Bass, and the accompanying introduction of white spot disease, are believed to have combined to eliminate *B. treurensis* from the Treur River; it now survives only in an isolated stretch of the Blyde bounded by two waterfalls, which is devoid of alien species (Pott, 1981).

Eerste River It has been claimed that the introduction of Smallmouth Bass to this river resulted in the elimination of *Barbus burgi*. De Moor and Bruton (1988), however, suggest that this record needs to be treated with caution.

Natal Smallmouth Bass have been less successful in Natal than in the Cape, possibly due to their inability to survive flash floods (Crass, 1964) or their intolerance of the higher water temperatures of Natal (De Moor & Bruton, 1988). In the few areas where *M. dolomieui* has managed to become established, Crass (1964) could find no evidence for any significant impact on native fishes such as *B. natalensis*, and the alien has not succeeded in colonizing rivers where *Clarias gariepinus* occurs. This may be due to predation on the young by catfish (De Moor & Bruton, 1988).

Summary In the Olifants (Clanwilliam), Berg and Breede Rivers, the introduction of Smallmouth Bass took place after that of Largemouth Bass, which were rapidly replaced as the dominant species (Harrison, 1963).

Many of the studies examined by De Moor and Bruton (1988) emphasize that it is the combined effect of environment degradation – in particular water extraction and siltation – and predation by Smallmouth Bass that has led to a severe decline in populations of native fishes. Loss of habitat, however, may exacerbate the impact of alien predators by rendering their prey more easily accessible (Gaigher *et al.*, 1980).

De Moor and Bruton (1988) suggest that the translocation of *Clarias gariepinus* might help to control *M. dolomieui*, but warn that the likely impact of the former species would have first to be assessed.

Zimbabwe

Although Toots (1970) said that Smallmouth Bass had failed to become established in Rhodesia (Zimbabwe), Robbins and MacCrimmon (1974) suggest that 'the species may be present in limited numbers in some waters'.

Unsuccessful Introductions

Robbins and MacCrimmon (1974) record the unsuccessful introduction in Africa of Smallmouth Bass to Swaziland and Uganda.

North America
Mexico

Rosas (1976b) refers to the rearing of Smallmouth Bass (for sport and as a food fish (Contreras & Escalante, 1984)) in Mexico, where Lee (1980) says they occur in the wild in parts of Amistad reservoir. Welcomme (1981, 1988) says they were first introduced to Mexico in 1975, and are currently 'established in some reservoirs in the northern part of the country'.

Transplantations

Outside its natural range in North America the Smallmouth Bass has been widely transplanted in the western States of the USA (Smith, 1896; Chapman, 1942) and in Canada, especially to British Columbia (Dymond, 1955; Carl & Guiguet, 1958).

South and Central America

Belize

Welcomme (1981, 1988) says that in 1969 Smallmouth Bass were imported as a sport fish to Belize (British Honduras), where they became established 'in a small area (Mountain Pineridge)'. Robbins and MacCrimmon (1974) say that the introduction was made in 1967, when 5000 fry were imported from the USA, followed by a second shipment of the same number of fry and 455 fingerlings in 1970.

Unsuccessful Introductions

In Central and South America, Robbins and MacCrimmon (1974) list failed introductions of Smallmouth Bass to Bolivia, Brazil, and possibly Costa Rica.

Oceania

Fiji

Devambez (1964) and Maciolek (1984) state that Smallmouth Bass may be established in Fiji, where Welcomme (1988) says they are reproducing. No details of their introduction are available.

Hawaiian Islands

In 1953, 74 Smallmouth Bass were imported from the California Department of Fish and Game's Elk Grove Hatchery to Oahu, where they were initially kept at the Fish and Game laboratory in Honolulu before being released in a lily pond near the Nuuanu reservoir, where spawning took place. Some of the original stock and their progeny were subsequently introduced to the reservoir, in which by 1956 they were said to be established. In May of that year, a second consignment of 1200 Smallmouth Bass fingerlings arrived from Elk Grove, and later in 1956 fish were planted in both forks of the Waialu River on Kauai and in Kaukonahua Stream above the Wahiawa reservoir on Oahu, and 2 years later in Puakea reservoir on Hawaii. Most of the fish used in these introductions came from Nuuanu, with some from the second 1956 shipment. By the late 1950s, Smallmouth Bass were naturalized in reservoirs and streams on Hawaii (e.g. at Kapaau near Hawi); Kauai (in a complex of tributaries of the Wailua River and Wailua reservoir); and Oahn (in the Wahiawa reservoir) (Brock, 1960; Robbins & MacCrimmon, 1974). (See also Randall, 1960; Hida & Thomson, 1962; Kanayama, 1968; Maciolek & Timbol, 1980; Maciolek, 1984.)

Ecological Impact   The Smallmouth Bass, which is a minor sport fish, appears to be a serious predator on native stream fauna. In the mid-section of the Wailua River on Kauai, Smallmouth Bass have displaced earlier populations of Largemouth Bass *M. salmoides* (q.v.) and Bluegills *Lepomis macrochirus* (q.v.) which had previously suppressed indigenous aquatic species, which Smallmouth Bass then eradicated. On Oahu, Smallmouth Bass have apparently eliminated all indigenous fishes and crustaceans in the Maunawili and Nuuanu Streams (Maciolek, 1984).

Unsuccessful Introductions

Attempts to establish Smallmouth Bass on the island of Guam in the Marianas have not been successful (Devambez, 1964; Robbins & MacCrimmon, 1974; Maciolek, 1984); nor have those on Mauritius (Robbins & MacCrimmon, 1974).

Spotted Bass
Micropterus punctulatus

The Spotted Bass has a much higher tolerance of silted and turbid waters and of drought and spate conditions than either *M. dolomieui* or *M. salmoides*, which has enabled it to become established in South African waters unsuitable for its congeners.

Natural Distribution

The central and lower Mississippi River system of the central and southeastern USA.

Naturalized Distribution

Africa: South Africa; ? Zimbabwe.

Africa

South Africa

In October 1939, 120 Spotted Bass fingerlings from Ohio, USA (where the subspecies is the nominate *M. p. punctulatus*), were imported to South Africa by the Cape Piscatorial Society and the Natal Provincial Administration, with the intention of stocking rivers unsuitable for *M. dolomieui* or *M. salmoides*, but only 29 survived the voyage. Six of these fish were retained in the Howick hatchery in Natal, while the remaining 23 were held in the Jonkershoek hatchery in the Cape, where spawning occurred in 1940 (Robbins & MacCrimmon, 1974). A hundred of their progeny were sent to reinforce the stock at Howick, while others were distributed to various localities in the eastern and southwestern Cape and in Natal (Harrison, 1964/1965a). De Moor and Bruton (1988: appendix VIII: 274–275) give details of the dates and locations of initial introductions of Spotted Bass (with references) in South Africa, and have summarized what is known of their past and present South African distribution.

In 1944–1945, Spotted Bass were released in the Cape, Orange Free State and Transvaal, and in 1949, 50 were freed in Natal (Robbins & MacCrimmon, 1974).

By the mid-1960s to early 1970s, Spotted Bass had become established in some acid, peat-stained waters in the eastern Cape (such as the lower Elandskloof River) and in Natal (Jubb, 1967), and in the Dunbar dam and the turbid Buffalo River which enters the Indian Ocean at East London (Harrison, 1954). Breeding populations were also to be found in the lower reaches of the Elandskloof River (Cape Agulhas) (Harrison, 1964/1965a); in the Klip River (Tugela system) at Ladysmith; and in the Umgeni River and its tributaries, the Lions and Karloof, at altitudes of between 900 and 1200 m asl (Crass, 1964). Although Spotted Bass became established with *M. dolomieui* and *M. salmoides* in some reservoirs and dams in Natal at between 1000 and 1300 m asl, in general only *salmoides* thrives in large impoundments (Robbins & MacCrimmon, 1974).

In most of the localities to which they were introduced, however, Spotted Bass failed to become established. It is possible that populations survive in the Buffalo River, and in 1988 some were taken in the Kubusi River (Kei system), 6 km below Stutterheim (De Moor & Bruton, 1988). In Natal, Spotted Bass are established in the Umgeni, Lions and Karloof Rivers, and a 'fairly large' population is naturalized in the Midmar dam; Spotted Bass are also found in the Wagendrift dam near Estcourt on the Bushmans River (Tugela system) (De Moor & Bruton, 1988).

Ecological Impact Introductions of Spotted Bass in the acidic south coastal streams of the Cape have been unsuccessful with the sole exception of the lower Elandskloof River which flows over limestone rocks (De Moor & Bruton, 1988). The pH in the lower reaches of this river is as high as around 8.5 (Harrison, 1964/1965a), and it thus seems that Spotted Bass are unable to tolerate low pH levels. This will exclude them from the south coast drainage rivers of the Cape, where several threatened endemic redfin minnows and other native species are found (De Moor & Bruton, 1988).

Zimbabwe

In 1944, a consignment of 414 Spotted Bass fingerlings was imported into Rhodesia (Zimbabwe) from the Jonkershoek hatchery, South Africa, and in the following year 254 were released in Sandy Spruit dam. These, and subsequent plantings in the Odzi River, were unsuccessful (Robbins & MacCrimmon, 1974).

Between 1945 or 1947 and at least 1959, fingerlings from Jonkershoek were imported almost annually by piscatorial societies and private individuals, but most of these also failed. In the late 1960s, there was only one known naturalized population of Spotted Bass in Zimbabwe, in Matopos dam (Toots, 1970).

Unsuccessful Introductions

Robbins and MacCrimmon (1974) list unsuccessful introductions of Spotted Bass to Swaziland (1946)

Botswana (Bechuanaland) and Lesotho (Basuto-land) in 1948, and to South West Africa (Namibia) in 1944–1945.

Translocations

Outside its natural range in the USA, the Spotted Bass has been translocated to several states, including North Carolina, Virginia, and California (Welcomme, 1988).

Largemouth Bass (Black Bass; Green Bass)
Micropterus salmoides

The Largemouth Bass is one of the most popular sport fishes in North America, where it has been widely translocated outside its natural range. Its success in northern latitudes, both in North America and elsewhere, is limited by its temperature requirements, since it prefers rather warmer waters than *M. dolomieui*. This higher thermal tolerance has enabled *M. salmoides* to become widely naturalized throughout the world except in some equatorial and northerly regions. In some countries established populations support commercial fisheries

(Welcomme, 1988). A predatory species, the Largemouth Bass feeds on crustaceans and other small invertebrates when young, progressing when adult to fish, frogs and larger invertebrates. It prefers the still waters of lakes and ponds to rivers and streams.

Natural Distribution

From northeastern Mexico to Florida, USA, most of the basin of the Mississippi northward to southern

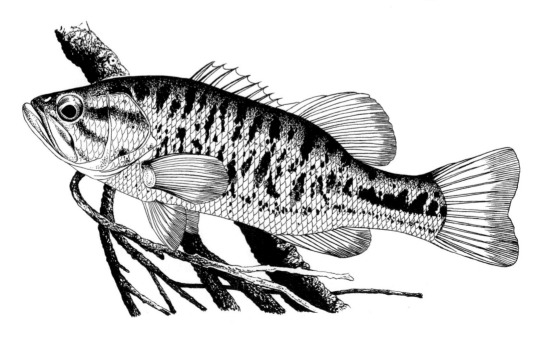

Ontario and Quebec, Canada, and the Atlantic slope north to southern–central South Carolina.

Naturalized Distribution

Europe: Austria; ? Belgium; British Isles (England); Cyprus; ? Czechoslovakia; France; ? Germany; Hungary; Italy; ? The Netherlands; Portugal; Sardinia; Spain; Switzerland; ? the former USSR; the former Yugoslavia. ***Asia:*** Japan; ? Korea; Malaysia; Philippines. ***Africa:*** Algeria; Botswana; Kenya; Lesotho; Madagascar; Malawi; Morocco; ? Mozambique; Namibia; South Africa; Swaziland; Tanzania; Uganda; Zimbabwe. ***South and Central America:*** Bolivia, Brazil; Colombia; Costa Rica; Cuba; Ecuador; ? El Salvador; Guatemala; Hispaniola (Haiti and Dominican Republic); Honduras; Panama; Puerto Rico. ***Oceania:*** ? Azores; ? Fiji; ? Guam; Hawaiian Islands; Mauritius; New Caledonia.

Europe

Robbins and MacCrimmon (1974) list the Largemouth Bass as fully naturalized in Europe only in France, Italy, Portugal and Spain, and to a limited extent also in Austria, the British Isles (England), Czechoslovakia, Hungary, Switzerland and the former USSR.

Austria

Max Von dem Borne (1894) records that he sent Largemouth Bass to various aquaculturists in Austria after 1885, specifically to Baron Von Washington at Schloss Poels in the Steiermark (Styria). The species was also introduced to the Worthersee, Carinthia, near the border with the former Yugoslavia, probably sometime after 1898 (Von Pirko, 1910); releases in the River Danube, however, proved unsuccessful (Robbins & MacCrimmon, 1974). In 1934, Largemouth Bass are known to have occurred in a lake near Kufstein, and in 1941 in the Mittersee near the Fernpass in the Tyrol.

Today, Largemouth Bass survive in considerable numbers in the Worthersee, and in such nearby waters as Keutschacher Lake, the Hafnersee, and the Forstsee which is joined to the Worthersee by a water pumpline (Robbins & MacCrimmon, 1974).

Belgium

Between 1885 and 1890, Max Von dem Borne shipped Largemouth Bass from his aquaculture

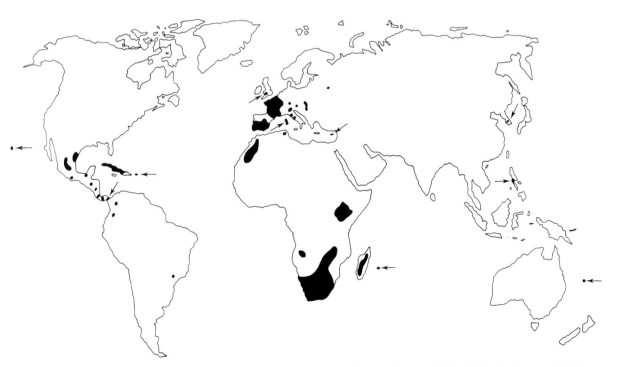

Naturalized distribution of Largemouth Bass *Micropterus salmoides* (adapted from Robbins & MacCrimmon, 1974).

centre at Berneuchen, Germany, to Belgium (Von dem Borne, 1894), where they became established in ponds at Groendaal and from where they were subsequently transferred to the main canals of Camp de Beverloo and the Ourthe and Semois Rivers, where in some localities they became the dominant species (Raveret-Wattel, 1900; Renault, 1951), and where Poll (1949) reported them to have survived 'in small numbers'.

Although Robbins and MacCrimmon (1974) say that 'at present the largemouth reproduces only in ponds ... near Brugge and in Hasselt ... it does not exist in any open waters', and Welcomme (1981) says it is not established in the wild in Belgium, Welcomme (1988) claims that 'some residual stocks probably still exist in some rivers'.

British Isles (England)

The earliest recorded introduction of Largemouth Bass to the British Isles appears to have taken place in 1879, when a number were imported from the USA by the Marquis of Exeter in a shipment of Smallmouth Bass *M. dolomieui* in 1881 or 1882 these were unsuccessfully released in Loch Baa, Scotland, by the Duke of Argyll.

In 1909, a specimen of the Largemouth Bass was seen in the Wherry Hotel on Oulton Broad in east Suffolk, England, 'exhibited as the only survivor(!) captured out of a consignment from Austria that had been deposited in local waters; the others, it is believed, were all devoured by the Oulton Pike' (quoted by Lever, 1977).

Fitter (1959) reported that:

> ... about thirty years ago renewed attempts ... were made with the large-mouthed species ... a German fish breeder had established a breeding stock of this form in Europe, and it was probably this acclimatised stock which was used for the second, more successful wave of introductions. In 1927 the Norwich Angling Club put 250 large-mouthed yearlings into a three acre [1.2 ha] lake, but they disappeared in the following year. At the same time fry from imported eggs were put into the River Ouse at Earith Bridge ...

In 1934 or 1935 D.F. Leney obtained some 10–18 cm Largemouth Bass for his Surrey Trout Farm and United Fisheries at Shottermill near Haslemere in Surrey, from a population established near Bourges in the River Loire, France. These were subsequently introduced to a gravel-pit at Send near Woking, Surrey, and to two ponds near Wareham in the Isle of Purbeck, Dorset and at Frome, Somerset; at Send several young fish were netted when the pit was cleared in 1937 or 1938 (indicating successful breeding), but before 1969 this population had died out (Wheeler & Maitland, 1973; Lever, 1977). Fitter (1959) said that Largemouth Bass from Shottermill were also introduced to a pond in northwestern Kent and to a lake in south Devon, where they grew 'to a fair size'.

Two further attempts to establish Largemouth Bass in England have been recorded; in 1935 a number were released in two reservoirs on the Lancashire moors where, although a few were subsequently caught, the remainder apparently disappeared. Before the Second World War, 100 yearlings were planted in the lake at Coombe Abbey, Warwickshire, where they similarly failed to become established (Wheeler & Maitland, 1973; Lever, 1977).

The status of the Largemouth Bass in England was reported by Lever (1977) as uncertain. It may be that the populations in an abandoned clay-pit near Wareham and in Somerset are the only ones extant. Fitter (1959) believed that the success of the fish at Wareham was due at least in part to the fact that the pit is well-sheltered and spring-fed, and held no predatory Pike *Esox lucius* but plenty of prey in the form of Roach *Rutilus rutilus* and Perch *Perca fluviatilis*. Pike and Perch have since been added to this pond, where both are believed to prey on the Largemouth Bass.

Maitland and Price (1969) reported the discovery of a North American monogenetic trematode, *Urocleidus principalis*, on the gills of Largemouth Bass from Wareham. They suggested that this species, previously unrecorded in Britain, was probably introduced with the original stock from the Loire, where the population is also presumably similarly infested.

It may be that the cold weather that frequently occurs during the breeding season of this early-spawning fish is one reason for its failure to become established in other apparently suitable waters in England, and indeed in continental Europe, which is at approximately the same latitude as the northern edge of its natural distribution in North America. (See also Robbins & MacCrimmon, 1974.)

Cyprus

In October 1971, about 400 Largemouth Bass fingerlings were sent from Ontario, Canada, to Cyprus, where they were placed in the Mia Milia dam, a fairly shallow reservoir with a considerable amount of aquatic vegetation into which Common Carp *Cyprinus carpio* (q.v.) and grey mullet (*Mugil* spp.) had been previously introduced (Robbins & MacCrimmon, 1974). Welcomme (1981, 1988) says that Largemouth Bass are established in two dams in Cyprus, where they are said to be 'very popular with anglers but very prolific and inclined to destroy population balance'.

Czechoslovakia

In the mid-1880s, Largemouth Bass from Max Von dem Borne's hatchery at Berneuchen, Germany, were sent to aquaculturists in Czechoslovakia, where in 1889 they were reported to be thriving in Baron Schwartzenburg's hatchery at Wittingau in Bohemia, from which they later escaped during a flood into the Moldau and Elbe Rivers. Before 1894, Largemouth Bass were also successfully reared at Weisenthal near Liberec, and it seems probable that they were also bred in other hatcheries elsewhere (Von dem Borne, 1894).

Although Robbins and MacCrimmon (1974) said that 'at present, the largemouth is not found in the natural waters of Czechoslovakia', Welcomme (1981, 1988) says that 'natural populations occur in the Danube, although it is rare; this species should also be recorded from other Danube countries due to its presence in that river'.

France

Although Largemouth Bass were first imported to France from North America around 1877, it was not until 1888–1889 that the species was successfully cultured in the country, near Versailles. The origin of these fish is not known, but it seems probable that they came from Max Von dem Borne's hatchery at Berneuchen in Germany, since Largemouth Bass from there were sent to Paris between 1885 and 1890 (Von dem Borne, 1894). Fish from the Berneuchen source were also successfully reared at Hueningen (Hueningue) in Germany (later Alsace, France) at around the same time (Von dem Borne, 1894). An attempt at around this time to introduce Largemouth Bass to Lac Annecy in the Savoie was unsuccessful (Hubault, 1955).

In 1903, 200 Largemouth Bass from Berneuchen were placed in a pond near Paris, followed by more

in the following year, which were reared successfully (Bastedo, 1904). In 1909, Largemouth Bass fingerlings, probably from Germany, were introduced to a pond at Courance, Seine-et-Marne, near Paris, by the Marquise de Ganay (Arenburg, 1911), and in later years large numbers of fry were released by piscatorial societies and private individuals in many *départements* throughout the country, some from hatcheries constructed in the 1930s (Renault, 1951). The spread of Largemouth Bass in France was assisted by natural dispersal through the canal system (Robbins & MacCrimmon, 1974). Roule (1925) said that the species was successfully established in suitable waters discontinuously throughout France, although it was self-maintaining in small numbers mainly around Redon in Brittany (Ille-et-Vilaine).

Vivier (1951b) reported the then distribution of Largemouth Bass in France as follows.

Rhône Basin   In the Saône, in the upper reaches of the Scey-sur-Saône, into which they escaped from a hatchery at Port-sur-Saône in 1931. Also in the Côte-d'Or and Saône-et-Loire, as well as in tributaries of the Saône such as the Brenne, l'Ognon and the Seille. In the lower reaches of the Saône, where they were spreading, Largemouth Bass appeared around 1945. They were apparently established in the Lac du Bourget near Brison, which joins the Rhône via the Savière canal. Small numbers were also to be found near Valence.

Loire Basin   In the Allier, in Lac Chambon (since 1930), and also in most of the tributaries of the Loire, such as the Nièvre, Vienne and their inflowing streams (the Gartempe, Clain and Vonne), the Thouet, Oudore, lower Sarthe, Mayenne, Loir and Cher. Largemouth Bass were, however, never very abundant. They did not occur in the upper reaches of the Creuse and only occasionally in the lower reaches, and were not found in the Loire itself. The fish seem to have come in part from a rearing facility at Aron near St Saulge (Nièvre) where they were introduced around 1920, and partially from the establishment of a fishery at Montayer, Maine-et-Loire, where they were cultured in about 1930.

Garonne Basin   Established for 20 years in the Lot, the middle Garonne, the Dordogne, but especially in the Isle near Montpon, and also in the Gers, the canalized Baïse, the Dropt, the Gelise and the Lède. Largemouth Bass were quite common in the Tarn and in the Aveyron, and very abundant in the

Gimone tributary of the Tarn. They were not yet established in the Corrèze. Some had been caught in the small rivers of Cantal.

Adour Basin Largemouth Bass had been reported in the Adour, the Gaves de Pau and the Oloron, and in such small rivers as the Luy de Béarn and the Luy de France.

Rhine Basin Only recorded in the Semoy, a tributary flowing into the right bank of the Meuse, where they were introduced in the 1930s. There had been little increase in numbers.

Inflowing Rivers of the Mediterranean Established in small numbers in the Hérault, in the Agde reservoirs in St-Thiberg village, where they were established in 1939. They were beginning to be found in the Midi canal, and in the lake formed by the damming of the St Ferréol.

In the *département* of Var, 1000 Largemouth Bass fry were in 1935 introduced to the Carcès dam where they bred successfully and became very numerous. In the outflowing Issole, however, a tributary of the Argens, Largemouth Bass did not occur.

Introductions of Largemouth Bass were made in 1948–1949 in the Basse Siagne, Alpes Maritimes, where early results seemed to be satisfactory.

Brittany Small numbers of Largemouth Bass had occurred in Ille-et-Vilaine since 1920, and in certain stretches of the Ille and Rance Canal. The species had thus not altered its distribution in Brittany since the report by Roule (1925).

Inflowing Rivers near the Bay of Biscay Largemouth Bass were found in the lower reaches of the Sèvre Niortaise and in the Charente where in some places they were doing very well, especially near Chanins and the Bel Ebat reservoir. They also occurred at Mimizan in the Landes, which does not flow into the sea.

Seine Basin Largemouth Bass were only naturalized in the River Eure at Louviers. Attempts had recently been made to rear them in hatcheries in the Loing near Moret; in the Eure between Chartres and Courville-sur-Eure; in the Burgundy canal (at St Florentin in the Yonne); in the Aisne at Soissons; in the Marne (a feeder canal of the Berry at Bac); in the Seine at Puisy; and in the Aube at Nogent-sur-Seine. These attempts had either not been very fruitful or were too recent for proper evaluation.

Largemouth Bass were not found in any inflowing rivers of the English Channel, and only sparsely in the Rhine basin.

Summary Although, following numerous introductions, Largemouth Bass were widely naturalized in France, they were reported by Vivier (1951b) to be nowhere abundant. The rivers where they appeared to be most successful were the Basse Saône (between Chalon and Lyon), the Charante, the Tarn, and much of the basin of the Garonne and the Midi canal.

Over 20 years later, Robbins and MacCrimmon (1974) reported Largemouth Bass to be established in the lower Ardour; the lower and middle Garonne; the Charente; both the Sèvres and the Landes lakes; most of the Loire tributaries; the Vilaine; the Brittany canals; the upper Saône to upstream of Lyons; the Rhône near Valence; and in lesser numbers elsewhere, such as the Seine canals and waters near the Mediterranean (Spillman, 1961). Populations are also established in ponds throughout France, and in most canal systems except those in the far north and east. In southern and central France, the Largemouth Bass is a popular sport fish (Robbins & MacCrimmon, 1974). Welcomme (1988) refers only to a 'slight presence' of the species in the south and east of France.

Germany

Max Von dem Borne, director of the Germany Fishery Society at Berneuchen, near Neudamm (now in Poland), was the most vociferous proponent of the introduction to Europe of North American *Micropterus* bass, on the grounds that they would provide an additional sport fish in European waters. In 1883, (Welcomme (1981, 1988) incorrectly says respectively in 1880 and 1888) he imported 45 Largemouth Bass from Greenwood Lake, New York, to Berneuchen (Von dem Borne, 1888); only 10 of these fish survived, but they produced 140 000 fry which Von dem Borne (1892, quoted by De Groot, 1985) considered would be enough 'to stock within 10–20 years all suitable German inland waters', an expectation never to be realized, despite the fact that before the turn of the century aquaculturists all over the country received stocks of Largemouth Bass, populations of which became widely naturalized, particularly in the south (Robbins & MacCrimmon, 1974). They became especially well established in Lake Barmsee in the Alps (890 m asl), which was first stocked in 1898, but eventually died out there due to suboptimal environmental condiions (Ladiges & Vogt, 1965). Other German waters which supported naturalized populations of Largemouth Bass in the early 20th century include those near Regensburg in Bavaria; the Naab, and other tributaries of the Danube; the Würmsee (Starnburgersee) in Bavaria; the Rhine; Bruehl near Cologne (Koln); the Main; and the Elbe (Robbins & MacCrimmon, 1974).

Thienemann (1950, quoted by Vooren, 1972) said that the Largemouth Bass was only 'locally acclimatized' in German waters. Although Robbins and MacCrimmon (1974) state that 'It seems most certain that the species has now died out in the natural waters of Germany', Welcomme (1981, 1988) says it is 'reproducing in one lake'.

Hungary

According to Von dem Borne (1892), Largemouth Bass from his hatchery at Berneuchen were sent to fish breeders in Hungary between 1885 and 1890 to prey on small Common Carp *Cyprinus carpio* (q.v.), thus enabling larger ones to achieve a higher weight-gain rate. Some were unsuccessfully released in fast-flowing waters (an unsuitable habitat) of northern Hungary, and in about 1895 successfully in more favourable still waters in the Tatra Mountains. At around the same time, Largemouth Bass escaped from rearing ponds at Sárdi in Somogy Province into Lake Balaton, from where although they failed to become established they subsequently dispersed through the Sío Canal into the River Danube and thence into the floodplains of the Drava near Peterhida, where they became widely distributed (Vutskits, 1913).

Thienemann (1950) and Vooren (1972) reported Largemouth Bass in Hungary to be numerous and well established in the Drava, where they were fished for commercially. Robbins and MacCrimmon (1974) said the species was 'reported sporadically in different waters . . . including the Danube where it is still present but rare. Also, a self-sustaining population is confined in a pond at a fish farm at Nagyated . . .'. Pinter (1978) records the species' distribution in Hungary as 'very limited'. According to Welcomme (1981, 1988), who says they were first introduced to Hungary in 1910 and again in the 1950s, Largemouth Bass occur only in 'very few localized self-sustaining populations in cooling ponds of power stations [and] also presumably in the Danube'.

Italy

Between 1886 and 1890, Largemouth Bass from Berneuchen were sent to fish breeders in various unspecified localities in Italy (Von dem Borne, 1892). The earliest recorded releases in the wild took place in about 1890, when they were successfully liberated in Lago Maggiore in Lombardy, where they continued to be stocked until around 1920 (Grimaldi, 1972). In 1897, 592 fish of mixed ages were planted in Lago Manate, where they initially became established but later declined. In 1900, 1008 fish were released in Lago Varano, where more were added in 1901–1902. In the early 1900s, Largemouth Bass were successfully released in Lago Monterosi (Besana, 1910; Thienemann, 1950). All these lakes are in Lombardy in northern Italy. Largemouth Bass were subsequently introduced with mixed results to other Italian waters.

Robbins and MacCrimmon (1974) said that the naturalized distribution of Largemouth Bass on the Italian mainland 'includes the majority of lakes in northern Italy which are located at a height not above 300 m, as well as the larger rivers and canals of the Padana (Po) Valley. . . . also in a few rivers and artificial reservoirs in Tuscany, and in the reservoirs of Monterosi near Viterbo and Salto Lake near Reiti'. In Lago Maggiore, Grimaldi (1972) said the fish 'show a very spotty distribution, as they concentrate along the very short stretches of shore line still lined with reeds'. Largemouth Bass in Italy are highly regarded as a sport fish and for the control of an excess of cyprinids, and have been widely stocked by both private anglers and local authorities (Alessio, 1984).

Ecological Impact In Lago Manate and Lago Varano, Vooren (1972) said that the local Bleak *Alburnus a. alborella* had been severely affected by the introduction of Largemouth Bass and Redbreast Sunfish *Lepomis auritus* (q.v.) and that local predators such as the Pike *Esox lucius* and Perch *Perca fluviatilis* had also declined in numbers. Alessio (1984), however, found that because of differing food preferences and feeding strategies, Largemouth Bass do not appear to come into competition with Pike. In Lago Monterosi, Largemouth Bass have reduced the population of the naturalized Sunfish (Besana, 1910; Thienemann, 1950).

The Netherlands

According to De Groot (1985), an unsuccessful attempt to introduce 18 'black bass' to The Netherlands in 1884 probably referred to *M. salmoides*. A second importation was more successful, and five so-called 'black bass' were kept in the aquarium at Amsterdam Zoo for a number of years. Von dem Borne (1892) reported shipping Largemouth Bass from Berneuchen to Holland between 1885 and 1890.

By 1900, Largemouth Bass in The Netherlands occurred only in the River Rhine, to which they had been introduced (Mulier, 1900). Those few caught in the River Meuse in The Netherlands are believed to have spread there by natural diffusion from Belgium (De Groot, 1985).

Portugal

Robbins and MacCrimmon (1974) have summarized what is known of the introduction of Largemouth Bass to Portugal.

The species was first introduced to the country in 1952, when the Director General of Forestry and Aquatic Services imported a shipment of fry from France; the purpose of the introduction was to provide an additional source of food and a sport fish in those waters unsuitable for trout and containing predominantly cyprinid species. The initial introduction was to Praiade Mira in central Portugal, where the fish quickly became established. Thereafter, they were released in the lower reaches of the River Vouga, which flows into the sea some 45 km south of Oporto; in Lake Pateira de Fermentelos; and in a number of neighbouring shallow ponds where cyprinids provide good forage.

Robbins and MacCrimmon (1974) reported the naturalization of the Largemouth Bass in Portugal in most still or slowly flowing waters south of the River Tejo (Tagus), which enters the Atlantic at Lisbon. It also occurred in canals and irrigation ditches and in the Mondego River in Beira Province, north–central Portugal. It was well established in many lentic waters in Alentejo and Algarve Provinces and in parts of the Provinces of Ribatejo and Estremadura, having been released as fry from natural spawning areas at Praiade Mira. North of the Tejo, Largemouth Bass are naturalized only in some small shallow ponds south of Aveiro; in Lake Pateira de Fermentelos; and in canals and irrigation ditches near the Rivers Mondego and Vouga. No attempts have been made to introduce Largemouth Bass into other northern waters where water temperatures are considered to be too low for successful spawning, and where they would come into competition with more important salmonids.

Sardinia

Cottiglia (1968) reports that in 1962 he introduced Largemouth Bass to a small lake near Cagliari on the island of Sardinia in the Mediterranean with such success that within 2 years the fish had consumed most of the available food, and the population had multiplied so rapidly that it had become stunted. In 1964, Cottiglia transferred 1700 adults from Cagliari to Lago Monte Pranu, where they also became established. In 1965–1966, Lago Mulargia was successfully stocked with 1000 mature Largemouth Bass together with *Atherina mochon* as forage. Robbins and MacCrimmon (1974) recorded Bass as 'abundant in the Pauliana Reservoir, in the lower Palmas River, in Monte Pranu Lake, in the River Monte S'Orcu, and Villa Peruccio Lake in the extreme south of the island'.

Spain

Largemouth Bass fingerlings from France were first introduced to Spain in 1955 or 1956 by the Inland Fisheries Service of the National Department of Game and Parks, for rearing in aquaculture at Aranjuez near Madrid and for sport fishing. This introduction proved so successful that culture was initiated at other hatcheries, from where fry were released in the lower reaches of most southward- and westward-flowing rivers devoid of salmonids (Robbins & MacCrimmon, 1974).

Vooren (1972) could find 'no records of the species being liberated in natural waters in Spain'. San Feliu (1973) said it occurred in the wild in some ponds and is abundant in one Mediterranean coastal lagoon. Robbins and MacCrimmon (1974), however, recorded the Largemouth Bass as naturalized in many lakes and rivers in Spain, where it reaches a good size. Elvira (1995a) lists it as established in the following drainages: North, Galicia, Douro, Tagus, Guadiana, Guadalquivir, South, Levant, Ebro and Eastern Pyrenees.

Switzerland

Von dem Borne (1892) records the introduction of Largemouth Bass from Berneuchen to Switzerland between 1885 and 1890. In 1921, fish from North America were successfully introduced to the 1.2-ha Lago Seeweid near Zürich, where Fehlmann (1930) said they were doing well. In 1964–1965, however, as a result of pollution of the lake by liquid manure, most of the fish were killed, the few survivors being transferred to the neighbouring Lago Lützel (9 ha), where Robbins and MacCrimmon (1974) reported them to be thriving. After Lago Seeweid had been restored, Largemouth Bass were reintroduced and quickly became re-established. They were also recorded by Robbins and MacCrimmon (1974) as naturalized in Lago di Lugarno; in a nearby river; and in the Swiss waters at the northern end of Lago Maggiore – all near Magadino.

USSR

Although Largemouth Bass from Berneuchen were apparently sent to the old Russian Empire between 1885 and 1890 (Von dem Borne, 1892), it was not until 1902 that the first release was made, in Lake Abrau on the shore of the Black Sea near Novorossiysk, where a naturally breeding population was reported in 1936. In the following year, four adult Bass were successfully transferred to the neighbouring Lake Limanchick. In 1948, Lake Ivan'kovskij (32 700 ha), 140 km northeast of Moscow, received 25 fish of various sizes. After 1948, Largemouth Bass (mostly from Lake Abrau) were stocked in the following waters; Lakes Beloje (northeast of the Caspian Sea); Dolgoje (in the southwest Ukraine); Krujloje (at the northeastern corner of the Sea of Azov); Sergilejevskoja (eastern Krasnodar); Pyalovskoje and Jackromckoje; the Novo-Matveevskei estuary (in Belorussia); and elsewhere (Berg, 1964–1965).

Robbins and MacCrimmon (1974) said that in the former USSR 'only in lakes Abrau and Limanchick is the largemouth reported in the commercial fish catch statistics but the quantities are small'.

Yugoslavia

Largemouth Bass, probably from either southern Austria or the Lake Balaton area of Hungary, were unsuccessfully imported to the former Yugoslavia in 1914 (Robbins & MacCrimmon, 1974). Welcomme (1981, 1988) says the date was 1920. 'At present', wrote Robbins and MacCrimmon (1974), 'negotiations are being undertaken by Yugoslav biologists with French authorities for the importation of a *Micropterus* species into Yugoslavia . . . an importation seems imminent'. This anticipated introduction presumably took place shortly afterwards, since Welcomme (1981, 1988) said that *M. salmoides* is reproducing in the wild in Yugoslavia where it is 'popular as a sport fish and for restricted aquaculture'.

Unsuccessful Introductions

In Europe, Largemouth Bass have failed to become established in Denmark (Jensen, 1987), Finland, Luxembourg, Poland and Sweden (Robbins & MacCrimmon, 1974).

Asia

Japan

In May 1925, 75 fingerlings and three adult *Micropterus* species (believed to be *salmoides*) were imported to Japan from the USA by the Faculty of Agriculture of Tokyo University, by whom they were released in Lake Ashino-Ko in Kanagawa Prefecture southwest of Tokyo. Here they became established, feeding mainly on the very abundant and valuable *Hypomesus olidus*. Small numbers of Largemouth Bass have also become established elsewhere, especially in waters around the nearby Mt Hakone (Okada, 1960, 1966). In 1962, some were transplanted from Lake Ashino-Ko to the Sagamiko River in the northeast, from where they dispersed naturally some 3 years later to Lake Tsukui. Robbins and MacCrimmon (1974) recorded them as naturalized in all the above waters. Chiba *et al.* (1989) said they are 'spreading rapidly over the Japanese main islands'.

Ecological Impact   After its initial introduction into Lake Ashino-Ko in 1925, for sport fishing according to Akaboshi (1959), and elsewhere, further translocations of the Largemouth Bass were forbidden in order to prevent predation of native fish populations. Its popularity as a sport fish, however, resulted in numerous clandestine and illegal movements of the species into many waters throughout Japan, where it has had the impact on indigenous species that had been anticipated (Chiba *et al.*,

1989). Nomura and Furuta (1977) found that the effect of Largemouth Bass on native fish populations appeared to vary with the depth of the water, the most severe predation tending to occur in shallow water; in deep waters, such as Lake Ashino-Ko, Largemouth Bass seem to have achieved an ecological balance with native species. Attempts to eradicate Largemouth Bass from Japanese waters have been unsuccessful (Kikukawa, 1980).

Korea

Introduced from the USA to Korea in 1963, the Largemouth Bass is, according to Welcomme (1981, 1988), reproducing, though whether in the wild or only in captivity is uncertain. Robbins and MacCrimmon (1974) do not mention this introduction.

Malaysia

According to Ang *et al.* (1989), the Largemouth Bass is the most recent species to have been introduced into Malaysian waters. In 1984, 1000 fry from Florida, USA, were imported for rearing in the hatchery of Boh Plantations, Ltd, where they were raised to fingerling size. These were planted to provide sport fishing in two lakes, each of around 3 ha in extent. There appear to have been no reports of the fish in the outflowing rivers downstream, and it is assumed that they remain confined to the two lakes.

Philippines

The earliest recorded introduction of an exotic fish species to the Philippines was in 1907, when 154 out of an original consignment of 175 Largemouth Bass fingerlings from Folsom, California, USA, imported by Alvin Seale on behalf of the Philippine Insular Government, were released for food and as a sport fish in high-altitude pools on the Trinidad River in the Montain Province (Baguio City), Luzon, and downstream in Trinidad Lake and in Caliraya Lake (a reservoir 700 m asl) in Laguna Province (Seale, 1915). The fish spawned successfully in Trinidad Lake, but did well only in Caliraya, where in 1985 more, of the subspecies *M. s. floridanus*, were released as a gift from California (Juliano *et al.*, 1989).

In 1909, large adult Largemouth Bass were transferred from Trinidad Lake to other nearby waters, including Cayman Lake at Los Baños (Laguna Bay

near Manila), apparently unsuccessfully (Seale 1915). Others were released in various unspecified reservoirs and lakes in 1958 and 1970. Herre (1953) said Largemouth Bass in the Philippines were naturalized only in the Taraka River and in Lake Lanao, Lanao Province, Mindanao. On Luzon, Robbins and MacCrimmon (1974) reported them only in the Ambuklao hydroelectric reservoir in Baguio, which was stocked with fish transferred from Burnham Park, Baguio City, during renovation.

Ecological Impact   Juliano *et al.* (1989) do not mention any ecological impact by Largemouth Bass in the Philippines, but Welcomme (1981) says they 'feed on local species'. In Lake Lanao these include several species of cyprinids (Payne, 1987).

Unsuccessful Introductions

In Asia, Largemouth Bass are reared in monoculture for human consumption but are not established in the wild in Taiwan (Liao & Liu, 1989), and are bred experimentally in the United Arab Republic (El Bolok & Labib, 1967) and in China (Yo-Jun & He-Yi, 1989).

Africa

Algeria

Welcomme (1981, 1988) says that in 1970 Largemouth Bass from France were introduced for aquacultural and angling purposes to Algeria, where they are established and breeding. Moreau *et al.* (1988) also give an earlier date of introduction, in 1956.

Botswana

In 1938, 110 Largemouth Bass were imported to Botswana (Bechuanaland) from a hatchery at Mbabane in Swaziland. In 1961, 500 fingerlings from the Jonkershoek hatchery in the Cape, South Africa, were also introduced to waters in Botswana, where the species has become established in ponds and reservoirs in the southeast of the country (Robbins & MacCrimmon, 1974).

Kenya

As early as 1910, Theodore Roosevelt, former president of the USA, while on his notorious big-game hunting safari to East Africa, offered some black bass o the Kenya Angling Association, a gift which for some unknown reason was declined (Harrison, 1934). In February 1929, however, 53 survivors out of a shipment of 104 Largemouth Bass fingerlings and 2-year-old fish from Europe were freed in the 260-km² Lake Naivasha, situated at an altitude of 1800 m in the Great Rift Valley, northwest of Nairobi (Harrison, 1936; Thompson, 1939). Here the Largemouth Bass flourished, partially as a result of the introduction in 1926 from two neighbouring rivers of a suitable forage fish, *Tilapia nigra.* Spawning was first noted in April 1930, and by the following year 'bass were being removed by the bagload' (Robbins & MacCrimmon, 1974). In spite of this initial success, prolonged droughts during the 1940s appear to have eradicated Largemouth Bass from Lake Naivasha and to have severely reduced the *Tilapia* population. The Largemouth Bass were only re-established as a result of a continuing restocking programme from 1952 (Robbins & MacCrimmon, 1974).

Today, Largemouth Bass are naturalized in a number of natural and artificial still waters in Kenya, mainly between 1200 and 2100 m asl, where the species is regarded as a valuable game fish. These include Lake Naivasha (where several previously introduced cichlids have been adversely affected); a smaller water nearby; the Ruthagati dam on the slopes of Mount Kenya; and the Sagana Fish Cultural Station, from which fish have been widely distributed throughout the country. Spawning seems to be most successful at an altitude of around 1800 m, where annual mean water temperatures are about 13°C (Robbins & MacCrimmon, 1974).

Lesotho

In 1937, 81 Largemouth Bass fingerlings were imported to Lesotho (Basutoland) from the hatchery at Mbabane in Swaziland. In later years, *M. salmoides* was stocked in a number of impoundments near Maseru and Leribe at the same time as Bluegills *Lepomis macrochirus* (q.v.) to provide forage, where a population explosion by the Largemouth Bass was followed by stunting; the species is, however, still present in this region (Robbins & MacCrimmon, 1974).

Madagascar

Largemouth Bass from France were originally introduced to fill an empty niche in the rivers and lakes of the island of Madagascar in 1951, specifically to Lake Alaotra in 1961 and to Lake Itasy in 1963 (Kiener, 1963; Moreau, 1979; Reinthal & Stiassny, 1991). Robbins and MacCrimmon (1974) said they were then established in several waters on the high plateaux and at middle elevations where water temperatures seldom exceed 25°C. In their survey of the island in 1988, Reinthal and Stiassny (1991) collected Largemouth Bass only in Lake Itasy on the central plateau west of Antananarivo.

Largemouth Bass in Madagascar have also provided a new species for commercial propagation, which has been undertaken on a considerable scale, and many were formerly also harvested from open plateaux waters (Therézein, 1960); they were regularly offered for sale in fish markets in the capital, Tananarive, and in Fianarantsoa, Moramanga and Ambatondrazaka (Robbins & MacCrimmon, 1974).

Ecological Impact Although Kiener (1963) cites widespread regional deforestation and concomitant siltation as a principal reason for the fall of populations of the endemic cichlid *Ptychochromoides betsileanus* in Lake Itasy, the introduction of the Goldfish *Carassius auratus* (q.v.), the Common Carp *Cyprinus carpio* (q.v.), the Redbreast Tilapia *Tilapia rendalli* (q.v.), the Nile Tilapia *Oreochromis niloticus* (q.v.) and Largemouth Bass also played an important rôle in the eventual extinction in the lake of *P. betsileanus* (Moreau, 1979).

Raminosa (1987), quoted by Reinthal and Stiassny (1991), found that Largemouth Bass and introduced Snakeheads *Channa striata* (q.v.) were together displacing the endemic cichlid *Paratilapia polleni* over much of its freshwater range on Madagascar.

In other open waters, however, Largemouth Bass have not eradicated their forage species, but have effectively kept in check populations of *Tilapia* spp. and Mosquitofish *Gambusia affinis* (q.v.). In closed waters, on the other hand, Largemouth Bass quickly eliminate all other species and eventually become cannibals, and often when a pond is drained only a few large Largemouth Bass individuals are found (Therézein, 1960).

Malawi

In 1937, the Swaziland Angling Association sent 11 Largemouth Bass fingerlings from the Mbabane hatchery to Malawi, where Robbins and

MacCrimmon (1974) were doubtful if the species had become established.

Some time after 1974, Largemouth Bass from Zimbabwe (Rhodesia) were imported for sport fishing to Malawi, where they have been 'successful in cooler areas in habitats too warm for trout and too cool for local species, for example Lujeri' (Welcomme, 1981, 1988).

Morocco

In 1934 and 1935 respectively, 13 adult Largemouth Bass and between 200 and 300 fry were imported to Morocco from France by the Société des Pêcheurs de Fès, in collaboration with the Administration des Eaux et Fôrets. The fish were first introduced to Lac d'Afourgah in the Moyen Atlas Mountains, followed by other releases in neighbouring montane lakes as well as in waters on the plains such as those around Dayèt-er-Roumi. Although the species at first did exceptionally well in Morocco, for some unknown reason, possibly disease, spawning virtually ceased between 1952 and 1960, apart from in three small rivers north of El-Arba-du-Rhab. As a result, Largemouth Bass disappeared from some lakes, such as Dayèt-Aouaoua, where they had previously been especially common, although even in the late 1950s others, for example the dam at Mechrâ-Homadi, continued to yield large numbers (Robbins & MacCrimmon, 1974).

Largemouth Bass are currently naturalized in Morocco in suitable waters from Lac Nfiss near Marrakech to the Arbaoua–Larache region in the Rif Mountains in the north, at altitudes ranging from near sea level to 1700 m. Siltation and the presence of introduced Barbel *Barbus barbus* (q.v.) and Common Carp *Cyprinus carpio* (q.v.) seem to be preventing any expansion of their range (Robbins & MacCrimmon, 1974).

Mozambique

In 1947, 50 Largemouth Bass fingerlings were sent from the Mbabane hatchery in Swaziland to the seaport of Lourenco Marques in extreme southern Mozambique, where Robbins and MacCrimmon (1974) considered that 'the presence of this bass in the border areas of Rhodesia [Zimbabwe], South Africa and Swaziland suggests that largemouth are now present in waters of the western part of Mozambique'. Welcomme (1981, 1988), however, was doubtful about the species' continued presence in the country, although Moreau *et al.* (1988) confirm its survival in some artificial waters.

Namibia

In June 1932, 40 Largemouth Bass fingerlings were despatched from the Jonkershoek hatchery in South Africa to a private individual at Narubus in South West Africa (Namibia), where it is believed that all died during a drought. In August 1934, Largemouth Bass were planted in a municipal dam at Windhoek, where their fate is unrecorded (Harrison, 1934, 1936; Robbins & MacCrimmon, 1974). There were numerous subsequent introductions between 1944 and 1949, again with uncertain results (Skelton & Merron, 1984).

Since 1983, Largemouth Bass have been supplied to stock numerous farm dams in Namibia where the species is now widely distributed, including in the fragile Omatako and Omuramba catchment system, and it thus poses a potential threat to the ecologically important Okavango delta (De Moor & Bruton, 1988). Fortunately, the extreme turbidity of seasonal rivers together with the species' specialized breeding requirements, limit the distribution of the Largemouth Bass in Namibia (Dixon & Blom, 1974; Schrader, 1985). The species also occurs in numerous stretches of the Swakops River, including the Von Bach dam (Skelton & Merron, 1984).

South Africa

Largemouth Bass which had been reared in The Netherlands were first imported to South Africa in 1927 for aquacultural purposes and the provision of recreational fishing. In that year, R.J. Neville of the Rand Piscatorial Association in Johannesburg acquired 49 fingerlings from D.F. Leney's Surrey Trout Farm at Haslemere in England, the 45 survivors being placed in February 1928 in a rearing pond at the Jonkershoek hatchery at Stellenbosch in the Cape, where in the following year they spawned successfully. In February 1930, a large number of fry were sent to the Rand Piscatorial Association in the Transvaal, and many more were released in the following waters of the Cape; the Berg River, by the Groot Angling Association; Brand Vlei, by the Worcester Trout Anglers' Association; Paarde Vlei, Somerset West; Caledon Reservoir; Brier Vlei, Caledon; and Oukloof Irrigation Board, Prince

Albert. In most of these waters the fish spawned successfully in 1931, and the Paarde Vlei was destined to become one of the best waters for Largemouth Bass fishing in the country. By 1933, the species had been released in numerous rivers, lakes, dams and reservoirs throughout the Cape (Harrison, 1934, 1936; Robbins & MacCrimmon, 1974).

Largemouth Bass were first sent to Natal in April 1931, when 25 yearlings were released in a reservoir near Nottingham Road where they spawned successfully in the following year, and from where fish were later transplanted to other waters in the province (Harrison, 1934).

In 1932, 110 Largemouth Bass fingerlings were placed in reservoirs at Bloemfontein, Kroonstad and Ficksburg in the Orange Free State with uncertain results (Harrison, 1934).

Between 1935 and 1939, large numbers of Largemouth Bass from the Mbane hatchery in Swaziland and from the Cape were successfully released in the Transvaal, Zululand and the Orange Free State. In 1939, a further 43 fingerlings from the Surrey Trout Farm were acquired by the Rand Piscatorial Association, by whom they were transferred to the Pirie hatchery of the Frontier Acclimatization Society at Kingwilliamstown. Here they seem to have done well, but by 1934 only a few had been freed in other waters in the Cape (Harrison, 1934).

Harrison (1936) recorded that at the time of writing

... the following rivers of the Cape Province ... have either been stocked direct with bass or receive the overflow of stocked enclosed waters: Olifants River (Bulshoek dam), Berg River, Hout Bay stream, Palmiet River; Bot River (possibly from Caledon and direct stocking projected); Klein River; Zonder End River (possibly from Vygeboom, or ascending Riversdale); Dwyka or Gamba River (from Oukloof dam, Prince Albert), and Kamanassie River (direct and from dam), and both possibly to the Gouritz River; Sundays River (from Graaff-Reinet and Caesar's dam, Lake Mentz); Kowie River; Fish River (from Cradock and Somerset East); Buffalo River; Golongi or Lower Toise; and possibly the Umzimvubu system from Matatiele.

Elsewhere, between 1935 and 1936 it is known that Largemouth Bass became naturalized in the Breede River, Princess Vlei (Heathfield) and Olieblom Vlei (Harrison, 1952). They failed to establish themselves in Sandvlei (Lakeside) (Harrison, 1976). By 1949,

Largemouth Bass are recorded as occurring in the Cape in the Clanwilliam and Bulshoek dams (Olifants system); the Lower Berg River; the Lower Breede system; Groenvlei Lake at Knysna; below the stretch occupied by introduced trout in the Buffalo River; the Lower Kubusie River at Stutterheim; the Nqabara River (Harrison, 1949, 1952); and in the lower reaches of the Sonderend River and the Palmiet and Kaffirkuils River (Harrison, 1940).

As a result of these introductions and transplantations, Robbins and MacCrimmon (1974) reported Largemouth Bass to be naturalized in rivers and lakes throughout the southern Cape, from the Olifants River south and east in the coastal rivers to Zululand, as well as in some rivers in the vicinity of Johannesburg. In addition, large numbers were regularly stocked in private waters throughout the country, but not in montane waters holding introduced Brown Trout *Salmo trutta* (q.v.) and Rainbow Trout *Oncorhynchus mykiss* (q.v.). Largemouth Bass were not doing well in the acidic peaty rivers of the southeast coast nor in the acid and frequently saline waters of the Orange River basin (but see below under Habitat Requirements). In Natal, Largemouth Bass were naturalized in many impoundments between 700 and 1300 m asl, especially in the Ixopo, Richmond, Pietermaritzburg and Umvoti districts (Crass, 1964). In the Transvaal, Largemouth Bass were widespread in dams and reservoirs in both the highveld and lowveld, being especially successful in the Phalaborwa area. The construction of artificial impoundments, which had helped Largemouth Bass to become established, had led to a corresponding decline in populations of native Tigerfish *Hydrocynus forskahlii/vittata*, whose natural migrations upriver they impeded.

Today, Largemouth Bass are widely naturalized in the southeastern, eastern and western Cape, and in Natal and the Transvaal. De Moor and Bruton (1988: appendix IX: 275–278), who have summarized recent records of the species in South Africa, and from whom much of the following account is derived, list the dates of first recorded introductions (with references) of Largemouth Bass to natural waters in various localities in South Africa.

Cape The species occurs in pools and dams throughout almost all of the Olifants system in the western Cape, such as Paardevlei (Somerset West), Kaffirkuils, Kromme, Van Staadens, Elands, Wit and

Kariega Rivers (Jubb, 1967). It is also found in the Tyume and Keiskamma Rivers (Mayekiso, 1986; quoted by De Moor & Bruton, 1988); in the Kubusie and Klipplaats Rivers (Great Kei system); and in Cintsa dam on the Cintsa River, north of East London (Jubb, 1967). Largemouth Bass occur but are rare in the Breede and Berg Rivers, but are found in most impoundments on both. They are present but are not flourishing in Brandvlei, where they are probably adversely affected by introduced Common Carp *Cyprinus carpio* (q.v.) (De Moor & Bruton, 1988). The species is established virtually throughout the Buffalo River apart from the estuary and the cold upper reaches (Jubb, 1967; Jackson, 1982). Until 1981, Largemouth Bass are known to have occurred in the Baakens River, where they may subsequently have been washed away by heavy floods (Heard & King, 1982). They are also found in the Swartkops River upstream from Despatch (Jubb, 1967); at Groenvlei near Knysna (Jubb, 1967; Van der Merwe, 1970); in every river in the Albany district apart from the Great Fish River, which is ecologically unsuitable (Coetzee, 1977); and in the large Rozendal farm dam near Stellenbosch (Van Schoor, 1969a). In 1986, Largemouth Bass were caught in the Groot River in the Gamtoos system (De Moor & Bruton, 1988).

In dams in the Riviersonderend and Eerste River catchments, especially in the Theewaterskloof, Largemouth Bass are widely distributed. Until the mid-1980s they did not occur in the Eerste River itself, but at around that time gained access to it through the Theewaterskloof tunnel. It would seem that the Eerste River provides only a marginally suitable habitat for the species, which must have entered it on numerous occasions by way of outflows from dams. Largemouth Bass are also found in Stettynskloof, upstream from Rawsonville, but do not occur in the Wemmershoek, Steenbras, Lakensvlei (outside Ceres), Keerom and Fairy Glen (outside Worcester) dams. Although Largemouth Bass are present in the Olifants, Berg and Breede Rivers, they are scarce there compared to introduced Smallmouth Bass *M. dolomieui* (q.v.). They do, however, occur in most of the impoundments in the catchments of the Berg and Breede which contain neither introduced Common Carp nor Bluegills *Lepomis macrochirus* (q.v.). The presence of Common Carp in Brandvlei lake is believed to have adversely affected the Largemouth Bass population there (De Moor & Bruton, 1988).

Orange Free State   Largemouth Bass are established in the dams of the Orange River catchment, but not in the rivers flowing into the Verwoerd dam (Jubb, 1972) nor in the dam itself (Hamman, 1980). Jackson *et al.* (1983) did not find Largemouth Bass in Lake le Roux, but they have been seen in the Bethulie dam on the Caledon River (De Moor & Bruton, 1988). It seems likely that the water in the main Orange River, particularly in the large dams, is too turbid to support the species (De Moor & Bruton, 1988); the dams referred to above by Jubb (1972) are all small farm dams.

Within the Orange/Vaal system, the only certain records of Largemouth Bass are from the upper Orange and Caledon Rivers (Skelton, 1986).

Transvaal   Although Largemouth Bass were stocked in catchment dams in the Vaal system well before 1956, it was not until that year that a specimen was caught near Ermelo (Du Plessis & Le Roux, 1965). Whether or not the species is still found in that locality is unclear, but by the mid-1980s Largemouth Bass were being recorded in numerous farm dams in the Amersfoort area of the Vaal catchment (De Moor & Bruton, 1988). Joubert (1984) reported their presence in the Vaal River at Vereeniging before 1983, apparently as a result of a dam collapsing, where they are today well established. Largemouth Bass are found in the Rietvlei dam in the Limpopo system, southeast of Pretoria (Smith, 1983), and in the Boskop dam on the Mooi River in the Vaal system at Potchefstroom the Ebenezer, Fanie Botha and Hans Merensky dams (Letaba system), the Mala Mala dam (Phalaborwra, Olifants River, Limpopo system) and the Braam Raubenheimer dam (Lydenburg, Crocodile River, Incomati system) (De Moor & Bruton, 1988). Largemouth Bass have been regularly stocked in the following dams in the Transvaal: Longmere, Klipkoppies, Da Gama and Primkop (Nelspruit/White River districts); Nooigedacht, Vygeboom, Westoe and Morgenstond (Carolina area); Jericho (Ermelo district); and the Witbank and Rondebosch (Witbank/Middleburg districts) (De Moor & Bruton, 1988).

Largemouth Bass are presently widely naturalized throughout the Transvaal. Batchelor (1974) recorded them as occurring but not thriving in the Doorndraai dam on the Sterk River in the Limpopo system near Potgietersrus, and they are also found in the Klipplaatdrift dam on the Elands River, a tributary of the Olifants River in the Limpopo system, as well as in dams in the Levubu catchment in the same system (De Moor & Bruton, 1988). Cochrane (1983) records their occasional occurrence in the Hartbeespoort dam.

Natal Crass (1966) reported the establishment of Largemouth Bass in the upper reaches of the Tugela system and in a coastal tributary of the Umhlatuzi River in Natal, while Bourquin and Sowler (1980) recorded them in natural pools in the Nyengelezi valley in the Umzinto River catchment in the Vernon Crookes Nature Reserve. 'Bass', believed by De Moor and Bruton (1988) to be *M. salmoides,* are known to be present in the Albert Falls dam and Craigieburn dam (midway between Greytown and Mooiriver). Bass in the Wagendrift dam (on the Bushmans River, Tugela system near Estcourt) (Smith, 1984) are certainly of this species, and those in the Midmar dam on the Umgeni River (Heeg, 1983) are probably so.

Habitat Requirements   After their initial stocking in the Breede, Berg and Olifants Rivers, Largemouth Bass quickly became naturalized and the dominant fish in each river. After the introduction of Small-mouth Bass, however, *M. salmoides* populations rapidly declined, and *M. dolomieui* assumed the dominant species role (Harrison, 1963). Although *M. salmoides* is somewhat more tolerant of turbid waters than *M. dolomieui,* neither species seems to flourish in extreme turbidness. *M. salmoides* is infrequent in the presence of dense populations of the Moggel or Mud Mullet *Labeo umbratus* (which tends to disturb bottom sediments, thus increasing turbidity), such as in the lower reaches of the Bushmans River, but occurs in tidal waters of the Kariega River from which *L. umbratus* is absent (Jubb, 1973).

   M. salmoides is said to be more tolerant of acid, peaty waters than introduced trout. It is, for example, common in the peat-stained waters of the Sonderend River (Elgin) and in the Kafferskuils system (Riversdale) (Harrison, 1940).

   Although in its native range in North America the Largemouth Bass lives sympatrically with the Bluegill *Lepomis macrochirus* (q.v.), in South Africa, for example in the upper Kariega River, it is often absent where high populations of Bluegills occur (Jubb, 1973). In Natal, Robbins and MacCrimmon (1974) noted that Largemouth Bass are often unsuccessful in waters occupied by the predatory Sharptooth Catfish *Clarias gariepinus* and the Chubbyhead Barb *Barbus anoplus.*

Ecological Impact   The Largemouth Bass has arguably had a greater negative impact on native species in South African waters than any other naturalized piscivore. This is especially so in those waters, such as parts of the Olifants River, devoid of large indigenous predators.

   Details of the ecological impact of Largemouth Bass in South Africa (derived from De Moor & Bruton, 1988) are given below.

Paardevlei Lake (Somerset West). Introduced in 1930, Largemouth Bass are held to be responsible for the extermination in this lake by January 1934 of the Cape Kurper *Sandelia capensis* (Anon., 1944; Harrison, 1952, 1954; Jubb, 1965). *S. capensis* is, however, alleged by Jubb (1965) to be less susceptible than many small native *Barbus* spp. to predation by *M. salmoides* and other introduced predators in a riverine environment.

Groenvlei Lake (Knysna). A survey of this lake carried out in 1940 (6 years after the introduction of *M. salmoides*) revealed that *Gilchristella aestuaria* (then believed to be the only abundant native species in the lake) still occurred in large numbers in shallow littoral waters. As a result of the poor condition of *M. salmoides,* however, which suggested a decline of the prey population, Bluegills and Mosquitofish *Gambusia affinis* (q.v.) were introduced to the lake (Harrison, 1976). It was later discovered that both *G. aestuaria* and another native species, *Atherina breviceps,* which superficially resembles *G. aestuaria,* are abundant in Groenvlei, despite intensive predation by *M. salmoides.* In the face of apparent preferential predation on *A. breviceps,* this species remains more plentiful in the lake than *G. aestuaria* (De Moor & Bruton, 1988).

Olifants River (Clanwilliam). Largemouth Bass were introduced to the lower Olifants in 1933 and to the upper Olifants on both sides of the Clanwilliam dam, in 1936–1937. Although in most cases environmental degradation has been the principal factor leading to the severe decline in the Olifants River system of the populations of 10 native species of catfish and minnows (*Austroglanis, Barbus* and *Labeo* spp.) (Bourquin *et al.,* 1984), predation by *M. salmoides* and *M. dolomieui* is believed to have played an important part (Skelton, 1987). Because the former is less well adapted than the latter to living in lotic waters, it is believed to have played only a minor predatory role (Bourquin *et al.,* 1984). (See also under *M. dolomieui.*)

Swartkops River. In the Swartkops River, where the date of their introduction is unknown, predation by Largemouth Bass is believed to be responsible for the decline of the minnow *Barbus afer* (Barrow, 1971).

Kromme, Van Staadens and *Elands River* (sections) (Eastern Cape). The introduction at apparently unrecorded dates of Largemouth Bass and Bluegills is thought to have been responsible for the decline of native *Barbus* spp. in these three rivers. In the southeastern Cape, rivers which are devoid of introduced predators contain flourishing populations of *Barbus* spp., whereas in the Kromme River *B. afer* is confined to a few shallow tributaries (Jubb, 1959b).

Breede River system. Although general habitat degradation in the Breede River system, where *M. salmoides* was introduced between 1930 and 1932, has been the principal factor in a severe decline in populations of the indigenous *Barbus burchelli*, the introduction of alien predators such as *M. salmoides, M. dolomieui* and *Oncorhynchus mykiss* (q.v.) probably played a contributory role (De Moor & Bruton, 1988). A survey of the Breede River system conducted between 1978 and 1983 revealed an absence of *B. burchelli* in those localities where bass and Rainbow Trout occurred, and that this threatened minnow is restricted to minor tributaries. Two native fishes, *Sandelia capensis* and *Galaxias zebratus*, were both found together with *B. burchelli* in various parts of the system, despite the fact that the former is a predatory species (Cambray & Stuart, 1985).

Keiskamma and *Tyume Rivers.* Three alien fishes (*M. salmoides, O. mykiss* and *L. macrochirus*) and two transplanted natives (the Sharptooth Catfish *Clarias gariepinus* and the Banded Tilapia *Tilapia sparrmanii*) occur in these two rivers, in the latter of which Mayekiso (1986) quoted by De Moor and Bruton, 1988), found them to be mainly distributed as follows:

Upper reaches:	*O. mykiss*
Upper/middle reaches:	*O. mykiss; T. sparrmanii*
Middle reaches:	*T. sparrmanii; C. gariepinus*
Lower/middle reaches:	*M. salmoides; L. macrochirus;*
	T. sparrmanii; C. gariepinus.

Two indigenes, *Barbus trevelyani* and *Sandelia bainsii*, live mainly in the upper and middle reaches respectively, in the former of which the presence of *O. mykiss* is believed to threaten *B. trevelyani*. The diet of *S. bainsii* closely corresponds to that of both *M. salmoides* and *O. mykiss*, but it does not come into competition with the two aliens because its range is limited by altitude and the aliens' distribution. Were *M. salmoides* and *O. mykiss* to become established in stretches of the Tyume where *S. bainsii* is abundant, they would probably have a serious impact, through predation and/or competition, on the native species. The effect of *C. gariepinus*, which was first reported in the Tyume as recently as 1985, on *S. bainsii* has yet to be assessed (Mayekiso, 1986; quoted by De Moor & Bruton, 1988).

Introduced Forage Species   De Moor and Bruton (1988) list the following alien species as having been introduced to South Africa primarily as forage for Largemouth Bass; Bluegills, Mosquitofish, Nile Tilapia *Oreochromis niloticus* (q.v.) and Redbelly Tilapia *Tilapia zillii*; and the translocated natives the Redtail Barb *Barbus gurneyi, T. sparrmanii* and the Mozambique Tilapia *O. mossambicus*. At least two of these forage species, the Bluegill and the Mozambique Tilapia, have had a significant negative ecological impact in those waters into which they have been introduced and translocated, which De Moor and Bruton (1988) reasonably describe as an 'indirect effect' caused by the presence of Largemouth Bass.

Swaziland

In May 1933, 18 Largemouth Bass fingerlings from the Pirie hatchery at Kingwilliamstown, South Africa, were successfully stocked in the clear and weedy Lake Adelaide at Mbabane, where they spawned in the following year. In 1935, some of the fry were translocated to other lentic and lotic waters in Swaziland (Harrison, 1936), of which the Sidokodo and Tubungu Rivers and some nearby reservoirs received a total of 414 (Robbins & MacCrimmon, 1974). Between 1935 and 1939, the Swaziland Angling Society distributed some 6000 Largemouth Bass from its hatchery at Mbabane to waters throughout Swaziland at a wide range of altitudes, and also to the Transvaal in South Africa (Robbins & MacCrimmon, 1974).

Once the species had become established other hatcheries were formed, and correct records of release programmes were no longer maintained. From 1952, however, a timber-exporting concern started properly controlled releases of Largemouth Bass in small and large dams and reservoirs – including, in 1966, the large reservoir on the

Lambongwenya River. From 1954, several waters were successfully stocked by the Usutu Pulp Company (Robbins & MacCrimmon, 1974).

George (1976) reported that Largemouth Bass were still being widely distributed in man-made waters in Swaziland where, however, floods and siltation have generally prevented their establishment in most rivers.

Ecological Impact Largemouth Bass have become the principal predator, especially of *Tilapia* spp., in those impoundments in which they are established in Swaziland (Robbins & MacCrimmon, 1974).

Tanzania

In about 1956, European settlers from Nairobi, Kenya, introduced Largemouth Bass as a sport fish to Tanzania, where they were stocked in the Ngwazi dam, Iringa, on the Rufiji River, northeast of Mahenge in Morogora Province. In late 1959 or early 1960, more were acquired from the Chief Fisheries Officer in Nairobi by A.W.S. Miller of the Colonial Pesticides Unit in Arusha, who placed them in the neighbouring Lake Duluti near Mount Kilimanjaro (Robbins & MacCrimmon, 1974).

Waters in the highland Mbozi (Mbeya) region south of Lake Rukwa near the border with Zambia have also been stocked with Largemouth Bass, although how they got there is apparently unrecorded (Robbins & MacCrimmon, 1974).

Largemouth Bass in Tanzania are today regarded primarily as a source of food for local people, and only secondarily as a sport fish.

Uganda

In 1960, Largemouth Bass from Kenya were imported to western Uganda by the Fisheries Department, by whom they were fairly successfully stocked in Lakes Mutanda and Bunyonyi (both around 1800 m asl) and less successfully in Lake Saka. Plantings made in waters at about 500 m asl all failed. The spread of Largemouth Bass in Uganda seems likely to be prevented by high water temperatures (Robbins & MacCrimmon, 1974).

Zimbabwe

Toots (1970) has traced the introduction of Largemouth Bass to Zimbabwe (Southern Rhodesia) where on 27 July 1932, 50 fingerlings, probably descended from the stock imported to the Jonkershoek hatchery in South Africa in 1928, were placed in the Matopos dam near Bulawayo by the Rhodesia Angling Society. Further shipments of Largemouth Bass fingerlings were acquired in 1936, 1937 and 1938, and between 1944 and 1959 the stocking of dams by local government authorities, angling societies, and private individuals with up to 5000 fingerlings per annum, obtained from Jonkershoek, the Mbabane hatchery in Swaziland, and elsewhere, proceeded apace. In 1967, the Fisheries Research Centre produced a total of 120 000 Largemouth Bass fingerlings which were successfully stocked in various suitable waters such as Lake Kyle (Ludbrook, 1974).

In most of the dams in which they were planted the fish did well, although in many waters regular restocking is required in order to make up for inadequate natural recruitment due to spawning failure and/or predation by native species.

Toots (1970) reported Largemouth Bass to be established in national parks, municipal reservoirs, and private farm dams throughout Zimbabwe. The larger reservoirs in the Bulawayo, Gwelo, Umtali and Salisbury (Harare) regions and two smaller dams near Fort William all support large naturalized populations (Harrison *et al.*, 1963; Robbins & MacCrimmon, 1974).

Jubb (1967, 1973) found that in Zimbabwe Largemouth Bass do not normally occur in rivers downstream of dams, particularly where Tigerfish *Hydrocynus forskahlii/vittatus* are found. They have, however, escaped from dams into a few rivers, where they have established populations which may naturally disperse into new impoundments constructed on those river systems.

Ecological Impact In some of the waters in Zimbabwe where Largemouth Bass have become established, they have eradicated local populations of the Rhodesian Mountain Catfish *Amphilius platychir*.

Unsuccessful Introductions

In Africa, Largemouth Bass have apparently failed to establish wild self-maintaining populations in Cameroun (Bard, 1960; Robbins & MacCrimmon, 1974), Congo, Egypt and Nigeria (Welcomme, 1981, 1988), and Malawi and Tunisia (Robbins & MacCrimmon, 1974).

North America

Translocations

The Largemouth Bass has been widely translocated outside its natural range in North America, in the USA, Canada, and Mexico. (For full details see Robbins & MacCrimmon, 1974.)

Ecological Impact

In Mexico, De Buen (1941) and Preciado (1955) said that predation by translocated Largemouth Bass had contributed to the disappearance of several native fish species, notably the Blanco de Pátzcuaro *Christoma estor/ ester* in Lago de Pátzcuaro, Michoacan, and *Goodea* spp. (Roots (1976) says that the translocation of 'bass' to Lago de Pátzcuaro has endangered the endemic eponymous salamander, but gives no further details or references.) Contreras (1969) referred to predation on and displacement of indigenous fishes and the possible replacement of another species of bass at Cuatro Cíenegas, while Contreras (1978) mentioned predation on and displacement of the El Potosí Pupfish *Megupsilon aporus* and to a lesser degree *Cyprinodon alvarezi* at El Potosí, Nuevo León. At Bustillos, Chihuahua, and Peña del Aguila and Río Tunal, Durango, translocated Largemouth Bass have replaced several native fish species (Contreras, 1975; Contreras *et al.*, 1976). (See also Contreras & Escalante, 1984.)

South and Central America

Bolivia

Introduced at an apparently unrecorded date to provide sport fishing in Bolivia, the Largemouth Bass is currently naturalized in various dams and small lakes, and in the Beri River, Santa Cruz (Welcomme, 1981, 1988). Robbins and MacCrimmon (1974) do not mention this introduction.

Brazil

Introductions of Largemouth Bass into Brazilian waters have been undertaken with caution because of the likely threat to indigenous predators, such as the Dourado *Salminus maxillosus*, which are important commercial species. *M. salmoides* may have been introduced as early as 1926, although the first certain importation, by an amateur aquaculturist to the state of Minas Gerais, was not made until 1935. Robbins and MacCrimmon (1974) said that the Fishing Institute of São Paulo was then rearing Largemouth Bass for release into waters holding native species such as lambaris *Astyanax* spp., and that Bass had become naturalized in several man-made waters where they provided useful recreational fishing.

Colombia

Largemouth Bass from the USA were introduced into Colombia by amateur anglers in 1956 for sport fishing. Although breeding success has only been moderate, the species has become naturalized in Antioquia Province in the northwest (Robbins & MacCrimmon, 1974).

Costa Rica

Introductions of Largemouth Bass have resulted in the establishment of self-maintaining populations in a small number of privately owned waters in Costa Rica. The only recorded attempt to establish them in a river, however, failed, probably as a result of interspecific competition from the native *Cichlasoma dovii* (Robbins & MacCrimmon, 1974).

Cuba

The earliest recorded introduction of Largemouth Bass to Cuba was made in 1915, when 1000 fish were sent by the United States Bureau of Fisheries to Las Indios Lake in Oriente Province (Johnson, 1915). In 1927 and 1935 respectively, 500 and 60 Largemouth Bass were imported to Cuba, as a result of which the species became naturalized around Havana and elsewhere (Leach & James, 1935), later spreading naturally to almost every watershed on the island, in a few places, particularly in the west, becoming quite common (Corral, 1936), before being subsequently eradicated from some waters (Rivero, 1936).

Holčík (1970) attributed the success of the Largemouth Bass in Cuba to an absence of native fish predators and an ideal environment for spawning and foraging. The species is currently established in lakes and lagoons all over the island, where it is fished for both commercially and for sport, and is also reared in aquaculture (Robbins & MacCrimmon, 1974).

Ecological Impact   Since no forage species such as Bluegills *Lepomis macrochirus* were introduced to

Cuba with Largemouth Bass, the latter has been a severe predator of native fishes (Rivero, 1936), and its predation in particular on the native *Gambusia* has been blamed for an increase in the incidence of malaria on the island (Robbins & MacCrimmon, 1974).

Ecuador

In 1960, Largemouth Bass from the USA were introduced to Ecuador where they were liberated in some ponds at an altitude of around 5800 m in the Andes on the Hacienda Guachalá at Cayambe in Provincia de Pichincha. From here, some were transferred by the Department of Pisciculture in 1964 and 1966 to Lago San Pablo, the largest lake in the Ecuadorean Andes, and to Lago de Colta in Provincia de Chimborazo. Although restocking of these waters takes place, the fish also reproduce in the wild (Robbins & MacCrimmon, 1974).

El Salvador

According to Welcomme (1981, 1988), Largemouth Bass from the USA were introduced for rearing in aquaculture in 1957 to El Salvador, where they are established in only two small lakes from which they are disappearing. Robbins and MacCrimmon (1974) do not mention this introduction.

Guatemala

In 1958, Largemouth Bass from the USA were introduced for sporting purposes to the landlocked Lago Atitlán, high in the mountains of southern Guatemala. In the following year more Largemouth Bass were imported and were released in waters less isolated than Atitlán (Robbins & MacCrimmon, 1974) – Welcomme (1981, 1988) specified Lago Calderas. They are currently established in both these lakes and possibly also in other waters.

Ecological Impact   Since at least the mid-16th century, Lago Atitlán provided a substantial subsistence fishery for small native species, probably the Liberty Molly *Poecilia sphenops* and the Convict Cichlid *Cichlasoma nigrofasciatum*, and crabs *Potamocarcinus guatemalensis*, which furnished the native Guatemalan Indians with both an important source of protein and a cash crop. After their introduction in 1958 (together with Black Crappies

Pomoxis nigromaculatus (q.v.), Largemouth Bass began to prey heavily on the small native fishes and crabs, which by the early 1970s had disappeared (Zaret & Paine, 1973; Miller, 1989; the latter misassigns Atitlán to Panama).

According to Robbins and MacCrimmon (1974) in Lago Atitlán 'the bass is said to have grown so large [up to 5 kg] that it now threatens the extinction of a rare grebe by preying on the goslings' (*sic*)! In fact, it was the loss of these small native fish and crabs, which were its staple diet, that contributed to the extinction of the rare endemic Atitlán Grebe *Podilymbus gigas*, a flightless species known locally as the *poc*, which was never known to number more than 100 pairs and which was already threatened by the destruction of the reedbeds in which it nested by local Indians to weave mats for the tourist trade (La Bastille, 1991; Shuker, 1993).

Hispaniola (Haiti and Dominican Republic)

The present status of the Largemouth Bass in Haiti and the Dominican Republic is uncertain. Welcomme (1981) says the species was introduced to the Dominican Republic for aquaculture purposes from Haiti in 1953 (but does not mention an introduction to Haiti), and that it was established 'in rivers and lakes ... populations much reduced despite widespread introductions'. Welcomme (1988), however, says the introduction was made from the USA in 1955, and that the species is reproducing only 'artificially'. Wetherbee (1989) says only that the species has been introduced from the USA to Hispaniola. Robbins and MacCrimmon (1974) do not mention this introduction.

Honduras

Largemouth Bass from the USA were in 1955 introduced to Honduras to provide a sport fishery, where they became established in the 200-km² Lago de Yojoa, some 75 km south of San Pedro Sula, and where by 1963 fish of up to 7.5 kg were being taken (Robbins & MacCrimmon, 1974).

Ecological Impact   According to Welcome (1981), Largemouth Bass have 'produced an eco-biological imbalance' in Lago de Yojoa.

Panama

Hildebrand (1938) records that in 1917, 1918 and 1925 respectively, 450, 1000 and 2250 fingerling and yearling Largemouth Bass were sent by the United States Bureau of Fisheries to Ancon (in the Canal Zone); these fish were introduced to the 42 315-ha Lago Gatún, where because of excessively high water temperatures and the presence of the predatory *Astyanax rubberimus* they apparently failed to become established.

According to Leach and James (1935), in that year 325 yearling and adult Largemouth Bass were despatched from the same source to Panama. Curiously, this introduction is not mentioned by Hildebrand (1938). González (1988) records that in 1955 Largemouth Bass were again imported to the isthmus, where they were successfully planted in the warm waters of volcanic lagoons in Chiriqui Province. Again, strangely, this introduction is not mentioned by Robbins and MacCrimmon (1974), so 1955 may be a literal for 1935. According to Robbins and MacCrimmon (1974), 'largemouth support fisheries in Los Lagunas de Volcán, two small lakes in the western part of the Country, some 400 km from Panamá City, and Goofy Lake, Cerro Azul, 46 km northeast of Panamá City'.

Puerto Rico

In 1915, and again in 1916, the United States Bureau of Fisheries unsuccessfully released 600 Largemouth Bass in the recently constructed Carite reservoir near Guayama, Puerto Rico (Johnson, 1915; Erdman, 1984). In 1946, the US Fish and Wildlife Service provided Félix Iñigo, superintendent of the Marícao Fish Hatchery, with 2170 fingerlings and mature fish, which became successfully established in a number of the island's reservoirs (Erdman, 1984). Largemouth Bass are now naturalized in all major Puerto Rican reservoirs, where they are the most successful predator, and in numerous farm ponds. Fish from Comerío reservoir have established themselves where the water flows slowly in the Río La Plata from Aibonito to as far upriver as Cayey. Introductions of Largemouth Bass to the Río Añasco, Cartagena, the Espíritu Santo, and to the Tortuguero lagoons proved unsuccessful, probably due to unsuitable environmental conditions (Erdman, 1984). Puerto Rican waters, where reservoir surface temperatures average around 27.5°C throughout the year and cooler streams tend to flow too swiftly, probably provide Largemouth Bass with only a marginal habitat (Robbins & MacCrimmon, 1974). (See also Meehean *et al.*, 1948; Iñigo, 1949; Soler, 1951; Bird, 1960; Schulte, 1974; Rivera-González, 1979.) Wetherbee (1989) does not mention this introduction.

Unsuccessful Introductions

In South and Central America, Largemouth Bass have apparently failed to establish viable populations in Argentina (Navas, 1987), and Nicaragua and Venezuela (Robbins & MacCrimmon, 1974).

Oceania

Azores

On 7 October 1898, Largemouth Bass fry from the USA were released in Lac Sete-Cidades on Île Saõ Miguel in the Azores by José Maria Raposo of Amaral. Initially the fish thrived and grew prodigiously, but following the excavation of a drainage ditch in 1937 which lowered the water level in the lake the population considerably declined (Robbins & MacCrimmon, 1974), and may now have disappeared (Moreira da Silva, 1977; Goubier *et al.*, 1983).

Fiji

Eldredge (1994), quoting Farman (1984), lists the Largemouth Bass as established in Fiji. Welcomme (1988) says the introduction took place in 1962 and was unsuccessful. Neither Robbins and MacCrimmon (1974) nor Maciolek (1984) mention an introduction to Fiji.

Guam (Mariana Islands)

Evidence of the status of Largemouth Bass in Guam is confusing and contradictory. Devambez (1964) says they were introduced in 1963 and became established. Robbins and MacCrimmon (1974) say that in September 1963, 56 juveniles were obtained from Hawaii and were planted in the Almagosa River and that in July 1965, a further 10 000 were acquired from the Miles City National Fish Hatchery in Montana, USA, the 3000 or more survivors being released in the same river and in the Fena reservoir; none of these introductions, Robbins and MacCrimmon (1974) add, was successful. Welcomme (1981) does not refer to the presence of Largemouth Bass on Guam. Maciolek (1984) lists the species as naturalized on Guam, whereas Welcomme (1988) says it is reproducing only 'artificially'.

Hawaiian Islands

In 1897, Largemouth Bass from California were imported to Hilo on the island of Hawaii, where the 21 survivors of the shipment were released in the Wailuku River. Largemouth Bass do not today survive in the Wailuku, but are well established in the spring-fed Waiakea mill pond in Hilo and in a large outflowing stream. In 1908, a second consignment of Largemouth Bass was imported from California, this time to the island of Oahu, where the fish were placed in the Wahiawa reservoir where they became established. Three years later a third shipment of 300 arrived from California, which were successfully released in Waiuli and Koloko reservoirs on Kauai (Brock, 1960). In about 1950, Largemouth Bass, mostly from the Lihue Plantation Company reservoir on Kauai, were planted in numerous man-made waters on Maui and Oahu (Brock, 1952).

Brock (1960) said that Largemouth Bass were then established in reservoirs on all the main islands. Robbins and MacCrimmon (1974) confirmed their presence in most reservoirs on Kauai, Oahu, Maui and Hawaii, whereas Maciolek (1984) omitted Maui from the list. Most of the waters containing Bass are used for the irrigation of sugar-cane and pineapple plantations; streams in the islands are too small or too fast-flowing for their survival (Robbins & MacCrimmon, 1974).

For details of a spontaneous hybrid between *M. salmoides* and *Lepomis macrochirus* on Kaui, see under the latter species.

Mauritius

In 1949, a dozen Largemouth Bass were shipped from the Jonkershoek hatchery at Stellenbosch, South Africa, to the island of Mauritius, for stocking in natural ponds and irrigation dams (Robbins & MacCrimmon, 1974). In the following year a further consignment arrived from the USA, and although Robbins and MacCrimmon (1974) expressed doubt about the result of these introductions, Welcomme (1988) said the species was 'present in small numbers'.

New Caledonia

Largemouth Bass have been introduced to and are established in New Caledonia (Devambez, 1960, 1964; Anon., 1963; Maciolek, 1984). Neither Robbins and MacCrimmon (1974) nor Welcomme (1981, 1988) refer to this introduction.

Unsuccessful Introductions

In Oceania, Largemouth Bass have failed to establish viable breeding populations in the wild in French Polynesia (Tahiti) (Maciolek, 1984).

White Crappie (Sac-A-Lai; Bachelor)
Pomoxis annularis

The White Crappie, a species of warm, slowly flowing streams, and lakes and ponds, is an important sport and food fish in North America.

Natural Distribution

Eastern–central North America, from southern Ontario, Canada and southwestern New York, USA, west of the Appalachians, south to the Gulf coast and westward to Minnesota, South Dakota and Texas.

Naturalized Distribution

North America: Mexico.

North America

Mexico

Introduced to Mexico for rearing as a food fish, White Crappies were released into the wild at an unknown date, but probably, according to Contreras and Escalante (1984), who do not recognize them as native in the Río Grande, in the early 1950s, in Marte R. Gómez reservoir, where they have become established (Barajas & Contreras, 1986). They have also been planted in Falcon reservoir, Tamaulipas (Lee, 1980); in the lower Río Yaqui (Hendrickson *et al.*, 1980); and in Baja California Norte (Follett, 1960), in all of which they have become naturalized. They are also established in the Río San Juan at Doctor Coss, Nuevo León (Contreras & Escalante, 1984).

Unsuccessful Introduction

Erdman (1984) refers to an unsuccessful attempt in 1957 to introduce White Crappies from the US Fish and Wildlife Service's hatchery at Welaka, Florida, to Puerto Rico.

Welcomme (1988) says the White Crappie has been successfully introduced to Panama; it is, however, its congener the Black Crappie *P. nigromaculatus* (q.v.), that is established in Panama (see González 1988).

Translocations

White Crappies have been widely translocated outside their natural range in the USA, where according to Welcomme (1988) they 'now occupy a good proportion of the country'. Specifically, Smith (1896) refers to transplants to California, Washington, and Idaho in 1890 and 1892, while Chapman (1942) mentions Oregon and Washington.

Black Crappie
Pomoxis nigromaculatus

The Black Crappie, which lives in warmwater ponds and shallow lakes and requires cleaner waters than the White Crappie, but is able to withstand somewhat lower temperatures, is also an important sport and food fish in North America.

Natural Distribution

The Atlantic slope from Virginia to Florida, and west along the Gulf coast to central Texas, northward to North Dakota and eastern Montana, east to the Appalachians in extreme southeastern Canada.

Naturalized Distribution

North America: Mexico. ***Central America:*** Guatemala; Panama.

Europe

Unsuccessful Introductions

In 1887, Black Crappies were imported into France, where according to De Groot (1985) they 'became a pest locally', implying their establishment in the wild. From France, Black Crappies were later exported to Germany, but in neither country are they established today.

North America

Mexico

Black Crappies were introduced to Mexico at an unknown date for rearing as a food fish (Contreras & Escalante, 1984); they are currently established in the lower Río Grande (Lee, 1980), and lower Río Colorado (Follett, 1960; Lee, 1980), and in Falcon reservoir, Tamaulipas (Contreras & Escalante, 1984).

Translocations

Black Crappies have been translocated to many places outside their natural range in North America. Hart (1934) refers to their appearance in British Columbia in the preceding year, since when they have been found in many other waters in the lower Fraser River system, to which they are believed to have dispersed naturally from translocations made to Washington and Oregon in 1890 and 1892 (Carl & Guiguet, 1958).

Central America

Guatemala

Zaret and Paine (1973) record the successful introduction in 1958 of Black Crappies to Lake Atitlán in the southern highlands of Guatemala.

Panama

Hildebrand (1938) says that in 1925 500 Black Crappies from the USA were released unsuccessfully in Lake Gatún, Panama. According to Dr Abelardo De Gracia, Jr (pers. comm. (1987) to González (1988)), between 1925 and 1935 they were successfully planted in Laguna de San Carlos and Mataahogado in Panama Province; today the Black Crappie is a popular sport fish in Laguna de San Carlos, where it is taken both by day and at night with live bait and artificial lures (González, 1988).

PERCIDAE

Common Perch (Redfin)
Perca fluviatilis

The Common Perch is a popular fish among European anglers, and the few introductions that have been made were for sporting purposes by European colonists. It favours lakes, ponds and slowly flowing rivers, where it lives among aquatic vegetation and submerged tree roots which blend well with the cryptic barred markings on its sides. It feeds mainly on zooplankton, bottom invertebrates, and other fish.

Natural Distribution

Europe, from Southeastern England east across Scandinavia and the former USSR to Siberia, southward to northern Italy and the Black and Caspian Seas. (The Yellow Perch *P. flavescens*, a fish of the middle Atlantic states and the Great Lakes of North America, has come to be regarded as a separate species from *P. fluviatilis*, which it closely resembles and to which it is extremely closely related.)

Naturalized Distribution

Europe: Cyprus; Spain. *Africa:* South Africa. *Australasia:* Australia; New Zealand. *Oceania:* Azores.

Europe

Cyprus

According to Welcomme (1981, 1988), Common Perch from Britain were introduced for sporting purposes to Cyprus in 1971, where they became established; they are 'very popular among anglers but prolific and voracious giving stunted populations in some waters'.

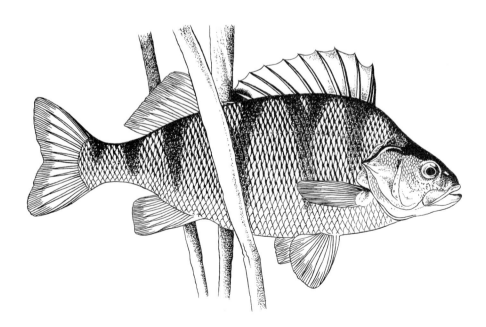

Spain

Between 1970 and 1979, Perch from France were introduced for angling purposes to Spain (Elvira, 1995b), where they have become established in the drainages of the Ebro and the Eastern Pyrenees (Elvira, 1995a).

Ecological Impact Because the Perch, when released in waters in which it does not naturally occur, tends to increase explosively and to form dense and stunted populations which adversely affect other species (Vooren, 1972), Brown Trout *Salmo trutta*, Common Carp *Cyprinus carpio* and minnows disappeared from the Gileppe dam after the Perch's introduction.

Translocations

Outside their natural range in Europe, Common Perch have been translocated to various parts of Italy (Welcomme, 1988), to the Gileppe dam at Verviers, Belgium (Thienemann, 1950), and to Scotland, Wales and Ireland.

In North America, the Yellow Perch has been widely translocated, in particular to the Pacific states of the USA (Smith, 1896; Chapman, 1942) and to British Columbia, Canada (Carl & Guiguet, 1958).

Africa

South Africa

De Moor and Bruton (1988) have traced what is known about the introduction of Perch to South Africa.

Perch were first imported to South Africa, to the Jonkershoek hatchery in the Cape, from England in 1915, in the hope that they would provide a sporting fish in waters too warm for trout. Although Perch proved difficult to rear in captivity, they were stocked in the Cape Province as follows.

In 1928, before the release of Largemouth Bass *Micropterus salmoides* (q.v.), Perch were introduced to Paardevlei in the western Cape, where they survived but did not do well, perhaps because of predation on their fry by Cape Kurper *Sandelia capensis*. After the introduction of Largemouth Bass in 1930 and of Bluegills *Lepomis macrochirus* (q.v.) in 1940, Cape Kurpers were eradicated, and Perch began to breed more successfully (Anon., 1944). In 1929, Perch were freed in the Berg River, also in the western Cape, where they failed to establish (Harrison, 1953). At an apparently unrecorded date Perch were also unsuccessfully planted in the Clanwilliam dam in the Olifants system of the western Cape (Scott, 1982). In 1936, a single Perch was caught in the Little Princess vlei (Heathfield), where the date of introduction is not known (Harrison, 1977).

Most introductions of Perch into South African waters proved unsuccessful (Hey & Van Wyck, 1950), but Harrison (1952) reported them to survive in Paardevlei and in some dams near Cathcart. In the mid-1980s they were recorded still to occur in Paardevlei, in the 'Downs Farm dam' near Cathcart, and in 1982 in the Roodepoort dam (De Moor & Bruton, 1988).

'It would be interesting' wrote De Moor and Bruton (1988), 'to establish the reasons why perch have not been able to colonise southern African waters, as this would shed light on the characteristics of an unsuccessful invader'. (See also Thorpe, 1977.)

Unsuccessful Introduction

Toots (1970) reported the failure of Perch to become established in Zimbabwe (Rhodesia) after their introduction into Ngamo dam in 1939, and subsequently elsewhere. Moreau *et al.* (1988) list Perch from Switzerland as released in some lakes in Morocco in 1941, where they are believed to have since died out.

Australasia

Australia

Curzon Allport (1874) describes the introduction by his brother, Morton, of the Perch to Australia:

> During four successive years prior to 1861 attempts were made to introduce the perch to Tasmania from England, but it was not until December of that year that a fifth attempt succeeded, when – as in prior attempts – at Mr Morton Allport's request and sole expense, the writer brought out under his personal charge and landed five fish, which were placed in ponds expressly built for them by Mr Allport. More fish were obtained by him the following year, from whence have sprung the immense supply now in Tasmania and Australia.

These fish were soon apparently breeding successfully, and their progeny were subsequently distributed to various suitable waters in Tasmania. *The*

Australasian newspaper of 12 May 1866 (quoted by Clements, 1988) proclaimed that 'Mr Morton Allport has in the corporation reservoirs, at Hobart Town, fully 30,000 English perch, which he is endeavouring to distribute'. In November 1866, Morton Allport presented 30 Perch to the Victoria Society, and in the following year a further 40 and a quantity of ova, which were placed in a reservoir at Royal Park. In 1868, six small Perch, again donated by Morton Allport, were released in Kirk's reservoir on the outskirts of Ballarat, Victoria, where they rapidly increased in numbers and from where, within a couple of years, their offspring were being transplanted to many other waters in the state, including Lake Wendouree, where they flourished. Ballarat became the principal centre from which Perch were later translocated on the Australian mainland (Clements, 1988).

Accounts differ as to when Perch were first introduced to New South Wales. According to Arentz (1966) the earliest attempt was made in 1882, when the state's Zoological Society ordered a shipment from Ballarat, all of which died *en route*. At around the same time, Perch were released in Lake George and in a lagoon near Canberra, and in 1884 Mr D.H. Campbell introduced Perch to dams on his property at Cunningham Plains. In the same year, seven Perch were freed in waters in the National Park, south of Sydney. In June 1888, Mr John Gale and Mr F. Campbell acquired 50 Perch from Ballarat, which they released in various waters near Queanbeyan (Clements, 1988), including the Queanbeyan, Molonglo, Naas and Cotter Rivers (Lintermans *et al.*, 1990).

According to the Department of Fisheries and Tourism's report for 1907, *The Fisheries of New South Wales* (quoted by Clements, 1988), the earliest introduction of Perch to that state

> was made by the Department of Fisheries in 1885, when two consignments of English perch were obtained from Ballarat at the instigation of the Zoological Society. ... these were liberated in the Kangaroos Creek, Port Hacking, the tributaries of the Lachlan River, and the waters of the Young district. Similar experiments were made in 1886.

Arentz (1966) also refers to this introduction. Roughley (1951) and Lake (1959) indicate that Perch were first introduced to New South Wales in 1888.

Today, Perch are abundant and widely distributed throughout the major rivers west of the Great Dividing Range (McKay, 1984). They are found in

suitable waters in most of Victoria, apart from Gippsland and streams of the northeast ranges where they are rare. They are, perhaps, commonest in the southern Murray–Darling Rivers system of New South Wales, but are rare in upland rivers and in the upper reaches of the Darling. They are also common in the larger slow-flowing rivers of South Australia and in Tasmania, and occur in southwestern Western Australia, where the date of their introduction is apparently unknown. They are occasionally reported from southern Queensland where, however, in general temperatures are too high for them (Clements, 1988; Twyford, 1991).

The main reason for the failure of Perch to become established in highland waters of southeastern Australia is that, in general, stream velocity is too high and suitable beds of aquatic plants to provide shelter are too scarce; it is significant that the only upland streams of New South Wales in which Perch occur are the Coolumbooka and Bombola – tributaries of the Snowy River – both of which are relatively slow-flowing and support extensive areas of aquatic vegetation (Weatherley & Lake, 1967).

Where they occur in the rivers of the southeast, the distribution of Perch tends to depend on habitat conditions. Normally, they are most common in still or slowly flowing water with plenty of vegetation. When severe flooding affects riverine habitats, however, and long stretches of river become ecologically more similar, Perch are distributed fairly evenly throughout river systems (Weatherley & Lake, 1967).

The discontinuous distribution of Perch in Tasmania closely resembles that of the Tench *Tinca tinca* (q.v.), both species only flourishing in slowly flowing rivers with an abundance of aquatic weeds. Perch are confined to the lower reaches of the Macquarie – South Esk Rivers system in the north and to the Jordan, Derwent and Coal Rivers system in the south. In the short and precipitous highland streams of the northeast, Perch are only found in those few into which they were introduced. Three apparently suitable western river systems – the Arthur, Pieman and Gordon – seem to be devoid of Perch, probably because of their isolation from colonized areas. There are apparently no Perch in the fast-flowing streams of the central highlands, and of the numerous lakes in this region Perch occur only in Lake Echo (where they were introduced in the 1920s), a mesotrophic water rather than the oligotrophic kind common in the Tasmanian highlands.

with an extensive shallow shoreline having an abundance of large weedbeds (Weatherley, 1963). (See also Whitley, 1951; Thorpe, 1977; Weatherley, 1977; Cadwallader & Backhouse, 1983; Merrick & Schmida, 1984; Arthington, 1986; Allen, 1989; McKay, 1989; Burchmore & Battaglene, 1990.)

Ecological Impact   The ecological impact of Perch in Australia is beginning to become evident, but is so far not fully understood. From an examination of the fish catches in the Kerang lakes between 1919 and 1949, Cadwallader (1978a) found that when the population of Perch was large those of native fish species were low, and did not increase until Perch began to decline. Fletcher (1986) has suggested that Perch, together with Brown Trout *Salmo trutta* (q.v.) and Chinook Salmon *Oncorhynchus shawytscha* (q.v.), may have affected populations of galaxiids, Pygmy Perch *Nannoperca australis* and Golden Perch *Macquaria ambigua* in southern rivers by selective predation. Similarly, Baxter *et al.* (1985) mention predation of fingerling Rainbow Trout *Oncorhynchus mykiss* (q.v.) by Perch, while McKay (1977) suspected them of being responsible for a decline in numbers of small *Galaxias nigrostriatus* in Western Australia.

Langdon (1986) described a new disease, epizootic haematopoietic necrosis (EHNV), of unknown origin, affecting Perch in the wild and which had spread to cultured Rainbow Trout. This virus has also proved extremely pathogenic to several native Australian fishes, including Silver Perch *Bidyanus bidyanus*, Mountain Galaxias *Galaxias olidus* and Macquarie Perch *Macquaria australasica*. (See also Tilzey, 1980; Arthington, 1986, 1991; Burchmore & Battaglene, 1990.)

New Zealand

Thomson (1922) has summarized what is known of the introduction of Perch into New Zealand.

In 1864, 200 Perch were imported by A.M. Johnson to Christchurch from England on the *British Empire*, but all were found to have died on the voyage. In 1868, 22 Perch were acquired from Hobart, Tasmania, by the Otago Acclimatization Society, which were released in the Ross Creek reservoir near Dunedin. In 1870, a further 18 were received from the same source, and the progeny of these fish were widely distributed throughout Otago, to Lawrence, Gore, Clydevale, Kaitangata, Otekaike, Elderslie, Tapanui, Waikouaiti, Waihemo, and elsewhere. Others were despatched to Ashburton, to the Canterbury Society, and to Nelson. The Otago Society's annual report for 1891 said that 'these fish are becoming very numerous; Kaitangata Lake and Lovell's Creek are simply swarming with them'.

In 1868, the Southland and Canterbury Societies acquired an unspecified number of Perch from Morton Allport (see above under Australia) of Hobart; the report for the latter society for 1871 said that the fish had 'successfully multiplied and no further importations are needed'. In 1883, the Canterbury Society received a report that 'perch in some numbers could be seen in some streams on the Wakanui road'.

In 1877, the Wanganui Acclimatization Society obtained around 600 Perch from Ballarat, Australia, some 24 of which were acquired in the following year by the Wellington Society, which released them in the Wellington reservoir. By 1886, they were reported to be 'very numerous', and some were transferred to lagoons in the Wairarapa and to lakes near Otaki.

In 1885, the Hamilton Domain Board received a consignment of 100 000 eggs from their *confrères* in Canterbury, and released the resulting fry in various waters in Waikato. In 1887, the Taranaki Society acquired some Perch from Mr Johnson of Opawa, and at around the same time some were released in Lake Mahinapua on the west coast.

For a short period after their introduction, Perch were a sought-after sporting fish in New Zealand, particularly in Taranaki, Manawatu, Wairarapa, Otago and North Canterbury, but interest in them soon declined, probably because of their inferiority to trout (McDowall, 1994b). As with the Tench *Tinca tinca* (q.v.), Perch were released illegally in the 1960s in and north of the Waikato, where they are now widely distributed. For a while, interest in Perch fishing was revived by research carried out in Hamilton Lake by the Ministry of Agriculture and Fisheries, but this renewed interest declined when the research was abandoned (McDowall, 1994b). McDowall (1990b) described the Perch in New Zealand as 'widespread but localised, occurring in the Auckland area, Waikato, Taranaki and Wanganui, Manawatu, Wairarapa, Canterbury, Westland, Otago and Southland. It is best known and most abundant in Canterbury and Otago'. (See also Thorpe, 1977;

McDowall, 1979, 1984; Druett, 1983; Clements, 1988; Lever, 1994.)

Oceania

Azores

On 8 November 1898, Perch from England were introduced to the volcanic Lac de Sete Cidades in the west of L'Île de Saõ Miguel in the Azores. Between the date of their introduction and 1913, Perch from Sete Cidades were transferred to Lac de Furnas in the east of the island, in both of which they remain established today (Goubier *et al.*, 1983).

Ruffe (Pope)
Gymnocephalus cernuus

The Ruffe is a species of lakes and slowly flowing rivers, and is also found in brackish waters. It breeds prolifically, and is a predator on the eggs and larvae of other fishes. It is of little or no value as food or for angling, and can be an ecological pest. Where it has been translocated outside its natural range in Europe, the Ruffe has tended to become extremely abundant and subject to stunting.

Natural Distribution

Eurasia, including England, northern and eastern Europe, rivers flowing into the Arctic Ocean, and the Caspian and Aral Seas.

Naturalized Distribution

North America: USA; Canada.

North America

United States; Canada

Pratt *et al.* (1992) have summarized what is known of the introduction of the Ruffe to the USA and Canada.

In 1986, a total of 66 Ruffe larvae were collected from three sites in the lower St Louis River, Minnesota, the westernmost tributary of Lake Superior. In 1987, 101 larvae were collected from four locations, and in July and August of the same year 31 small Ruffe were caught at three sites, followed by more in September.

The most likely vector for this alien species was the dumped ballast water of a sea-going vessel. The Duluth/Superior harbour is an international grain shipping port situated at the mouth of the St Louis River, and is regularly visited by many in-ballast grain ships originating in north European sea ports, which are frequently situated in the estuaries or lower reaches of rivers where Ruffe are common.

Between 1988 and 1991, the abundance of Ruffe in the St Louis River greatly increased, and the species has slowly extended its range into southwestern Lake Superior and into the estuaries of the Amnicon, Brule and Iron Rivers, which are some 25, 40 and 50 km respectively east of the St Louis River along the southern shore of Lake Superior in Wisconsin. In August 1991, a Ruffe was taken on rod and line in the Kaministiquia River at Thunder Bay, Ontario, Canada, where six more were caught in September and October. Since Thunder Bay is some 300 km northeast of the St Louis River, it seems unlikely that the fish arrived by natural dispersal but rather by translocation from the St Louis River in the ballast of intralake shipping. (See also Courtenay *et al.*, 1991; Courtenay, 1993.)

Ecological and Economic Impact At the time of writing the Ruffe has not occurred for a long enough period nor in sufficient numbers for any ecological impact to have become apparent in North America. Pratt *et al.* (1992) point out, however, that for the past 40 years or more the Great Lakes fisheries have been seriously affected by the invasion of exotic fish species, and 'we are greatly concerned that ruffe will have negative impacts on the native fishes of the Great Lakes, of other lakes and rivers throughout North America, and on the fisheries they support'. This fear has been echoed by the US Fish and Wildlife Service which has suggested that the Ruffe 'poses the greatest threat to Great Lakes fisheries of any of the new exotic species' (quoted by Laycock, 1992), and by Mills *et al.* (1994) who say that the species 'has the potential to disperse throughout much of North America and become a major competitor with other fish species throughout the Great Lakes'. Were the Ruffe to colonize Lake Erie, site of the world's largest fishery for Yellow Perch *Perca flavescens* and Walleye *Stizostedion vitreum*, the potential annual economic loss has been estimated at some $90 million (Anon., 1992, quoted by Mills *et al.*, 1994 and Courtenay, 1995).

Zander (Pike–Perch)
Stizostedion lucioperca

Throughout its range, the Zander is regarded as an important commercial and sporting asset, qualities that have led to its introduction to several European countries. A large and important commercial fishery is established in the lower reaches of the River Volga. A voracious piscivore, the Zander favours murky water in lakes and slowly flowing rivers, where it is a crepuscular hunter, remaining at other times near the bottom. Populations living in brackish waters such as the Baltic Sea migrate up in-flowing rivers to spawn. Because of its predatory nature, international introductions of Zander have not been generally regarded with favour by conservationists.

Natural Distribution

Central and eastern Europe, from Sweden and Finland south to the former Yugoslavia and the Black and Caspian Seas, and eastward to the Ural Mountains in the former USSR.

Naturalized Distribution

Europe: British Isles (England); Denmark; France; Italy; The Netherlands; Spain. ***Asia:*** Turkey. ***Africa:*** Morocco.

Europe

British Isles (England)

As early as 1861, the recently formed Society for the Acclimatization of Animals, Birds, Fishes, Insects and Vegetables within the United Kingdom was inveighing against the proposed introduction to British waters of so potentially dangerous a predator as the Pike–Perch, as it was then generally known (Lever, 1994). Nevertheless, many unsuccessful attempts, some of them, like those of the Wels or European Catfish *Silurus glanis* (q.v.), irresponsibly sponsored by *The Field* magazine, were made to introduce Zander to Britain before they were at last successfully imported (Sachs, 1878):

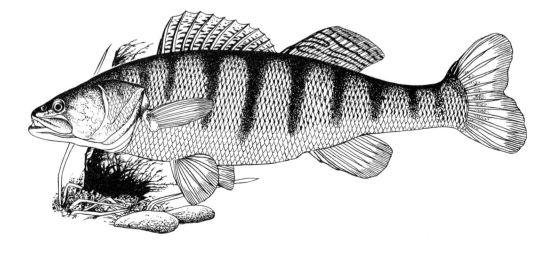

Mr Dallmer, chief fishing master of Schleswig-Holstein relates ... that he had the honour of being requested ... to supply His Grace the Duke of Bedford with about one hundred small Zander ...

Mr Dallmer came to the conclusion to get the Zander from shallow lakes, as he knew such fish were better to transport, and in winter they lived long in a fish box. He therefore wrote ... that he thought the undertaking possible, and would select fish of a large sort: first, because small fish were weak; second, they would more easily fall prey to larger fish; third, larger fish would soon after their arrival in England produce a family of English Zander.

Mr Dallmer selected 24 two-pounders [900 gm] – twelve male, twelve female [from the lake at Bothkamper], so as to procure as many marriages as possible. ... in London they were met by servants of His Grace the Duke who took charge of the carriers, and conveyed them to the railway station. On arrival [at Bedford] carriages were waiting to convey the fish to the estate [at Woburn Abbey, Bedfordshire], ... where His Grace received them in person at the door of his mansion. 'Are the fish alive?' he called out ... twenty-three were very lively, the twenty-fourth being rather doubtful. His Grace had him cooked for dinner; four fish, two male and two female, were placed in one sheet of water, the others in a lake of about twenty-three acres [9 ha], which was full of small fish but no pike, the gravelly bottom being eminently suitable for Zander.

This excellent result has succeeded ... after this who will say Zander will not bear transporting? Henceforth, England possesses the celebrated aristocratic choice-eating fish, *Luceo perca zandra*, or pike–perch, thanks to His Grace the Duke of Bedford.

Three previous attempts were made by my friend, Alexander Seydell of Stettin, and self to procure Zander from Stettin by sea to London. On the first voyage all the fish died; the second produced three small live Zander. They all died in Mr Frank Buckland's museum at South Kensington. The third voyage nearly proved a shipwreck ... and all the gear ... the large Brighton Aquarium iron carrier included, were tossed into the North Sea.

In 1910, the Duke of Bedford reinforced his stock of Zander with a second consignment of fish from Germany. In 1947, the Leighton Buzzard Angling Club was given about 50 Zander from Woburn Abbey, most of which were liberated in Firbank's Pit at Leighton Buzzard, a few being introduced into the River Ouzel and into an artificial watercourse that runs into the Grand Union Canal near Tring. No Zander were reported from any of these waters until

1950, but thereafter a number were caught on rod-and-line until 1966 when the pit was filled in.

In 1950, the Leighton Buzzard anglers were presented with a further 30 young Zander from Woburn, which they placed in Claydon Lake near Steeple Claydon, south of Buckingham. Three years later these fish were breeding, and became the source of new stock for at least one other water nearby (Fitter, 1959; Lever, 1977).

Wheeler and Maitland (1973) and Cawkwell & McAngus (1976) have documented the extension of the Zander's range from the early 1960s. In the winter of 1959–1960 the lakes at Woburn Abbey were again netted, and 97 Zander were removed; these fish were either released directly into the Ouse Relief Channel of the Great Ouse river system at Stow Bridge, 5 km downstream of Downham Market, or, after being kept for a time in a stock-pond at Hengrave Hall near Bury St Edmunds, Suffolk, their progeny were freed in the Channel in 1963 (two versions of this stocking exist). This planting was a success, and in 1969 it was reported that 'zanders are now prevalent in the Relief Channel between Denver [south of Downham Market] and King's Lynn and also in the cut-off channel upstream of Denver sluice'.

In July 1965, in order to reinforce the existing stock, 100 Zander from Sweden were released in the Relief Channel at Stow Bridge by the Great Ouse River Authority. (Another account states that 500 fish were freed.) There is also evidence of an attempt in 1969 to introduce Zander to 'a private fishery' in the Fenland district of Lincolnshire; this may have taken place at the Landbeach Lakes or in Mepal Pit near Cambridge, where in 1965 it was reported that 'a few pike–perch were put in as fingerlings three seasons ago'. By the early 1970s, Wheeler and Maitland (1973) reported that in the Relief Channel Zander 'are now numerous, breeding and firmly established ... and have shown a rapid growth rate'.

In times of flood the Relief Channel connects with the Great Ouse river system, while a second channel runs into the Rivers Little Ouse and Wissey. In this last river the *Angling Times* of 30 January 1969 reported that 'zander have now been seen ... as far upriver as Stoke Ferry [11 km south-east of Downham Market]'. Wheeler and Maitland (1973) mention the following records of Zander caught in East Anglian waters other than the Relief Channel in 1969 and 1970: the Rivers Cam, Old Bedford and Nene, and at

Brownshill Staunch (1969); in the Great Ouse River, the Middle Level Drain, and at Sutton Gault (1970).

In 1974, it was reported that Zander were 'showing up in good numbers throughout the whole length of the River Delph ... the Old Bedford River ... holds fair numbers ... the Great Ouse itself holds zander ... the advance is getting dangerously close to the Nene system too'. In the following year 'Zander are now being taken in rapidly increasing numbers in the Forty Foot, Sixteen Foot and Pophams Eau. They have also appeared for the first time in the Twenty Foot and Old River Nene as they continue their spread in a westwards direction'.

In the winter of 1973–1974, Zander are known to have been illegally released in a flooded gravel-pit at Maxey, 13 km northwest of Peterborough, Northamptonshire. Some 2 years later, the Anglian Water Authority removed a total of 791 Zander of different ages from this water.

Away from East Anglia, Zander were reported in the late 1960s from the River Mease near Tamworth, Staffordshire; from a lake at Stoke Poges near Slough, Buckinghamshire; and from a small lake near Sevenoaks in Kent, where Zander from Woburn are said to have been stocked in 1952 (Maitland, 1969).

'There seems to be little doubt', Wheeler and Maitland (1973) conclude:

> that the zander will spread through at least the rivers of East Anglia. Its spread has been considerably assisted by the stocking of fishing waters, and the presence of zander in numerous accessible waters will probably lead to misguided attempts to introduce the species elsewhere for angling purposes. It is noticeable that not until it was released into a river system (the Ouse Relief Channel) did the expansion in its range become uncontrolled, and apparently uncontrollable.

This prediction proved prophetical. Although by the mid-1970s the Zander had still only partially colonized the extensive system of interconnecting waterways of East Anglia, it was steadily increasing its range, and had been further illegally introduced by anglers to other localities in England (Linfield & Rickards, 1979; Klee, 1981; Linfield, 1984), including those in the Rivers Severn and Trent water system and the River (Lower) Avon (Hickley, 1987).

By the mid-1980s, Zander had travelled further upstream in the River Cam; had invaded the Reach and Burwell Lodes; had fully colonized the main channels of the Middle Level drainage system, including numerous small interconnecting drains; and had spawned successfully in the River Stour in Suffolk, where they are believed to have arrived as a result of a water transference scheme in the mid-1970s (Cawkwell & McAngus, 1976; Linfield & Rickards, 1979) and where a decade later they had become well established (Linfield, 1984). In 1980, large numbers of Zander fry were removed from Abberton reservoir which draws water from the lower Stour and is listed as of international conservation importance by the Ramsar Convention (Linfield, 1984).

While the spread of the Zander within East Anglia has, since the last stocking in the Great Ouse Relief Channel in 1963, been mainly by natural dispersal, its appearance elsewhere has resulted from illegal releases by anglers, each of which has provided a fresh locus for the natural expansion of the species' range, and this process is continuing in many parts of England (Linfield, 1984). Rickards and Fickling (1979) listed no fewer than 15 counties in England in which Zander were considered to be established, and the presence of fish in waters as far apart as the Midlands Canal system near Coventry and the Rivers Severn and Thames suggested that widespread colonization of much of at least central and southern England may occur, as has already taken place in most of East Anglia (Linfield, 1984).

As mentioned above, by 1976 it had become clear that Zander had been illegally stocked in waters in the catchment of the Rivers Severn and Trent; Hickley (1987) has traced the history of their colonization of this region.

In 1976, Zander were first reported from Coombe Abbey Lake, Warwickshire (where 14 from East Anglia are believed to have been stocked in 1973 and 1974) and a short stretch of the Oxford Canal, both about 30 km east of Birmingham. By 1978, they had spread to the Ashby Canal, and in the following year to gravel pits at Wanlip near Leicester. During the early 1980s, Zander began to spread into the catchment's river systems, where they were first reported from the lower Severn in 1980, the lower Avon region and the confluence with the River Teme (1981), the middle Avon and River Soar (1982), and the middle Severn in 1983. In 1984, isolated reports of Zander came from the Grand Union and Stratford Canals.

Recent records suggest that the Zander is continuing to extend its range in England both within and without its East Anglian stronghold. The achievement of Zander in England is in marked contrast to the remarkable claim by Vooren (1972) that success 'has so far been marginal ... the species has not spread of its own accord'.

Ecological Impact   Linfield (1984) has summarized what is known of the ecological impact of Zander in East Anglia.

Between 1963, when Zander were freed in the Great Ouse Relief Channel, and 1978, reactions to the fish in England were mixed, ranging from ardent enthusiasm (by anglers) to total opposition (by conservationists). Linfield and Rickards (1979), Klee (1981) and Linfield (1981), have described the impact that Zander have had on other fish species in East Anglia.

It was revealed that, in the early 1970s, there was relatively poor recruitment to cyprinid stocks, which resulted in generally lower than expected biomass levels in 1978–1980. Furthermore, stocks of cyprinids in rivers infested by Zander had generally fallen to a lower level than in Zander-free waters, and the extent of the decline appeared to correlate with the period during which Zander had been present. The earliest open waters to be infested, the Relief Channel and the Ely Ouse, had fallen to exceptionally low biomass levels, and comparison of the composition of diminished stocks in those waters occupied by Zander with those as yet uncolonized clearly suggested a reason for degeneration. A very strong 1975 year-class for most cyprinids in East Anglia, especially Roach *Rutilus rutilus*, was subdued in those waters holding Zander, and many young fish were apparently dying. A comparison of the combined Pike *Esox lucius* and Zander biomass with that of cyprinids showed an imbalance in the predator/prey relationship, and the degree of this imbalance seemed loosely related to both the duration of the presence of Zander and the amount by which the cyprinid biomass had declined.

The impact of Zander on fisheries in the Severn/Trent river system will depend on the rate of expansion of the populations and the ability of native species to adapt to the presence of an alien predator. There is anecdotal evidence that in Coombe Abbey Lake, Roach have declined since the introduction of Zander (Hickley, 1987). (See also Kell, 1985.)

Methods of Control   The principal threat of Zander becoming established outside their present range is posed by continued illegal translocations by anglers, about which little can be done. Electro-fishing and chlorination and dechlorination of water have been used to some effect to prevent the further expansion of Zander by natural means (Linfield, 1984).

In contrast to the experience of the Zander in England, elsewhere in Europe it is a valued sport species, as is its homologue in North America, the Walleye *S. vitreum*. Why, then, should the Zander have increased so greatly and have had such a deleterious effect in England? The answer, given by Linfield (1984), seems to be that until the early 1980s English anglers, in contrast to their European or North American counterparts, generally employed a 'catch and release' policy with Zander, thus reducing mortality through angling to a minimum. To counteract this, anglers were encouraged to kill all Zander caught, rather than to return them alive to the water. Since this policy proved ineffective in controlling either the spread or increase of Zander, it has since been abandoned in East Anglia.

Denmark

Welcomme (1981, 1988) and Jensen (1987) say that Zander were introduced to Denmark from Germany between 1879 and 1913, and from Sweden between 1915 and 1936, and that they are now established in lakes and rivers throughout the country. They also occur in brackish lagoons off the Baltic Sea.

Ecological Impact   Dahl (1962) noted that after the introduction of Zander to Danish lakes, changes occurred in the yield of other fish species, including Eel *Anguilla anguilla*, Pike *Esox lucius* and Perch *Perca fluviatilis*.

France

According to Welcomme (1981, 1988), Zander from an unknown source were introduced to France accidentally in 1912 and later deliberately in 1958, and the species is now 'abundant in the Rhône and its tributaries where it is highly appreciated for its taste and sporting qualities'. Lever (1977) records an introduction to Huningue on the Rhine in the early 1890s.

Ecological Impact   Goubier (1975), quoted by Fickling and Lee (1983), poses the question 'Has the zander any negative effects on the resident fish species?' in France. He notes that in the valley of the

Seine, anecdotal evidence implicates the species with the eradication of some bottom-feeding herbivores, but that this is not confirmed by contemporary scientific observations.

Italy

Welcomme (1981, 1988) records the introduction to Italy of Zander from France and eastern Europe between 1964 and 1966 and in 1975, and that the species is 'established in the River Tiber and is considered very successful'.

The Netherlands

The Zander entered Dutch waters in the late 19th century by natural diffusion from the River Rhine in Germany (to which it had been translocated in 1883), and was subsequently assisted to expand its range by human intervention. For example, in 1901 the Heidemaatschappij imported 50 000 eyed ova from a hatchery at Wittingen, Germany, to their hatchery at Vaassen, and more fertilized eggs were acquired from the same source in the following year. Today, Zander are naturalized in slowly flowing rivers and large low lying lakes throughout The Netherlands, being particularly numerous in Lake Yssel where they prey primarily on Smelt *Osmerus eperlanus*; they are the country's second most important commercial species, and their range in The Netherlands is believed to be expanding (Vooren, 1972; De Groot, 1985).

Ecological Impact   In some Dutch waters there appears to be a correlation between the arrival of Zander and a decline in the number of Pike *Esox lucius* (Vooren, 1972).

Spain

Between 1970 and 1979, Zander from France were introduced for sporting purposes to Spain (Elvira, 1995b), where they are only established but are locally abundant in the drainage of the Eastern Pyrenees (Elvira, 1995a).

Translocations

In Europe, Zander have been widely translocated in Germany (especially in Schleswig-Holstein and to Lake Constance, and in the Rivers Elbe, Main, Mosel and Rhine) (Haack, 1893; Mylius, 1894; Lever, 1977; De Groot, 1985); Sweden (Fickling & Lee, 1983); and the Eurasian former USSR (Drapkin, 1968; Fickling & Lee, 1983).

Ecological Impact

Zander translocated to Lake Ymsen in Sweden in 1911 reduced populations of three important prey species, Perch *Perca fluviatilis*, Roach *Rutilus rutilus*, and Bream *Abramis brama*, while in Lake Erken they affected populations of Pike *Esox lucius* and Perch (Fickling & Lee, 1983). Translocated to Lakes Balkash and Alakul in Siberia in the former USSR, the Zander has, according to Roots (1976) who gives no references or further details, 'conflicted with the Balkash perch'.

Asia
Turkey

In 1955, 10 000 young Zander (known locally as *sudak*) were introduced from Austria into each of Lakes Mermere and Egridir in the western Taurus lakes area of the Isparta Province in western Turkey. In Lake Mermere, owing to the warm water and an abundance of food, the Zander flourished, and in the first fishing season (1959–1960) yielded 40 t of fish (Aksiray, 1961).

Ecological and Economic Impact   After the introduction of Zander to Lake Mermere, the populations of several native fish species of low commercial value, such as Eels *Anguilla anguilla* and Bitterling *Rhodeus sericeus*, began to decline, while the economically important Common Carp *Cyprinus carpio* started to increase in weight. There was thus an increase in the total biomass of commercial fish in the lake (Aksiray, 1961; Bard, 1977).

According to Roots (1976), who again gives no reference or further details, 'the cicek, an endemic fish of Lake Egridin [Egridir] in Anatolia, is rare owing to direct predation by the pike–perch or zander, introduced in 1953'. The species referred to is *Acanthorutilus handlirschi*.

Africa
Morocco

Moreau *et al.* (1988) list Zander as having been introduced in 1944 and 1972 from France to Morocco, in some waters of which they became established.

SCIAENIDAE

Bairdiella
Bairdiella icistia

Orangemouth Corvina
Cynoscion xanthulus

Both the Bairdiella and the Orangemouth Corvina are coastal marine drums, the former reaching a length of around 30 cm and the latter about 80 cm and a weight of up to 15 kg.

Natural Distribution

Both these species are natives of the Pacific coast of Mexico, where the Bairdiella ranges north to Almejas Bay, east of Isla Santa Margarita, Baja California Sur, while the Orangemouth Corvina is found from Acapulco northward to the Gulf of California. Both are said to be particularly abundant in the vicinity of Mazatlán in the Gulf of California.

Naturalized Distribution

North America: USA.

North America

United States

Many native and alien species have been transplanted and introduced into the Salton Sea (73.5 m bsl) in Imperial and Riverside counties, California (Walker *et al.* 1961), but only two of the exotics, the Bairdiella and the Orangemouth Corvina, have established breeding populations.

In October 1950, a total of 67 Bairdiellas were introduced to the Salton Sea (Walker *et al.*, 1961; Shapovalov *et al.*, 1981) by the California Department of Fish and Game as a sport fish and as forage for the Orangemouth Corvina (Courtenay *et al.*, 1984), while between 1950 and 1955 around 272 of the latter species were released as a game fish by the same organization (Walker *et al.*, 1961).

Both the Bairdiella and the Orangemouth Corvina

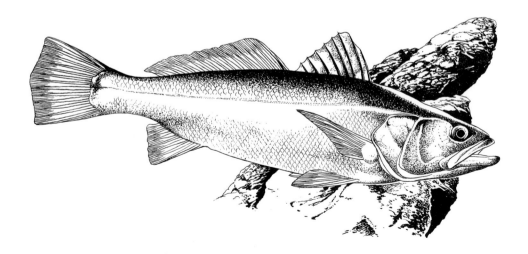

multiplied explosively in the Salton Sea, where Walker *et al.* (1961) recorded each as occurring in millions. Moyle (1976a) listed the Bairdiella as one of the three most dominant fishes in the Salton Sea.

The success of the Bairdiella and the Orange-mouth Corvina (which is today an important game species) is, as Lachner *et al.* (1970) point out, of particular interest because of the nature of the environment in the Salton Sea, the water of which is not only saline and subject to considerable variation in temperature, but also differs from sea water in the rela-

tive amounts of various salts present. After the Salton Sea was formed in 1906 by the diversion of the Colorado River, the existing freshwater fishes gradually died out, and the Common Carp *Cyprinus carpio* (q.v.), which was the most abundant species in 1916, disappeared by 1929 due to the increase in salinity. 'The Salton Sea' wrote Lachner *et al.* (1970), 'is an excellent example of an area where there was good chance to gain and nothing to lose or harm by trial and error introductions.' (See also Courtenay *et al.*, 1991; Courtenay, 1992, 1993.)

CICHLIDAE

Astatoreochromis alluaudi

This small (10 cm) tropical fish is a molluscivorous member of the cichlid species flocks in several of the East African Lakes.

Natural Distribution

Lake Victoria (Kenya, Uganda and Tanzania) and Lakes Edward, Nakachira and Nakivali, Uganda.

Naturalized Distribution

Africa: Cameroun; Central African Republic; Congo; Zaire; Zambia.

Africa

Cameroun; Central African Republic; Congo; Zaire; Zambia

This species has been introduced from Uganda to aquacultural installations and lakes in the above African countries in an attempt to control the snail vectors of schistosomiasis (bilharzia). The dates of introduction were *Cameroun* (1960s), *Central African Republic* (1969), *Congo* (1960), *Zaire* (1960s) and *Zambia* (1971) (Welcomme, 1981, 1988). The fish have established naturally breeding populations both in captivity and in the wild.

Ecological Impact   Jhingran and Gopalakrishnan (1974) recorded reductions in snail populations of between 64 and 98% in those countries to which *Astatoreochromis alluaudi* has been introduced.

Oscar
Astronotus ocellatus

The Oscar has been widely distributed outside South America as a popular tropical aquarium fish. Welcomme (1988) believes that it probably occurs in the wild much more frequently than has been reported.

Natural Distribution

The Orinoco, Amazon and La Plata river systems of South America, including the Parana, Paraguay and Negro Rivers.

Naturalized Distribution

Africa: ? Ivory Coast. **North America:** USA. *Oceania:* Guam (Mariana Islands); Hawaiian Islands.

Africa

Ivory Coast

The status of Oscars which have escaped from aquaria in the Ivory Coast, where they were introduced from Hong Kong in 1976, is uncertain (Welcomme, 1988).

North America
United States

Rivas (1965) recorded the presence of Oscars in canals and other waterways in Miami, southeastern Dade County, Florida, where they had been deliberately released from an ornamental aquarium fish farm in the late 1950s, and where they became established in the Miami Lakes Estate section, in the Little River area, and in the Snapper and Black Creeks canal system (Courtenay *et al.*, 1974). Further introductions in the Miami area were subsequently made by anglers. Hogg (1976a) said that Oscars were established in most canal systems to the north of and including Black Creek, Snake Creek, Miami and Tamiami Canals, and in Coral Gables Waterway. In the late 1970s/early 1980s Oscars began to expand their range in Florida and became naturalized in Dade, Broward, Glades and Palm Beach counties, and probably in Hendry county (Courtenay *et al.*, 1984). Courtenay and Stauffer (1990) recorded them as established and breeding also in Collier and Monroe counties.

Oscars have also been collected but are not naturalized in Massachusetts, Mississippi, Pennsylvania and Rhode Island, all apparently as a result of releases from aquaria (Courtenay & Stauffer, 1990). (See also Courtenay & Robins, 1973; Courtenay *et al.*, 1986, 1991; Courtenay, 1993.)

Oceania
Guam (Mariana Islands)

Best and Davidson (1981), Shepard and Myers (1981) and Maciolek (1984) record the successful establishment of Oscars on Guam in the Mariana Islands.

Hawaiian Islands

In 1952, a small stock of Oscars was imported from the Steinhart Aquarium in San Francisco, California, and bred successfully in tanks of the Fish and Game Department laboratory in Honolulu on Oahu. Sixty-four were subsequently released in Nuuanu reservoir number 2, and in 1958 some of their progeny were translocated to Wahiawa reservoir on Oahu and to Waialua reservoir on Kauai (Brock, 1960). Maciolek (1984) recorded Oscars as established and increasing on both islands. (See also Brock, 1952; Kanayama, 1968; Lachner *et al.*, 1970.)

Unsuccessful Introductions

In 1971, Erdman (1984) found a non-breeding colony of Oscars in a farm pond at Aibonito, Puerto Rico. (See also Wetherbee, 1989.)

McKay (1986–1987, 1989) said that Oscars have been introduced to several urban creeks in Queensland, Australia (such as Enoggera, Maryborough), but that it was doubtful if breeding populations occurred, since they were unlikely to survive winter temperatures.

Tucunare (Pavon; Peacock Bass; Peacock Cichlid)
Cichla ocellaris

The large (up to 50 cm and 2 kg) and voracious predatory Tucunare (Portuguese) or Pavon (Spanish) is renowned both as a game fish and for the quality of its flesh.

Natural Distribution

The River Amazon and its tributaries in South America.

Naturalized Distribution

North America: USA. ***Central America:*** Hispaniola (Dominican Republic); Panama; ? Puerto Rico. ***Oceania:*** Guam (Mariana Islands); Hawaiian Islands.

North America

United States

Moe (1964) and Ogilvie (1966), quoted by Lachner *et al.* (1970), say that in 1964, 10 000 juvenile Tucunares from Venezuela were released around Fort Lauderdale, Dade County, Florida, by the Florida Game and Freshwater Fish Commission, where they were apparently killed during the cold winter of 1964–1965 (Courtenay & Robins, 1973). Courtenay (1992) says that Tucunares were 'recently' introduced to extreme southeastern Florida, where they have become established (Courtenay *et al.*, 1991; Courtenay, 1993).

In the USA Tucunares have also been taken in the wild, but are not established in Texas (Courtenay *et al.*, 1991).

Ecological Impact According to Courtenay (1992), in Florida the Tucunare 'is reported to prey mostly, but not exclusively, on other introduced cichlid fishes. It was, in part, introduced to feed on previously introduced cichlids that had become dominant, but instead probably has added additional "cichlid pressures" to native fishes'.

Central America

Hispaniola (Dominican Republic)

In 1976, Tucunares from Colombia were introduced to the Dominican Republic on Hispaniola, where they have become naturalized in rivers, dams and lagoons (Welcomme, 1981, 1988). This introduction is not mentioned by Wetherbee (1989).

Ecological Impact Although of uncertain geographic distribution, Tucunares in the Dominican Republic are said to have adversely affected native fish populations (Welcomme, 1981, 1988).

Panama

Zaret and Paine (1973) and Zaret (1974) have documented the introduction of the Tucunare to Panama.

In the first decade of the 20th century, a dam was built in the lower reaches of the Chagres River, which resulted in the flooding of the river basin and the formation of an extensive (42 315 ha) though shallow freshwater lake, Lago Gatún, which is now an integral part of the Panama Canal.

In 1965, a local businessman, with the approval of the Panamanian government, arranged for the importation of 100 Tucunare fingerlings (Zaret & Paine, 1973) or only 60 (Zaret, 1984) from the fish culture station at Buga, Colombia, to Panama. These fish he placed in a small dam, with the intention of providing both food and sport for his employees and for local residents. During the rainy season this impoundment, formed by the damming of the

Quebrada Ancha creek, overflows into the Río Gatuncillo, a small tributary at the northern end of the Chagres River. It is surmised that at the time of the rains in late 1966, some Tucunares gained access to this tributary where they spawned, and from where by the following year they had spread the 8 km to the main Chagres River. Here they flourished, and by early 1970 had dispersed down the Chagres, and catches were reported near Gamboa, where the river debouches into the eastern end of Lago Gatún. By June 1970, Tucunares were observed near the small settlement of Frijoles, having apparently spread along the eastern shore of the lake. In the spring of the following year, Tucunares appeared off Barro Colorado Island opposite Frijoles, from where by 1972 they had spread northwest to the township of Gatún near the northern end of the lake, where their presence may have been partially due to translocations by local anglers. By the mid-1970s, Tucunares had spread into the Trinidad Arm in the extreme southwest of the lake, and the whole of Lago Gatún had been overrun.

Since their initial introduction in 1965, Tucunares have been frequently, and often successfully, introduced to rivers and lakes throughout the isthmus of Panama; specifically, Welcomme (1981, 1988) mentions the Bayano hydroelectricity reservoir (in 1972) and Lakes Alajuela and La Yaguada.

Ecological Impact   Zaret and Paine (1973) and Zaret (1974) have analysed the ecological effect of Tucunares on the native fish fauna of Lago Gatún. They found that the introduction of this piscivore has caused dramatic alterations in the biotic community of the lake which has affected the entire lake ecosystem, including species populations from tertiary consumers down to primary consumers, and probably also primary producers, and thus the entire lake's food web.

As Tucunares extended their range in the lake they had a devastating effect on indigenous fish species. Off Barro Colorado Island, Zaret and Paine (1973) found that six of the eight previously abundant species – *Astyanax ruberrimus, Roeboides guatemalensis, Aequidens coeruleopunctatus, Gobiomorus dormitor, Gambusia nicaraguagensis* and *Poecilia mexicana* – had apparently been effectively eliminated, and the population of a seventh, *Melaniris chagresi*, appeared to have been reduced by 50%.

'The apparently drastic perturbation produced in the Gatun Lake community, due solely to the introduction of this single species of top-level predator, is', Zaret (1974) claims, 'especially striking when contrasted with the relative absence of these effects in the Great Lakes of East Africa following the introduction of the predatory Nile Perch, *Lates niloticus*, [q.v.], which some authors predicted would create problems'. This claim is not borne out by events since, as has been shown, the Nile Perch has had a marked impact on the fish fauna of Lake Victoria in East Africa since its introduction in 1954. Indeed, the effect of Tucunares on the fishes of Lago Gatún is not dissimilar to that of Nile Perch in Lake Victoria.

'Why', Zaret (1974) continues, 'since Gatun Lake has other fish predators present, has the *Cichla* population been able to take over so completely in a way that the others have not?'. The answer, Zaret concluded, lay in differing hunting strategies. The two principal indigenous predators, *Hoplias microlepis* and *Gobiomorus dormitor*, both depend on their cryptic coloration to hide from their intended prey, and emerge suddenly from 'ambush' to capture unsuspecting victims. Characteristically, these species are primarily crepuscular hunters. Brightly coloured Tucunares, on the other hand, are strictly diurnal 'pursuit' predators, prepared to chase their prey for up to 10 m or more (*cf.* the hunting strategies of the Leopard *Panthera pardus* and Cheetah *Acinonyx jubatus* respectively). Furthermore, non-breeding Tucunares hunt in schools of up to 15 individuals, instead of singly like the native species. The success of Tucunares over indigenous predators can thus be ascribed to the inability of native prey species to combat new hunting techniques to which they are not preadapted.

The impact of Tucunares on fish populations in Lago Gatún has, as pointed out by Zaret and Paine (1973) and Zaret (1974), in turn had a knock-on effect at other trophic levels. Thus populations of piscivorous birds have declined, and the elimination of important planktivores such as *Melaniris chagresi* has resulted in alterations in the zooplankton and phytoplankton communities. Specifically, populations of such tertiary consumers as Tarpon *Tarpon atlanticus* (which regularly enter Lago Gatún via the Panama Canal locks and, once in the lake, consume large numbers of *M. chagresi*), Black Terns *Chlidonias niger* (which follow Tarpon hunting *Melaniris* and feed on those which leap from the water to escape Tarpon), kingfishers and herons, all of which feed on small fishes, have declined in numbers. There also appears to have been a revival in the populations of malaria-causing mosquitoes in the lake, as a result of the decline or elimination of such insectivores as *A. ruberrimus*, *R. guatemalensis*, *M. chagresi* and *G. nicaraguagensis*, and a change in the predominant type of malaria in the Canal Zone from *Plasmodium vivax* as the disease agent to the more dangerous *P. falciparum*.

It has been suggested that, where planktivores impact on the formation of zooplanktonic communities, corresponding alterations also occur at primary producer level, and that this may be a principal control of the abundance of algae. In Lago Gatún, there does appear to be a lower crop of phytoplankton in those areas infested with Tucunares than where Tucunares are absent.

Zaret (1974) pointed out that although in most areas of the lake, Tucunares had eradicated native fishes, where the Chagres River flows into the lake some of the latter had survived. It appears, Zaret concluded, that the increase in water flow together with a decrease in visibility due to the resulting turbidity during the rainy season, allows the reintroduction and re-establishment of fishes from neighbouring tributaries. During the following 9 months of the dry season, numbers of these fishes are gradually reduced through predation by Tucunares. Since, however, the predator is not well established in the tributaries, the river systems serve as a source of further supplies of prey fishes. This annual resurgence of the prey species provides an element of long-term stability for the whole system, and highlights the ecological importance of the Chagres River and its inflowing tributaries on the lake ecosystem.

Since the reports by Zaret and Paine (1973) and Zaret (1974), it appears that their findings may have been unduly pessimistic, and that even more prey species than had been suspected have survived in refugia at the mouths of inflowing streams. The position is further complicated by the introduction of other alien species, such as Common Carp *Cyprinus carpio* (q.v.) and Nile Tilapia *Oreochromis niloticus* (q.v.), which provide Tucunares with an alternative prey (Welcomme, 1988). It may thus be that the return to an ecological equilibrium in Lago Gatún, as predicted by Zaret and Paine (1973), whereby a balance is struck between predator and prey, has been realized. (See also Zaret, 1982, 1984; Swartzmann & Zaret, 1983; Payne, 1987; Miller, 1989.)

Puerto Rico

On 27 January 1967, a consignment of 200 Tucunare fingerlings arrived in San Juan, Puerto Rico, from the fish culture station at Buga, Colombia, as a governmental gift. Only about 50 arrived alive, and of these only 30 survived the 4-hour journey by road to Maricao. Spawning first occurred in late May 1968, but the fry soon disappeared, and the first successful spawning did not take place until the following August. Although it was found that self-maintaining populations could occasionally be established for short periods, long-term establishment in farm ponds was not achieved, and this stocking programme was abandoned. In reservoirs, such as those at Toa Vaca and La Plata, the fish tended to fare better, but still did not come up to expectations, and their numbers were much reduced by uncontrolled fishing (Erdman, 1984). Welcomme (1981, 1988), however, says that Tucunares are established and breeding in the wild in Puerto Rico, where they are 'regarded as an excellent sport fish'. (See also Wetherbee, 1989.)

Oceania

Guam (Mariana Islands)

Best and Davidson (1981) and Maciolek (1984) list the Tucunare as naturalized on Guam in the Mariana Islands, where it was first introduced in 1966 (Courtright, 1970).

Hawaiian Islands

In 1961, a shipment of Tucunares from New York, USA, was despatched to the islands of Kauai and Oahu, where only two individuals arrived alive. (Welcomme (1981) says incorrectly that this introduction took place in 1966.) From these are descended the breeding populations which are today abundant in reservoirs on the two islands, where the Tucunare is a popular sport fish. (See also Kanayama, 1968; Lachner *et al.*, 1970; Randall, 1980; Maciolek, 1984.)

Unsuccessful Introduction

In 1970, Tucunares were introduced to control stunted populations of *Tilapia* spp. and *Sarotherodon* spp. in Kenya, apparently without success (Welcomme, 1981, 1988).

Black Acara (Two-spotted Cichlid)
Cichlasoma bimaculatum

The Black Acara is a favourite tropical aquarium species which has been widely distributed throughout the world. Its high water-temperature requirements render its establishment in temperate countries, which are the main importers of tropical fishes, unlikely (Welcomme, 1988).

Natural Distribution

South America, from eastern Venezuela, Trinidad, Guyana, Surinam, French Guiana, Brazil, ? Ecuador, Bolivia, Paraguay, and Uruguay to northern Argentina.

Naturalized Distribution

North America: USA.

North America

United States

The Black Acara was one of the earliest fish species to be cultivated in Florida, possibly in the 1930s, and was an important element of the trade in aquarium fish until the late 1950s, when the arrival of jet cargo aircraft facilitated the importation of more varied and colourful cichlids from further afield. There may have been accidental escapes from fish-rearing facilities, but there are known to have been several deliberate releases to get rid of surplus stock (Courtenay & Stauffer, 1990).

Rivas (1965) first reported Black Acaras (which he misidentified as *Aequidens portalegrensis*, a common error, according to Courtenay *et al.* (1974) in aquacultural literature) in the wild in Florida in the early 1960s. This misnomer was repeated by Lachner *et al.* (1970), who reported them to be well established in the canal system along the southern coast of Florida, and by Kushlan (1972), who recorded their discovery in an Alligator *Alligator mississipiensis* (*sic*) pond in the Big Cypress Swamp of Monroe County, 80 km west of Miami – the first record of the species' penetration into the swamps and marshes of the interior.

Courtenay *et al.* (1974), who reported Black Acaras to be established from west of Lantana, Palm Beach County, south through Broward and Dade Counties, and west into northeastern Monroe County, described them as having 'the widest distribution of all exotic fishes in southeast Florida', and as the sole alien fish species then to be established in the Everglades National Park. In some of the main canals west of Fort Lauderdale, the Black Acara was the dominant species, comprising up to 64% of the biomass of all fishes present. In one canal, near the site

of the suspected original introduction in Broward County, the Black Acara accounted for some 80% of the total fish biomass.

Hogg (1976a,b) said that the Black Acara was 'the most abundant and widespread cichlid in Dade County and has an extensive range also in Palm Beach, and adjoining Broward and Collier counties'. Although it was quite abundant in the northern two-thirds of Dade County, where it was first recorded in 1972, the species had apparently been unable to colonize canals in the southeast of the county. Many of the waterways in that locality are prone to constant infiltration by sea water, and support mainly brackish-water communities. Black Acaras in captivity have been found to be relatively intolerant of saline conditions compared to some other cichlid species.

Courtenay *et al.* (1984) recorded the naturalization of Black Acaras in Florida in Broward, Dade, Collier, Hendry, probably Glades, Monroe and Palm Beach counties. Courtenay *et al.* (1986) extended this range to include waters bordering State Road 84 and Naples, and possibly Lake Okeechobee. To this list Courtenay and Stauffer (1990) added possibly Lee County. (See also Courtenay & Robins, 1973; Courtenay & Hensley, 1979; Courtenay *et al.*, 1991; Courtenay, 1993.)

Ecological Impact   As early as 1972, Kushlan was expressing concern at the appearance in the wild of

Black Acaras within 16 km of the Everglades National Park, where he feared that they might compete with native species and have a detrimental impact on populations of aquatic organisms in the wetlands of southern Florida. Hogg (1976a,b) reported seeing male and female introduced Spotted Tilapia *Tilapia mariae* (q.v.) defending their nest and young in 1974 from intrusions by *Cichlasoma bimaculatum.*

The Black Acara is a very territorial species, reacting aggressively to potential intruders. It tends to make use of spawning sites similar to those of indigenous centrarchids, with which it thus comes into resource exploitation competition, and where such sites are in short supply it utilizes marginal habitats such as small ledges in soil borrow pits. Its high fecundity, the degree of care of its young shown by parent fish, a lengthy spawning season compared to that of native species, and its adaptability in the choice of spawning sites, have together enabled the Black Acara to establish and maintain itself in high population densities (Courtenay *et al.*, 1974). Taylor *et al.* (1984) attribute the success of Black Acaras in becoming established in Florida in part to the fact that they are less prone to desert their nests than, for example, native sunfishes *Lepomis* spp., and their eggs are thus not so susceptible to predation. (See also under *C. meeki*).

Midas Cichlid
Cichlasoma citrinellum

Although the Midas Cichlid is not one of the most popular of warmwater aquarium fishes, it has been quite widely distributed by aquaculturists. In spite of its relative tolerance of low temperatures, its establishment in the waters of temperate countries, which are the principal importers of tropical fishes, is unlikely (Welcomme, 1988).

Natural Distribution

The Atlantic slope of Nicaragua and Costa Rica.

Naturalized Distribution

North America: USA.

North America
United States

Courtenay *et al.* (1984, 1986) reported the discovery in May 1981 by members of the Non-Native Fish Research Laboratory of the Florida Game and Fresh Water Fish Commission of Midas Cichlids in Black

Creek Canal northeast of Homestead, in Dade County, Florida, where they occurred in 4.8 km of that waterway. By late 1982/early 1983 they had spread, via the interconnected canal system, to other nearby channels, and in 1984 were recorded within the Everglades National Park, Dade County. They were believed to be derived from the deliberate release or escape of surplus aquarium stock. Welcomme (1988) said they had been recorded 'from one canal' only. (See also Courtenay & Stauffer, 1990; Courtenay *et al.*, 1991; Courtenay, 1993.)

Chanchito (Chameleon Cichlid)
Cichlasoma facetum

The aggressive Chanchito is, for a warmwater species, unusually tolerant of low water temperatures.

Natural Distribution

Brazil, Paraguay, Uruguay and northern Argentina, especially the Mar de Solís and the Rivers Paraná and Uruguay in Argentina.

Naturalized Distribution

Europe: Portugal; Spain. *South America:* Chile.

Europe
Portugal

Maitland (1977) says the Chanchito 'has apparently been successfully introduced to southern Portugal', where his map indicates the eastern Algarve.

Spain

Between 1980 and about 1986, *C. facetum*, probably from Brazil, was introduced by aquarists to Spain (Elvira, 1995b), where it occurs in the wild only in the basin of the River Guadiana (Elvira, 1995a).

South America
Chile

According to De Buen (1959), who gives no further details or date, the Chanchito is naturalized in some central Chilean waters as a result of introductions for ornamental purposes from Argentina.

Jaguar Guapote (Green Guapote; Guapote Tigre)
Cichlasoma managuense

The Jaguar Guapote is one of the principal local fishes reared in aquaculture in Central America where, as a voracious predator, it is used in lakes for controlling young tilapias and other forage species, and where in consequence it is now of value in commercial fisheries. Its intolerance of low water temperatures makes it unlikely to become established in the wild in temperate countries which are the main importers of tropical fishes (Welcomme, 1988).

Natural Distribution

Endemic to Nicaragua and Costa Rica.

Naturalized Distribution

North America: USA. *Central America:* Cuba; El Salvador; Guatemala; Honduras; Panama.

North America

United States

In 1988, Jaguar Guapotes that had been released by aquarists were found to be established in a thermal spring in St George, Washington County, Utah (Courtenay & Stauffer, 1990), where their continued presence was confirmed by Courtenay *et al.* (1991) and Courtenay (1993); the former list them as having also been caught, but not established, in Ontario, Canada.

Central America

Cuba; El Salvador; Guatemala; Honduras; Panama

Welcomme (1981, 1988) lists introductions of Jaguar Guapotes in Central America as shown below.

Translocation

Villa (1971) reported the translocation in the mid- to late-1960s of Jaguar Guapotes from Lago Managua to Lago Xiloá, both in Nicaragua, although the species was already present in the latter water.

DATE	COUNTRY	ORIGIN	PURPOSE	REMARKS
1956	Honduras	Nicaragua	Aquaculture	Established in Tegucigalpa D.C. Under study for polyculture
1958	El Salvador	Nicaragua	Aquaculture and repopulation of lakes	Established in almost all lake and rivers. Used as a biological control in aquaculture ponds. Has displaced local predators
1958	Guatemala	El Salvador	Aquaculture	Widespread in ponds, rivers and lagoons. Highly appreciated for its flavour and quality. Used as a mild predator
1972	Panama	Costa Rica	Aquaculture	Established in some small reservoirs. Useful in control of forage species
1983	Cuba	Nicaragua	Aquaculture	Reproducing

Firemouth Cichlid
Cichlasoma meeki

The Firemouth Cichlid is a popular warmwater species that has been widely distributed around the world by aquarists. Although relatively resistant to cold temperatures, it requires water of at least 24°C in which to breed successfully.

Natural Distribution

The Atlantic Slope drainages from the Río Tonala in Veracruz and Tabasco States, Mexico, south through the Yucatán peninsula, southern Belize, and the upper Usumacinta basin in Guatemala.

Naturalized Distribution

North America: USA. ***South America:*** Colombia. ***Oceania:*** Hawaiian Islands.

North America

United States

Hogg (1976a) reported the establishment of Firemouth Cichlids in Comfort Canal in Miami, Florida, and westward into several smaller waterways south of the Tamiami Canal, Dade County. Although he said that this was the first record of the species in any of the state's open waters, Courtenay *et al.* (1974) had previously mentioned the presence of apparently non-breeding Firemouth Cichlids in a canal west of Lantana, Palm Beach County, close to a fish farm from which they had escaped, as well as the establishment of a reproducing population in a limestone sinkhole on private property in South Miami, Dade County.

Courtenay *et al.* (1984, 1986) say that although the status of the population in the sinkhole was unknown, that in Palm Beach County had died out. Firemouth Cichlids were then known to be naturalized in Comfort Canal, and probably in the interconnecting Tamiami and Snapper Canals, and had been collected from a channel in southeastern Dade County. A large and well-established population in flooded soil borrow pits south of Fort Lauderdale, Broward County, had been eradicated by members of the Florida Game and Fresh Water Fish Commission in July 1981. The populations in the canal systems in Dade County are believed to have originated in escapes or release by aquarists, while that in Broward County seems to have been the result of a deliberate release (Courtenay & Stauffer, 1990).

Taylor *et al.* (1984) attribute the success of Firemouth Cichlids in Florida in part to the fact that they are less prone to desert their nests than, for example, native sunfishes *Lepomis* spp., and their eggs are thus not so susceptible to predation. (See also under *C. bimaculatum.*)

Minckley (1973) reported the capture of a single Firemouth Cichlid from a canal in Mesa, Maricopa County, Arizona, where, *pace* Welcomme (1988), the species has never been established. (See also Shafland, 1979; Courtenay *et al.*, 1991; Courtenay, 1993).

South America

Colombia

Welcomme (1988) records the introduction for ornamental purposes of Firemouth Cichlids to Colombia, where he says they are reproducing but gives no further details.

Oceania

Hawaiian Islands

Brock (1960) said of the Firemouth Cichlid only that it was 'established in Nuuanu Reservoir [on Oahu] by a release in 1940. Reported to occur also in a drainage canal in the McCully district of Honolulu', also on Oahu. Maciolek (1984) confirmed its marginal success, but reported that the species' hold on Oahu was 'tenuous'. (See also Brock, 1952 and Kanayama, 1968.)

Convict Cichlid (Zebra Cichlid)
Cichlasoma nigrofasciatum

The Convict Cichlid is a favourite fish among tropical aquarists, by whom it has been widely distributed outside Central America. Like many other cichlids, its high water-temperature requirements make its establishment in temperate countries, which are the main importers of tropical fish, improbable, except in naturally or artificially heated waters (Welcomme, 1988).

Natural Distribution

The Pacific coast drainages of Central America from Guatemala to Costa Rica, and the Atlantic Slope drainages of Costa Rica.

Naturalized Distribution

North America: USA. ***Australasia:*** Australia. ***Oceania:*** ? Hawaiian Islands.

North America

United States

In March 1961, a breeding population of Convict Cichlids was discovered in Rogers Spring, near the Overton Arm of Lake Mead, in Clark County, Nevada (Deacon *et al.*, 1964). Hubbs and Deacon (1964) reported the discovery in June of that year of a second population, 145 km north–northwest in Ash Spring. Subsequently, a third population was found in the outflow of Crystal Spring, 8 km further north (Courtenay & Deacon, 1982, 1983). These two latter sites are both in the Pahranagat Valley, near Alamo, in Lincoln County. The fish are believed to have been spread deliberately by man from Rogers Spring to Ash Spring, and thence to have dispersed naturally (or again possibly by man) to Crystal Spring through an intermittently-flowing connecting stream (Hubbs & Deacon, 1964).

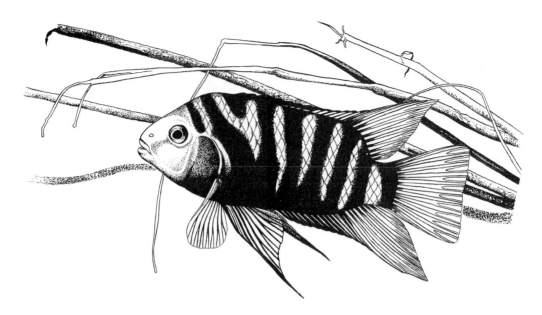

In September 1985, Courtenay *et al.* (1987) collected 19 Convict Cichlids from an established population in Barney Hot Spring, in the Little Lost River Valley, 67 km north–northwest of Howe, in Custer County, southern Idaho.

Courtenay (1993) lists the Convict Cichlid as also naturalized in Wyoming, but gives no further details.

All these populations are said to be derived from the deliberate release of unwanted aquarium pet fish (Courtenay *et al.*, 1986), originally imported from Guatemala (Welcomme, 1988). (See also Lachner *et al.*, 1970; Courtenay *et al.*, 1984.)

Ecological Impact   Deacon *et al.* (1964) voiced their concern about the possible impact of Convict Cichlids in Rogers Spring on the sport fishery in neighbouring Lake Mead, and listed local races of such species as *Pantosteus intermedius*, *Gila robusta*, *Rhinichthys osculus* and *Crenichthys baileyi* as among those that might be adversely affected by introduced exotics; a fifth native species, *Lepidomeda altivelis*, had apparently already been eradicated.

The prophecy by Deacon *et al.* (1964) was soon to be fulfilled, at least in the case of one species, though not in the way that might have been anticipated. During 1966–1967, the native endemic Hiko White River Springfish *Crenichthys baileyi grandis* was eradicated from its type locality, Hiko Springs, by introduced exotics. This federally listed endangered fish was reintroduced into Hiko Springs early in 1984; shortly thereafter, Convict Cichlids were found for

the first time in the springs, to which Baugh *et al.* (1985) and Courtenay and Stauffer (1990) believed they had been maliciously transferred from neighbouring Crystal Spring in a deliberate attempt to prevent the re-establishment of the native species.

Unsuccessful Introductions   Populations of Convict Cichlids previously established in Florida (Rivas, 1965), Arizona (Minckley, 1973), and Alberta, Canada (McAllister, 1969) have all since died out (Courtenay *et al.*, 1991).

Australasia

Australia

In 1978, Convict Cichlids (and Spotted Tilapia *Tilapia mariae* (q.v.)) were taken in eel-nets from the cooling-water discharge of the Hazelwood power station near Morwell, Victoria (Cadwallader *et al.*, 1980), and subsequently also from the Eel Hole Creek downstream, where water temperatures are around 10°C above ambient creek temperatures. Although heated water from Eel Hole Creek flows into the Morwell River, the latter does not support any exotic species (Arthington, 1986). The fish appear to be confined to areas with aquatic vegetation to provide shelter (Merrick & Schmida, 1984), and are believed to have originated in the release of unwanted aquarium stock (Allen, 1989). (See also Courtenay *et al.*, 1984; McKay, 1986–1987, 1989; Clements, 1988; Arthington, 1989, 1991.)

Oceania

Hawaiian Islands

Eldredge (1994) lists the Convict Cichlid as established on the island of Oahu since 1983.

Jack Dempsey
Cichlasoma octofasciatum

The aptly named pugnacious Jack Dempsey is another tropical fish popular with warmwater aquarists, by whom it has been widely transported around the world, but which, because of its high temperature requirements, is unlikely to survive in the wild in those temperate countries to which it has been most frequently introduced (Welcomme, 1988).

Natural Distribution

The Atlantic Slope drainages from the Río Chachalacas basin, Veracruz, Mexico, south through the Yucatán peninsula to the basin of the Río Ulua in Honduras, Central America. Some authors extend the species' range south through Costa Rica to the basins of the Río Negro and the Amazon in South America.

Naturalized Distribution

Asia: ? Thailand. **North America:** USA. *Australasia:* ? Australia.

Asia

Thailand

According to Piyakarnchana (1989), 'the Jack Dempsey (*Cichlasoma biocellatum*) was first imported from South American countries (Paraguay, Bolivia and Argentina) and in the 1950s reintroduced from Hong Kong and Japan. It breeds successfully in the natural habitats of Thailand'.

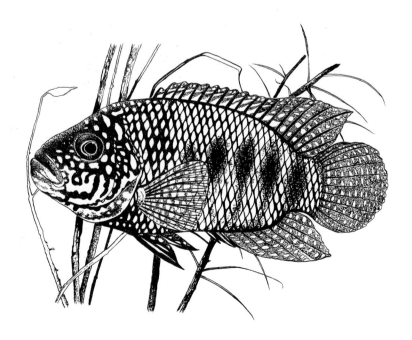

North America

United States

Courtenay *et al.* (1974) recorded the establishment of the Jack Dempsey in a canal west of Lantana in Palm Beach County, Florida, and in a number of roadside ditches in Ruskin, Hillsborough County, as a result of accidental escapes from neighbouring tropical fish farms. Hogg (1976a,b) reported the species' spread to an 8-km stretch of the Snapper Creek Canal and to Black Creek, both in Dade County. The population west of Lantana has since been eradicated, as has another (by the Florida Game and Fresh Water Fish Commission) formerly established in a rockpit in Levy County (Levine *et al.*, 1979; quoted by Courtenay *et al.*, 1984).

Dial and Wainright (1983) reported the naturalization of the Jack Dempsey in four Florida counties; in ditches on the campus of the University of Florida at Gainesville, Alachua County; in canals near the Satellite Beach Civic Center and in others between Satellite Beach and Canova Beach, Brevard County; and in the sites in Dade and Hillsborough Counties. The population in Brevard County is believed to result from releases of unwanted pet fish. Courtenay (1993) gave the date of the earliest introductions as the 1960s. (See also Courtenay & Robins, 1973; Courtenay *et al.*, 1986, 1993; Courtenay & Stauffer, 1990.)

Australasia

Australia

In March 1979, a single Jack Dempsey was caught in an eel-net in the cooling-water discharge of the Hazelwood power station near Morwell, Victoria, where 'there is no evidence to suggest that a self-maintaining population has become established' (Cadwallader *et al.*, 1980). Since then, reports on the species' status in Victoria are confusing and contradictory. Whereas McKay (1984) said it 'may not have become established', Merrick and Schmida (1984), inaccurately quoting Cadwallader *et al.* (1980), say 'it has established wild populations in a power station cooling pond in Victoria'. Allen (1989) records the species as 'not well established', while McKay (1989) again incorrectly quoting Cadwallader *et al.* (1980), says it is 'reported to be established (now declining population?) in thermal waters of a power station in Victoria'. Finally, Arthington (1991) lists the Jack Dempsey as among those 'introduced freshwater fish that have established self-maintaining populations in Australian inland waters'.

Mayan Cichlid (Mojarra Criolla)
Cichlasoma urophthalmus

The Mayan Cichlid is another species popular with warmwater aquarists, who have transported it to many countries. Like most other cichlids, it is unlikely to become established in the cool waters of most of the countries to which it has been introduced (Welcomme, 1988).

Natural Distribution

The Atlantic Slope, from the Río Coatzacoalas basin in Veracruz, Mexico, and the Río Usumacinta, south through the Yucatán peninsula (including Isla Mujeres and Isla Cozumel) to Nicaragua.

Naturalized Distribution

North America: USA.

North America

United States

In 1984, specimens of the Mayan Cichlid were collected in Snook Creek, Joe Bay, Everglades National Park, Dade County. Florida (Courtenay *et al.*, 1986), where the then status of this euryhaline cichlid was unknown. Courtenay and Stauffer (1990) said that it had probably been deliberately released by aquarists.

The species' present disjunct range in the park indicates two release sites, one of which is extremely saline, in both of which it has become established (Loftus, 1987, 1989; Courtenay & Stauffer, 1990). Its continued presence is confirmed by Courtenay *et al.* (1991) and Courtenay (1993).

Translocation

Contreras and Escalante (1984) say that in 1968 what appears, from a photograph, to be atypical *C. urophthalmus*, derived from the wild in Tabasco, Mexico, were translocated to the Río Papeloapan near Temazcal, where their present status is unknown.

Allied Species

According to Welcomme (1981, 1988), in 1960 the Mojara Azul *Cichlasoma guttulatum*, a native of southern Mexico and Guatemala, was introduced from the latter country to try to populate some lakes in neighbouring El Salvador, where it became established in one lake only in which its numbers are declining.

The same author says that *Cichlasoma motaguense*, which is native to the Amazon basin, has been introduced for aquacultural purposes from Brazil to Guatemala, where it is 'reproducing in ponds', though whether in the wild or only in aquaculture is not stated.

Courtenay *et al.* (1991) and Courtenay (1993) list the Blue-eyed or Cutler's Cichlid *Cichlasoma spilurum* as having become successfully naturalized in the Hawaiian Islands, where Eldredge (1994) says it was first reported in 1984 and that it occurs solely on the island of Oahu.

Pearl Spot (Red Tilapia)
Etroplus suratensis

This tropical brackish cichlid readily adapts to living in fresh water.

Natural Distribution

Brackish waters of Sri Lanka.

Naturalized Distribution

Asia: India; Malaysia.

Asia

India

According to Welcomme (1981, 1988), Pearl Spots from Sri Lanka were in 1950 'very successfully transferred to fresh water [in India], where they fill a vacant niche'. They have been used both for rearing in aquaculture and for stocking lakes.

Malaysia

In 1975, the Pearl Spot was introduced as a potential source of food to Sarawak and Sabah, where it is established in the wild in small numbers but has a slow rate of growth which makes it uncompetitive with other faster-growing species (Ang *et al.*, 1989).

Unsuccessful Introductions

Welcomme (1988) says the Pearl Spot was introduced from Malaysia to Indonesia in 1979, where it is reproducing only 'artificially'. Muhammad Eidman (1989) does not mention this introduction. The species has also been imported to the island of Mauritius, with uncertain results (Welcomme, 1981, 1988).

Translocations

Since 1910, Pearl Spots have been successfully translocated to freshwater rivers and reservoirs within Sri Lanka, where the species forms the basis for 'important [commercial] catches in certain reservoirs' (De Silva, 1987).

Redstriped Eartheater
Geophagus surinamensis

A large tropical aquarium species that is also caught by artisanal fisheries (Welcomme, 1988).

Natural Distribution

The Guianas and the basin of the River Amazon in Bolivia, Brazil, Colombia and Peru.

Naturalized Distribution

North America: USA.

North America

United States

In 1982, a population of Redstriped Eartheaters was found by staff of the Non-Native Fish Research Laboratory of the Florida Game and Fresh Water Fish Commission to be established in Snapper Creek Canal, Dade County (Courtenay *et al.*, 1984, 1986).

Although the species was not collected from the Snapper Creek Canal system between 1983 and 1986, its continued survival there was confirmed in 1989 (Courtenay & Stauffer, 1990). The original stock is assumed to have escaped from a nearby fish farm. As is the case with other cichlid species naturalized in southern Florida, it has the potential considerably to expand its distribution (Courtenay *et al.*, 1986). (See also Courtenay *et al.*, 1991; Courtenay, 1993.)

Allied Species

McKay (1986, 1989) says that the Pearl Cichlid *Geophagus brasiliensis*, which is endemic to eastern Brazil, has been 'introduced to the Bajool area, Queensland, and in an ornamental pond, Rockhampton', in Australia, where the species' status is unclear. The Pearl Cichlid has also occurred, but is not known to be established, in natural waters of Florida, USA (Courtenay *et al.*, 1991).

African Jewelfish
Hemichromis bimaculatus/letourneauxi

This small tropical cichlid is, as its name would imply, a species popular with warmwater aquarists.

Natural Distribution

Rivers and lakes of west Africa, from the basins of the Rivers Chad and Nile south to the River Congo.

Naturalized Distribution

North America: ? Canada; USA. *Australasia:* Australia. *Oceania:* ? Hawaiian Islands.

North America

Canada

In 1976, African Jewelfish were found to be established in the marshy outflow area below Cave and Basin Hot Springs in the Banff National Park, Alberta, where they are believed to have been released by local aquarists. Although Crossman (1984) and Nelson (1984) recorded their presence there in 1981, and Courtenay *et al.* (1986) confirmed their continued survival, Welcomme (1988) says they have since disappeared. (See also Crossman, 1984.)

United States

Rivas (1965) recorded the establishment of African Jewelfish in the Hialeah Canal in the Miami River system, Dade County, Florida. Lachner *et al.* (1970) and Courtenay *et al.* (1974) found them to be abundant in waterways along the western perimeter of Miami International Airport, and believed that they might also be established in Eureka Springs, Hillsborough County. Hogg (1976a,b) said that they had recently spread southward into the Comfort Canal, the channelized South Fork of the Miami River, and that in the vicinity of the airport the Jewelfish ranked second in abundance among naturalized exotics only behind the Black Acara *Cichlasoma bimaculatum* (q.v.). Courtenay *et al.* (1984) recorded the probable establishment of Jewelfish in a canal east of Goulds, and also possibly in Snapper Creek, north of the Tamiami Canal, and said that they had been collected from, but were not known to be established in, a canal close to an aquarium fish farm near Micco in Brevard County. Courtenay and Stauffer (1990) believed that the original stock might have escaped from aquaculturists alongside the canal system, and that the species might also have been deliberately released, while in transit, from holding facilities at the airport, which is a major port of entry for aquarium fishes to the USA (Courtenay *et al.*, 1986). See also Courtenay & Robins, 1973; Courtenay *et al.*, 1991; Courtenay, 1993.)

Australasia
Australia

McKay (1986, 1987) says that the African Jewelfish has escaped from ornamental fish ponds in Cairns on the coast of northern Queensland, where it has become 'established in urban drains'.

Oceania
Hawaiian Islands

Eldredge (1994), quoting Miyada (1991), says the African Jewelfish has been reported from Lake Wilson on the island of Oahu.

Three-spot Tilapia (Kafue Tilapia)
Oreochromis andersoni

The Three-spot Tilapia is one of the most important food fishes in the Zambezi River catchment.

Natural Distribution

The upper reaches of the Zambezi, Cunéné and Ngomi Rivers, portions of the upper Zaire, the Okavango drainage system, and Lake Liambezi.

Naturalized Distribution

Africa: Tanzania.

Africa
Tanzania

In 1968, Three-spot Tilapias from Zambia were introduced for aquaculture purposes to Tanzania (Lema *et al.*, 1975), where according to Moreau (1979) they have hybridized in the wild with local *Tilapia* species.

Translocations

Toots (1970) records unsuccessful transplantations of Three-spot Tilapia within Rhodesia (Zimbabwe).

De Moor and Bruton (1988) list successful translocations to the Shashi dam on the Shashi River in the Limpopo system, Botswana, and to northern South West Africa (Namibia), probably for sporting purposes or as a forage fish for introduced bass *Micropterus* spp., and a failed transplantation in the region of the confluence of the Vaal and Orange Rivers in South Africa. Moreau *et al.* (1988) say that in 1956–1957 Three-spot Tilapia were successfully translocated to the River Kipopo in Zaire.

Allied Species

In 1953, 1959 and 1960 *Oreochromis alcalicus grahami*, which is a native of Lakes Natron (Tanzania) and Magadi (Kenya), was successfully translocated to fill a vacant ecological niche in Lake Nakuru, Kenya, where it grew more rapidly than in its natural waters (Moreau *et al.*, 1988).

Blue Tilapia (Jordan St Peter's Fish; Blue Mouthbrooder; Israeli Tilapia)
Oreochromis aureus

The Blue Tilapia is a popular fish for rearing in aquaculture, and has been widely distributed for this purpose, particularly in Central and South America. It is tolerant of highly saline water (and of temperatures as low as 6.5°C), and can thus be bred in estuarine conditions. In introductions and for experimental hybridization programmes it has often been misidentified as the Nile Tilapia *O. niloticus* (q.v.) (Welcomme, 1988).

Natural Distribution

Africa: the Senegal River, the middle Niger as far south as Bussa, upper tributaries of the Benue River, Lake Chad and the lower Chari and Logone Rivers, and the lower Nile from Cairo to the delta lakes. Asia: the Na'aman River, the Yarkon River near Tel Aviv, Lake Hulah, Israel, the Jordan River system, the Asraq marshes, and the heated pools in the oasis of Ein Fashka on the shore of the Dead Sea.

Naturalized Distribution

Europe: ? Cyprus. *Asia:* Philippines; ? Singapore; ? Taiwan; ? USSR. *Africa:* South Africa. *North America:* Mexico; USA. *South and Central America:* ? Antigua; Brazil; ? Costa Rica; Cuba; ? El Salvador; Guatemala; Hispaniola (Haiti and Dominican Republic); ? Nicaragua; ? Panama; ? Peru; Puerto Rico.

Europe

Cyprus

In 1976, Blue Tilapia from Israel were introduced to dams to fill an empty niche in the Cypriot ichthyofauna, where they established 'flourishing self-breeding populations in the wild until 1979 when all were lost due to a winter kill' (Welcomme, 1981, 1988). Since then the dams have been restocked, but with what result is uncertain. (See also under Europe in *O. niloticus.*)

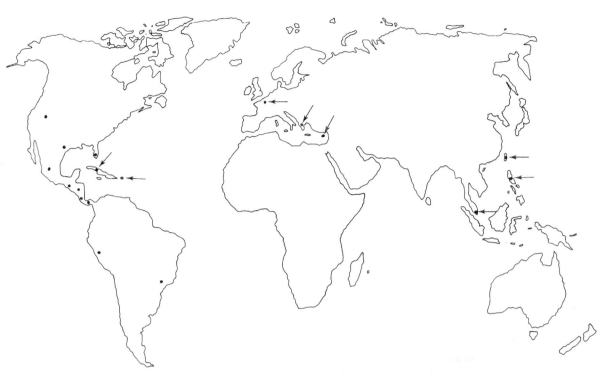

Principal introductions outside Africa of Blue Tilapia *Oreochromis aureus* (adapted from Philippart & Ruwet, 1982).

Asia

Philippines

In 1977, Blue Tilapia from Auburn University, Alabama, USA (Juliano *et al.*, 1989) and Singapore (Welcomme, 1981, 1988) were introduced to the Philippines, where although Juliano *et al.* (1989) do not indicate their current status, Welcomme (1981, 1988) says they are 'established in lakes and rivers'. Philippart and Ruwet (1982) say they were introduced for hybridization purposes.

Singapore

From their introduction to the Philippines from Singapore, Welcomme (1988) deduced their importation to the latter where, however, he says their status is unknown.

Taiwan

In 1974, 83 male Blue Tilapia from Israel were introduced to Taiwan, with the intention of obtaining male offspring when crossed with female Nile Tilapia (Chen, 1976; Liao & Chen, 1983). The resulting hybrid has become the principal tilapia used in aqua-

culture in Taiwan, and 'provides a fine example of success achieved with exotic aquatic species. Annual production of this hybrid currently exceeds 50 000 tonnes' (Liao & Liu, 1989). Whether the species occurs only in aquaculture or also in the wild is uncertain.

USSR

Welcomme (1988) lists the Blue Tilapia as having been introduced in 1984 for aquacultural purposes to the former USSR, where its present status is unknown.

Unsuccessful Introduction

In 1981 Blue Tilapia were introduced for hybridization purposes to China (Yo-Jun & He-Yi, 1989). Chiba *et al.* (1989) list the Blue Tilapia as among those species in Japan only 'being reproduced in certain experimental or natural ponds'.

Translocation

Within Israel, *O. aureus*, known locally as the Jordan St Peter's Fish, has been successfully translocated from Lake Hulah into Lake Kinneret (Sea of Galilee/Lake Tiberias), and also to fish ponds in

Galilee, the Beth Shean Valley, and the coastal plain where, as in Taiwan, it has been crossed with *O. niloticus* (and *O. vulcani*) to produce all male offspring (Ben-Tuvia, 1981).

Ecological Impact

Opinions vary regarding the ecological impact, if any, of *O. aureus* in Lake Kinneret. Ben-Tuvia (1981) stated that the stocking of the Jordan St Peter's Fish, the Silver Carp *Hypophthalmichtys molitrix* (q.v.) (which does not breed in the lake), and mullets (Mugilidae) 'added about 400 tonnes to the annual catch in recent years without any apparent negative effect on other stocks'. Gophen *et al.* (1983), however, said they had 'evidence that the increase of stocked species has occurred at the expense of the Galilee St Peter's Fish *O. galilaeus* population with whom the stocked species compete for plankton resources'. Davidoff and Chervinski (1984) agree with Ben-Tuvia (1981), and state that the catch of *O. galilaeus* appears to be cyclical, and declines can be attributed to poor year-class survival rather than to competition with *O. aureus*.

Africa

South Africa

De Moor and Bruton (1988) have summarized what is known about the introduction of Blue Tilapia to South Africa.

The species was first imported from Israel in 1910 for experimental rearing in the Jonkershoek hatchery at Stellenbosch. It was subsequently introduced to Natal, where it was bred by the Natal Park Board in their hatchery at Umgeni and elsewhere (Bourquin *et al.*, 1984). De Moor and Bruton (1988) list the three following more recent introductions of Blue Tilapia to South Africa.

In 1965, Blue Tilapia and Galilee St Peter's Fish were stocked in experimental ponds at the Jonkershoek hatchery. Although there is no record of either species ever having been released in private farm dams, in 1967 a dam on Rozendal farm near Jonkershoek 'already contained large flourishing populations of largemouth bass and *Tilapia aurea* and a smaller population of *Tilapia mossambica*' (Van Schoor, 1969a).

In 1978, both Blue Tilapia and Nile Tilapia *O. niloticus* (q.v.) were transferred from the Fisheries Development Corporation Hatchery at Amatikulu to a small dam in northern Natal, where after hybridizing they were all later destroyed.

In 1982, Blue and Nile Tilapia were liberated by the Tongaat Sugar Company in the Dudley Pringle dam in the Wewe River catchment in Natal (Bruton & Van As, 1986).

Today, Blue Tilapia are believed to survive in the Wewe River catchment, and in dams on the Somerlust and Vergenoeg–Lynedoch farms, and possibly also still at Rozendal (De Moor & Bruton, 1988). Welcomme (1981, 1988), who incorrectly gives the year of introduction as 1976, claims that the species is 'established in fish culture only'.

Ecological Impact

Although it is possible that Blue Tilapia might hybridize with Mozambique Tilapia *O. mossambicus* (q.v.), which occur naturally in the Wewe catchment, it is improbable that the alien or any hybrids could spread further without additional translocations (Bruton & Van As, 1986). Nevertheless, De Moor and Bruton (1988) expressed the fear that *O. aureus* 'could become a major pest if its range is extended', and that it 'should be regarded as a potentially deleterious alien fish which should not be translocated into natural waterways ... South Africa and Mozambique probably have the only genetically pure stocks of *O. mossambicus* in the world and every effort should be made to prevent hybridization between alien tilapias, such as *O. aureus*, and the indigenous species'.

Unsuccessful Introduction

According to Moreau *et al.* (1988), Blue Tilapia from Israel were introduced for rearing in the Kajanzi Farm in Uganda in 1962. C. Nugent (in Pullin, 1988) says that in 1981 Blue Tilapia of Israeli origin were unsuccessfully introduced to the Ivory Coast from Tihange, Belgium.

North America

Mexico

Blue Tilapia from Auburn University, Alabama, USA, were introduced to Mexico as a food fish in 1964 (Delgadillo, 1976), and releases were probably made in the lower Río Colorado, where the species is now naturalized, shortly thereafter (Minckley, 1973). Specimens from the lower Río Colorado (south from the Palo Verde Valley, California, USA) are most likely hybrids between *O. aureus* and probably the Nile Tilapia *O. niloticus* (q.v.) or perhaps the Redbelly Tilapia *Tilapia zillii* (q.v.) (Contreras & Escalante, 1984).

United States

Courtenay *et al.* (1984) summarized the then known range of the Blue Tilapia in the USA.

For many years, Blue Tilapia have been stocked annually in ponds at Auburn University in Alabama, where they have normally overwintered in heated buildings (Smith-Vaniz, 1968). *O. aureus* (or an *O. aureus* × *O. niloticus* hybrid) is established and locally dominant in the lower Colorado River in extreme southwestern Arizona and southeastern California below Laguna dam, and north of Yuma on the Gila River. Blue Tilapia have been used to control growths of algae near Gila Bend, Arizona, and *O. aureus* (or a closely related species or a hybrid) has been reared commercially on a catfish *Ictalurus* sp. farm and on a pig farm near Alamosa, San Luis Valley, Colorado, where its success is said to be due to thermally heated water. In the Colorado River, Blue Tilapia appear to be established as far north as above Parker dam (Lake Havasu).

In Florida, where they were first introduced from Auburn University in August 1961 (Courtenay & Robins, 1973), Courtenay *et al.* (1984) recorded Blue Tilapia as naturalized in 18 counties: Alachua, Brevard, Dade, De Soto, Hardee, Hernando, Hillsborough, Lake, Manatee, Marion, Orange, Osceola, Palm Beach, Pinellas, Polk, Sarasota, Seminole, Volusia, and in the saline waters of Tampa Bay. There were also unconfirmed reports by Foote (1977) for Broward, Charlotte, Glades and Pasco Counties. Courtenay *et al.* (1986) reported the presence of overwintering populations in a number of ponds in Duval County in 1983. Courtenay and Robins (1973) and Courtenay *et al.* (1974) said that Blue Tilapia were initially translocated by anglers from experimental ponds of the Florida Game and Fresh Water Fish Commission at Pleasant Grove, Hillsborough County, and from Lake Parker, Polk County, to neighbouring counties in central Florida and along the Gulf Coast.

Blue Tilapia have been reported, but unproven, to be established in ponds on a golf course at Sea Island and St Simons Island, Glynn County, Georgia. In Iowa, the species has been stocked in ponds to assess its growth potential, where although it reproduced successfully it was unable to survive winter temperatures (Pelren & Carlander, 1971). Blue Tilapia were introduced in 1965 to Lake Julian, a heated reservoir of the Carolina Power & Light Company in Buncombe County, North Carolina, where they became established. In 1977, the species was found by Pigg (1978) to be reproducing in the North Canadian River between Lakes Overholser and Eufaula (a distance of some 383 km), northwest of Harrah in Oklahoma County, Oklahoma.

In Texas, Blue Tilapia were reported by Noble *et al.* (1975) and Germany (1977) to be the dominant species in Trinidad reservoir, Henderson County, where they apparently died out in about 1978 when the power plant discharging excess heated water was turned off during cold weather. Meanwhile, other populations became established in the following waters: Braunig reservoir, Bexar County (Hubbs *et al.*, 1978); Lake Nasworthy, Tom Green County; Amistad reservoir, Val Verde County (Hubbs, 1976); Falcon reservoir, Zapata County, and for 60 km downstream in the Río Grande, where they were the dominant species (Courtenay *et al.*, 1984); Canyon reservoir, Comal County (Hubbs *et al.*, 1978); Lake Calaveras; Tradinghouse Creek reservoir, McLennan County; Lake Creek reservoir; Lake Colorado City, Mitchell County; and Lake Fairfield, Freestone County (Hubbs, 1982). Many of these waters possess thermal generation stations. Falcon reservoir is far enough south to make the risk of winter kills improbable. Amistad and Canyon reservoirs are in aquifer discharge localities, where low winter temperatures are likely to be moderated by stenothermal outflows. The widespread populations of Blue Tilapia in Texas provide, as Hubbs (1982) points out, ample opportunities for gradual natural selection to increased cold tolerance, which would enable the species to become even more widely disseminated. The northern edge of the Blue Tilapia's natural distribution is around 33°N, which approximates to that of the city of Dallas.

Skinner (1984) reported the establishment of Blue Tilapia near power plant thermal effluents in the lower Susquehanna River, Pennsylvania, while Courtenay *et al.* (1987) recorded the species' probable establishment near natural thermal outflows in the Snake River, Idaho, following escapes from aquaculture centres in Hagerman Valley, Twin Falls County.

Blue Tilapia were introduced to the USA for a variety of reasons. Introductions to Alabama by Auburn University were made for experimental research purposes (Smith-Vaniz, 1968); to Arizona by the state's Department of Fish and Game to control growths of algae; to Colorado privately for breeding in aquaculture; to Florida initially by the state's Game and Fresh

Water Fish Commission for research at their Pleasant Grove Research Station in Hillsborough County into the species' possible use in the control of aquatic weeds, and latterly by private individuals (Crittenden, 1962; Buntz & Manooch, 1968, 1969a,b; Courtenay & Robins, 1973; Courtenay *et al.*, 1974; Harris, 1978); to Georgia possibly privately to control growths of aquatic vegetation; to North Carolina as a potential sport fish; to Oklahoma possibly by the Oklahoma Gas & Electric Company for experiments in aquaculture (Pigg, 1978); and to Texas for use in aquaculture and to dispose of surplus bait fish. Courtenay *et al.* (1984) believed that the then distribution in the lower Colorado River in Arizona, California, and possibly in southern Nevada, was likely to expand because of the release of unwanted bait fish.

Courtenay *et al.* (1991) listed the Blue Tilapia as naturalized in Arizona, California, Florida (where it is now the most widespread exotic species), Georgia, North Carolina, Oklahoma and Texas, and to have been collected in the wild but not known to be established in Alabama, Colorado, Idaho and Pennsylvania. Courtenay (1993) queries the species' survival in Georgia. (See also Stanton Hales, 1991).

Ecological Impact   Evidence for the ecological impact of Blue Tilapia in the USA is largely anecdotal and contradictory. Although Noble *et al.* (1975) attributed increases in turbidity in some ponds to roiling by Blue Tilapia, where they exist in eutrophic waters in Florida and Texas there is, according to Taylor *et al.* (1984), no direct evidence that their presence has resulted in increases in turbidness or growth of algal blooms.

Opinions on their impact on native species also vary. According to Buntz and Manooch (1968), Blue Tilapia compete directly with native species for spawning sites, living space, and food. Taylor *et al.* (1984) quoted evidence to suggest that competition for food between native shad *Dorosoma* spp. and Blue Tilapia may cause the displacement of the indigenous species in eutrophic lakes and reservoirs in Florida and Texas, although as they point out the evidence is inconclusive. Courtenay and Robins (1973) said that in parts of Hillsborough County, Florida, where the Blue Tilapia is abundant, streams are devoid of most aquatic vegetation and nearly all native fishes, and 'the ecological disruption it causes is high. ... Florida acquired a highly undesirable exotic fish'. Taylor *et al.* (1984), however, said that

changes in water quality caused by tilapia have, as in the case of Grass Carp *Ctenopharyngodon idella* (q.v.), yet to be linked directly to alterations in the populations of native fish species.

Blue Tilapia are capable of explosive increases in population, and thus in their percentage of the total biomass of fish in many waters, particularly in eutrophic lakes. In a survey of Lake Parker, in the city of Lakeland in Polk County, Florida, Buntz and Manooch (1968) found that Blue Tilapia comprised 20% by weight of all fish present. In another Florida water, they increased within 2 years from 7.4% biomass to 69.1%, while in Lake Effie, near Lake Wales, Polk County, Blue Tilapia increased from nil biomass in 1966 to 0.66% in 1968, and to a staggering 93.26% in 1972 (Courtenay *et al.*, 1974). In some waters in Florida and Texas, Blue Tilapia are now so numerous that they are adversely affecting breeding of more valuable native centrarchid sport species.

South and Central America

Brazil; Costa Rica; Cuba; El Salvador; Guatemala; Nicaragua; Panama; Peru

Welcomme (1981, 1988) lists the introduction of Blue Tilapia for aquacultural purposes to the above South and Central American countries and their status as follows:

COUNTRY	YEAR	SOURCE	REMARKS
Brazil	1965	USA	Established in Pentecoste, Ceara
Costa Rica	1965	El Salvador	Well established in fish culture stations. Mainly useful for production of hybrids
Cuba	1968	Mexico	An advantageous introduction
El Salvador	1963/ 1979	USA	Widely diffused ... cultivated in artificial ponds and dams
Guatemala	1974	El Salvador	Widespread ... cultivated in ponds, rivers and lagoons. Easy adaptation to temperate climate reproduction and feeding habits
Nicaragua	1978	Costa Rica	—
Panama	1987	Puerto Rico	—
Peru	1983	Cuba	—

Puerto Rico; Hispaniola (Haiti and Dominican Republic); Antigua

According to Pagan-Font (1973), quoted by Erdman (1984), Blue Tilapia from Auburn University, Alabama, USA, were introduced to Puerto Rico in 1971 for experimental rearing in aquacultural ponds as a potential source of food.

Wetherbee (1989) lists the Blue Tilapia as among the various *Tilapia* introduced and naturalized in the West Indies [which] are ubiquitous in both fresh and brackish waters'. He modifies this somewhat extravagant statement by referring specifically to

their presence on the island of Hispaniola, but does not mention Puerto Rico.

Welcomme (1981) says that the Blue Tilapia 'has also been recorded from Antigua', but gives no further details.

Unsuccessful Introductions

Blue Tilapia may have hybridized in Australia with Mozambique Tilapia *O. mossambicus* (q.v.), but no pure *O. aureus* are known to occur there in the wild. *O. aureus* has been unsuccessfully introduced to Fiji (Andrews, 1985; Nelson & Eldredge, 1991; Eldredge, 1994).

Ngege
Oreochromis esculentus

This species has virtually disappeared from its native range due to a combination of over-fishing, predation from introduced Nile Perch *Lates niloticus* (q.v.), and competition from introduced Nile Tilapia *O. niloticus* (q.v.) (Welcomme, 1988).

Natural Distribution

Endemic to Lakes Victoria (Kenya, Uganda and Tanzania), and Kyoga and Nabugabo (Uganda).

Naturalized Distribution

Africa: Rwanda.

Africa

Rwanda

According to Welcomme (1988), *O. esculentus* was introduced in the 1950s from Lake Victoria to small lakes in the Kagera River system in eastern Rwanda, where it has become established and forms the basis for a valuable commercial fishery.

Translocation

Welcomme (1981, 1988), who treats the movement as an introduction rather than as a translocation, says that *O. esculentus* was also transferred in the 1950s from Lake Victoria to a number of small dams in Tanzania, from which it escaped to Lake Nyumba ya Mungu where it became established. It has proved a 'highly successful species for stocking into lakes and small dams where it fills a planktonophage niche. Provides up to 90% of the catch of Lake Nyumba ya Mungu'. Moreau *et al.*, (1988), who say that the translocation occurred in about 1960, name one of the Tanzanian waters in which *O. esculentus* has become established as Lake Kitangiri, where it both competes and hybridizes with the endemic *Sarotherodon amphimelas*.

The same authors also state that *O. esculentus* was unsuccessfully released in lakes and dams throughout Uganda between 1940 and 1945. Lowe-McConnell (1958) and Philippart and Ruwet (1982) specify Lakes Koki and Nkugute. In the latter, *O. esculentus* has hybridized with introduced *O. niloticus* (q.v.) (Lowe-McConnell, 1958).

Oreochromis leucostictus

O. leucostictus is a fairly small species which is resistant to low dissolved oxygen tensions (Welcomme, 1988).

Natural Distribution

Endemic to Lakes Edward, Albert and George, Uganda.

Naturalized Distribution

Africa: Kenya; Tanzania; ? Zaire.

Africa

Kenya; Tanzania; Uganda

Welcomme (1967) lists the following introductions and translocations of *O. leucostictus* into the Kenyan, Tanzanian and Ugandan waters of Lake Victoria:

YEAR	LOCALITY	NUMBER	SOURCE
1951/ 1953	Entebbe, Uganda	?	Lake Albert via Kidetok dam, Teso
1953	Kavirondo Gulf, Kenya	6	Uganda via Kisumu, Kenya
1955	Mara Bay, Musoma and Mwanza, Tanzania; Kavirondo Gulf and Mfwangu Island, Kenya	*c.* 12 000 (including *Tilapia zillii*)	?
1955/ 1957	Pilkington Bay, Bovuma, Entebbe, Uganda	?	?
1956/ 1957	Lake Victoria generally	*c.* 1500 (including *O. niloticus* and *T. zillii*)	?
1958	Entebbe, Uganda	490	?
1959	Mwanza, Tanzania	2173	Butimba prison
1960	Smith Sound, Mwanza, Tanzania	1751	?

According to Welcomme (1981, 1988), in Lake Victoria *O. leucostictus* occupies marginal deoxygenated lagoon habitats not inhabited by other species, but is not a favoured food fish because of its muddy taste. It has also been naturalized since 1955 or 1956 in Lake Naivasha, Kenya, where it forms the basis for a commercial fishery (Muchiri & Hickley 1991; Muchiri *et al.*, 1995 and has been stocked in dams in various localities in East Africa (Welcomme 1988). Craig (1992) says that in the early 1950s/1960s it was translocated to Lake Kyoga, Uganda. (See also under *O. spilurus*.)

Ecological Impact   After 1964, *O. leucostictus* became the dominant species in lagoons around Lake Victoria, and in the vicinity of papyrus beds and in shallow muddy bays (Lowe-McConnell, 1982) where Ogutu-Ohwayo (1990a) suggested some competition with indigenous *O. variabilis*. (See also under *O. niloticus* and *Tilapia zillii*.)

In Lake Kyoga, Moreau *et al.* (1988) say that *O. leucostictus* has been eliminated by *O. niloticus*.

In Lake Naivasha, *O. leucostictus* hybridized with another alien, *O. spilurus nigra/niger* (q.v.). Hybrids were abundant in 1961 but within a decade had disappeared, since when only *O. leucostictus* has survived (Lowe-McConnell, 1982).

Zaire

In 1955, *O. leucostictus* was introduced from Uganda to Zaire, where although it has been reported in the Lualaba River its status there is uncertain (Moreau 1979; Moreau *et al.*, 1988).

Longfin Tilapia (Greenhead Tilapia)
Oreochromis macrochir

The Longfin Tilapia has been widely stocked in ponds and dams in southern Africa, and has also been introduced for aquacultural purposes to many other African countries. As Welcomme (1988) points out, the majority of records are for artificial ponds, and it is difficult to assess the extent to which Longfin Tilapia have become established in the wild.

Natural Distribution

From Central Africa south to the Cunéné, Okavango, and upper Zambezi River systems, including Lakes Bangweulu and Mweru.

Naturalized Distribution

Africa: ? Algeria; ? Benin; ? Burundi; Cameroun; ? Central African Republic; ? Congo; ? Gabon; ? Ghana; ? Ivory Coast; Kenya; Madagascar; Rwanda; Zaire; Zambia. *Oceania:* Hawaiian Islands; Mauritius; Wallis Island.

Africa

Algeria; Benin; Burundi; Cameroun; Central African Republic; Congo; Gabon; Ghana; Ivory Coast; Kenya; Rwanda; Zaire; Zambia

Welcomme (1981, 1988) and Moreau *et al.* (1988) have listed introductions of Longfin Tilapia in Africa (other than Madagascar) as follows:

COUNTRY	DATE	SOURCE	REMARKS
Algeria	1961	Zaire	Useful at low altitudes
Benin	?	?	?
Burundi	1948/1950s	Zaire	—
Cameroun	1950s	Congo	Rivers Loum and Noum (rare)
Central African Republic	1953	Zaire	Attempts at culture with this species have been discontinued
Congo	1953	Zaire	—
Gabon	?	?	Libreville and Franceville
Ghana	1962	Kenya	—

(cont'd . . .)

COUNTRY	DATE	SOURCE	REMARKS
Ivory Coast	1957–1958	Cameroun	Hybridization experiments
Kenya	?1955	Zambia	Self-propagating, can be found in ponds. Generally not popular, and its culture has been abandoned
Rwanda	1948–1949	Zaire	Lakes Akagera and Kivu
Zaire	1954–55	Zambia	River Lualaba
Zambia	1959 & 1961	?	Lakes Kariba and Lusiwashi

Madagascar

Longfin Tilapia from the Djoumouna Fish Culture Centre in Brazzaville in the Congo were introduced to Madagascar in 1955 (Welcomme, 1981, 1988) or 1951 (Reinthal & Stiassny, 1991, quoting Kiener (1963) and Moreau (1979)), specifically to Lake Alaotra in 1958. Lamarque *et al.* (1975) and Philippart and Ruwet (1982) say they became established throughout the island in mangrove swamps, for example at Nemakia. Welcomme (1988) says that Longfin Tilapia in Madagascar are 'used for aquaculture and fisheries at low altitudes. Declining because of competition with *O. niloticus*' (q.v.). According to Moreau (1979), the culture of Longfin Tilapia on the island has been abandoned, but it has been an important species in open water fisheries at all altitudes' (J. Moreau, in Pullin, 1988). None was collected in the wild in 1988 by Reinthal and Stiassny (1991).

Ecological Impact Welcomme (1984) draws attention to an interesting result of hybridization of tilapia on Madagascar. Although in many instances disturbed sex ratios among hybrid tilapia offspring are frequent, self-breeding stocks of hybrids can apparently emerge in the wild, resulting in the formation of new genotypes where more than one species has become established. In Lake Itasy, parental stocks of introduced Nile Tilapia *O. niloticus* (q.v.) and Longfin Tilapia all but disappeared, and were replaced by a hybrid known locally as 'Tilapia Trois Quarts'. Subsequently, an equilibrium was achieved between Nile Tilapia and the new hybrid, in favour of the former, but Longfin Tilapia have failed to recover.

Translocations

Toots (1970) records the translocation of Longfin Tilapia to numerous dams in Rhodesia (Zimbabwe) where in many cases they have done well, and also less successfully to Lakes Kyle and Kariba. (See also Bell-Cross & Bell-Cross, 1971; Marshall, 1979.) Philippart and Ruwet (1982) mention the species' success in Lake McIlwaine, Zimbabwe. De Moor and Bruton (1988) list translocations to Shashi dam or the Shashi River in the Limpopo system, Botswana, in 1977, and also to the Hardap Dam, Namibia. (See also B. Marshall, in Pullin, 1988.)

Ecological Impact

Although the spread of Longfin Tilapia southward from the Shashi dam in Botswana is likely to be checked by their relative lack of tolerance to low temperatures, the area available for invasion – the warmer stretches of the Limpopo, Incomati, Usutu Phongolo and Mkuze systems – would still be considerable. There is a risk that Longfin Tilapia could hybridize with such commercially valuable native species as Mozambique Tilapia *O. mossambicus*, and they would also be likely to compete for food with detritivores such as other native tilapia and mullet (Mugilidae) (De Moor & Bruton, 1988).

Oceania

Hawaiian Islands

In 1957, 52 4–5 cm Longfin Tilapia from Leopoldville in the then Belgian Congo (Zaire) were imported to Oahu by the Fish and Game Laboratory. From the following year, their progeny were stocked in Wahiawa reservoir on Oahu, and also in reservoirs on Maui, in both of which they became established (Brock, 1960). (See also Hida & Thomson, 1962; Kanayama, 1968; Maciolek, 1984; Courtenay *et al.*, 1991; Courtenay, 1993.)

Mauritius

Moreau *et al.* (1988) say that in 1955 Longfin Tilapia from Madagascar were introduced for aquacultural purposes to the island of Mauritius, where they are widespread and expanding.

Wallis Island

Eldredge (1994) lists the Longfin Tilapia as having been successfully introduced between 1967 and 1970 to Lake Kikila on Wallis Island in the South Pacific, from where it has spread naturally to other freshwater areas.

Unsuccessful Introduction

In Asia, Chiba *et al.* (1989) list an unsuccessful introduction of Longfin Tilapia in 1964 from the USA to Japan.

Mozambique Tilapia (Peter's Tilapia (Israel); Mozambique Mouthbrooder; Peter's Mouthbrooder; Largemouth Kurper, Bream, Mudfish (Africa); Java Tilapia, Ikan Mudjair (Indonesia); Common Tilapia; Wu-Kuo yu (Taiwan); Makau (Papua New Guinea))
Oreochromis mossambicus

The Mozambique Tilapia was the earliest tilapiine cichlid to be introduced outside its natural range, first appearing on Java, probably as a result of escapes or releases from aquaria, before the outbreak of the Second World War. Since then, following introductions, translocations, accidental escapes and deliberate releases from captivity, and natural diffusion, it has become almost cosmopolitan. Originally highly regarded for use in aquaculture, its fecundity and thus tendency to stunt both in captivity and in the wild has caused its value to be reassessed. Large numbers are, however, fished for commercially in rivers, ponds, canals, ditches and rice paddy-fields, and even, since it is extremely hardy and tolerant of high salinities, in estuaries and marine habitats. Although in Asia the Mozambique Tilapia is still used as a major aquacultural species, elsewhere it is largely confined to the breeding of monosex hybrids, especially with Nile Tilapia *O. niloticus* (q.v.) (Welcomme, 1988).

Natural Distribution

Eastward-flowing rivers of Africa debouching into the Indian Ocean, from the lower Zambezi in Mozambique southward to the Brak River and coastal drainages to Algoa Bay, South Africa.

Naturalized Distribution

Europe: ? Malta; ? USSR ***Asia:*** ? Bangladesh; China; Hong Kong; India; Indonesia (Bali, Banjak (Pulau Pulau Islands), Banka, Borneo (Kalimantan), Celebes, Java, Lombok, Molucca Islands, Sumatra); ? Israel; Japan; ? Laos; Malaysia; ? Pakistan; Philippines; ? Saudi Arabia; ? Singapore; Sri Lanka; Taiwan; ? Thailand; ? Vietnam. ***Africa:*** ? Algeria; ? Benin; Madagascar; Namibia; Tanzania; ? Tunisia; ? Uganda; Zaire. ***North America:*** ? Mexico; United States. ***South and Central America:*** ? Bolivia; Brazil; Colombia; Costa Rica; El Salvador; Guatemala; ? Guyana; Honduras; Nicaragua; Panama; Peru; Venezuela; West Indies (Barbados, Cuba, Dominica, Grenada, Hispaniola (Haiti and Dominican Republic), Jamaica, Martinique, Puerto Rico, St Lucia, Trinidad). ***Australasia:*** Australia; Papua New Guinea. ***Oceania:*** Hawaiian Islands; Line Islands (Fanning Atoll and Washington Island); Melanesia (Fiji, New Caledonia, Solomon Islands, Vanuatu); Polynesia (Cook Islands, Western Samoa, Tonga, American Samoa, Society Islands (Tahiti), Wallis Island, Niue, Tuvalu); Micronesia (Nauru, Kiribati, Mariana Islands, Caroline Islands).

Europe
Malta

Welcomme (1988) lists the Mozambique Tilapia as reproducing on the island of Malta, but whether in the wild or only in captivity is unclear. Philippart and Ruwet (1982) record the introduction as a 'complete failure'.

USSR

According to Welcomme (1988), Mozambique Tilapia from Vietnam were introduced for aquacultural purposes to the former European USSR in 1962, where their present status is unknown.

Unsuccessful Introduction

Welcomme (1988) says the Mozambique Tilapia has been introduced for 'experimental biology' to Britain where it is reproducing 'artificially'.

Principal introductions outside Africa of Mozambique Tilapia *Oreochromis mossambicus* (adapted from Philippart & Ruwet, 1982).

Asia

Bangladesh

Welcomme (1981, 1988) states that in 1954 the Mozambique Tilapia was introduced from Thailand to Bangladesh for rearing in captivity, where it became 'established in aquacultural ponds throughout the country . . . useful for food but tends to disappear spontaneously'. Philippart and Ruwet (1982) classify the species as a 'failure' in Bangladesh.

China

In 1957, Mozambique Tilapia were imported from Vietnam to China for breeding in aquaculture (Yo-Jun & He-Yi, 1989). They are a useful source of food, and have become 'widespread throughout the southern part of China in ponds and natural waters' (Welcomme, 1981, 1988). Borgstrom (1978) and Philippart and Ruwet (1982) recorded Mozambique Tilapia as occurring in brackish and marine coastal waters and in rivers of the central and southern provinces such as Fukien. (See also below under Taiwan.)

Hong Kong

According to Tubb (1954), Mozambique Tilapia from the USA and Singapore were introduced to Hong Kong before 1940 in consignments of other tropical aquarium fishes. After 1945, several shipments were made to the colony for stocking aquaria, at least one of which came from the USA, the remainder arriving from Singapore. None of these early introductions appears to have been made for aquacultural purposes. Tubb (1954) recorded the following importations from 1948:

Mozambique Tilapia are now widely distributed in ponds and reservoirs throughout Hong Kong, in spite of their fecundity and thus tendency to stunt and their inability to tolerate temperatures below around 10°C (Hodgkiss & Man, 1978).

Ecological Impact 'One of the best-studied examples of the role of a tilapia in an exotic fish community is that of *O. mossambicus* in Plover Cove Reservoir, Hong Kong' (Lowe-McConnell, 1982), where a freshwater storage reservoir has been formed from a marine inlet (Hodgkiss & Man, 1977; Man & Hodgkiss, 1977). The initial 20 species of fish in the reservoir included marine ones which survived but did not breed, rice-field and river fish from the drainage basin, Chinese carps stocked between 1967 and 1971, and Mozambique Tilapia which, having escaped from an aquaculture centre, established themselves and were before long the principal species in gill-net catches. Mainly a detritivore, *O. mossambicus* in Plover Cove reservoir had the widest of food spectra, thus showing itself to be well adapted to reservoir life since nearly all available food sources were utilized, when some items were in short supply others being exploited. The Mozambique Tilapia also helped to control both chironomids and algae, and to remove nutrients and decaying organic detritus from the mud, thereby indirectly assisting in the reduction of algal growth (Lowe-McConnell, 1982).

India

Mozambique Tilapia were first introduced to India, to Tamil Nadu, on 7 August 1952, when a shipment of 50 fingerlings from Bangkok, Thailand, was imported by the Central Marine Fisheries Research

DATE	NUMBERS	SOURCE	IMPORTER	REMARKS
1948	100 fry	Singapore	Department of Agriculture, Fisheries & Forestry (DAFF)	Introduced to a pond at Kam Tin in January, 1949, where all died during cold weather
1953	4 young	Kowloon, China	Fisheries Research Unit (FRU)	Transferred to outdoor pool at the DAFF, Laichikok, and in 1954 to private lotus pond at Sha Tin, where they reared some 300 young
1954	250 young	Bangkok, Thailand	DAFF	Introduced to pool at Laichikok
1954	6 young	Kowloon, China	FRU	Spawned unsuccessfully
1954	10 adults	Bureau of Fisheries, Bangkok, Thailand	FRU	Spawned successfully

Institute to Mandapam (Panikkar & Tampi, 1954; Jhingran, 1989; Sukumaran & Tripathi, 1989). In the same year, a second consignment was brought in from Ceylon (Sri Lanka) by the Madras Fisheries Department (Devadas & Chacko, 1953; Jhingran, 1989). In December 1953, a small stock was transferred from Mandapam to the Pond Culture Division of the Central Inland Fisheries Research Institute at Cuttack, for experimental research and study of the species' biology and culture and of its likely impact on the Indian 'major carps' (Jhingran, 1989; Sukumaran & Tripathi, 1989; Yadav, 1993). Since then, in spite of governmental advice that it should be restricted to waters south of the Vaigai River, it has been widely distributed throughout southern India (Sreenivasan & Chandrasekaran, 1989), and elsewhere (Sukumaran & Tripathi, 1989). In Tamil Nadu, it is especially numerous in Amaravathy, Vaigai, Bhavanisagar, Tirumoorthy, Upper Dam and Krishnagiri reservoirs (Shetty *et al.*, 1989).

In Madras State, Mozambique Tilapia were first introduced to a fish farm in South Kanara in 1954 (Sreenivasan, 1967). At around the same time an aquarist in Mangalore began selling Mozambique Tilapia, some of which may have been deliberately released in natural waters in Karnataka. Also in the early 1950s, large consignments of Gangetic carp eggs were flown to Madras from Calcutta, Bengal, and were planted in reservoirs and other impoundments in the state; it is believed that some Mozambique Tilapia may have been accidentally included in these shipments (Chandrasekharaiah, 1989). The species is now naturalized in Karnataka in water tanks and reservoirs at Krishnarajasagar, Kabini and Hemavathi and probably in some other impoundments. It also occurs in the River Cauvery and its tributaries. It is flourishing in all perennial water tanks connected to sewage and with an abundance of algae and aquatic weeds around the cities of Bangalore and Mysore, such as Shettykere, Kukkarahalli, Ulsoor, Hebbal and Bellandur, where it accounts for up to 80% of the annual catch (Chandrasekharaiah, 1989).

In the mid- to late-1950s, Mozambique Tilapia from neighbouring Bangladesh were introduced by private individuals into West Bengal (Shetty *et al.*, 1989), where they became established, even in such arid areas as Ramanathapuram, Salem, Coimbatore, and in Madurai (Sreenivasan, 1991). From West Bengal, despite governmental opposition, they were

successfully transferred in the early 1960s to ponds at Banas-pokhar, Ranchi and Hazaribagh in south Bihar (Banerji & Satish, 1989). Mozambique Tilapia are today widely distributed in many parts of India.

Ecological and Economic Impact   'The Java tilapia [see below under Indonesia] *Oreochromis mossambicus* ... is the most controversial among the exotics in India' (Shetty *et al.*, 1989). 'No other species [in India] has caused such an "ecological explosion" as has the tilapia' (Sreenivasan, 1991).

Mozambique Tilapia were introduced to India to provide a cheap and readily available source of protein (the species is referred to locally as the 'aquatic chicken') for local people. As such, it has aroused considerable controversy, which still continues, regarding its value and ecological significance.

'Introduction of tilapia into the open waters', wrote Jhingran (1989), 'should be averted at all costs'. Ravichandra Reddy *et al.* (1990) point out that although the Mozambique Tilapia may have made a significant contribution to the development of freshwater fisheries in India, its introduction into some reservoirs has been followed by a decline in the numbers of several native carp species. In the Nelligudda reservoir near Bangalore, for example, *Labeo boggut*, *Puntius sarana* and *P. dorsalis* were the dominant native species before the advent of Mozambique Tilapia. Since the arrival of the exotic, the two *Puntius* spp. have been superseded as dominants by *Oreochromis*, which now comprises the greater part of the catch. In the absence of sufficient phytoplankton, the omnivorous (or even on occasion carnivorous) (Shetty *et al.*, 1989) diet of *Oreochromis* seems to coincide with that of the two *Puntius* spp. and with that of *L. boggut*. Moreover, the spawning of the latter species is seasonal, whereas that of *Oreochromis* occurs throughout the year, giving the alien an in-built competitive advantage. Chandrasekharaiah (1989) reported the disappearance of some *Puntius* spp. in Krishnarajasagar reservoir in Karnataka. Similarly, Banerji and Satish (1989) say that the introduction of Mozambique Tilapia to ponds occupied by Indian 'major carps' in south Bihar resulted in a gradual diminution in the numbers and size of the native species, which are now confined to a few municipal waters only. In Tamil Nadu, Sreenivasan and Chandrasekaran (1989) say that the species soon stunted following precocious breeding and fecundity, and that as a result of interspecific competition the 'Chanos fishery was ruined by Tilapia'. In those reservoirs inhabited by 'major carps' the impact of Mozambique Tilapia was mixed. When 'proper sizes' (presumably fairly large) of 'major carps' were stocked, they were unaffected by the presence of Mozambique Tilapia, which were, however, themselves adversely affected by the occurrence of native *Rhinomugil corsula* and *Wallago attu*. It is possible to manage Mozambique Tilapia populations in order to increase 'major carp' production in reservoirs. The revenue produced by reservoirs in which Mozambique Tilapia is the dominant species is less than that yielded in those dominated by 'major carps' and catfish. For small, seasonal reservoirs, Mozambique Tilapia seems the ideal fish for commercial production.

In a rather remarkable *volte face* from the view expressed in the 1989 joint paper quoted above, Sreenivasan (1991) says that the Mozambique Tilapia's fecundity and breeding precocity in a wide variety of aquatic habitats and at all seasons render it available as a source of food for the poor throughout the year. Although it may have adversely affected the Milkfish *Chanos chanos* fishery in some watersheds, Sreenivasan continues, the negative impact is by no means universal, and the highest yield of Milkfish in some ponds and reservoirs comes from those where *O. mossambicus* is the dominant species. In the past 15 years, Mozambique Tilapia have contributed to almost 30% of the total catch of freshwater fishes in Tamil Nadu, and a species that has supplied so much nutrition to poor people must be of socioeconomic benefit. 'Taking this into consideration', Sreenivasan (1991) concludes, 'the introduction of this fish in 1952, should be considered beneficial'. Chandrasekharaiah (1989) says that the culture of Catla *Catla catla*, Rohu *Labeo rohita* and Mrigal *Cirrhina mrigala* in small tanks and ponds is unaffected by the presence of Mozambique Tilapia, and that in waters where Mozambique Tilapia and Indian 'major carps' and introduced Common Carp *Cyprinus carpio* (q.v.) coexist, it is the growth of the exotics that is retarded.

Indonesia (Bali; Banjak (Pulau Pulau Islands); Banka; Borneo (Kalimantan); Celebes; Java; Lombok; Moluccas; Sumatra)

In Indonesia, Mozambique Tilapias (three males and two females) were first discovered in 1938 or 1939 in a lagoon at the mouth of the River Serang, in a remote and uninhabited region at Blitar, Residency Kediri, on the south coast of East *Java*, by an employee of the Fishery Extension Service, Pak Mudjair, in whose honour the species was named 'Ikan Mudjair' (*ikan* = fish) by the Department of Inland Fisheries. From Blitar, Mozambique Tilapias were transferred via Surabaja in East Java to Bojor (Bogor) by the Laboratory for Inland Fisheries; to Tasikmalaja by a private aquaculturist; and later to Djakarta for the control of malarial mosquitoes – all in West Java (Hofstede & Botke, 1950; Schuster, 1950, 1952; Vaas & Hofstede, 1952; Atz, 1954; Muhammad Eidman, 1989). How Mozambique Tilapias arrived in Java, the first place outside Africa where the species has definitely occurred in the wild, has apparently never been determined.

The outbreak of the Second World War in 1939 caused havoc among Indonesian fish farmers; in particular, coastal fishing for young Milkfish *Chanos chanos* to stock brackish aquaculture ponds was greatly curtailed, and the transportation of fry of other species by coastal shipping was seriously disrupted. Because of the ease with which it could be reared and the speed of its growth, albeit accompanied by stunting, together with the urgent necessity to produce an additional supply of protein, the Mozambique Tilapia was widely distributed throughout Java during the War by the occupying Japanese forces (Atz, 1954), becoming established in a variety of natural and artificial waters, including lakes, ponds, streams, swamps, drainage-ditches, irrigation reservoirs (*waduks*), and in brackish pools (*tambaks*).

In some waters, such as Lake Ranu Lamongan in eastern Java, Mozambique Tilapia and the translocated Tawes *Barbus javanicus/gonionotus* (q.v.) support a flourishing commercial fishery (Payne, 1987). During the War, the Japanese (who named the species 'Ikan Nippon') (Hofstede & Botke, 1950) spread the Mozambique Tilapia in Indonesia outside Java to *Sumatra* in 1939, *Lombok* (1941), *Bali* (1943) and *Celebes* (1944) (Atz, 1954). After the War, Mozambique Tilapia were introduced to other Indonesian islands, including *Amboina* in the *Moluccas* in 1949–1950, *Banka* in 1950, and *Banjak* (*Pulau Pulau Islands*) and *Borneo* (*Kalimantan*) also in 1950 (Welcomme, 1988). They are now widely naturalized throughout Indonesia, as a result of accidental escapes or deliberate releases into natural waters (Atz, 1954). Although regarded with disfavour by some aquaculturists, who consider them competitors with more valuable food species, they are an important source of protein for local people and useful for the control of mosquitoes (Muhammad Eidman, 1989).

Israel

In 1975, Mozambique Tilapia from Natal, South Africa, were imported to Israel by the Fish Breeding Research Station at Dor for experimental hybridization purposes. Others were imported to Arava north of Elat to try to improve the yield of *Chlorella* cultivated in seawater tanks by the Israel Oceanographic & Limnological Research Company, where fish excreta fertilize the water and stimulate the growth of the unicellular green algae (Ben-Tuvia, 1981). Whether Mozambique Tilapia are established in the wild in Israel is uncertain.

Japan

In and after 1954, Mozambique Tilapia were introduced for aquacultural purposes to Japan, where they have colonized polluted brackish estuarine waters around Okinawa Island (Imai, 1980; Chiba *et al.*, 1989).

Laos

Riedel (1965) and Welcomme (1988) list the Mozambique Tilapia as reproducing in Laos, to which it was introduced, presumably for rearing in aquaculture, from Thailand in 1955, but whether in the wild or only in captivity is unclear.

Malaysia

Mozambique Tilapia were first introduced to Malaysia from Java by the Japanese in 1944 during the Second World War (Mohsin & Ambak, 1983), quickly becoming established in many pools, impoundments, and even in some coastal regions such as the Johore Straits between Malaysia and Singapore (Ang *et al.*, 1989). Initially popular with aquaculturists, Mozambique Tilapia were widely reared throughout Malaysia during the 1950s, and were introduced to Sabah (northern Borneo) in 1951 and to Sarawak in 1960 (Welcomme, 1988). Thereafter, however, the species declined in favour, since its precocious breeding, fecundity, and year-round reproduction resulted in population explo-

sions and severe stunting, and caused competition for food and living space with more valuable species, whose growth rate was reduced. Since the early 1960s, Mozambique Tilapia have been rarely cultured in Malaysia (except for experimental breeding with *O. urolepis hornorum* to produce male monosex hybrids), where the species is now naturalized in both fresh and brackish waters, including ponds, canals, tin-mining pools, and streams (Ang *et al.*, 1989). Chen Foo Yan (in Pullin, 1988), however, reported an introduction of Mozambique Tilapia from Pietermaritzburg, South Africa, to Batu Berendam, Malacca.

Pakistan

According to Atz (1954), Mozambique Tilapia were introduced from Java to Sabang, Pakistan, presumably for rearing in aquaculture, in 1951, where their current status is unknown.

Philippines

In 1950, three male and one female Mozambique Tilapia, the sole survivors of a larger shipment, were imported from Thailand to the Philippines by the Bureau of Fisheries and Aquatic Resources, where in brackishwater fish farms they competed for food with the more valuable Milkfish *Chanos chanos*. Mozambique Tilapia are now established in brackishwaters throughout the Philippines, and in freshwater lakes, reservoirs, dams, rivers, rice paddy-field ponds and swamps in upland regions (Atz, 1954; Pillai 1972; Juliano *et al.*, 1989), although they are said to be disappearing from some localities around Laguna de Bay (R.D. Guerrero III, in Pullin, 1988). Because of the very small founder stock, Mozambique Tilapia in the Philippines are highly inbred, and hybridization with Nile Tilapia *O. niloticus* (q.v.) has occurred both in the wild and among farm stocks (R. D. Guerrero III, in Pullin, 1988).

Saudi Arabia

Ross (1984) describes his discovery in the previous year of a breeding population of Mozambique Tilapia in irrigation canals near wells in the Hofuf Oasis in the Eastern Province of Saudi Arabia, where he believed they had either been released by aquarists as surplus stock or that an attempt had been made to establish a commercial venture.

Singapore

Mozambique Tilapia from Indonesia were introduced to Singapore by Japanese forces during their occupation of the island in the Second World War between 1942 and 1945 (Harrison & Tham, 1973). The species is presently naturalized in both freshwater and low-salinity waters (Chou & Lam, 1989), including those uninhabited by native species due to heavy pollution (De Silva, 1987). Chen Foo Yan (in Pullin, 1988) refers to populations in the Botanical Garden and in the lake at the Seramban Sports Club. He also mentions the presence in a Singapore reservoir of an 'interesting strain', which he describes as 'a short, deep-bodied fish, rather disc-like in appearance like a pompano [*Trachinotus* sp.]. It can be easily recognized from other populations of *O. mossambicus* . . . This is a good example of the plasticity of tilapias in different environments and fish communities – in this case, heavy predation by *Channa micropeltes*'.

Sri Lanka

Fernando (1971) reported the introduction of Mozambique Tilapia from East Africa to Ceylon (Sri Lanka) for aquacultural purposes in 1952, where De Silva (1987) records them as having been stocked in 'man-made lakes', and Welcomme (1981) says they are 'very abundant especially in estuarine areas'.

De Silva (1987) attributed the success of cichlids in Sri Lanka more to the lack of purely lacustrine native species able to colonize a new artificial environment such as reservoirs than to an absence of indigenous herbivores. Fernando (1971) also suggested that a high degree of pressure by predators and a restriction of spawning sites has enabled Mozambique Tilapia to attain a large size in perennial impoundments. Blue-green algae (Cyanobacteria) are a major source of food for *O. mossambicus*, and the abundance of algae in reservoirs may also have aided the species' success (De Silva, 1987). According to Payne (1987), before the introduction of Mozambique Tilapia a mature impoundment in Sri Lanka yielded an annual average of 1.07 kg of fish per hectare. By 1974, 22 years after the introduction of Mozambique Tilapia, the yield had risen to 445 kg per hectare, a figure that by 1987 had more than doubled to 918 kg per hectare, and Mozambique Tilapia formed 99% of the catch.

Ecological and Economic Impact   The significance of Mozambique Tilapia in Sri Lanka appears to have been mostly beneficial. Fernando (1971) pointed out that as well as itself providing a new and valuable commercial catch, the alien has been indirectly responsible for increasing the take of native species which, in the absence of the exotic, would not have been commercially fished. More recently, De Silva and Fernando (1980) have suggested that Mozambique Tilapia have created a new source of food for native species, in the form of partially digested blue-green algae, which has resulted in improved growth rates of indigenes. Being largely planktonivores, Mozambique Tilapia in reservoirs do not compete for food with native fishes (Payne, 1987).

Taiwan

Mozambique Tilapia were first, unsuccessfully, introduced from Indonesia to Taiwan (Formosa) in 1944, when the island was under Japanese occupation during the Second World War. In 1946, after the surrender of Japan, two Taiwanese named Wu and Kuo, who had received commissions in the Japanese army, returned to Taiwan taking with them from Singapore a number of Mozambique Tilapia, of which 13 survived the journey. In honour of the species' importers it became known as Wu-Kuo yu (*yu* = fish). In the aftermath of the War, when food in Taiwan was in short supply, the Mozambique Tilapia was readily accepted as a cheap and available source of protein (Chen, 1976; Liao & Liu, 1989). In the rice paddies of Taiwan, Mozambique Tilapia show slow rates of growth and are subject to high winter mortality (Philippart & Ruwet, 1982).

Thailand

In 1949, Mozambique Tilapia from Malaysia were introduced to Thailand where, after experimental breeding for 2.5 years in a government field station, large numbers were transferred, in a programme sponsored by the Food & Agriculture Organization of the United Nations, to thousands of ponds, ditches, canals, swamps and rice paddy-fields (Atz, 1954).

Evidence on the present status of the species in Thailand is contradictory. Whereas Welcomme (1981, 1988) says it was 'well-established in 1950–60 but has now almost disappeared due to a spontaneous decline in population', and Philippart and

Ruwet (1982) classify it as a 'failure', Piyakarnchana (1989) reports Mozambique Tilapia to be 'well established in the ecosystem of Thailand. They are completely viable . . . capable of reproducing in reservoirs and in natural waters'. M. Tangtrongpiros (in Pullin, 1988) reported that the species is declining in Thailand.

Ecological Impact   Piyakarnchana (1989) regarded the Mozambique Tilapia as a 'nuisance species' in Thailand, and believed it to compete for food in aquaculture ponds with the bottom feeding brackish-water shrimp *Penaeus merguiensis*.

Vietnam

According to Welcomme (1988), Mozambique Tilapia from the Philippines were introduced to Vietnam in 1955, where they are reproducing, though whether in the wild or only in aquaculture is uncertain.

Unsuccessful Introduction

Welcomme (1981, 1988) records the introduction in 1953 of Mozambique Tilapia from Thailand to Korea, where the species is unable to withstand low winter temperatures. Riedel (1965) says the introduction, to South Korea, took place in 1955.

Africa

Algeria

In 1957, Mozambique Tilapia from an unrecorded source were introduced to open waters at Ain Skrouna, Algeria, where they are also naturalized in waters between 1000 and 1500 m asl (Moreau *et al.*, 1988).

Benin

In the 1980s, Mozambique Tilapia from an apparently unrecorded source were introduced to Lake Nokoué, Benin, to try to establish a species tolerant of saline conditions and superior to the native Blackchin Tilapia *Sarotherodon melanotheron* (q.v.) (Welcomme, 1988).

Madagascar

Mozambique Tilapia from Mauritius and Mozambique were first introduced to Madagascar in

1956, specifically to Lake Alaotra in 1960 and Lake Itasy in 1961–1962, for aquaculture and rearing in rice-paddies (Kiener, 1963; Moreau, 1979; Reinthal & Stiassny, 1991). *O. mossambicus* was at one time widespread and abundant at low elevations and in brackish waters of the Malagasy Republic, but following hybridization with other *Oreochromis* species the current distribution on the island of pure *mossambicus* is difficult to determine. In 1972, a population was discovered in a small isolated freshwater body, known locally as 'Lac Sacré', near Mahajungaha City airport on the northwest coast, where around 200 fish of various ages were living sympatrically with introduced *Heterotis niloticus* (q.v.) Because of its remote situation and relative inaccessibility, this population of Mozambique Tilapia probably remains pure (J. Moreau, in Pullin, 1988). The species is presently established in small numbers of uncertain purity in a few coastal lagoons and rice-fields, and in some rivers, such as a small freshwater tributary of the Mangoro, but nowhere dominates catches. In 1988, Reinthal and Stiassny (1991) collected Mozambique Tilapia from the following Malagasy waters; small outflowing streams from Lac Vert in the northeast central rain forest; Antsirabé market (fish said to have been caught in the River Tsiribihina) and Lake Itasy on the central high plateau; the Rivers Namorona, Fanolafana and Fotobohitra in the Ranomafana National Park; and in a feeder of the River Mangoro in the central east coastal area. (See also Lamarque *et al.*, 1975; Philippart & Ruwet, 1982.)

Namibia

Dixon and Blom (1974) record the presence of introduced Mozambique Tilapia in six localities in the Kuiseb, Omaruru and Ugab Rivers in central Namibia (South West Africa), while Skelton (1981), quoted by De Moor and Bruton (1988), refers to them in the Hardap dam on the Fish River in the Orange system. Schrader (1985) found them to be widely distributed in Namibian farm dams, including those in the Omatako Omuramba catchment, and in the Von Bach dam in the Swakop system and in the Omatoko dam in the Okavango system (Skelton & Merron, 1984).

Ecological Impact   Although there are believed to be no certain records of hybridization in natural waters in Namibia between Mozambique Tilapia and native species, it is possible that the exotic could interbreed with either or both the Longfin Tilapia *O. macrochir* or the Three-spot Tilapia *O. andersoni* in the country's northern river systems (De Moor & Bruton, 1988). Schrader (1985) believed that interspecific competition with native cichlids for food and spawning sites was also probable.

As Mozambique Tilapia are established in the Omatoko dam and also in numerous isolated pans in the Omatoko Omuramba drainage, which in time of flood is joined to the Okavango River, it seems likely that they will eventually spread to the Okavango drainage. If this does indeed take place, hybridization with Three-spot Tilapia, and thus genetic dilution of the indigene in the Okavango system, seems probable (Skelton & Merron, 1984).

Tanzania

In 1958, *O. mortimeri*/*O. mossambicus mortimeri* (see below under Zaire) was successfully introduced, probably from the Kafue River in Zambia, to impoundments at Mwadingusha, Koni and Nzilo in Tanzania (Moreau *et al.*, 1988).

Tunisia

Moreau *et al.* (1988) say that Mozambique Tilapia from Zaire (presumably *O. mortimeri* or *O. mossambicus mortimeri*) (see below under Zaire) were in 1966 introduced to Tunisia, where they became established in warm brackish waters. Welcomme (1988), however, says that it is unknown whether the species is reproducing in Tunisia. Philippart and Ruwet (1982) say they were introduced 'into the reservoir of a southern Tunisian oasis' to develop a new fishery.

Uganda

Moreau *et al.* (1988) record the introduction of Mozambique Tilapia for hybridization purposes from Zanzibar to Uganda between 1962 and 1966.

Zaire

Oreochromis mortimeri is taxonomically so close to *O. mossambicus*, with which it shares the same natural distribution, that in spite of differences in coloration it is sometimes regarded as only a subspecies of *mossambicus*. In 1957, *O. mortimeri*/*O. mossambicus mortimeri* was imported from the Kafwe River in Zambia and suc-

cessfully stocked in dams, reservoirs, and the Rivers Kipopo and Lufira in Zaire (Moreau *et al.*, 1988).

Translocations

De Moor and Bruton (1988) say that Mozambique Tilapia have been widely translocated in South Africa, especially to the Cape, for aquacultural purposes, to provide sport fishing, as a forage fish for introduced black bass *Micropterus* spp., and for the biological control of chironomids and macrophytes (Begg, 1976; Ashton *et al.*, 1986). Welcomme (1981, 1988), quoting Moreau (1979), refers to the species' successful introduction in 1951 for aquacultural purposes to Lake Sibaya.

Mozambique Tilapia have been the subject of widespread translocations within Zimbabwe (Rhodesia), particularly around 1920 to the Mazoe River, where they were released by local farmers. *O. mortimeri*/*O. mossambicus mortimeri* has been stocked in Lake Kariba (between Zimbabwe and Zambia) and elsewhere in Zimbabwe (Toots, 1970).

Ecological Impact

In the Jan Diesels River, South Africa, translocated Mozambique Tilapia compete with native *Barbus* spp. (Crass, 1969), while in rivers in the Cape Gaigher *et al.* (1980) noted that they had an adverse effect on sport fisheries and probably also on native species. De Moor and Bruton (1988) could trace no detailed studies on the impact of Mozambique Tilapia in those areas to which they have been translocated within South Africa.

Unsuccessful Introduction

In Africa, El Bolok and Labib (1967) refer to an unsuccessful attempt between 1934 and 1962 to establish Mozambique Tilapia in Egypt. Riedel (1965) says that the first introduction, from Thailand, took place in 1954. Moreau *et al.* (1988) list a failed introduction of the species to Rwanda in 1962.

North America

Mexico

The status of Mozambique Tilapia in Mexico is unclear. According to Contreras and Escalante (1984), the official records of introduced species of the Departmento de Pesca, which specifically list *O. mossambicus*, *O. niloticus* and *O. melanopleura* (? *zillii*), are all otherwise based on tilapia regardless of species. Thus records may apply to any species of introduced tilapia. Contreras and Escalante (1984), who specify *O. mossambicus* as established and reproducing, mention unidentified tilapia as naturalized in the following Mexican localities: Marte R. Gómez reservoir (1974); Falcon reservoir, Tamaulipas (1975); Peña del Aguila and elsewhere in the upper Río Mezquital, Durango (1981); the Río Nazas at El Salvador, Durango; the coastal plains of Sinoloa (1978); and La Boca reservoir, Monterrey (1979). Welcomme (1981) says Mozambique Tilapia were first introduced to Mexico in 1964, and are established in Miguel Alemar, Benito Juarez and La Anostara reservoirs, where they form the basis for important fisheries.

United States

Courtenay *et al.* (1974) reported the collection of Mozambique Tilapia in Florida from the Lake Worth Drainage District Canal L-15 in Palm Beach County before 1969, and subsequently from the Comfort Canal (the channelized southern fork of the Miami River), Dade County; the Civic Center Pond in Satellite Beach, Brevard County; near a fish farm north of Vero Beach, Indian River County; in drainage ditches on a fish farm in Micco, Brevard County; and in Six-Mile Creek, Hillsborough County. Populations in each of these locations originated from separate introductions. Courtenay *et al.* (1974) said that Mozambique Tilapia were then considered to be naturalized in the south fork of the Miami River; in areas near a fish farm outside the town of Fellsmere, Indian River County; in Six-Mile Creek; and possibly in the Banana River, Cocoa Beach, Brevard County.

Hogg (1974, 1976a,b) reported the dispersal of Mozambique Tilapia in a westerly direction through the Airport Lakes to their confluence with the Tamiami Canal, and the sighting of isolated individuals in the Opa Locka Canal and in a small canal joining the Snapper Creek Canal.

Dial and Wainwright (1983) and Courtenay *et al.* (1984) confirmed the species' establishment in the saline waters of the Banana River, and at five sites in Dade County; in ponds at the Aventura condominium community; in the Snapper Creek Canal for around 9 km above its confluence with the Tamiami Canal; in the Tamiami Canal at US Highways 41 and

27; in the Comfort Canal; and in two canals east of Goulds.

Courtenay *et al.* (1974) considered that the species' high fecundity; the parental care of young; its adaptability to the acceptance of a variety of spawning habitats; and its high tolerance of saline waters had combined to help the Mozambique Tilapia's establishment in the waters of southern Florida. The fish are believed to have originated in releases by aquaculturists. (See also Courtenay & Robins, 1973; Harris, 1978.)

Mozambique Tilapia observed by St Amant (1966) on 3 January 1964 in a small pond and an outflowing stream near the Hot Mineral Spa, Imperial County, some 8 km north of the Salton Sea 'represent the first verified record of free-living tilapia in California'. The fish are believed to have escaped or been released from a tropical fish farm, Del Rancho El Sargent, owned by Mr D. J. Sargent. Although more than 5000 Mozambique Tilapia, weighing around 350 kg, were subsequently removed from the pond alone, the attempt to eradicate them was unsuccessful.

On 2 July 1968, further populations of Mozambique Tilapia were discovered in California, in the Araz Drain and Reservation Main Drain – both irrigation drains near Bard, Imperial County, where they were found over a length of around 24 km. They are believed to have dispersed naturally to the Bard Valley from Yuma, Arizona, where they were introduced by the Arizona Game and Fish Department, or from illegal releases (Hoover & St Amant, 1970).

Before 1976, Mozambique Tilapia in California also became established in the lower Colorado River (Moyle, 1976a,b). They were later also released in agricultural drains, and by 1978 were invading the Salton Sea, where they (or a hybrid) are now the dominant species in terms of biomass (Courtenay *et al.*, 1984).

Mozambique Tilapia and Wami Tilapia *O. urolepis hornorum* (q.v.) (and possibly hybrids between them) were introduced into Coyote Creek, a tributary of the San Gabriel River (Legner *et al.*, 1980), and to waters in Long Beach, Los Angeles County, and the Santa Ana River, Orange County, in 1973 (Courtenay *et al.*, 1984). Knaggs (1977) reported the collection of Mozambique Tilapia from marine waters at Seal Bay, Orange County and Cerritos Lagoon, Long Beach, Los Angeles County, to which it is believed they were translocated by children from the San Gabriel River.

In January 1983, Mozambique Tilapia were released in High Rock Spring in Lassen County (Courtenay *et al.*, 1984), who reported them to be established in California in Imperial, Los Angeles, Orange and Riverside Counties.

Brown (1961) reported the capture on 19 October 1959 of a single Mozambique Tilapia in the San Marcos River, San Marcos, Hays County, Texas, near the A.E. Wood State Fish Hatchery. Although this was the first known record of the species in the wild in Texas, Brown (1961) had been aware of its presence in the headwaters of the San Antonio River at San Antonio, Bexar County, for several years, and was able to trace the history of its introduction to the State.

Prior to 1956, the San Antonio Zoo acquired some Mozambique Tilapia from the Steinhart Aquarium in San Francisco, California, which it is believed had imported them direct from Africa. In the San Antonio Zoo the fish spawned successfully, and some were placed in a wire retaining pen in the headwaters of the San Antonio River. During the spring of 1956 some 40 8–12-cm fish escaped into a canal adjacent to the river, where they rapidly increased in size and were soon breeding successfully.

In June 1958, the Texas Game and Fish Commission acquired an unrecorded number of Mozambique Tilapia from the San Antonio River, and planted them in an experimental rearing pond on the A. E. Wood State Fish Hatchery at San Marcos. Some were later transferred to Spring Lake in the headwaters of the San Marcos River, where they spawned. In 1958 and 1959, large numbers of fry and fingerlings are believed to have escaped from the pond into a drain which flows into the San Marcos River, and more probably entered the river in subsequent years.

Brown (1961) recorded Mozambique Tilapia in Texas to be naturalized in the headwaters of the San Antonio River from the San Antonio Zoo downstream possibly to its confluence with the Medina River, and in the headwaters of the San Marcos River from Spring Lake for about 8 km downstream. Hubbs *et al.* (1978) confirmed the species' survival at San Antonio, and Hubbs (1982) and Courtenay *et al.* (1984) reported its continued existence also in the San Marcos River.

Minckley (1973) records the establishment of Mozambique Tilapia in agricultural drains and mitigation ponds near Yuma, Yuma County, in Arizona since the early 1960s, where they now coexist with Blue Tilapia *O. aureus* (q.v.). *O. mossambicus* also previously occurred in drains and in some stretches of the Gila River between Phoenix and north of Yuma, but has apparently now been supplanted by *O. aureus* or by a hybrid with the Nile Tilapia *O. niloticus* (q.v.) (Courtenay *et al.*, 1984). According to Minckley (1973), other populations established in Arizona at Warm Springs on the San Carlos River, Gila and Graham Counties, and in the Salt River in Tempe, Maricopa County, succumbed to flooding.

In 1985, Courtenay *et al.* (1987) collected 142 Mozambique Tilapia from Barney Hot Spring and the upper end of Barney Creek, Custer County, in the Little Lost River Valley north northwest of Howe, Butte County, Idaho; in the following year 10 juvenile tilapia (probably hybrids between *O. mossambicus* and *O. urolepis hornorum*) were taken in the Bruneau River at Bruneau Hot Springs, below the Blackstone Grasmere Road Bridge in Owyhee County, Idaho.

Courtenay *et al.* (1984) said that Mozambique Tilapia might be established in ponds on golf courses on Sea Island and St Simons Island in Glynn County, Georgia.

Between 1965 and the early 1970s, Mozambique Tilapia were established for sport fishing in Lake Julian, Buncombe County, North Carolina, and they are or have been reared and/or stocked annually, but are not naturalized, in Alabama, Colorado, Illinois and Montana (Courtenay *et al.*, 1984).

Mozambique Tilapia have been introduced to the USA with a variety of motives; to Arizona by the Arizona Game and Fish Department and the Arizona Cooperative Fishery Research Unit of the US Fish and Wildlife Service to control aquatic vegetation (Minckley, 1973); to California for the control of aquatic weeds, mosquitoes and chironomid midges by various agencies, including the Orange County and Southeast Mosquito Abatement Districts (Knaggs, 1977; Legner *et al.*, 1980); those in the Salton Sea are probably descended from releases into irrigation drains after 1976. In Florida, most of the populations in Dade County are believed to originate in escapes from tropical aquarium fish farms, although in one locality they were apparently released by a building developer to control aquatic weeds; in Brevard County, the release of pet fish at Satellite Beach seems to have been the origin of present day populations (Courtenay *et al.*, 1974, 1984).

Welcomme (1981, 1988) includes Tennessee and

North Carolina in his list of States where Mozambique Tilapia are established, although they are known to have died out in the latter in the early 1970s, and says they may also be established 'possibly in Alabama', despite the fact that Courtenay *et al.* (1984) say 'there is no evidence of overwintering there'.

Courtenay and Stauffer (1990) list *O. mossambicus* as naturalized in Arizona, California, Florida, Idaho and Texas. Courtenay *et al.* (1991) confirm this list, and say that the species has been collected but is not known to be established in Alabama, Colorado, Georgia, Illinois, Montana, Nevada, New York and North Carolina. Courtenay (1992, 1993) omits Idaho from the states in which the species is breeding naturally in the wild. (See also Lachner *et al.*, 1970; Courtenay & Hensley, 1980b; Courtenay *et al.*, 1986.)

Ecological Impact Little research has been carried out into the ecological impact, if any, of Mozambique Tilapia in the USA. Knaggs (1977) indicated a decline in populations of native fishes in the San Gabriel River, California, which he attributed to an increase in the number of *O. mossambicus*, and predicted a similar future impact on coastal marine species.

South and Central America

West Indies (Barbados; Cuba; Dominica; Grenada; Hispaniola (Haiti and Dominican Republic); Jamaica; Martinique; Puerto Rico; St Lucia; Trinidad)

Atz (1954) has traced the introduction to that date of Mozambique Tilapia to the West Indies.

The West Indian islands support very few suitable freshwater food fishes. To augment their meagre stocks, shortly after the Second World War Dr C. F. Hickling, Fisheries Adviser to the British Colonial Office, suggested the importation of tilapia. Accordingly, in 1949 Mr Swithin Schouten, Agricultural Superintendent of *St Lucia*, imported 450 *O. mossambicus* by boat from Malaya via New York through Paramount Aquarium, Inc., the largest importers of tropical fish in the USA. Only two fish failed to survive the voyage. In St Lucia the Mozambique Tilapia thrived and multiplied, and some of their progeny were transferred from St Lucia to *Trinidad* and *Grenada* in 1949 and to *Barbados,*

Dominica, Jamaica and *Martinique* in 1950; from Jamaica to *Hispaniola (Haiti)* in 1951; and from Haiti to the *Dominican Republic (Hispaniola)* in 1953. In St Lucia and Grenada the fish were deliberately released into the wild, but in Jamaica they accidentally escaped into a river where they became established. In Haiti, within 2 years of their introduction, Mozambique Tilapia had become naturalized in several rivers and a brackish water lake, and had become an important source of protein to the local people. (See also Wetherbee, 1989, and under *O. aureus*.)

Post-1954 introductions of Mozambique Tilapia in the West Indies have been made to *Puerto Rico* and to *Cuba*. On 11 July 1958, fish from Auburn University, Alabama, USA, were introduced to Puerto Rico, primarily to control growths of algae that were clogging canals used to irrigate sugar-cane plantations. 'This tilapia now has the widest distribution on the island' (Erdman, 1984), and in Cartagena Lagoon has displaced the introduced Bluegill *Lepomis macrochirus* (q.v.). (See also under the Redbreast Tilapia *Tilapia rendalli*.)

Welcomme (1981) says that Mozambique Tilapia from Mexico were imported to Cuba in 1967 for 'cultivation for commercial fishing', and that they are 'established throughout the country'. Welcomme (1988) advances the date of their introduction from Mexico to Cuba to 1868 (31 years before the species' first recorded exportation from Africa and nearly a century before it was introduced to Mexico) and 1883, and says they were introduced for 'aquaculture, forage, sport, mosquito control'.

Nicaragua

On 22 December 1959, 100 Mozambique Tilapia fry, imported from El Salvador, were released in Lake Moyua, which lies at an altitude of 420 m some 80 km north of the capital, Managua. The extent of the lake varies considerably according to the time of year; in the wet season it measures up to 4 km × 2.5 km (Riedel, 1965).

Ecological Impact In Lake Moyua the Mozambique Tilapia flourished and multiplied rapidly, soon becoming the staple food of resident Least Grebes *Podiceps dominicus* and Pied-billed Grebes *Podilymbus podiceps*, whose predation partially governs both the growth rate and increase in numbers of Tilapia.

Riedel (1965) considered that it was highly unlikely that Mozambique Tilapia would ever become established in coastal waters of Nicaragua, although he believed that were they ever to do so considerable economic advantages might accrue. Similarly, he believed that if Mozambique Tilapia were to spread to or be stocked in freshwater bodies in Nicaragua, they could prove of economic value not presently provided by native *Cichlasoma* spp.

Welcomme (1981, 1988) lists the introduction of Mozambique Tilapia, all for aquacultural purposes, to other countries in Central and South America, as shown below.

Riedel (1965) says the introduction to Guatemala was from Haiti, and took place in 1958, and that to Guyana (British Guinea) was in 1954. Infante (1985) refers to a population in Lake Valencia, Venezuela.

Unsuccessful Introductions

Welcomme (1981) says that the Mozambique Tilapia has been introduced to, but is not established in, Ecuador and Surinam.

Australasia

Australia

Despite the fact that the importation to Australia of Mozambique Tilapia had been forbidden since 1963,

in 1969 three specimens were found in Western Australia and destroyed (Jubb & Petrick, 1970), and in 1977 aquarists in Brisbane and Townsville, Queensland, were openly offering the species for sale (McKay, 1977). Mozambique Tilapia were first reported in the wild also in 1977, when some were caught in Tingalpa reservoir (Leslie Harrison dam), Brisbane, and in the North Pine dam (Lake Samsonvale), Brisbane, in 1979, and others were found in inflowing creeks, including Ferny Creek, at Palm Beach, north of Cairns, and in 1978 in ornamental ponds in Anderson Park Botanic Gardens, Townsville. In Western Australia, the species was discovered and eradicated in ornamental ponds in Geraldton in 1978, but in 1981 it was found to be widespread in the Gascoyne River, Carnarvon, and in the following year in river pools near Carnarvon (Arthington *et al.*, 1984). McKay (1984) says the species may also be established in the Chapman River in Western Australia.

These reports initiated surveys of waters near Brisbane and Townsville in 1981–1982, which revealed naturalized populations in Tingalpa reservoir and North Pine dam, Brisbane, and in Townsville in freshwater ornamental ponds and in portions of the Southern Townsville drainage system that receives tidal waters from Ross Creek. The discontinuous distributional records and various dates

COUNTRY	DATE	SOURCE	REMARKS
Bolivia	1983	Brazil	Reproducing
Brazil	1960s	?	Used in management of dams and reservoirs in the northeast, where it makes a significant contribution to catches
Colombia	1960	USA	Reproducing below 1000 m. Have adversely affected some native species
Costa Rica	1960 *et seq.*	?	Established in fishponds and small streams in Huila Province
El Salvador	1958 [but see under Guatemala]	Guatemala	Tilapias contribute over 50% of commercial catch. Eliminated from aquaculture farms and substituted by *O. aureus*. State of wild stocks not clear
Guatemala	1955 [but see under El Salvador]	El Salvador	Widespread cultivated in ponds and established in natural waters
Guyana	1951	Haiti	Reproducing
Honduras	1956 [but see under El Salvador]	El Salvador	Established Tegucigalpa, D.C. and Lago de Yojoa. Little use because of precarious reproduction, stunting, and low yield
Panama	1950	?	Established in a few impoundments. Of certain importance, but now superseded by *O. niloticus*
Peru	1981	Panama	Reproducing. Positive introduction
Venezuela	1958	Trinidad	Only free-living population breeds in sewer-system of Cumana (Mago). Unpopular species. Should be used with care

of first records suggest several separate introductions, probably from the release of surplus aquarium stock or unwanted pet fish, and/or the deliberate stocking of ornamental ponds, although the near proximity of the waters occupied by the species in Townsville also suggest the possibility of natural dispersal from park ponds through overflow drains into the Townsville drainage system and thence into Ross Creek (Arthington *et al.*, 1984; Arthington, 1986; Arthington & Milton, 1986; Blühdorn & Arthington, 1989, 1990). McKay (1984) says that a recent report indicates the species' presence in the Burdekin River.

Blühdorn *et al.*, (1990) reported in detail the then known distribution of Mozambique Tilapia in Australia:

Tingalpa Reservoir, Brisbane. Widely distributed, though principally near aquatic vegetation. Fewer in number and smaller in size than in North Pine dam, probably due to a less suitable environment.
Tingalpa Creek, Brisbane. Apparently confined to the spillway base area.
North Pine Dam, Brisbane. Widespread and abundant throughout, and specimens grow to the maximum recorded size for the species. Some have been removed by private individuals to stock ornamental pools.
North Pine River, Brisbane, downstream from the dam. Many tonnes of large fish were caught during and after flooding in 1988.
South Pine River, Brisbane. Catches as for the North Pine River. Breeding populations believed to have become established.
Dowes Lagoon, Sandgate. Presence confirmed by Brisbane City Council.
Ross River Estuary and above Aplin Weir, Townsville. Captures have been reported.
Woolcock Street stormwater drain, Townsville. Fish are established and breeding in this frequently hypersaline water, and may also enter via the tide gate.
Artificial lake development upstream of the Woolcock Street drains, Townsville. A thriving population has existed in this optimum habitat since the lake's construction. Inflow screening has not prevented external recruitment.
Healey Creek, north of Townsville. A confirmed population is established.
Kewarra Beach/Palm Cove, Cairns. A population is naturalized in an interconnecting system of stormwater drains, freshwater creeks, *Melaleuca* spp. swamps, and

estuaries, where the water ranges from saline to fresh. Both normal and stunted *O. mossambicus* occur, and also Spotted Tilapia *T. mariae* (q.v.).
Port Douglas, north of Cairns. A large but overcrowded and stunted population is established in an ornamental irrigation pond in the Mirage Resort golf course.
Gascoyne/Lyons River system, Western Australia. A stunted population is established in numerous small pools of the discontinuous Gascoyne River, and also in pools in the Lyons River some 300 km upstream from Carnarvon, where the population originated.

Blühdorn *et al.* (1990) also reported unconfirmed but probable populations of Mozambique Tilapia in Lake Kurwongbah, Brisbane, and in the South Townsville drainage system and in the Townsville Common. They considered that there has been very little natural expansion of the species' range in Queensland, and that most new records of its occurrence are the result of human agency, since the majority of populations are near urban conurbations and public roads. All the northern Queensland populations appear to be derived from ornamental ponds that have overflowed in times of flood, or have been permitted to drain into nearby watercourses.

In Western Australia, however, the population in the Gascoyne River has dispersed naturally far upstream in this intermittent river system, where the species is making use of its tendency to stunt to cope with the adverse environmental conditions.

Blühdorn *et al.* (1990) considered it a 'virtual certainty' that many other man-made waters in Australia support breeding populations of Mozambique Tilapia, and that 'only by maintaining control (however tenuous) of its translocation is there any prospect of preventing its widespread introduction throughout Queensland'. (See also Merrick & Schmida, 1984; McKay, 1986–1987; Allen, 1989; Blühdorn & Arthington, 1990.)

Genetics of Tilapia spp. in Australia Research has shown that there are two separate genetic strains of tilapia in Australia; one is a relatively pure form of *O. mossambicus*, the other based on this species but with hybrid genetic variations from one or more of the Nile Tilapia *O. niloticus* (q.v.), the Wami Tilapia *O. urolepis hornorum* (q.v.), and the Blue Tilapia *O. aureus* (q.v.). The former 'southern' strain occurs in Brisbane, Townsville, and the Gascoyne/Lyons system, while the latter 'northern' strain is found only around Cairns. (For further details see Arthington, 1989, 1991; Blühdorn *et al.*, 1990; Mather & Arthington, 1991; Arthington & Blühdorn, 1994.)

Ecological Impact Probably because of the species' relatively recent naturalization, there is little if any firm evidence regarding the ecological impact of *O. mossambicus* in Australia. In North Pine dam and Tingalpa reservoir it has been shown that with adult fish there is no overlap between the diet of the alien and that of two common native species, the bottom-feeding Eel-tailed Catfish *Tandanus tandanus* and the omnivorous Spangled Perch *Leiopotherapon unicolor*. It is, however, not known how the young of these and other species interact in Australian waters, and competition for food may be more important among juveniles than among mature fish. It has been suggested that there may be competition for breeding sites between *O. mossambicus* and *T. tandanus*, and that the former may prey upon small forage fish (Melanotaeniidae and Eleotriidae) in subtropical impoundments, and that their consumption of fish and aquatic invertebrates may be greater in waters with low primary productivity, but research on these aspects has yet to be carried out (Blühdorn *et al.*, 1990); Arthington & Blühdorn, 1994). McKay (1984) has postulated the theory that the presence in the Brisbane reservoirs of the voracious and aggressive *L. unicolor* could check the population expansion of Mozambique Tilapia, but that as the impoundments mature the population of the native predator is likely to decline, possibly resulting in an explosive growth in numbers of *O. mossambicus*.

Papua New Guinea

Glucksman *et al.* (1976) have traced the introduction of Mozambique Tilapia to Papua New Guinea, and thence to Bougainville in the Solomon Islands (see below).

In December 1954, a consignment of 250 fry from Malaya was introduced to ponds at the Bomana Gaol, Port Moresby in the Central District, where they became established. West and Glucksman (1976) say that 200 were subsequently transferred to ponds at Goroka in the Eastern Highlands District. Sometime before 1959 progeny from Bomana were transferred 90 km to the Kemp Welch River, where they also established breeding populations.

Experimental rearing of Mozambique Tilapia (and

other species) in highland areas was conducted at Dobel, Mount Hagen, where Mozambique Tilapia were first introduced in October 1955. Here the Mozambique Tilapia increased rapidly and, the ponds becoming grossly overcrowded, stunted severely (West & Glucksman, 1976).

In lowland areas, Mozambique Tilapia were first recorded in the wild in a stream near Port Moresby in 1956. In the following years large numbers were translocated all over the island on the grounds that they were more efficient at controlling mosquitoes than introduced Mosquitofish *Gambusia affinis* (q.v.) and, being larger, were more desirable for human consumption. In the Port Moresby area, Mozambique Tilapia caught in the Waigani Swamp were first offered for sale, at Koki market, in June 1961.

In June 1957, Mozambique Tilapia were introduced to a natural pond at Maprik from where, after a flood in 1959, they were washed into the Screw River, a tributary of the Sepik, where by 1966 they were abundant (West & Glucksman, 1976).

Berra *et al.* (1975) included the Mozambique Tilapia in their list of exotic species naturalized in Papua New Guinea in the Laloki, Brown and Goldie Rivers. Referring to the Laloki, the same authors commented that 'it is sad commentary on man's propensity for juggling freshwater fish faunas that the species collected at the most localities was an introduced one, *Tilapia mossambica*. This species also ranked second in total number of individuals . . .'. Glucksman *et al.* (1976) list Mozambique Tilapia as 'widespread' in all districts apart from the Southern Highlands, and say that they occur in both fresh and estuarine waters, while West and Glucksman (1976) said they were an important source of food in the Sepik and Central provinces. Allen (1991) recorded them as abundant in the Lower Ramu and Middle and Lower Sepik Rivers, and said that in the latter they have become the most important human food fish, forming 50% of the total catch (Coates, 1987).

West and Glucksman (1976: table 5) list numerous movements of Mozambique Tilapia within Papua New Guinea, mostly from Goroka and Bomana, and one introduction from Singapore in October 1956.

Oceania

Hawaiian Islands

In 1951, some 60 small Mozambique Tilapia from Singapore were imported to Honolulu on the island of Oahu, where all but 14 subsequently became diseased and died. The survivors bred successfully in the laboratory tanks of the Fish and Game Division on Oahu, the progeny being used for stocking purposes on all the main islands where, following further introductions, they are now naturalized in reservoirs, dams, streams, estuaries and low-lying wetlands. They have been used as live-bait for Skipjack Tuna *Katsuwonus pelamis* fishing; to control aquatic vegetation in irrigation systems; as a human food fish; and for sport fishing (Brock, 1952, 1954, 1960; Maciolek, 1984). (See also Atz, 1954; Randall, 1960; Hida & Thomson, 1962; Neill, 1966; Kanayama, 1968; Maciolek & Timbol 1980; Courtenay *et al.*, 1991; Courtenay, 1993.)

Ecological Impact   Mozambique Tilapia in the Hawaiian Islands are suspected of competing aggressively and advantageously with the Striped Mullet *Mugil cephalus* (Eldredge, 1994).

Fanning Atoll and Washington Island (Line Islands)

Lobel (1980) has traced the introduction of Mozambique Tilapia from the Hawaiian Islands to Fanning Atoll and Washington Island in the Line Islands, central Pacific Ocean, in 1958 (his map indicates the date as 1955). The fish were released by the US Bureau of Commercial Fisheries from their research vessel, the *Charles H. Gilbert*, in October or November 1958 in a saltwater pond on the southern side of English Harbor and in a tidepool on the northern side; on the west coast of Fanning Atoll; and in a freshwater lake on nearby Washington Island.

During a survey of Fanning Atoll in the summer of 1978, Lobel (1980) found that Mozambique Tilapia had dispersed over about 16 km of coastline from the site of their original introduction, but had not yet succeeded in colonizing the entire atoll. As the fish are not used by the resident islanders either for bait or for food, their spread had been entirely by natural dispersal, and the rationale for their introduction is a mystery. Lobel also found Mozambique Tilapia to be abundant throughout the lake on Washington Island.

Ecological and Socio-economic Impact   On Fanning Atoll, Mozambique Tilapia occupy estuaries at low

tide, retreating to shallow areas as the tide rises. Fish trapped in exposed shallows during high or low tides are preyed on by seabirds, and also by piscivorous fishes such as jacks *Caranx* spp. (mostly *C. ignobilis*) and juvenile Blacktip Sharks *Carcharhinus melanopterus*. Local fishermen claim that Mozambique Tilapia have caused a decrease in stocks of native mullet *Mugil cephalus* and *M. engeli*, Milkfish *Chanos chanos* and Bonefish *Albula vulpes*, which use the estuaries as nursery grounds, and on which the islanders depend for food and bait. Any decrease in the numbers of these species could have a serious negative impact on both subsistence fishing and also on any potential future commercial fishery (Lobel, 1980).

Solomon Islands

In about 1957, Mozambique Tilapia from an unknown source were introduced to ponds on Guadacanal in the Solomon Islands (Anon., 1958), and thence subsequently from Honiara to Lake Tenaggano on Rennell Island in the southern Solomons (Wolff, 1969); they later also became naturalized on the islands of Malaita and Santa Anna (Nelson & Eldredge, 1991).

In 1959, an apparently unrecorded number of Mozambique Tilapia were air-lifted from the Kemp Welch River on Papua New Guinea to the Arawa Plantation on Bougainville Island in the Solomons, some 1000 km to the east, where they were placed in an ornamental pond by the owner, a Mr McKillop, who also gave some to several of his employees, by whom they were released in their village ponds. In some regions of Bougainville Island there is a long tradition of aquaculture, either indigenous or acquired from Chinese coolies on German plantations before the First World War.

In or before 1970, some Mozambique Tilapia on Bougainville Island were freed in ponds at Musinau on the Toyo River, a tributary of the Jaba River, from which in September 1972 a severe flood washed the fish into the Toyo, and a number became established in a dam at the Jaba Pumping Station and in Tum Creek, another tributary of the Jaba.

In May 1974, a large population of Mozambique Tilapia was observed in a billabong on the Pagana River just downstream of the Moratana Mission, to which they had probably been carried by human agency from the pumping station (Glucksman *et al.*, 1976).

Ecological Impact On Rennell Island, Mozambique Tilapia introduced into Lake Tenaggano, the island's only large body of fresh water, were responsible for the elimination by 1959 of the local subspecies of the Grey Teal *Anas gibberifrons remissa*, which was only described in 1942 (Kear & Williams, 1978), and also perhaps for the extinction of the local race of the Grey Duck *A. superciliosa* (Eldredge, 1994). The Osprey *Pandion haliaetus melvillensis*, on the other hand, only spread from the coast to the lake after the introduction of Mozambique Tilapia had provided it with an additional and abundant source of prey (Diamond, 1984).

Melanesia; Polynesia; Micronesia

Nelson and Eldredge (1991), from whom much of the following account is derived, have traced the widespread introduction of Mozambique Tilapia elsewhere in the South Pacific (Melanesia and Polynesia) and Micronesia.

Because of their value as biological controlling agents (mainly of mosquito larvae and aquatic vegetation) and their adaptability to rearing in aquaculture, several species of tilapia, including the Blue Tilapia *Oreochromis aureus* (q.v.), the Nile Tilapia *O. niloticus* (q.v.), the Wami Tilapia *O. urolepis hornorum* (q.v.), the Redbelly Tilapia *Tilapia zillii* (q.v.), but predominantly the Mozambique Tilapia, have been introduced since the 1950s to numerous islands of the tropical Pacific. On many of these they, but most frequently *O. mossambicus*, have become naturalized in a variety of freshwater and coastal marine habitats (including lakes, impoundments, rivers, streams, mangrove swamps, and lagoons) on both coral atolls and volcanic islands, in some of which they have had a serious ecological impact, whereas in others they have caused an improvement in local fisheries.

Melanesia

In 1954, Mozambique Tilapia from Malaya were introduced to *Fiji* (Holmes, 1954), where they are established in fresh waters on Viti Levu (Chimits, 1957; Anon., 1958; Devambez, 1964; Ryan, 1980; Maciolek, 1984; Andrews, 1985), from where they have since been successfully translocated to Vanua Levu (Eldredge, 1994). In 1954 or 1955, 40 Mozambique Tilapia from Manila in the Philippines were introduced to ponds at the Port Laguerre Farm School in *New Caledonia* (Van Pel, 1956; Anon.,

1958), where they have become naturalized in freshwater habitats (Van Pel, 1959; Maciolek, 1984; Eldredge, 1994). (Riedel (1965) says the introduction was from Fiji.) By 1956, Mozambique Tilapia from Noumea, New Caledonia, had been introduced to Efate and Tanna Islands of *Vanuatu* (Chimits, 1957; Anon., 1958; Devambez, 1964; Maciolek, 1984). The rationale for these introductions in Melanesia was for aquaculture and the improvement of fish stocks.

Polynesia

In 1955, *Western Samoa, Tonga (Tongatapu),* and the *Cook Islands* acquired Mozambique Tilapia fingerlings from the Sigatoka Agricultural Station in Fiji (Chimits, 1957; Devambez, 1964; Van Pel, 1961; Maciolek, 1984). From the Sopu area of Tongatapu, they were translocated in the 1970s to Nomuka; in 1982 to Lake Vailahi on Niuafo'ou; at an unrecorded date to Vava'u (Eldredge, 1994); and in 1982 to Lake Ava Ano in Nomuka, H'Aapi (Nelson & Eldredge, 1991). From Western Samoa, Mozambique Tilapia were introduced in 1957 into a brackish water swamp on Aunu'u Island, and later into Leone Creek on Tutuila, in *American Samoa* (Anon., 1958; Van Pel, 1959; Devambez, 1964; Maciolek, 1984; Eldredge, 1994). At some time in the 1950s, Mozambique Tilapia were introduced to Tahiti in the *Society Islands* (Devambez, 1964; Maciolek, 1984). In 1966, Mozambique Tilapia were introduced to two caldera lakes on Wallis Island, *Wallis* and *Fortuna* (Hinds, 1969), and at unknown dates to *Niue* (Cook Islands), (Devambez, 1964), and *Tuvalu* (Funafuti and Niutao Atolls and Nanumanga Island) (Uwate *et al.*, 1984). As in Melanesia, the purpose of all but two of these introductions was for aquaculture to supply an additional source of protein to the islanders. The exceptions were in Tonga, where the objective was the biological control of mosquito larvae and in Lake Ava Ano, to provide additional stocks for fishing (Nelson & Eldredge, 1991).

Micronesia

In eastern Micronesia, Mozambique Tilapia were introduced to brackishwater ponds on *Nauru* in the early 1960s to control mosquitoes (Devambez, 1964; Maciolek, 1984). In 1963, the Department of Agriculture imported some, probably from Fiji, to some of the islands, such as Tarawa, in the Gilbert archipelago in *Kiribati*, where they were placed in

Milkfish *Chanos chanos* ponds to provide an additional source of food, especially when rough weather made sea-fishing impracticable (Nelson & Eldredge, 1991).

In western Micronesia, Mozambique Tilapia have been introduced intermittently since the late 1930s (Ikebe, 1939) and especially in the early 1950s. In 1954 or 1955, the Department of Agriculture on Guam in the *Mariana Islands* imported some from the Philippines (Brock & Takata, 1956; Chimits, 1957; Devambez, 1964; Kami *et al.*, 1968; Best & Davidson, 1981; Shepard & Myers, 1981; Maciolek, 1984), which were released in Fena reservoir in southern Guam to control aquatic vegetation. Since 1973, *O. mossambicus* × *O. niloticus* hybrids have been periodically introduced from Taiwan to Guam (FitzGerald & Nelson, 1979). In 1955, 309 Mozambique Tilapia from the Philippines were released in the 10-ha Lake Susupe in Saipan in the northern Mariana Islands. In the same year, 200 fingerlings from Saipan were liberated in each of two lakes on the island of Pagan (Nelson & Eldredge, 1991). The purpose of these introductions was presumably to supply an additional source of food. Finally, in the 1970s Mozambique Tilapia from an unrecorded source were introduced for aquaculture to mangrove swamps on Yap in the *Caroline Islands* (Nelson & Hopper, 1989).

Naturalization in the South Pacific and Micronesia
Although many introductions of Mozambique Tilapia were originally restricted to artificial impoundments, they soon became established in a variety of natural waters through deliberate releases or accidental escapes. As a result, in numerous places on many islands in the South Pacific and Micronesia, Mozambique Tilapias have become naturalized in artificial impoundments (including tanks, ponds, dams and reservoirs), streams, rivers, lakes, natural ponds, mangrove swamps, and even in brackish and marine waters (Nelson & Eldredge, 1991).

Freshwater Ecosystems   Nelson and Eldredge (1991) traced populations of naturalized Mozambique Tilapia in freshwater ecosystems in the South Pacific and Micronesia as follows: in the Cook Islands; American and Western Samoa; Wallis Island (Hinds, 1969); the Mariana Islands (Brock & Takata, 1956; Best & Davidson, 1981); Fiji (Ryan, 1980; Andrews, 1985); New Caledonia (Van Pel, 1959);

Nauru (Ranoemihardjo, 1981; quoted by Nelson & Eldredge, 1991); the Solomon Islands; Tonga (Maciolek & Yamada, 1981); on Washington in the Line Islands (Lobel, 1980); and on Yap in the Caroline Islands (Nelson & Hopper, 1989).

Brackishwater and Marine Ecosystems   As a result of their tolerance of high salinities, Mozambique Tilapia have become naturalized in brackish and marine waters in several islands in the South and Central Pacific, where Nelson and Eldredge (1991) traced populations as follows. As a result of their escape from a small fishpond, they have become the dominant species in brackishwaters in the Sopu area of Tongatapu, Tonga (Uwate *et al.*, 1984); they have also become naturalized in brackishwaters at Tuvalu (Uwate *et al.*, 1984); in mangrove swamps on Yap Island in the Caroline archipelago (Nelson & Hopper, 1989); on Fanning Atoll in the Line Islands (Lobel, 1980); and in brackishwater ponds and lagoons of Nauru and Tarawa in Kiribati (Uwate *et al.*, 1984).

Ecological and Socio-economic Impacts   Impacts of Mozambique Tilapia in Melanesia, Polynesia and Micronesia – whether ecological or socio-economic – are believed to be numerous but are difficult to quantify, since in the majority of cases their impacts on the native fauna and flora and on aquaculture or commercial fisheries are virtually unknown.

One of the species' best documented impacts has been, not as might be supposed on other fishes or on insects, but on native birds (see also above under Solomon Islands). On Niuafo'ou, Tonga, for example, the algae-coloured green water of Crater Lake cleared after the introduction of Mozambique Tilapia, and the resident duck population thereafter markedly decreased (Scott, 1993). In Lake Susupe on Saipan, in the Mariana Islands, competition with Mozambique Tilapia is believed to have been partially responsible for the decline of the endemic form of the Common Moorhen or Gallinule *Gallinula chloropus guami* (Stinson *et al.*, 1991), and also possibly for that of the endemic Marianas Mallard *Anas 'oustaleti'*.

Maciolek (1984) and Allen (1991) drew attention to the general tendency of naturalized Mozambique Tilapia to become pests, and the former recorded the species' unpopularity as a food fish. Nelson and Eldredge (1991) have traced what is known of the impact of Mozambique Tilapia on aquaculture in the region.

In some localities, Mozambique Tilapia are believed to have displaced more highly valued species in shallow inshore fisheries. On Tongatapu, Tonga, for example, they hinder the culture of Striped Mullet *Mugil cephalus*, and are regarded as a pest. On Fanafuti, Tuvalu, the stunted population of Mozambique Tilapia in the island's lagoon impedes the rearing of valuable Milkfish *Chanos chanos* (Uwate *et al.*, 1984).

Naturalized Mozambique Tilapia are also a nuisance to traditional aquacultural practices. On Nauru, for example, they have almost destroyed the traditional culture of highly valued Milkfish in freshwater ponds and lagoons (Ranoemihardjo, 1981; quoted by Nelson and Eldredge, 1991), and Milkfish are now imported to Nauru from Guam and Kiribati where they are cultured commercially (Nelson & Eldredge, 1991). Similarly, stunted Mozambique Tilapia on Tarawa and elsewhere in Kiribati compete with adult Milkfish for food and prey on their fry (Ranoemihardjo, 1981; quoted by Nelson & Eldredge, 1991).

Although on most islands in the region Mozambique Tilapia were originally introduced primarily for aquacultural purposes, there are few places in which they are at present being cultured. The only commercial rearing of Mozambique Tilapia of any significance takes place in Guam, where *O. mossambicus* and *O. mossambicus* × *O. niloticus* hybrids, imported from Taiwan, are cultured in ponds and sold alive in local markets (Nelson, 1988; quoted by Nelson & Eldredge, 1991). The domestic rearing of Mozambique Tilapia on Guam has put an end to the importation of fish from the Philippines, and the annual production of Mozambique Tilapia in Guam has progressively increased in recent years to the pre-1991 level of 125 000 kg which more than satisfies the local market (Nelson & Eldredge, 1991). Maciolek (1984), who lists the hybrids as *O. mossambicus* × *O. niloticus*, said that on Guam they 'were marketed under the name "Cherry Snapper" . . . to avoid the local prejudice against tilapias as a food fish. (Locally, the term "tilapia" refers to *O. mossambicus* which is considered an undesirable fish by most residents.)'

Mozambique Tilapia have also been cultured on a small scale in Western Samoa where, however, marketing the fish is said to have proved difficult (Uwate *et al.*, 1984).

In some places, the introduction of Mozambique Tilapia has been of benefit to both commercial and subsistence freshwater fisheries. Thus, over 16 t of Mozambique Tilapia per annum have been taken from Lake Tenaggano on Rennell Island in the Solomon group. In Lake Vailahi on the island of Niuafo'ou in Tonga, Mozambique Tilapia provide the principal source of protein for the islanders when the weather is too severe for sea-fishing; They are also fished on the islands of Vava'u and Nomuka in Tonga; where they occur in American Samoa; in the past in the Cook Islands; and on Niutao Atoll to provide pig food (Uwate *et al.*, 1984).

In few, if any, places have naturalized Mozambique Tilapia been successfully eradicated. In the Cook Islands, where they were originally introduced to control mosquito larvae and as an additional source of human food, they initially multiplied explosively, but are now so reduced in numbers that fishing is no longer worthwhile (Nelson & Eldredge, 1991). They are now mainly caught only to provide bait for barracuda *Sphyraena* sp. fishing. Popper and Lichatowich (1975), who found that Mozambique Tilapia also bred profusely in saltwater ponds in Fiji with salinities of up to 49‰, conjectured that their expansion there was largely contained through predation by the native *Elops hawaiiensis*. Attempts to eradicate Mozambique Tilapia in lagoons on Tarawa in Kiribati and brackishwater ponds of Nauru, where as mentioned above they have been accused of adversely affecting the traditional culture of Milkfish, have so far been unsuccessful (Ranoemihardjo, 1981, quoted by Nelson & Eldredge, 1991); Uwate *et al.*, 1984; Eldredge, 1994).

General Ecological Impact

See under Redbreast Tilapia *Tilapia rendalli*.

Hybrid Species

For details of *O. mossambicus* × *O. niloticus* hybrids (so-called 'Red Tilapia') see under *O. niloticus*.

Nile Tilapia
Oreochromis niloticus

Naturally the most widespread of tilapiine ciclids, the Nile Tilapia is one of the most popular for rearing in aquaculture and for stocking impoundments. Because it shows a good rate of growth and is less likely to produce stunted populations, it is often preferred for these purposes to the Mozambique Tilapia. It is widely crossed with other tilapias, especially *O. mossambicus*, to produce male hybrid offspring. It has been widely distributed throughout the world, but as a subtropical species is unable to reproduce in the wild in many of the temperate countries to which it has been imported (Welcomme, 1988).

The Nile Tilapia has the reputation of being the hardiest of tilapia species. It is physiologically well adapted to both fresh and brackish waters and to a wide pH range. It is also eurythermal, and can live in waters with a temperature range of between 14 and 33°C, though for maximum growth rate it has an optimum range of between 23 and 28°C. It is mainly a phytoplanktonivore and prefers planktonic filamentous algae. It also feeds on higher plants, but not to such an extent that it can be used effectively for the control of aquatic weeds. Its hybrid offspring with *O. mossambicus* are said to be more resistant than pure-bred progeny to cold conditions (Jhingran, 1989).

Natural Distribution

The basins of the Rivers Niger, Chari, Benue, Volta, Gambia, Senegal and Chad in West Africa, eastward through Lake Chad, Jebel Marra between Lake Chad and the River Nile, through the Congo basin to East Africa, and the Nile downstream of the Albert Nile to the delta in Egypt. The species may also be native to the River Yarkon in Israel (but see under Israel) and the Jordan Valley.

Naturalized Distribution

Europe: ? Cyprus. ***Asia:*** ? Bangladesh; China; ? Hong Kong; ? India; ? Indonesia; ? Israel; Japan; Malaysia; ? Philippines; Sri Lanka; ? Taiwan; ? Thailand; Vietnam. ***Africa:*** Benin; Burundi; Cameroun; ? Central African Republic; ? Guinea; Ivory Coast; Kenya; Madagascar; Mali; Rwanda; Sierra Leone; South Africa; Tanzania; Tunisia; Uganda; Zaire. ***North America:*** ? Mexico. ***South and Central America:*** ? Bolivia; ? Brazil; ? Colombia; ? Costa Rica; Cuba; ? Hispaniola (Haiti and Dominican Republic); Ecuador; ? El Salvador; Guatemala; ? Honduras; ? Nicaragua; Panama; Paraguay; ? Peru; Puerto Rico; Venezuela. ***Oceania:*** ? Cook Islands; Fiji; Mauritius; Réunion.

Principal introductions outside Africa of Nile Tilapia *Oreochromis niloticus* (adapted from Philippart & Ruwet, 1982).

Europe

Cyprus

Welcomme (1981) lists the Nile Tilapia as having been introduced from Israel to one dam in Cyprus for angling purposes and to fill a vacant ecological niche, but does not indicate if it became established.

Unsuccessful Introductions

In Belgium, Nile Tilapia imported from Israel are 'farmed in heated power station cooling waters with a production of 240 t (1986)' (Welcomme, 1988). In 1957, Nile Tilapia from Ghana were introduced 'for experimental aquaculture in cooling ponds' in the then Federal Republic of Germany (Welcomme, 1988).

In 1995, hybrids between Nile Tilapia and Blue Tilapia *O. aureus* (q.v.) were reared in heated water used to cool the plant of the Courtauld's textile factory at Spondon on the River Derwent, in Derbyshire, England, and marketed under the name 'St Peter's Fish'.

Asia

Bangladesh

Welcomme (1981, 1988) says that in 1974 Nile Tilapia from Thailand (see below) were imported for aquacultural purposes to Bangladesh, where the species became 'widespread all over the country where it is regarded as an excellent food fish'.

China

Nile Tilapia from Africa were introduced to mainland China in 1978 (Yo-Jun & He-Yi, 1989), where they are at present 'established mainly in aquaculture ponds in Hubei Province' (Welcomme, 1981, 1988), and are a useful source of food.

Hong Kong

Welcomme (1981, 1988) says that Nile Tilapia from Taiwan (see below) were in 1972 imported to Hong Kong, where they became widely distributed in fish ponds and where Welcomme (1981) quotes Tubb (1954) as saying they were stunting and hybridizing with other tilapia species. Tubb's paper, written 18 years before the date quoted for the introduction, deals exclusively with *O. mossambicus* and does not mention *O. niloticus*.

India

According to Jhingran (1989), the Nile Tilapia was one of several 'exotic species which have been suggested as potential candidates for introduction into the country'. Shetty *et al.* (1989) reported that '*O. niloticus* has already been introduced through private trade in West Bengal and is being cultured in sewage-fed farms'. Its status in the wild is unknown.

Indonesia

In 1969, Mr Rustami Djajadiredja imported Nile Tilapia from Taiwan to Indonesia, where they are established in both aquaculture and in the wild (Philippart & Ruwet, 1982; Muhammad Eidman, 1989).

Israel

Although some authorities include southern Israel within the natural range of the Nile Tilapia, Ben-Tuvia (1981) considers that the species' restricted distribution and rarity make this possibility doubtful, and the fact that Nile Tilapia have been introduced to Israel for aquacultural purposes since the 1960s from both East and West Africa seems to confirm Ben-Tuvia's opinion. Fish that have escaped or have been released from captivity – both pure-bred *O. niloticus* and hybrids with *O. aureus* (q.v.) – have become established and are breeding in the wild. Pure *O. niloticus* occur in two coastal streams, the Yarkon near Tel Aviv and the Alexander (Fishelson, 1962; Trewavas, 1965) and are believed to be the result of releases by aquarists (Ben-Tuvia, 1981). (See also G. Hulata, in Pullin, 1988.)

Japan

Nile Tilapia from Egypt were introduced to Japan in 1962, where the species is a useful food fish and is widely reared in naturally hot springs or in the cooling water discharge from factories or power plants (Chiba *et al.*, 1989). Nile Tilapia have also colonized estuarine waters in the Okinawa Islands (Imai, 1980).

Malaysia

Nile Tilapia were first imported to Malaysia in the 1960s, but as a result of deliberate hybridization with *O. mossambicus* pure-bred *O. niloticus* eventually disappeared. In 1979, 400 more were imported from Chengmai, Thailand (see below) by the Department of Fisheries, who reared them at their Jitra hatchery, Kedah. These fish comprised the broodstock for the Department's hatcheries to the present day. In Malaysia, Nile Tilapia are mono- or polycultured in ponds, dams, and reservoirs, and are established in the wild in mining pools (Ang *et al.*, 1989).

Philippines

Since 1970, Nile Tilapia from various sources, including Thailand and Israel, have been introduced to the Philippines, where the species 'is now an important freshwater fish cultured in pens, cages and ponds' (Juliano *et al.*, 1989). Welcomme (1981, 1988) says that Nile Tilapia are 'established in Laguna de Bay area', where hybridization with *O. mossambicus* has occurred.

Sri Lanka

Nile Tilapia from East Africa were first imported to man-made lakes in Sri Lanka in 1956 and again in 1975 (Mendis, 1977), where they are now a common and popular food fish in certain reservoirs (De Silva, 1987).

Taiwan

In 1966, 56 Nile Tilapia from Japan were introduced to Taiwan by Mr T. C. Yu and Mr Imai (Chen, 1976; Liao & Chen, 1983). *O. niloticus* became a popular aquacultural species, and also as a hybrid with *O. mossambicus* (q.v.) and especially with *O. aureus* (q.v.) (Liao & Liu, 1989). The status of Nile Tilapia in the wild in Taiwan is unclear.

Thailand

Nile Tilapia from Japan were first introduced for aquacultural purposes to Thailand on 25 March 1965, when Prince Akihito sent 50 fingerlings to the King of Thailand. Subsequent introductions were made in 1972 (20 000), 1973 (15 000), and in 1981 (35 400) (De Iongh & Van Zon, 1993), and in 1983 (1000 fingerlings from Israel). Nile Tilapia 'are now well established in the ecosystem of Thailand. They are completely viable. *O. niloticus* is more popular as food than [*O. mossambicus*] . . . [they] are capable of reproducing in reservoirs and in natural waters' (Piyakarnchana, 1989). According to Welcomme (1981, 1988), Nile Tilapia are apt to decline in numbers some 4–5 years after stocking in natural waters. Philippart and Ruwet (1982) classify the Nile Tilapia as a 'failure' in Thailand.

Vietnam

Welcomme (1981, 1988) lists the Nile Tilapia as 'widespread in lakes' in Vietnam, but gives no date or source of the introduction.

Unsuccessful Introduction

Haas and Pal (1984) list an attempt to establish Nile Tilapia for the control of mosquito larvae in Turkey, where they were unable to survive low winter temperatures.

Africa

Kenya; Tanzania; Uganda

Although *O. niloticus* occurs naturally in Lakes Albert, Edward and George in Uganda, and in most of the Rift Valley lakes in Kenya, transplantations into Lake Kyoka, Uganda, and to Lake Victoria, which borders on Kenya, Tanzania and Uganda, and to other waters in the three countries, are treated here as introductions.

Welcomme (1967) listed the following introductions to date of Nile Tilapia to Lake Victoria and affluent waters:

DATE	LOCALITY	NUMBER	REMARKS
1954	Kagera River, Uganda	?	Via Koki Lakes which were first stocked in 1936 from Lake Bunyoni
1956–1957	Lake Victoria generally	c.1500 (including *T. zilli* & *O. leucostictus*)	—
1957	Kisumu, Kenya	Under 200	—
1958	Entebbe, Uganda	c.1290	—
1959	Mwanza, Tanzania	4153	Mwalogwabagole and Butimba prison ponds
1960	Smith Sound, Mwanza, Tanzania	414	—
1961–1962	Entebbe, Uganda	1474	—

The rationale for these introductions was the decline in the lake's fisheries due to overcropping and a desire to augment the surviving stocks (Welcomme, 1964, 1966; Craig, 1992).

After the introduction of Nile Perch *Lates niloticus* (q.v.) to Lake Victoria in the late 1950s, the large flock of more than 300 endemic haplochromine cichlids and many other non-haplochromine species greatly declined in numbers and distribution. Stocks of the native pelagic cyprinid *Rastrineobola argentea* and of the introduced *O. niloticus*, however, increased (Witte *et al.*, 1992), and by 1959 the latter was reported to be spreading rapidly throughout the lake; in the following year the species featured for the first time in commercial fishing records (Welcomme 1964, 1967).

Moreau *et al.* (1988) list the Nile Tilapia as established in Lake Kitangiri, Tanzania, where it has hybridized naturally with the endemic *O. amphimelas*; in Lake Naivasha, Kenya, where it was introduced in 1965 and died out, according to Muchiri and Hickley (1991), in 1967; and to Lakes Koki (from Lake Albert in 1935) and Bunyoni, (in 1932), Uganda: in Lake Bunyoni, *O. niloticus* has hybridized with and eradicated the previously introduced *O. spilurus nigra/niger* (q.v.). Worthington (1932) says that the introduction of *O. niloticus* to Lake Bunyoni took place in 1927 and 1930, where Trewavas (1933) recorded the species as still very uncommon. It subsequently increased and became a valuable food fish in the lake (De Vos *et al.*, 1990). Philippart and Ruwet (1982) refer to the species' successful establishment in Lake Nkugute, Uganda, where it has hybridized with *O. esculentus* (Lowe-McConnell, 1958).

Ecological Impact 'The introduction of Nile perch and Nile tilapia, as well as other alien fishes, into Lake Victoria, combined with overfishing for the indigenous cichlid species, has resulted in marked changes to the fish communities and the fisheries that depend on them. The most important impacts of . . . the tilapias include hybridization, overcrowding, competition for food and possibly the introduction of parasites and diseases' (Bruton, 1990).

Examining the impact of Nile Tilapia in greater detail, Ogutu-Ohwayo (1990a,b) suggested possible competition for living space in offshore waters with the native *O. esculentus* in many places *O. niloticus* became the dominant species to the virtual exclusion of *O. esculentus* (Welcomme, 1966; Lowe-McConnell, 1982). Since it is a mouthbrooder and a phytoplanktonivore, adult Nile Tilapia also compete with the native species for nursery sites (Lowe-McConnell,

1982) and possibly when juvenile also for food (Ogutu-Ohwayo, 1990a).

Among the tilapiine species, only *O. niloticus* has remained abundant in Lakes Victoria and Kyoga. The parallel success of *O. niloticus* and the predatory *Lates niloticus* may, Ogutu-Ohwayo (1990a) suggests, be related to their similar natural distribution. *O. niloticus* is less likely to be vulnerable to predation by *L. niloticus* than native tilapiines in the lakes since it has evolved in the presence of the latter. Furthermore, *O. niloticus* attains a larger size, grows and breeds more rapidly, has greater longevity, is more catholic in its diet, and has less limiting habitat requirements than any of the native tilapiine species (Fryer & Iles, 1972). Finally, *O. esculentus* (and another native, *O. variablis*) may have become less fit than *O. niloticus* as a result of competition or through genetic deterioration after hybridization with an introduced tilapiine (Welcomme, 1964, 1966; Ogutu-Ohwayo, 1990a).

Madagascar

Nile Tilapia were first introduced to the island of Madagascar in 1956, specifically to Lakes Alaotra in 1960 and Itasy in 1962 (Reinthal & Stiassny, 1991), and in 1970–1972 to Lakes Mantasoa and Tziazompaniry. Moreau *et al.* (1988) say the introduced stock came from the Egyptian Nile and Mauritius, and that spontaneous hybridization has occurred in Lake Itasy, presumably with introduced Mozambique Tilapia *O. mossambicas* (q.v.).

During their survey of Malagasy waters in 1988, Reinthal and Stiassny (1991) collected the species from the following waters: the River Mangoro at an altitude of 850 m in the northeast central rainforest; the River Namorona at an altitude of around 800 m in the Ranomafana National Park in the southeast central rainforest; and in a small stream by Soaindrana, south of Fianarantsoa, at an altitude of about 1300 m on the central high plateau. Moreau *et al.* (1988) say that Nile Tilapia also occur in waters near the coast. (See also Lamarque *et al.*, 1975; Philippart & Ruwet, 1982.)

For the Nile Tilapia's ecological impact in Madagascar see under Common Carp *Cyprinus carpio* or Largemouth Bass *Micropterus salmoides*.

Rwanda

Between 1935 and 1938, some European farmers in northeastern Rwanda, with the official approval of the Belgian colonial administration, obtained young Nile Tilapia from Lake Bunyoni over the border in neighbouring Uganda, which they released in Lakes Bulera and Luhondo in Rwanda. With the Nile Tilapia were accidentally introduced a small number of a *Haplochromis* sp. Both exotics adapted well to conditions in Lake Luhondo, in which for many years *Oreochromis* considerably outnumbered the less fecund **Haplochromis**, and the population of the latter only began to exceed that of the former after the early 1950s (De Vos *et al.*, 1990).

Welcomme (1981, 1988), who mentions introductions of Nile Tilapia to Rwanda in 1951 and 1952, says the species is now 'successful and very widely distributed in lakes'.

Ecological Impact   Before the introduction of *Oreochromis niloticus* and the *Haplochromis* sp. to Lake Luhondo, the principal, if not the only, autochthonous fish species were three cyprinids: a small and abundant barbel *Barbus neumayeri* (*luhondo*) and two larger ones, *B. microbarbis* (which may not have been a valid species but rather a hybrid between another *Barbus* and a *Varicorhinus* species) and *Varicorhinus ruandae*. Since the introduction of the two exotics, which are now the dominant species in the lake, and certainly since 1952, the two larger indigenes seem to have disappeared, while the smaller *B. neumayeri* has been reduced to a relict population in some small effluent tributaries. De Vos *et al.* (1990) believed that the introduced *Oreochromis*, and to a lesser degree the *Haplochromis*, played an important part in the eradication and decline of the three native species, though in what form they were unable to say.

Information on the impact of the Nile Tilapia in other Rwandan waters seems to be unavailable.

South Africa

Nile Tilapia from Israel were first imported to the Jonkershoek hatchery in the Cape before 1955 and again in 1959 as a forage fish for introduced black bass *Micropterus* spp. and for experimental aquacultural purposes (Van Schoor, 1966). The objective of the latter was the production of a species (or a hybrid) with a wider range of temperature tolerance than the native Mozambique Tilapia *O. mossambicus* (q.v.) and a more rapid rate of growth than the native Banded Tilapia *Tilapia sparrmani* (De Moor & Bruton, 1988). In about 1959–1960, Nile Tilapia were placed in 15 farm dams in the valleys of the Eerste and Lourens Rivers, where surveys carried out in 1964 revealed thriving populations in 11 of these dams (Van Schoor, 1966).

In 1978, Nile Tilapia were stocked with Blue Tilapia *O. aureus* (q.v.) from the Fisheries Development Corporation Hatchery at Amatikulu in a small dam in northern Natal, where the fish were subsequently destroyed (De Moor & Bruton, 1988).

In 1982, both Nile and Blue Tilapia were released in the Dudley Pringle dam in the Wewe River catchment, Natal, by the Tongaat Sugar Company (Bruton & Van As, 1986).

According to De Moor and Bruton (1988), Nile Tilapia are today probably established in waters in the Wewe River catchment and in farm dams in the catchments of the Eerste and Lourens Rivers, though confirmatory evidence is lacking.

Ecological Impact Since Nile Tilapia produce fertile hybrids when crossed with Mozambique Tilapia, the genetic integrity of the native species is likely to be contaminated wherever the two occur sympatrically. Although in densely populated ponds in the Jonkershoek hatchery, Nile Tilapia effectively cleared thick growths of *Limnanthemum* vegetation, in less heavily populated waters they had no significant impact on aquatic plants (Van Schoor, 1966).

Other introductions

Welcomme (1981, 1988) and Moreau *et al.* (1988) list the following introductions (with additions) of Nile Tilapia, mainly for aquacultural purposes, elsewhere in Africa, though given the species' known natural distribution many, if not most, may be domestic transplantations. Countries not listed by Welcomme include *Benin* (1979) and *Mali* (1982), both from Bouaké, Ivory Coast (C. Nugent, in Pullin, 1988).

Burundi; Cameroun; Central African Republic; Guinea; Ivory Coast; Sierra Leone; Tunisia; Zaire

COUNTRY	DATE	SOURCE	REMARKS
Burundi	1951–1952	Zaire	Established in lakes of Akagera. Basis of local fishery

COUNTRY	DATE	SOURCE	REMARKS
Cameroun	1958 & 1975	Bangui, Central African Republic	Established in Noum, Djerem, Segana Ayamé (since 1962) and Kossou (since 1971) marshes, and later in other (artificial) waters. Has replaced native species in the wild
Central African Republic	1957/1963	Congo	From Djoumouna
Guinea	1978 [and 1983]	?	—
Ivory Coast	1957	Upper Volta [Burkina Faso]	Has hybridized in captivity. Very successful fishery. Appeared in rivers by diffusion.
Sierra Leone	1978 [and 1970]	Upper Volta via Bouaké, Ivory Coast	—
Tunisia	1966	?	Reproducing
Zaire	1975	Bangui, Central African Republic	—

North America

Mexico

The current status of the Nile Tilapia in Mexico is uncertain. Welcomme (1981, 1988) lists the species as introduced in 1978 from Costa Rica, to which he says it was imported from Panama in 1979 (*sic*), and that it is 'established in one farm only'. Philippart and Ruwet (1982) state that Nile Tilapia are 'cultivated in numerous fish stations and established in numerous artificial lakes in the center and south of the country'. Contreras and Escalante (1984) say only that official records show that Nile Tilapia were one of the tilapiines imported by the Departamento de Pesca, without indicating any releases or establishment in the wild.

Unsuccessful Introduction

Crittenden (1962) described an attempt in the previous year to establish Nile Tilapia in Florida, USA, where they subsequently died out.

South and Central America

Puerto Rico

'The Nile Tilapia was introduced from Brazil in 1973 for experimental purposes (production of all-male hybrids) by the University of Puerto Rico' (Erdman, 1984), where it was established only in 'experimental ponds' and is regarded as a 'controversial' introduction. (Elsewhere, Erdman (1984) gives the date of introduction as 1974.) (See also Wetherbee (1989) and under *O. mossambicus.*) Welcomme (1981, 1988) lists the introductions shown in the table below of Nile Tilapia elsewhere in South and Central America. Countries not listed by Welcomme include **Paraguay** (1968) and **Venezuela** (1971), both from Bouaké, Ivory Coast (C. Nugent, in Pullin, 1988).

Oceania

Cook Islands

In 1993, two shipments of Nile Tilapia from Fiji were imported to Rarotonga in the Cook Islands; no fish survived from the first consignment, and it is too early to assess the success or failure of the second (Eldredge, 1994).

Fiji

Nile Tilapia from Israel were introduced to Viti Levu, Fiji, in 1968, for subsistence aquaculture, and between 1988 and 1990 were transferred to Vanua Levu (Maciolek, 1984; Andrews, 1985; Nelson & Eldredge, 1991; Eldredge, 1994). Populations of fish that presumably escaped from captivity have become established in local freshwater streams in Fiji (Nelson & Eldredge, 1991).

Mauritius

In 1957, Nile Tilapia from Tanzania were successfully introduced for sporting purposes to the island of Mauritius (Moreau *et al.*, 1988), where Welcomme (1988), who says that they were introduced from Madagascar, reported them to be 'widespread and expanding'.

COUNTRY	DATE	SOURCE	REMARKS
Bolivia	1977	Brazil	Established Methodist mission Altobeni
Brazil	1971–1972	Ivory Coast	Excellent quality for rearing but tendency to stunt makes it useless. *O. urolepis hornorum* × *O. niloticus* hybrid used to avoid this
Colombia	1980	Panama	Established in aquaculture installation in Huila Province. (This introduction may be *O. aureus.*)
Costa Rica	1979	Panama	Well established in fish culture stations. Important for hybrid production
Cuba	1967	Peru	Established all over the country. Acclimatized
Dominican Republic	1979	?	Reproducing artificially
Ecuador	?	?	*Oreochromis* sp. have become established in the Chone river where they have invaded the traditional 'chamera' fishery and compete with *Dormitator latifrons* reared there
El Salvador	1979	USA [probably Auburn University]	Tilapias contribute over 50% of commercial catch
Guatemala	1974 [but see under El Salvador]	El Salvador	Widespread in fish farms . . . and also found in many small lakes
Haiti	1977–1978	?	Reproducing
Honduras	1978	USA [probably Auburn University]	—
Nicaragua	1964 [but see under El Salvador]	El Salvador	Has disappeared (Welcomme, 1981). Reproducing (Welcomme, 1988)
Panama	1976	Brazil	Established in ponds and reservoirs. Very useful, rapid growth and excellent food value
Peru	1978	Brazil	Established in two fish stations
	1979	Ecuador; Israel	

Réunion

In 1957, Nile Tilapia from Madagascar were imported as a sport fish to the island of Réunion, where they have become established (Welcomme, 1981, 1988).

Hybrid Species

An *O. niloticus* × *O. mossambicus* hybrid (so-called 'Red Tilapia'), apparently first bred in Taiwan, has been widely distributed for experimental aquacultural purposes, especially in South and Central America and in Indonesia, as follows (Welcomme, 1988):

COUNTRY	DATE	SOURCE
Indonesia	1981	Philippines
Cuba	1979	Philippines
Dominican Republic	1981	Taiwan
Honduras	1978	Panama
Panama	1978	Taiwan

General Ecological Impact

See under Redbreast Tilapia *Tilapia rendalli*.

Oreochromis spilurus

The young and adults of *O. spilurus* feed on phytoplankton and snails respectively.

Natural Distribution

Eastward-flowing coastal rivers of Kenya and Somalia, including the Webi, Shebeli, Ewaso Nyiro, Tana, Athi, Tuchi and Voi.

Naturalized Distribution

Africa: Madagascar; Uganda; Zaire.

Africa

Madagascar

In 1955, *O. spilurus nigra/niger* was introduced from Kenya to the island of Madagascar (Kiener, 1963), where M.M.J. Vincke (pers. comm. to Pullin, 1988) reported that it still occurs in some rice fields around Antsirabé. Reinthal and Stiassny (1991) make no mention of this introduction.

Uganda

Welcomme (1988) says that in 1958 *O. spilurus* was unsuccessfully introduced from Kenya to Lake Bunyoni near the border with Rwanda in extreme southwestern Uganda. Moreau *et al.* (1988), who list the subspecies involved as *O. s. nigra* and give the

date of this introduction as 1927, say that *O. spilurus* disappeared after hybridizing with *O. niloticus* (q.v.) which was introduced in 1932. Between 1962 and 1966, the species was successfully introduced to Lake Batadi, where Moreau *et al.* (1988) say it is the basis for a fishery. Welcomme (1988) says that *O. spilurus* has been stocked in dams elsewhere in Uganda.

Zaire

O. spilurus nigra/niger from Kenya was introduced in 1946–1947 to Zaire, where it became established in Yangambi and the River Maniema, but has since almost disappeared (Moreau *et al.*, 1988).

Translocations and Ecological Impact *O. spilurus nigra/niger* has been stocked in farm dams in Kenya since 1922, and in 1925 was translocated to fill a vacant ecological niche in Lake Naivasha in the Rift Valley northwest of Nairobi; here, after initially increasing, it hybridized with another exotic, *O. leucostictus* (q.v.), finally disappearing in 1971 (Moreau *et al.*, 1988). Introductions and translocations of all three subspecies of *O. spilurus* with other *Oreochromis* spp. has ensured widespread hybridization (Welcomme, 1988).

Unsuccessful Introductions Welcomme (1981, 1988) and Moreau *et al.* (1988) list the following attempts to establish *O. spilurus* elsewhere in Africa which are believed to have been unsuccessful (dates and sources where known): Ivory Coast; Cameroun (1950, Kenya); Congo (about 1965, Kenya); Egypt (Kenya); Benin (to Lake Nokoué to try to find a saline-resistant species superior to the native Blackchin Tilapia *Sarotherodon melanotheron*); Mozambique (Jhingran & Gopalakrishnan, 1974). (Welcomme (1988) misplaces Lake Nokoué in both Benin and the Ivory Coast!)

Allied Species

Courtenay *et al.* (1991), Devick (1991), Courtenay (1993) and Eldredge (1994) list the Rainbow Krib *Pelvicachromis pulcher* as introduced from Nigeria to Oahu in the Hawaiian Islands, where it has been established since 1984.

Wami Tilapia
Oreochromis urolepis hornorum

Two subspecies of the Wami Tilapia are known, the nominate *O. u. urolepis* and *O. u. hornorum*. The latter form, under the designation '*O. hornorum*', has been widely introduced, especially in South and Central America, for hybridization purposes, since matings between males of '*O. hornorum*' and females of several other *Oreochromis* spp. normally produce all-male offspring (Balarin, 1979).

Natural Distribution

The basins of the Wami (and perhaps the Rufiji) River, Tanzania. (See also under Zanzibar.)

Naturalized Distribution

Africa: Ivory Coast; Zanzibar. **North America:** ? Mexico; USA. **South and Central America:** Brazil; Colombia; Costa Rica; Cuba; Hispaniola (Dominican Republic); El Salvador; Guatemala; Honduras; Nicaragua; Panama; Peru; Puerto Rico.

Asia

Unsuccessful Introduction

De Silva (1987) lists an unsuccessful attempt in 1969 to introduce the Wami Tilapia for experimental

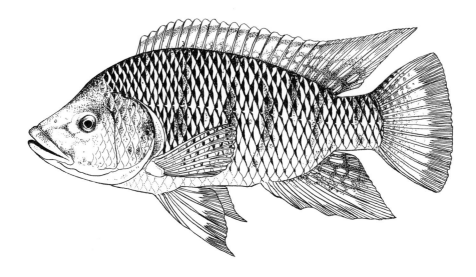

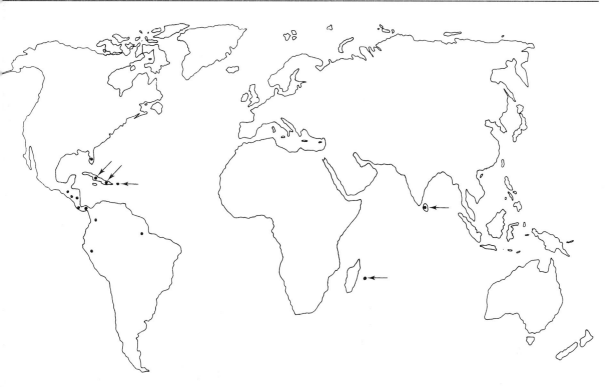

Principal introductions outside Africa of Wami Tilapia *Oreochromis urolepis hornorum* (adapted from Philippart & Ruwet, 1982).

aquaculture in Sri Lanka. Chiba *et al.* (1989) say that in 1981 the species was introduced (under the name *Tilapia macrocephala*) from Israel (to which it was imported in 1977 from Bouaké, Ivory Coast, via Pentecoste, Brazil) to Japan, where it is 'being reproduced in certain experimental or natural ponds'. Liao and Liu (1989) list the Wami Tilapia as introduced from Costa Rica to Taiwan in 1981 for the production of all-male hybrids in aquaculture.

Africa

Ivory Coast

Moreau *et al.* (1988) and C. Nugent (in Pullin, 1988) say that in 1967 40 juvenile Wami Tilapia from the Fish Culture Research Station at Batu Berandam, Malaysia were introduced for hybridization purposes to the Ivory Coast, where they became established at Bouaké, some 280 km northwest of Abidjan.

Zanzibar

Welcomme (1981, 1988) and Moreau *et al.* (1988) incorrectly place the Wami River on Zanzibar,

thereby including the island in the Wami Tilapia's natural distribution. It is now generally accepted that the species was originally introduced to Zanzibar from the Wami River system before records of fish introductions were begun.

Unsuccessful Introduction

Moreau *et al.* (1988) list an apparently unsuccessful attempt between 1962 and 1966 to introduce Wami Tilapia from Zanzibar to Uganda.

North America

Mexico

Welcomme (1981, 1988) lists the Wami Tilapia as successfully introduced to Mexico from Costa Rica in 1978. Contreras and Escalante (1984), however, make no mention of the species' introduction to Mexico.

United States

Hauser *et al.* (1976), Legner and Pelsue (1977), Legner (1979) and Legner *et al.* (1980) found Wami

Tilapia to be established and breeding in drainage channels in Imperial and Riverside Counties, California, and also in Coyote Creek, Long Beach, Los Angeles County. Knaggs (1977) reported the species, or a hybrid with *O. mossambicus* (q.v.) to be established in the Cerritos flood control channel, Cerritos Lagoon, and confirmed its presence in the Coyote Creek–San Gabriel River drainage. Wami Tilapia are also probably naturalized in the Bolsa Chica flood control channel in Huntington Beach, Orange County, California (Courtenay *et al.*, 1984, 1986). Courtenay *et al.* (1991) and Courtenay (1993) confirm the species' (or hybrid's) survival in California, and say it was first introduced in the 1970s.

Wami Tilapia were released in California at the University of California, Riverside, and by the Southeast and Orange County Mosquito Abatement Districts, to control excess growths of aquatic vegetation and to reduce populations of mosquitoes and chironomid midges (Legner & Pelsue, 1977).

South and Central America

Puerto Rico

Wami Tilapia from the Cooperative Fishery Research Unit in Tucson, Arizona, USA, were introduced to Puerto Rico in November 1963. Male *O. u. hornorum* were successfully crossed at Maricao with female *O. mossambicus* to produce all-male hybrids, which were reared in a number of island ponds (Erdman, 1984). To ensure the maintenance of pure-bred Wami Tilapia stock, D.S. Erdman planted around 100 fingerlings in a small (0.04-ha) pond on Mona Island (between Puerto Rico and Hispaniola) where the species is still established and breeding. Because of the small size of the pond, the fish never exceed 67 mm in length (Erdman, 1984). (See also Wetherbee, 1989.)

Brazil; Colombia; Costa Rica; Cuba; Hispaniola (Dominican Republic); El Salvador; Guatemala; Honduras; Nicaragua; Panama; Peru

Welcomme (1981, 1988) lists below the introduction of Wami Tilapia elsewhere in South and Central America.

Unsuccessful Introduction

Nelson and Eldredge (1991) and Eldredge (1994) list an unsuccessful attempt in 1985 to establish *O. u. hornorum* imported from Taiwan, on Viti Levu, Fiji.

COUNTRY	DATE	SOURCE	REMARKS
Brazil	1971–1972	Ivory Coast	Established in reservoirs. Excellent quality for intensive rearing but high proliferation leads to useless population of *O. hornorum* × *O. niloticus* hybrids. A positive introduction
Colombia	?	?	Reproducing
Costa Rica	1960+	?	Useful species for hybrid production
Cuba	1976 & 1983	Mexico & Nicaragua	Reproducing. Positive introduction
Hispaniola (Dominican Republic)	1980	Panama	Reproducing artificially
El Salvador	1979	USA	Mainly used for hybridization. Tilapias contribute over 50% of commercial catch
Guatemala	1974 [but see under El Salvador]	El Salvador	Established mainly in fish ponds but also in small natural ponds
Honduras	1979	Panama	Reproducing
Nicaragua	1974 [but see under El Salvador]	El Salvador	Reproducing
Panama	1976	Brazil	Reproducing. Undesirable in natural waters
Peru	1978 & 1980	Brazil & Panama	Established in one fish station. Monoculture experiments

Blackchin Tilapia
Sarotherodon melanotheron

The Blackchin Tilapia is a popular species not only with warmwater aquarists but also for laboratory experimentation. Its tolerance of high salinities makes it ideal for rearing in coastal lagoons, although it has seldom been actively cultured (Welcomme, 1988).

Natural Distribution

Coastal river delta lagoons from central Liberia to southern Cameroun.

Naturalized Distribution

North America: USA. *Oceania:* Hawaiian Islands.

North America

United States

In the late 1950s or early 1960s, Blackchin Tilapia became established in Hillsborough County, Florida, from Lithia Springs to the mouth of the Alafia River, and southward along the east coast in the brackish waters of Tampa Bay as far as Cockroach Bay, Manatee County (Springer & Finucane, 1963; Finucane & Rinckey, 1964; Buntz & Manooch, 1969b; Lachner et al., 1970; Courtenay et al., 1974, 1984, 1986). Springer and Finucane (1963) reported that in Tampa the species was marketed as the 'African Sunfish', and said that between 13 and 17 December 1962 a commercial catch weighing 1553 lbs (704 kg) had been taken in Bullfrog Creek, near Old Tampa Bay. The same authors suggested that the fish originated from a deliberate release or an escape from an aquarium fish farm on the eastern shore of Tampa Bay.

In the late 1970s, Blackchin Tilapia became naturalized in canals near Satellite Beach and in the neighbouring saline Banana and Indian Rivers, Brevard County, Florida, from Merritt Island to the south of Canova Beach, a distance of some 27 km, possibly following the release of aquarium fish into the reflecting pool at the Satellite Beach Civic Center (Dial & Wainright, 1983).

Courtenay and Stauffer (1990) recorded the Blackchin Tilapia's range in Florida as being from near Titusville, Brevard County, southward to near Vero Beach, Indian River County. (See also Courtenay & Robins, 1973; Hogg, 1974; Courtenay et al., 1984, 1986, 1991; Courtenay, 1993.)

Ecological Impact   Courtenay et al. (1974) reported that in Lithia Springs, where they comprised around 90% of the total fish biomass, Blackchin Tilapia had been observed grazing on green and blue-green algae (Cyanophyta). Aquatic vegetation of all kinds had been much reduced, apparently as a result of grazing by Tilapia. The only conspicuous native fishes in Lithia Springs, Bluegills *Lepomis macrochirus* and Largemouth Bass *Micropterus salmoides*, appeared diseased and ill-fed.

Oceania
Hawaiian Islands

In 1962, Blackchin Tilapia were imported from New York, USA (Eldredge, 1994) to the Hawaiian Islands, where they have become established in freshwater reservoirs and marine waters of the island of Oahu (Maciolek, 1984; Courtenay et al., 1991).

Unsuccessful Introduction

Welcomme (1988) lists an unsuccessful attempt to introduce Blackchin Tilapia to the former USSR.

Allied Species

In 1959, Israeli Tilapia *S. galilaeus*, natives of Israel, Jordan, and much of Africa north of the Zambezi, were imported from Israel, together with Redbelly Tilapia *Tilapia zillii* (q.v.) and Nile Tilapia *Oreochromis niloticus* (q.v.), to South Africa, where they were bred in the Jonkershoek hatchery in the Cape. Fry were subsequently stocked in farm dams around Stellenbosch (Van Schoor, 1966), where their descendants may survive (De Moor & Bruton, 1988).

Ecological Impact

Were *S. galilaeus* to become widely established in the wild in South Africa, it would probably compete for food with such native detritivores as *O. mossambicus* and *O. andersoni* (De Moor & Bruton, 1988).

Banded Largemouth (Nembwe)
Serranochromis robustus

A large and aggressive predatory cichlid, which lives in weedy areas or in the vicinity of submerged trees or rocks in the quieter stretches of rivers, from which it emerges to hunt its prey.

Natural Distribution
The catchment of the Zambezi River in Africa.

Naturalized Distribution
Africa: South Africa; Swaziland.

Africa
South Africa

In the early 1960s, the subspecies *S. r. jallae* (Angola, Okavango, Upper and Middle Zambezi, and Bangweulu) was introduced from Zimbabwe to the Umgeni hatchery in Natal in the hope that it would augment or replace Largemouth Bass *Micropterus salmoides* (q.v.) as a game fish (Bourquin *et al.*, 1984). Fish reared at Umgeni were introduced to the Uvonga River above the falls in 1965–1966 and at Margate quarries in 1968; to Cornhill farm dam, Port Edward, and to Seldon Park farm dam, Margate, in 1968; and at an unrecorded date to the Dreadnought hatchery and to a farm dam in the Eshowe district, and to farm dams in Empangeni (Pike, 1980; De Moor & Bruton, 1988).

In 1965, *S. r. jallae* were sent from Umgeni to the Jonkershoek hatchery at Stellenbosch in the Cape, where in the following year some were released in four farm dams at Uitkyk, Warwick and Woodlands in the Eerste River basin; in 1967 others were planted in the Rozendal farm dam, Stellenbosch, and in 1968 in the dam at Plant Quarantine Station, Stellenbosch (Van Schoor, 1969b).

De Moor and Bruton (1988) recorded the continued presence of this species in the Eshowe, Empangeni and Port Edward districts of Natal, and probably also in Stellenbosch in the Cape, although there had been no recent confirmed records of the species in the latter area.

Swaziland

In 1975, the nominotypical subspecies *S. r. robustus* was introduced from Lake Malawi (where it is endemic) to the Sand River dam in Swaziland, where it became established (Welcomme, 1981, 1988; De Moor & Bruton, 1988).

Translocations

The Banded Largemouth has been stocked in farm dams within its native range in Malawi, Zambia and Zimbabwe, mainly for sporting purposes and to control stunted populations of tilapia (Welcomme, 1988).

Gilmore (1978), quoted by De Moor and Bruton (1988), believed that the presence of this species in the Shashi dam on the Shashi River in the Limpopo system, Botswana, was probably as a result of their translocation in 1973–1974 from the Okavango River system by the Fisheries Section of the Ministry of Agriculture.

Ecological Impact

Because large native predatory freshwater fishes are absent from western and southern Cape rivers of South Africa, the impact of *S. r. jallae* on rare endemic fish communities in these waters, were they ever to become widely established, would probably be more serious than in Swaziland, Natal, and the Limpopo system of Botswana, where large native predators such as *Clarias gariepinus* and, in some localities, *Hydrocynus forskahlii*, are found (De Moor & Bruton, 1988).

Allied Species

Eldredge (1994), referring to Devick (1991), lists the establishment in fresh waters on Oahu in the Hawaiian Islands since 1982 of a *Pterophyllum* sp. of angelfish, a native of the Amazon basin in South America and a popular aquarium fish.

Welcomme (1988) says that *P. scalare*, which is very intolerant of low temperatures and favours fairly acidic water which limits its establishment outside its natural range, has been caught in the wild, but is unlikely to be reproducing, in Guyana and Surinam.

Natarajan and Ramachandra Menon (1989) list *Pterophyllum* spp. as having been introduced into Tamil Nadu, India, but do not indicate the outcome.

Spotted Tilapia (Black Mangrove Cichlid; Niger Cichlid; Tiger Cichlid)
Tilapia mariae

A popular but aggressive warmwater aquarium cichlid, the Spotted Tilapia occurs in both lentic and lotic waters; in rocky areas; underneath overhanging banks; and, where there is no cover, over muddy bottoms.

Natural Distribution

Fresh and brackish waters in lowlying coastal areas from central Ivory Coast to southwestern Ghana, and from southeastern Benin to southwestern Cameroun.

Naturalized Distribution

North America: USA. ***Australasia:*** Australia.

North America

United States

In April 1974, adult Spotted Tilapia were first collected in the wild in the USA in Snapper Creek Canal, and subsequently in Little River Canal, Tamiami Canal, and Coral Gables Waterway, all of which are indirectly interconnecting, in Dade County, South Miami, Florida. In early June fry and attendant parents were collected, and spawning pairs were observed. It was believed that the fish had either escaped from local aquarium fish farms or were discarded pets (Hogg, 1974). Two years later, Hogg (1976a) reported that the population had spread rapidly 8–16 km to the north, west and south, where its numbers were increasing. Hogg also collected Spotted Tilapia in lesser numbers over an area of some 64 km², including canals and lakes in the Little River, Biscayne and Opa Locka canal systems. A small discrete satellite population found in a disused quarry lake near Perrine, 12 km northeast of Homestead in southern Dade County, showed such strikingly different coloration, body shape, and fin size to those of the canal populations that Hogg (1976a) considered that they almost certainly originated from a separate source.

A survey of waters in southeastern Florida carried out in 1978 showed that Spotted Tilapia had become the dominant fish species in many lakes and canals in eastern Dade County (with population sizes in excess of 50% of the total fish biomass), and that they occurred in an area of some 2000 km² from near Homestead in southern Dade County north as far as the South New River Canal in southeastern–central Broward County. They were also collected in a pond south of Copeland in south–central Collier County, suggesting a westward expansion from Dade County through the Tamiami Canal (Courtenay & Hensley, 1979, 1980b).

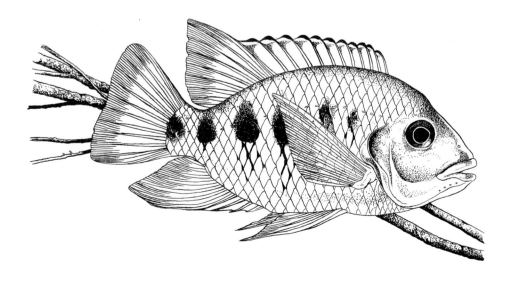

Courtenay *et al.* (1986) reported that a population discovered near Micco in Brevard County in 1979 had since died out, but that in southeastern Florida the Spotted Tilapia had continued rapidly to extend its range, occurring throughout the New River Canal system in central Broward County; in the Tamiami Canal in western Dade County; in the Aerojet Canal system, Dade County; and at the eastern entrance to the Everglades National Park. It had also been found below salinity dams in three canals in Dade County leading into Biscayne Bay.

The presence in Florida of such native predators as *Micropterus salmoides* and *Lepisosteus platyrhincus* has not prevented the Spotted Tilapia's spread in Florida (Courtenay & Deacon, 1983), which has been further helped by the species' ability to breed throughout the year, whereas native species have strictly delimited breeding seasons (Hogg, 1976b; Taylor *et al.*, 1984).

On 18 October 1980, Spotted Tilapias were collected in Rogers Spring, 2 km southwest of Blue Point Spring and some 68 km northeast of Las Vegas, above the Overton Arm of Lake Mead in the Lake Mead National Recreational Area, Clark County, Nevada, where both numerically and in terms of biomass they were found to be the dominant fish species; they were believed to have originated in the release of pet fish (Courtenay & Deacon, 1982, 1983).

Most of the Spotted Tilapia in Rogers Spring, even large individuals, had the banded pattern of juveniles, only relatively few large fish displaying the typical adult pattern of spots on their flanks. Furthermore, most of the juvenile-patterned individuals were missing parts of their pelvic fins, and other fins (especially the soft dorsal and upper caudal) were frequently damaged. The only fish present able to inflict such injuries were typically-coloured adults, none of which showed damaged fins. Populations of Spotted Tilapia in Florida have never reached the numerical density of the one in Rogers Spring. It thus seems likely that the trophic and spatial capacities of Rogers Spring for the species have been reached, and that this social/behavioural hierarchy, led by a small number of extremely aggressive individuals, has evolved to prevent excessive overpopulation (Courtenay & Deacon, 1982, 1983). (See also Hogg, 1976b; Courtenay *et al.*, 1984, 1986; Courtenay & Stauffer, 1990; Courtenay *et al.*, 1991; Courtenay, 1993.)

Ecological Impact   Courtenay and Hensley (1979) expressed their concern at the potential ecological impact of Spotted Tilapias in Florida. Nearly all the canals occupied by *T. mariae* were also inhabited by other naturalized exotics, including the Oscar *Astronotus ocellatus* (q.v.), the Black Acara *Cichlasoma bimaculatum* (q.v.), the African Jewelfish *Hemichromis bimaculatus* (q.v.), the Mozambique Tilapia *Oreochromis mossambicus* (q.v.) – all cichlids – and the Pike Killifish *Belonesox belizanus* (q.v.) – a poeciliid – as well as by native centrarchids and other fishes. In canals containing Spotted Tilapia and other aliens, native species were in decline. In some waters, populations of other previously flourishing exotics, such as Oscars and Pike Killifish, appeared to have decreased. Hogg (1974) observed Spotted Tilapia parents aggressively defending their nests and young against other fish, especially Black Acaras, while Courtenay and Hensley (1979) noted them preventing the same species from occupying its usual shallow water spawning areas and confining it to the margins.

Although in its native West African range the Spotted Tilapia is an omnivore, in Florida it has shown a preference for green algae. When the species was discovered in Nevada, the bottom of Rogers Spring was found to be barren of green algae, presumably as a consequence of grazing by *T. mariae*. Courtenay and Deacon (1982, 1983) expressed fear that, were Spotted Tilapias to spread elsewhere, such a trophic preference, together with an omnivorous diet, could prove serious for a number of small endemic fish species and subspecies in the southwest, where in desert springs green algae are a favourite food. Some of these endemics also feed on small snails and ostracods (freshwater mussels and seed shrimps), for which green algae are their main food source. Furthermore, a fish capable of inflicting damage on its own species, as observed in Rogers Spring, would seem likely to prey on other fish species.

Australasia

Australia

During 1978, Spotted Tilapia (and Convict Cichlids *Cichlasoma nigrofasciatum* (q.v.)) were caught in eel-nets in the 480-ha cooling pond of the Hazelwood power station near Morwell, Victoria, and in Eel Hole Creek, an intermittent tributary of the Morwell River, below the pondage. Spotted Tilapia were found to be widely distributed, common, and breeding throughout the Hazelwood pondage and in Eel Hole Creek, occurring in both still and flowing water; in rocky and detrital areas; underneath overhanging banks; and in shallow mud-bottomed bays with little or no cover (Cadwallader *et al.*, 1980). Subsequently, Spotted Tilapia were also reported, by Cadwallader and Backhouse (1983), from the warm waters of La Trobe River where it receives heated effluent from the Yallourn power station, some 10 km north of Hazelwood.

Clements (1988) was informed of the probable presence since around 1971 of Spotted Tilapia from two localities near Cairns in northern Queensland – Ferny Creek at Palm Grove and the creek associated with the Australian Bird Park north of Cairns. McKay (1989) confirms the species' presence in this area. (See also Merrick & Schmida, 1984; McKay, 1984, 1986–1987; Allen, 1989; Arthington, 1989, 1991.)

Ecological Impact Cadwallader *et al.* (1980) expressed concern about the establishment of Spotted Tilapias and Convict Cichlids in Victoria. Although it seems possible that both species, and the Jack Dempsey *C. octofasciatum* (q.v.), could establish self-maintaining populations in the warmer stretches of La Trobe River, they would be unable to survive the much lower winter temperatures (<15°C) in most natural Victorian waters, and thus in the opinion of Arthington (1986) pose no threat to the state. Indeed, Arthington reported recent evidence that numbers at Hazelwood had recently declined 'to very low levels'. Because of their known ability to become naturalized in other parts of the world, however, these cichlids are all regarded as potentially destructive species in Australia (McKay, 1984), where their importation is prohibited (Arthington, 1986).

Redbreast Tilapia
Tilapia rendalli

T. rendalli and *T. zillii* (q.v.) are virtually indistinguishable morphologically and ecologically, and are frequently confused when introductions and distribution are described. The position is further confused because where they occur sympatrically the two species appear to hybridize freely. Both fish, to a greater or lesser degree, graze on higher plants, and have thus been extensively introduced for the control of aquatic vegetation as well as for aquacultural purposes. Both *T. rendalli* and *T. zillii* have become widely naturalized and have proved somewhat controversial; in some localities they are valued for aquatic weed control and form the basis for useful commercial fisheries, while in others their ability to control vegetation has been questioned and they have achieved pest status (Welcomme, 1988). In some places *T. rendalli* is highly regarded as an angling fish and as a source of human food (De Moor & Bruton, 1988). (*T. melanopleura* = either *T. rendalli* or more probably *T. zillii*.)

Natural Distribution

Tropical Africa south to the lower Phongolo River in the east and the Cunéné River in the west, including the Katanga and Lualaba Rivers, Lakes Tanganyika and Nyasa, the Zambezi, coastal waters of Mozambique and Natal, and the Okavango.

Naturalized Distribution

Asia: ? Sri Lanka; ? Taiwan; Thailand. ***Africa:*** Burundi; Kenya; Madagascar; Rwanda; Tanzania; Uganda. ***North America:*** ? USA. ***South and Central America:*** Brazil; Colombia; Cuba; El Salvador; Peru; Puerto Rico; West Indies (Antigua). ***Australasia:*** Papua New Guinea. ***Oceania:*** Hawaiian Islands; Mauritius; Wallis Island (Wallis and Futuna).

Asia
Sri Lanka

Evidence for the introduction to, and present status of, the Redbreast Tilapia in Sri Lanka is contradictory and confusing. Welcomme (1981) says the species was introduced from Malacca [Malaysia] in 1969 'to fill a vacant niche in aquaculture', and that it became 'very important economically'. Philippart and Ruwet (1982) describe it as 'well established in natural

waters'. De Silva (1987) listed the Redbreast Tilapia as 'introduced from East Africa for stocking as a food fish in man-made lakes'. Welcomme (1988) states that it was introduced from Zaire in 1955 and is 'not popular and although species can breed it is disappearing as it cannot compete with local species'.

Taiwan

In 1981, 25 Redbreast Tilapia from South Africa were imported to Taiwan 'for the mere purpose of adding a new species to the local gene pool' (Liao & Chen, 1983; Liao & Liu, 1989).

Thailand

In 1955, Redbreast Tilapia from Zaire via Belgium were imported into Thailand, where although they initially became well established in the wild they then progressively declined, due probably to competition from local species (Welcomme, 1981, 1988; Philippart & Ruwet, 1982). Piyakarnchana (1989) does not mention this introduction.

Unsuccessful Introductions

In Asia, Redbreast Tilapia have been unsuccessfully introduced to Malaysia (Jhringran & Gopalakrishnan, 1974; Ang *et al.*, 1989) and Vietnam (Jhringran & Gopalakrishnan, 1974).

Africa

Kenya; Tanzania; Uganda

The decline in the fisheries of Lake Victoria (Kenya, Tanzania and Uganda) and Kyoga (Uganda), as a result of overfishing, led to the introduction in the early 1950s and 1960s of exotic tilapiines to augment the surviving stocks of native fishes (Welcomme, 1964, 1966; Ogari, 1990) which they eventually replaced (Craig, 1992).

In 1955, Redbreast Tilapia from an apparently unrecorded source were introduced for aquaculture and fishing to the Kenyan waters of Lake Victoria (where they have hybridized with *T. zillii* (q.v.)) and to the Tana River, in both of which they became established (Welcomme, 1981, 1988).

In 1952, 3 years before they say the species was introduced *to* Kenya, Moreau *et al.* (1988) say the Redbreast Tilapia was introduced *from* Kenya to Uganda (presumably to Lake Kyoga), from which it 'escaped into Lake Victoria' (Welcomme, 1981, 1988) where it has hybridized with the introduced *T. zillii* (q.v.) (Moreau *et al.*, 1988).

In 1962, Redbreast Tilapia, again from an apparently unrecorded source (probably Kenya), were introduced to stock dams in Tanzania where the species 'contributes to the fisheries' (Welcomme, 1988). The Redbreast Tilapia also occurs in the Tanzanian waters of Lake Victoria by natural diffusion from the waters of Kenya and/or Uganda. Trewavas (1966) reported the species' escape from aquaculture into the Pangani River in Tanzania, where it became established.

Madagascar

In 1955, Redbreast Tilapia were introduced to Lakes Alaotra and Itasy in Madagascar (Reinthal & Stiassny,

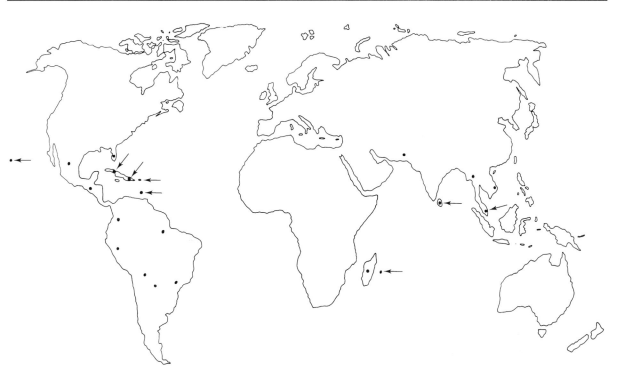

Principal introductions outside Africa of Redbreast Tilapia *Tilapia rendalli* (adapted from Philippart & Ruwet, 1982).

1991). (According to De Moor & Bruton (1988) the introduction was accidental.) Welcomme (1981, 1988) and Moreau *et al.* (1988) say that the introduction took place in 1951 (from the Djoumouna Fish Culture Centre in Brazzaville in the Congo) and that the species is 'widespread' on the island, where it is 'important for aquaculture and fisheries in natural waters including rivers in the highlands. Useful for weed control and supports high altitude and high salinity in lakes such as Lake Ihotry'. According to J. Moreau (in Pullin, 1988), '*T. rendalli* exhibits great cold tolerance in Madagascar (surviving down to 8–10°C) and is the only *Tilapia* for which reproduction has been recorded in natural waters above 1600 m (Kiener, 1963) ... It can still be found as a pure species in high altitude lakes ... It also exhibits exceptional salt-tolerance (almost up to full strength seawater) in the Ihotry Lake near Toliary in south-west Madagascar ... It remains the most widespread tilapia in Madagascar and can still be found as a pure species in this lake'. Lamarque *et al.* (1975) and Philippart and Ruwet (1982) say the Redbreast Tilapia is found in numerous rivers and lakes, and has seriously disrupted the ecological balance of Lake Kinkony (see below). Moreau *et al.* (1988) say it

does well in those waters without phytophagous species. Reinthal and Stiassny (1991), in their survey of 1988, collected Redbreast Tilapia in Lake Itasy on the central high plateau west of Antananarivo, and in a small feeder stream (about 1 m wide and a maximum of 2 m deep, at an altitude of around 50 m asl) of the River Mangoro on the central east coastal plain.

Ecological Impact Within 3 years of its introduction to Lake Kinkony, Madagascar, the Redbreast Tilapia had devastated nearly 3000 ha of *Ceratophyllum* and *Nymphaea* beds, which resulted in the virtual disappearance of a valuable native fish, *Paretroplus petiti* (Lamarque *et al.*, 1975; Philippart & Ruwet, 1982). (See also under *Cyprinus carpio* and *Micropterus salmoides*.)

Burundi; Rwanda

In 1956, Redbreast Tilapia from Zaire were introduced to both Burundi and Rwanda. In the former, the species 'forms the basis of a local fishery in the Akagera Lakes'. In the latter, it was 'stocked in Lake Kivu and other lakes. It is common in the Kagera River and its lakes where it contributes to commercial fisheries. In Lake Kivu its history is less certain as the species has disappeared' (Welcomme, 1981, 1988). Moreau *et al.* (1988), who say that the introductions to both countries took place in 1948–1949, list the species as established in Burundi in Lake Bugesera and in Rwanda in Lakes Akagera and Kivu.

Unsuccessful Introductions

In Africa, the Redbreast Tilapia has been unsuccessfully introduced or translocated in the Central African Republic (in 1953); to control aquatic vegetation in Sudanese irrigation canals (1960); in the Ivory Coast (1957–1958); Liberia (1960); the Congo (1953); Malawi; and Lake Kyle, Zimbabwe (1957–1962) (Moreau *et al.*, 1988). In about 1961, Redbreast Tilapia from Chilanga, Zambia, were planted in the waters of Lake Kariba (between Zambia and Zimbabwe) where, however, '*T. rendalli* already occurred in the system and there is no evidence to show that it would not have become abundant without stocking' (Marshall, 1979).

Successful Translocations

The subspecies *T. r. swierstrae* (the Limpopo system south to the Mkuzi, and also the Incomati system to around 1000 m) has been successfully translocated outside its natural range in South Africa in the Transvaal and Natal for the control of aquatic macrophytes and mosquito larvae (De Moor & Bruton, 1988).

T. rendalli has also been successfully established since 1949 in the Rivers Djerem and Noum in Cameroun; in Lake Lushiwashi in Zambia, to which it spread after the rupture of a dam; and since 1948 in Lakes Kivu and Munkumba, and elsewhere in Zaire (Moreau *et al.*, 1988).

Ecological Impact

In the Transvaal and Natal transplanted Redbreast Tilapia have had a major ecological impact as a result of their destruction of aquatic vegetation, which has deprived other fish, such as the Mozambique Tilapia *Oreochromis mossambicus*, and aquatic birds, such as the African Black Duck *Anas sparsa*, of shelter, nest-building material, and food (Batchelor, 1978). The eradication of the Nile Crocodile *Crocodylus niloticus* from many waters may have contributed to the population explosion there of Redbreast Tilapia (For further details see De Moor & Bruton, 1988.)

North America
United States

T. rendalli is 'taxonomically liable to be confused with *T. zillii* and may be in continental US waters but at present all individuals of the *T. zilii/T. guineensis/T. rendalli* group are treated as *T. zillii* (Welcomme, 1981, 1988).

Unsuccessful Introduction

Welcomme (1988) lists the introduction of Redbreast Tilapia to Mexico in 1974. This is not mentioned by Contreras and Escalante (1984).

South and Central America
Brazil

In 1953, Redbreast Tilapia from Zaire were introduced to Brazil, where they were intensively reared in parts of the northeast and central–south of the country. Escaped or released fish became established in the wild, notably in Lago Pinheiro in Brazilia and in many hydroelectricity reservoirs in São Paulo State (Nomura, 1976, 1977; Philippart & Ruwet, 1982).

Ecological Impact   Nomura (1976, 1977) documented numerous accounts of stunting of Redbreast Tilapia in Brazil as a result of explosive population increases and consequent destruction of aquatic macrophyte vegetation.

Colombia

Introduced from the USA to Colombia since 1960, Redbreast Tilapia were initially cultivated in ponds from which they later escaped or were released into natural waters in the Valle del Cauca, Huila and Atlantico Provinces, between 1000 and 1400 m asl (Philippart & Ruwet, 1982; Welcomme, 1988).

Puerto Rico

On 27 July 1963, 19 juvenile Redbreast Tilapia from Auburn University, Alabama, USA, were imported to Puerto Rico to control such rooted macrophytes as 'southern naiad'; the species is also an efficient predator of the pulmonate snail *Biomphalaria glabrata*, which is common in farm ponds in Puerto Rico where it acts as an intermediate host for the blood fluke *Schistosoma mansoni*, the vector of bilharzia.

From time to time, particularly in flood conditions,

tilapia are commercially fished for sale in the market at Río Piedras and elsewhere. Although less abundant than *O. mossambicus*, *T. rendalli* is more readily caught on hook and line, and thus contributes to the hundreds of kilograms resulting from this fishery (Erdman, 1984). (See also Wetherbee, 1989.)

Ecological Impact In 1966, blue-green algae (Cyanophyta) began infesting the Loiza reservoir near Carraizo dam, where they were expensively treated with copper sulphate. In April of that year, Redbreast and Mozambique Tilapia were introduced to the reservoir, where shortly afterwards the algae disappeared.

Other Introductions

Welcome (1981, 1988) has summarized the introduction of the Redbreast Tilapia elsewhere in South and Central America:

COUNTRY	DATE	SOURCE	REMARKS
Cuba	1968 & 1970	Mexico	Present throughout the country but its use is limited by overpopulation and stunting
El Salvador	1960	USA	This and other tilapias contribute over 50% of commercial catch
Peru	1966	Brazil	Established in some lakes
West Indies (Antigua)	?	?	Reproducing

Unsuccessful Introductions

In South and Central America, Redbreast Tilapia were in 1979 introduced from Mexico to the Dominican Republic, where they are reproducing only in aquacultural installations, and in 1977 from Puerto Rico to Panama where they were eradicated because of their poor growth rate (Welcomme, 1981, 1988). Philippart and Ruwet (1982) also list the species as cultivated in ponds in Bolivia and Paraguay.

Australasia

Papua New Guinea

Eldredge (1994), quoting Osborne (1993), states that the Redbreast Tilapia is a 'recent introduction'

to the Sepik and Ramu Rivers in Papua New Guinea, where in 1991 150 000 fingerlings were imported from the UK 'to enhance stock, rapidly spreading'.

Oceania

Hawaiian Islands

Maciolek (1984) received 'unconfirmed reports of other established fishes such as the hybrid red tilapia and *Tilapia rendalli*' in the Hawaiian Islands, where Welcomme (1988) said the species was introduced in 1956 and was reproducing only 'artificially'. No mention of the species is made by several other authorities (e.g. Brock, 1960.)

However, Courtenay *et al.* (1991) and Courtenay (1993) say the Redbreast Tilapia is naturalized in the Hawaiian Islands, where the latter gives the date of its introduction as 1957.

Mauritius

Welcomme (1981, 1988) lists the Redbreast Tilapia as introduced in 1956 from Madagascar to establish a commercial fishery in Mauritius, where it escaped into reservoirs and rivers and is now 'widespread and expanding'.

Ecological Impact George (1976) and Philippart and Ruwet (1982) say that Redbreast Tilapia in Mauritius have from time to time seriously affected the native fauna and macrophyte flora of those waters in which they have become established.

Wallis Island (Wallis and Futuna)

Eldredge (1994) lists the Redbreast Tilapia as established in Lake Kikila on Wallis Island in the South Pacific following introductions (as *T. melanopleura*) from an unknown source between 1967 and 1970.

General Ecological Impact

As mentioned above, the naturalization of the Redbreast Tilapia has been somewhat controversial and has evoked mixed reactions. Even the species' efficacy in the control of aquatic vegetation has been disputed. 'Redbreast Tilapia and Redbelly Tilapia (*Tilapia zillii*) [q.v.] for some reason have gained an undeserved reputation as weed eaters. True, these species will eat weeds, but they will eat anything else as well from detritus to young fish, and resulting population explosions have not generally endeared the species to those in whose waters it has been introduced' (Welcomme, 1984). 'Numerous fishes are reported to consume aquatic vegetation. However, only a few of these fish have been investigated and show promise for weed control [including cichlids *O. mossambicus, O. niloticus, T. rendalli* and *T.*

zillii]. Many of these species may not be suitable for weed control because the individual has insufficient consumption (high stocking rates needed), they are prolific spawners (often cause overcrowding), or they are restricted to warm climates (must be overwintered in controlled environments)' (Shireman, 1984).

Banded Tilapia
Tilapia sparrmani

The Banded Tilapia is not a favoured angling fish but is fairly popular with warm water aquarists. It is a useful forage fish for predators such as black bass *Micropterus* spp. It has a wide range of habitats, favouring shallow, sheltered areas with plenty of marginal plant cover in rivers, lakes or swamps.

Natural Distribution

From the southern tributaries of the Zaire River system south to much of the Orange River system, including the basins of the Congo and Zambezi and Lakes Nyasa, Bangwelu and Moero.

Naturalized Distribution

Asia: Japan. *Africa:* Madagascar.

Asia
Japan

According to Imai (1980) and Chiba *et al.* (1989), Banded Tilapia have become established in estuarine waters in the Okinawa Islands of Japan, following attempts to rear them elsewhere in industrial effluents (Philippart & Ruwet, 1982).

Africa
Madagascar

In their survey of Malagasy waters in 1988, Reinthal and Stiassny (1991) collected Banded Tilapias

from the following localities: the River Mangoro in the northeast central rainforest; Lake Itasy on the central high plateau west of Antananarivo; the River Namorona (in three places) in the Ranomafana National Park area in the southeast central rainforest; and in Bay Lake on the central east coastal plain. They were unable to trace any record of the introduction of the species to Madagascar.

Unsuccessful Introduction

In Africa, Banded Tilapias have been unsuccessfully introduced to Tanzania (Moreau *et al.*, 1988).

Translocations

The Banded Tilapia has been translocated to southern and eastern rivers of the Cape outside its natural range in South Africa as a forage fish for black bass *Micropterus* spp., where it has become widely distributed (De Moor & Bruton, 1988).

In Namibia (South West Africa), the Banded Tilapia has been taken in the Von Bach dam in the Swakop system (Skelton & Merron, 1984) and in the Friedman dam in the Kuiseb River system, probably following translocations (De Moor & Bruton, 1988).

Toots (1970) reported the frequent translocation of Banded Tilapia within Rhodesia (Zimbabwe).

Ecological Impact In the Oliphants River, South Africa, the presence of Banded Tilapia is believed to be responsible for a decline of several native fish species, due to competition with adults for food and predation on juveniles (Gaigher, 1981). In the Gourits River, Banded Tilapia are considered by Skelton (1987) to be one of the factors that threaten the 'rare' native *Barbus tenuis*.

North America

Unsuccessful Introductions

Several unsuccessful attempts have been made to introduce Banded Tilapias to the USA (see e.g. Pelzman, 1972).

Redbelly Tilapia (Zill's Cichlid; Striped Tilapia; Jordan St Peter's Fish)
Tilapia zillii

T. zillii and *T. rendalli* (q.v.) are virtually indistinguishable morphologically and ecologically, and are frequently confused when introductions and distribution are described. The position is further complicated because where they occur sympatrically the two species appear to hybridize freely. Both fish, to a greater or lesser degree, graze on water plants, and have thus been extensively introduced for the control of aquatic vegetation as well as for aquacultural purposes. Both *T. zillii* and *T. rendalli* have become widely naturalized and have proved somewhat controversial; in some localities they are valued for aquatic weed control and form the basis for useful commercial fisheries, while in others their ability to control vegetation has been questioned and they have achieved pest status (Welcomme, 1988).

In its native range, *T. zillii* usually occurs sympatrically with such large predators as *Lates niloticus* and *Hydrocynus forskahlii*, and is normally found in shallow pools, lagoons, and the margins of rivers and floodplains over rock, sand or muddy substrates, where dense aquatic macrophytes provide shelter (Lowe-McConnell, 1982). (*T. melanopleura* = either *T. zillii* or possibly *T. rendalli*.)

Natural Distribution

From West Africa eastward through the basins of the Chad and Nile Rivers to Lakes Albert and George, northward into Israel and the Jordan valley.

Naturalized Distribution

Europe: ? British Isles (England). *Asia:* Japan. *Africa:* Ethiopia; Kenya; Madagascar; ? South Africa; Tanzania; Uganda; *North America:* Mexico; USA. *Central America:* West Indies (Antigua). *Oceania:* Guam; Hawaiian Islands; Mauritius; New Caledonia.

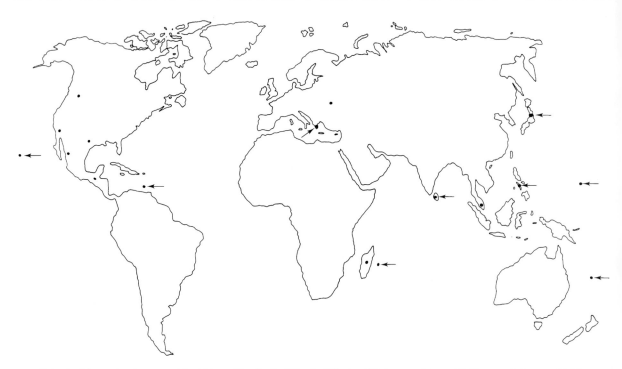

Principal introductions outside Africa of Redbelly Tilapia *Tilapia zillii* (adapted from Philippart & Ruwet, 1982).

Europe

British Isles (England)

Wheeler and Maitland (1973) reported the capture in 1967 of three cichlids, subsequently identified as *T. zillii*, in the St Helen's Canal, Lancashire. The fish were confined to the Church Street stretch of the canal (Welcomme (1981, 1988) wrongly refers to it as the Church Street Canal), where the water was heated by effluent discharged from the glassworks of Pilkington Brothers (Wheeler, 1974). The fish were believed to be derived from stock dumped in the canal when a nearby tropical-fish shop closed in 1963 (Lever, 1977). This population has since apparently died out. (See also under Guppy *Poecilia reticulata*.)

Asia

Japan

Imai (1980) and Chiba *et al.* (1989) record the establishment of Redbelly Tilapia in estuarine waters in the Okinawa Islands of Japan.

Unsuccessful Introductions

In 1969, Redbelly Tilapia from East Africa were unsuccessfully introduced to Sri Lanka (De Silva,

1987). Juliano *et al.* (1989) reported the failure of Redbelly Tilapia imported from Israel around 1977 (and others introduced from Taiwan in 1973) to become established in the Philippines. Introduced Redbelly Tilapia have also failed to establish themselves in the former USSR (Welcomme, 1988). Philippart and Ruwet (1982) refer to unsuccessful attempts to establish Redbelly Tilapia in Malacca (Malaysia) and Taiwan, while Yo-Jun and He-Yi (1989) say the species occurs in China only in 'experimental hatcheries.'

Translocations

Trewavas (1965) recorded the successful transplantation of Redbelly Tilapia within Jordan into the Asraq springs, some 120 km southeast of Amman.

Ben-Tuvia (1981) reported that the species had been successfully transplanted within Israel to Ein Fashka sometime after 1967.

Africa

Ethiopia

In 1974, Fisseha Haile Meskel, with the co-operation of the Ministry of Agriculture, imported Redbelly

Tilapia from Uganda to Ethiopia, to stock ponds and for hybridization with native Nile Tilapia *Oreochromis niloticus* (q.v.). Fry maintained in a pond at Doukam (Dukam), some 30 km east of Addis Ababa, showed excellent growth rate, and subsequently reproduction, before the project was discontinued in 1976. In 1975, the Ministry of Agriculture imported more Redbelly Tilapia from the same source, which were reared in ponds near Sebeta, 30 km southwest of Addis Ababa. In the late 1970s, Redbelly Tilapia were released by the Ministry in the following Ethiopian waters: River Teltele (Showa Administrative Region), 'micro-dam in children's village', Lake Langano, Lake Zwai, Koka dam, and Lakes Kuriftu and Babogaya (both near Debre-Zeit) (Tedla & Meskel, 1981).

Africa

Kenya; Tanzania; Uganda

Welcomme (1967) listed the following introductions and/or translocations of Redbelly Tilapia to Lake Victoria, which lies between Kenya, Tanzania and Uganda:

DATE	LOCALITY	NUMBER/REMARKS
1953	Kavirondo, Kenya	14 000 from Lake Albert stock reared in 10 fish ponds at Kisumu
1954	Musoma & Mwanza, Tanzania	200 in Capri Bay, Mwanza; 50 in Mara Bay, Musoma
1955	Mara Bay & Mwanza, Tanzania; Entebbe, Uganda; Kavirondo Gulf & Mfwangu I., Kenya	c. 12 000 mixed *T. zillii* and *Oreochromis leucostictus*
1955	Entebbe Harbour	128 from Kajansi Fish Farm
1956–1957	Pilkington Bay, Bovuma, Entebbe	—
1956–1957	Lake Victoria generally	c. 1500 *T. zillii*, *O. niloticus* and *O. leucostictus*
1958	Entebbe	c. 440
1959	Mwanza	908 at Mwalogwabagole & Butimba prison ponds
1960	Smith Sound, Mwanza	991

Since Welcomme's (1967) report, it has been generally accepted that Redbelly Tilapia from Lake Albert (between Uganda and Zaire) were first introduced, into Kenyan waters, in Lake Victoria in 1954. They spread rapidly in the north of the lake, first appearing in Tanzanian waters in the south and east in 1960; by 1964 they were virtually ubiquitous and abundant in the lake (De Moor & Bruton, 1988).

In 1954–1956, the species was introduced for commercial and sport fishing to Lake Naivasha, northwest of Nairobi, Kenya (a water containing no phytophagus competitors) and at an unrecorded date to the Tana River, where it has been less successful. Since 1962, Redbelly Tilapia have been widely and successfully stocked in impoundments in Tanzania (Moreau *et al.*, 1988). (See also Welcomme, 1964, 1966, 1981, 1988; Ogari, 1990; and Craig, 1992, under *T. rendalli*.)

Ecological Impact In Lake Victoria, *T. zillii* eventually supplanted *Oreochromis variabilis* as the dominant species in those areas where the latter had previously predominated, with fry and juveniles found mainly on sheltered shelving, sandy shores. There seemed to be little competition between adults for food, the principal rivalry being apparently for breeding sites. Both *T. zillii* and *O. variabilis* require a clean, firm substrate for spawning, and the more aggressive alien out-competed the smaller native for these areas (Philippart & Ruwet, 1982; De Moor & Bruton, 1988). In Lake Victoria, *T. zillii* has hybridized freely with another exotic species, *T. rendalli* (q.v.) (Welcomme, 1966). (See also under *O. leucostictus* and *O. niloticus*.)

Madagascar

Redbelly Tilapia from Nairobi, Kenya, were first introduced to the island of Madagascar in 1955 (Reinthal & Stiassny, 1991), where Lamarque *et al.* (1975) and Philippart and Ruwet (1982) reported them to be established in many rivers, lakes (in Lake Itasy since 1961–1962) and ponds, being especially abundant in the basin of Lake Kinkony (J. Moreau, in Pullin, 1988). Although Welcomme (1988) said that their importance in aquaculture and fisheries was declining, Reinthal and Stiassny (1991), in their survey of Malagasy waters in 1988, collected Redbelly Tilapia in the following localities: a small stream with a maximum width and depth of about 1 m around

1 km west of Manjakandriana east from Antananarivo in the northeast central rainforest; the River Namorona north of Ambatolahy in the Ranomafana National Park area of the southeast central rainforest; at sea level in the Canal des Pangalanes (see Glossary) 7 km north of Menagerie on the southeast coastal plain; the River Mania on the central high plateau; and in Bay Lake on the central east coastal plain.

South Africa

In 1959, *T. zillii*, known in South Africa as Jordan's St Peter's Fish, were imported from Israel for experimental aquacultural purposes in the Jonkershoek hatchery in the Cape. In or around the following year they were stocked in 15 farm dams in the Eerste and Lourens River basins to control aquatic vegetation (Van Schoor, 1966). At about the same time, *T. zillii* were introduced with 'Taiwan red tilapia' (presumably *Oreochromis niloticus* × *O. mossambicus* hybrids) for experimental purposes to Natal (Bourquin *et al.*, 1984). There are no records of the species being released into the wild in Natal, where it was probably only maintained in captivity.

In 1964, only one of the farm dams in the Cape previously stocked with *T. zillii* was found to contain a flourishing population, although the species survived in a further seven (Van Schoor, 1966). Although there had to date been no further records, De Moor and Bruton (1988) believed that populations may still have existed in some other farm dams in the area.

Ecological Impact   Although there are no native cichlids in the Jonkershoek area with which *T. zillii* could compete as it does with those in Lake Victoria, its possible impact on indigenous *Barbus* spp. in the western Cape is impossible to assess. If it were ever to gain access to natural waters in Natal it might hybridize and/or compete with such autochthonous species as Redbreast Tilapia *T. rendalli* (q.v.) and Banded Tilapia *T. sparmani* (q.v.), almost certainly to the natives' disadvantage.

Unsuccessful Introductions

In Africa, Redbelly Tilapia have failed to become established in the wild in Algeria (Welcomme, 1981, 1988), Egypt (Bayoumi, 1969), and the Ivory Coast, Zimbabwe and Cameroun (Moreau *et al.*, 1988).

North America

Mexico

Redbelly Tilapia were first imported to Mexico in 1964 (as *T. melanopleura*) from Auburn University, Alabama, USA (Delgadillo, 1976) to try to establish a new source of food. They may have been first released into the wild in the mid-1960s (Minckley, 1973), and are now established in the lower Río Colorado (Hensley & Courtenay, 1980; Contreras & Escalante, 1984). Welcomme (1981), who gives the date of introduction as 1974, says the species is 'widely established in many states'. (See also under *Oreochromis mossambicus*.)

United States

In 1972, *T. zillii* (or possibly a hybrid with *T. guineensis*) was introduced, with *Oreochromis mossambicus*, to control the excessive growth of aquatic vegetation in irrigation drainage ditches in the Coachella, Imperial and Palo Verde valleys in southern California, where it became widely distributed (Pelzman, 1973; Moyle, 1976b). The species was released in some 20 ponds, lakes and creeks in Kern, Los Angeles, Orange, Riverside and Santa Clara counties, with uncertain results (Moyle, 1976b), and was reported by Legner and Pelsue (1977) to have died out in Los Angeles County. Knaggs (1977) collected Redbelly Tilapia from coastal marine waters of Orange County where, however, the species was not reproducing. In the Napa Valley, Pelzman (1973) reported Redbelly Tilapia to have survived successfully for 2 years before succumbing to winter chill. In January 1983, the species was released with *Oreochromis mossambicus* in High Rock Spring in Lassen County (Courtenay *et al.*, 1984, 1986).

In 1980, Redbelly Tilapia were found in a pond on a golf course in Pahrump Valley, Nye County, Nevada, where although they again managed to survive for at least two winters (Courtenay & Deacon, 1982) they later died out (Courtenay *et al.*, 1984, 1986).

In 1978, M. Mangini and Peter Rubec collected some Redbelly Tilapia from the spring-fed waters of the San Antonio Zoo, Bexar County, Texas, which flow into the San Antonio River, where the fish are established and breeding (Hubbs, 1982).

Smith-Vaniz (1968) reported that Redbelly Tilapia were annually stocked by the Alabama Department of

Conservation and by Auburn University in lakes and farm ponds in Alabama where, however, they were not reproducing.

In 1975, six Redbelly Tilapia were collected in a small quarry lake near Perrine, Southern Dade County, Florida, where no breeding or young were observed (Hogg, 1976a,b). This population was reported by Courtenay *et al.* (1984, 1986) to have subsequently been eradicated by the Florida Game and Fresh Water Fish Commission.

In North Carolina, Redbelly Tilapia were reported by Courtenay *et al.* (1984, 1986) to be stocked to control aquatic weeds by Texasgulf, Inc., near Aurora in Beaufort County, and by the Carolina Power & Light Company in the heated waters of a cooling reservoir at Wilmington, Brunswick County. The same authors also reported the species to be cultured to control aquatic plants and as a food-fish in an artificially heated area of the Santee–Cooper reservoir in South Carolina, and they received unconfirmed reports that it was being reared for human consumption in Idaho.

In Arizona, *T. zillii* (or again possibly a hybrid with *T. guineensis*) became established in ponds in Papago Park, Scottsdale, and in other waters near Phoenix, Maricopa County, and possibly elsewhere (Minckley, 1973).

In California, where introductions were made by a number of agencies, including the Goleta, Orange County, Northwest and Southwest Mosquito Abatement Districts, the University of California, Riverside, and privately, the reasons were the control of aquatic vegetation and of mosquito larvae and chironomid midges (Legner & Pelsue, 1977; Courtenay *et al.*, 1984, 1986). The release in Nevada was made illegally by a private developer (Courtenay & Deacon, 1982). In Texas, the fish appear to have been released by, or to have escaped from, the San Antonio Zoo (Hubbs, 1982). In Arizona, they were introduced to control aquatic vegetation jointly by the State Department of Game and Fish and the University of Arizona (Minckley, 1973).

The population formerly present but not breeding in Florida appears to have originated in the release of unwanted pets or from an aquaculture farm (Hogg, 1976a,b). In the late 1960s, the Florida Game and Fresh Water Fish Commission was experimenting with *T. zillii* or a closely related species (probably *T. rendalli*) as a potential agent for the control of excessive growths of aquatic weeds; these experi-

ments were abandoned and the surviving fish destroyed in the early 1970s, when it was recognized that the fish could do more harm than good in the waters of Florida (Courtenay *et al.*, 1984, 1986).

Courtenay *et al.* (1991) and Courtenay (1993) confirmed the Redbelly Tilapia's naturalization in Arizona, California, North Carolina and Texas.

Ecological Impact Evidence for the value or otherwise of Redbelly Tilapia as biological controlling agents of surplus aquatic vegetation in the USA is contradictory. Legner *et al.* (1973), Legner *et al.* (1975) and Shireman (1984) reported that when stocked at sufficiently high densities, they were successful in suppressing macrophytes and other noxious aquatic plants in Californian irrigation canals. The species has, however, the potential to disrupt aquatic ecosystems by changing plant communities (Moyle, 1976a), and has been called by W. R. Courtenay Jr. (quoted by Hogg, 1976a) 'the most destructive fish to submerged vegetation known next to the grass carp' [*Ctenopharyngodon idella*] (q.v.).

No effects on autochthonous faunal communities have as yet been attributed to the removal of aquatic vegetation by Redbelly Tilapia, although concern has been expressed about the possible destruction of habitats vital for the survival of threatened aquatic species in the southwestern USA were this fish or other cichlids to gain access to natural waters (Deacon, 1979; Courtenay & Deacon, 1982, 1983; Taylor *et al.*, 1984).

See also in General Ecological Impact under the Redbreast Tilapia *Tilapia rendalli*.

Central America
West Indies (Antigua)

Welcomme (1981) says that the Redbelly Tilapia was introduced in 1945 for sport fishing and mosquito control to the island of Antigua, West Indies, where it has become established. Welcomme (1988) gives the date as 1943, and lists the control of mosquitoes as the only reason for the species' introduction. This introduction is not mentioned by Wetherbee (1989).

Australasia
Unsuccessful Introduction

According to Allen (1989), a breeding population of adult and juvenile Redbelly Tilapia was discovered in

1975 in a series of small ponds and drains flowing into the upper reaches of the Swan River estuary in the Bayswater district of Perth, Western Australia. This population was subsequently eradicated by the Western Australia Department of Fisheries and Wildlife.

Oceania

Guam (Mariana Islands)

In 1956, Redbelly Tilapia, probably from the Hawaiian Islands, were imported by the Guam Department of Agriculture to the Fena reservoir in southern Guam in the Mariana Islands for the control of aquatic weeds (Van Pel, 1958; Devambez, 1964; Best & Davidson, 1981; Shepard & Myers, 1981; Maciolek, 1984; Nelson & Eldredge, 1991; Eldredge, 1988, 1994). The species is today common in Fena reservoir, where it forms the basis for a small recreational fishery but where, although it was introduced to control aquatic vegetation, the shallows remain choked with growths of *Hydrilla verticillata* (Leith *et al.*, 1984; Nelson & Eldredge, 1991).

Hawaiian Islands

In 1955, 19 Redbelly Tilapia were imported from the island of Antigua in the West Indies to Oahu, where they bred successfully in the Fish and Game Laboratory tanks in Honolulu until 1957, when 1500 were liberated in plantation reservoirs on Maui; further plantings were made in the two succeeding years in reservoirs on Hawaii and Oahu. In these waters Redbelly Tilapia proved remarkably effective in clearing unwanted aquatic vegetation (Brock, 1960). The species is currently established and spreading on all three islands (Maciolek, 1984). (See also Randall, 1960; Hida & Thomson, 1962; Kanayama, 1968.)

Maciolek (1984) also lists the introduction in 1957 of 50 *T. melanopleura* from the Congo to Kauai, Maui and Oahu. This introduction probably involved *T. zillii*.

Mauritius

In 1957, Redbelly Tilapia from Madagascar via Kenya were imported for aquacultural purposes to the island of Mauritius, where they are at present widespread and expanding (Moreau *et al.*, 1988).

New Caledonia

In about 1955, fingerling Redbelly Tilapia from the Hawaiian Islands were released in ponds constructed by the Port Laguerre Farm School in New Caledonia (Van Pel, 1959), where they later became established in the wild (Devambez, 1964; Maciolek, 1984, Nelson & Eldredge, 1991; Eldredge, 1994).

Unsuccessful Introduction

Andrews (1985), Nelson and Eldredge (1991) and Eldredge (1994) list the Redbelly Tilapia as unsuccessfully imported from the Hawaiian Islands to Fiji in 1957. Welcomme (1981, 1988) and Maciolek (1984) do not refer to this introduction.

GOBIIDAE

Yellowfin Goby (Oriental Goby; Japanese Goby)
Acanthogobius flavimanus

Gobies are benthic fish with characteristically fused pelvic fins that form a suctorial pad, which is used to secure the fish to boulders in fast-flowing fresh water or in waves along rocky shorelines. Gobies are generally robust species that prefer brackish estuaries or lagoons and adapt to habitats unacceptable to other species (Jude *et al.*, 1992).

Natural Distribution

Marine and brackish estuarine waters and freshwater rivers of southern Japan, South Korea, and eastern China.

Naturalized Distribution

North America: ? Mexico; USA. ***Australasia:*** Australia.

North America

Mexico; United States

The first Yellowfin Goby to be taken in the wild in the USA was collected on 18 January 1963 by Arnold B. Albrecht and Vincent A. Catania off Prisoners' Point on the southwestern shore of Venice Island in the Sacramento–San Joaquim River delta, California. On 29 March of the same year, a second specimen was caught by Catania, E.G. Gunderson and Armand P. Croft, Jr, in the Stockton Deepwater Channel just upstream of the confluence with the Calaveras River and downstream of Port Stockton. Since both these individuals were small in size, it appeared likely to Brittan *et al.* (1963) that a local breeding population was in the process of becoming established. They believed that the form and source of this, and possibly later introductions, was as eggs on fouling organisms growing on the hulls of, or in the ballast water pumped from, trans-Pacific shipping.

Thereafter, the spread of the Yellowfin Goby in California was nothing less than explosive. Brittan *et al.* (1970) listed the localities at which it was collected, sometimes in considerable numbers, as follows:

DATE	LOCALITY
Late 1964	Palo Alto Yacht Harbor; Leslie Salt Company evaporation pond, Alviso
August 1965	Off Marin Island near San Rafael, Marin County; Newman wasteway, Delta–Mendota Canal

DATE	LOCALITY
1966	Around Treasure I., adjacent to Yerba Buena I. between San Francisco and Oakland; San Rafael Channel, Plummer Creek near Newark, Richardson Bay, Angel I., Dunbarton bridge – all in San Francisco Bay; San Pablo Bay at McNear Beach; Napa Slough; Carquinez Strait between San Pablo and Suisun Bay at Benicia; the Delta at Antioch; the Tracy Pumping Plant
1967	Between Angel and Treasure I., east of San Rafael; Foster City Lagoon near Mateo Bridge; Suisun Bay and Montezuma Slough; Lake Merritt; Belvedere Lagoon; Aquatic Park, Berkeley; Paradise Cay, Tilburon peninsula; elsewhere in San Francisco Bay
1968	San Pablo and Suisun Bays, and the Delta, San Francisco Bay; Snodgrass Slough, a tributary of the Sacramento River near Walnut Grove; Sacramento Ship Channel south of Sacramento.

Thus, by the end of the 1960s, Yellowfin Gobies had become naturalized in the San Francisco Bay area, the Sacramento Delta, the Delta–Mendota Canal, the San Luis reservoir, Suisun Bay, and the Bolina Lagoon in Alameda, Contra Costa, Marin, ? Napa, San Francisco, San Mateo, Santa Clara, Solano and Sonoma Counties (Brittan *et al.*, 1970). Kukowski (1972) reported a southerly extension of range when in 1970 and 1971 Yellowfin Gobies were caught in the Elkhorn Slough, Monterey County, while Miller and Lea (1972) recorded the species in Tomales Bay and Estero Americano, Marin County. In 1977–1978, Yellowfin Gobies were seen or caught in Los Angeles and Long Beach Harbors; Upper Newport Bay, 37 km southeast of Los Angeles Harbor; the San Gabriel River, between the Westminster Avenue and Seventh Street Bridges, Long Beach, Los Angeles County, 13 km east of Los Angeles Harbor; and the Long Beach Swimming Lagoon, near the mouth of the Los Angeles River (Haacker, 1979). C.A. Usui (pers. comm. to Courtenay *et al.*, 1984, 1986) reported Yellowfin Gobies in the Bolsa Chica Flood Control Channel in Orange County, and by 1980 as far south as San Diego and perhaps Baja California Norte, Mexico. (See also Lachner *et al.*, 1970; Moyle, 1976a, b; Courtenay & Hensley, 1980b, Courtenay *et al.*, 1991; Courtenay, 1993.)

Ecological Impact   Moyle (1976a) expressed the fear that if Yellowfin Gobies were to continue to spread into streams, lagoons and bays down the coast of California they might eradicate local populations of Tidewater Gobies *Eucyclogobius newberryi*, while Brittan *et al.* (1970) considered that the Staghorn Sculpin *Leptocottus armatus* might be displaced. If Yellowfin Gobies were to be collected with such bait fish as the Longjaw Mudsucker Goby *Gillichthys mirabilis* for fishing in the Salton Sea, Lake Mead, and the Colorado River, the alien could be transported to, and become naturalized in, further marine and freshwater habitats (Haacker, 1979).

Australasia

Australia

In June 1971, a Yellowfin Goby was collected in the waters of Sydney Harbour, New South Wales, where the species soon became widespread and abundant and from where it later spread into the lower reaches of the Parramatta River and subsequently to other areas. The means of entry to Australia is believed to have been, as to the USA, as eggs attached to the hull and/or as adults carried in ballast tanks of transoceanic vessels (Hoese, 1973; Middleton, 1982; Clements, 1988).

Round Goby
Neogobius melanostomus

Tubenose Goby
Proterorhinus marmoratus

For characteristics see under *Acanthogobius flavimanus*.

Natural Distribution

Littoral fresh and slightly brackish waters of the Caspian Sea and of the north, west and east of the Black Sea, and throughout the Sea of Azov. The range of *Neogobius melanostomus* extends westward into the Sea of Marmara and the northern Aegean.

Naturalized Distribution

North America: Canada; USA.

North America

Canada; United States

The earliest published reference of the Round Goby in Canadian waters was probably by Crossman (1991), who included it in a list of species possibly accidentally introduced in ballast water discharged from a foreign tanker. On 28 June, 18 July and 23 September 1990, three Round Gobies were caught by anglers fishing the St Clair River, which runs for 63 km between Lakes Huron and St Clair, near Sarnia, Ontario, Canada, some 28 km north of the Detroit Edison Company's Belle River Power Plant, Michigan, USA, where on 11 April of the same year a Tubenose Goby had been collected. In November and December 1990, a total of 34 Round and Tubenose Gobies were trawled from a small cove south of the Belle River Power Plant, and between April 1990 and the winter of 1991 31 Tubenose Gobies and 14 Round Gobies were collected at or near the power plant in the St Clair River, of which nine of the latter were believed to be young of the year (Crossman *et al.*, 1992; Jude *et al.*, 1992). It was thus apparent that Round Gobies occurred throughout the St Clair River, including the Canadian side of the river, and in Lake St Clair (Crossman *et al.*, 1992).

Since then, Jude *et al.*, (in press) have summarized the distribution and spread of the two gobies in the St Clair and Detroit Rivers; in the latter, which flows between Lake St Clair and Lake Erie, Round Gobies were caught in 1993 on the east side off Pêche Island near Lake St Clair, north of Windsor, Ontario. Six Round Gobies were also reported to have been taken from the Grand River which flows into Lake Erie

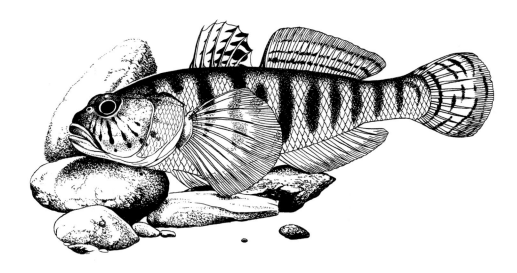

near Cleveland, Ohio, which is extensively used by tankers and freighters transporting sand, gravel and other cargoes. In January 1993, Round Gobies were caught by anglers live-bait fishing with minnows through holes in the ice for Yellow Perch *Perca flavescens* in the Grand Calumet River, Illinois, where vessels in ballast with water from the St Clair River pump it out while in transit. The Grand Calumet River joins, via the Calumet Sag Channel, the Chicago Sanitary and Ship Canal and thence the Mississippi River system. Jude *et al.* (1996) found that Round Gobies have spread rapidly, and are now nearly as common in this area of Lake Michigan as in the St Clair River. In September 1994, Round Gobies were recorded in southern Lake Michigan. Populations of Tubenose Gobies have remained small and the species has spread very little.

Ecological Impact   Crossman *et al.* (1992) were informed that a number of native species had declined in areas where Round Gobies had become established, and their consumption of fish suggested to Jude *et al.* (1992) that they might have a negative impact on indigenous species through competition for food and/or predation on their eggs and young. Since Round Gobies grow much larger than native Mottled Sculpin *Cottus bairdi*, Logperch *Percina caprodes* and Rainbow Darters *Etheostoma caeruleum*, Jude *et al.* (1992) considered that Round Gobies might eat young of the year and yearlings of these species.

Jude *et al.* (in press), who examined the impact of Round and Tubenose Gobies on autochthonous benthic species in greater detail, found that Round Gobies have well-developed lateral line systems, and reactive distances to prey in the dark as long as, or longer than, those of Mottled Sculpin, giving them an inbuilt competitive feeding advantage over them and most other benthic feeders in poor lighting conditions. In the St Clair River, populations of Mottled Sculpin especially and those of other benthic species,

particularly the Logperch, have sharply declined since the gobies' arrival, almost certainly as a result of spatial overlap and thus competition with the aliens for food and space, especially spawning sites, and predation on their eggs and young.

Since Tubenose Gobies are approximately the same size as, rather than larger than, native Mottled Sculpins, are almost entirely benthic feeders, and seldom venture outside beds of macrophytic vegetation, they will probably never become abundant in the St Clair River, and will thus not have the same negative impact as that of Round Gobies which are ecologically quite similar to, but larger than, Sculpins. They feed at night, favour rocky habitats with large interstitial spaces, are aggressively territorial, and both reclusive and cryptic when threatened (Jude *et al.*, in press).

A further reason why Round Gobies are increasing so rapidly in the St Clair and Grand Calumet Rivers is probably the presence of alien Zebra Mussels *Dreissena polymorpha*, introduced to the Great Lakes in the 1980s (Mills *et al.*, 1993, 1994), which serve as food but which had hitherto been relatively little used as a food source by larger fish.

An additional potential threat posed by Round Gobies is to the Deepwater Sculpin *Myoxocephalus thompsoni* where the two species overlap.

Jude *et al.* (in press) expressed their belief that Round Gobies will continue to expand their range within the basin of the Great Lakes by way of freighters *en route* to the Duluth–Superior Harbor in western Lake Superior, where their ecological impact seems likely to increase.

Predators   Jude *et al.* (in press) found that predators of Round Gobies in the St Clair River included Walleyes *Stizostedion vitreum*, Smallmouth Bass *Micropterus dolomieui*, Rock Bass *Ambloplites rupestris*, Yellow Perch (which consume young of the year), Stonecats *Noturus flavus* and Tubenose Gobies.

Tridentiger bifasciatus

For characteristics see under *Acanthogobius flavimanus*.

Since the occurrence of *T. bifasciatus* in the USA is its first record in the wild outside Asia, no common name for the species exists. Matern and Fleming (in press) have proposed modifying and adopting its Japanese name, '*shimofuri shimahaze*'. '*Shimofuri*' refers to fat-marbled beef, and aptly describes the pattern of white spots on the species' head. '*Shimahaze*' = Striped Goby, and is to be abandoned to avoid confusion and redundancy. Matern and Fleming (in press) accordingly suggest that *T. bifasciatus* be known as the Shimofuri Goby, but at the time of writing this vernacular name has yet to be accepted.

Natural Distribution

The estuaries of Asian rivers flowing into the Pacific Ocean. *T. bifasciatus* occurs commonly in fresh water, and is not found in salinities greater than 22‰. The salinity ranges of *T. bifasciatus* and *T. trigonocephalus* (q.v.) do, however, overlap, and in the estuaries of some Japanese rivers the two species exist sympatrically. Both species favour habitats with structurally complex substrates.

Naturalized Distribution

North America: USA. ***Australasia:*** ? Australia.

North America

United States

The following account of the establishment of *T. bifasciatus* in the USA is derived from Matern and Fleming (in press).

After more than half a century of synonymy with the Chameleon Goby *T. trigonocephalus* (q.v.) it is only recently that *T. bifasciatus* has been reinstated as a separate and valid species (Akihito & Sakamoto, 1989).

Until 1985, when fish believed to be Chameleon Gobies were collected in the low salinity and fresh waters of the upper San Francisco Bay estuary,

California, *T. trigonocephalus* had only been recorded in the higher salinity waters of the lower estuary. Meng *et al.* (1994) hypothesized that increasing levels of salinity, decreasing outflows, and drought conditions had interacted with the flushing flows of 1986 to create conditions that had enabled Chameleon Gobies to extend their range upstream into the area of Suisun Marsh. Based on Akihito and Sakamoto's (1989) criteria, Matern and Fleming (in press) re-examined *Tridentiger* specimens collected in 1985 and 1987 from Suisun Marsh and near Byron, Contra Costa County (Raquel, 1988), and found them to be not *trigonocephalus* but *bifasciatus*. Having subsequently examined several hundred *Tridentiger* specimens from the San Francisco Bay estuary, Matern and Fleming (in press) believed that all from the limnetic zone and high salinity area of the estuary were *bifasciatus.*

By 1989, *T. bifasciatus* was the commonest adult fish species in Suisun Marsh, and the third commonest larval fish in the upper estuary. By the following year it was the most abundant larva and had been transported via the Californian Aqueduct some 513 km south to Pyramid reservoir (Swift *et al.*, 1993). In 1992, it was collected downstream of that reservoir in Piru Creek (Swift *et al.*, 1993), but by 1995 had yet to reach Piru Lake further downstream. Specimens in and below Pyramid Reservoir were originally identified as *T. trigonocephalus*, and were only correctly named as *T. bifasciatus* in 1994.

As in the case of the Yellowfin Goby *Acanthogobius flavimanus* (q.v.), *T. bifasciatus* probably arrived in the Sacramento–San Joaquim estuary in the ballast-water of transoceanic vessels from Japan.

Ecological Impact   In Suisun Marsh *T. bifasciatus* feeds principally on benthic invertebrates such as amphipods (e.g. water lice) and mysid shrimps. Dietary and habitat preferences of *T. bifasciatus* potentially place it in direct competition for food and spawning sites with the federally endangered Tidewater Goby *Eucyclogobius newberryi*, on which it may also prey (Swenson & Matern, in prep). Although the two species seem to be allopatric, the potential for interaction between them may soon be realized, either through natural extension of range by the alien or through the State Water Project System. This could happen when water from the California Aqueduct is diverted into Lake Cachuma in the south of the state. Water from this reservoir flows into the Santa Ynez River, and the larvae and adults of *T. bifasciatus* could thus gain access to the coastal brackish waters occupied by *E. newberryi*. Matern and Fleming (in press) considered it vital that, as is planned, water diverted into Lake Cachuma is previously treated. The populations of *E. newberryi* in the lower Santa Clara River are even more seriously at risk, since *T. bifasciatus* is established upstream in Piru Creek. Downstream dispersal of the alien species throughout this drainage has been predicted by Swift *et al.* (1993), a diffusion

which may have been accelerated by the severe early winter floods of 1995.

Australasia
Australia

Since 1973, *T. trigonocephalus* (q.v.) has been established in Sydney Harbour, New South Wales. In view of the morphological similarities between the two species and their recent taxonomic separation, Matern and Fleming (in press) consider that *T. bifasciatus* may well also occur undetected in Sydney Harbour as it did for several years in San Francisco Bay, especially if the *Tridentiger* populations are found in harbour waters of low salinity.

Chameleon Goby (Striped Goby; Trident Goby)
Tridentiger trigonocephalus

For characteristics see under *Acanthogobius flavimanus*.

Natural Distribution

Brackish and marine waters of the southern Amur River, the Suifen River, the mouths of rivers debouching into Peter the Great Bay, the Tumen'-ula River, the east coast of Korea as far south as Pusan, Chemulpo, the Liao River, Lusbun (Port Arthur), China at least as far south as Guangzhou (Canton) and Japan. It seldom occurs in salinities below 22‰.

Naturalized Distribution

North America: USA. ***Australasia:*** Australia.

North America
United States

On 1 June 1960, J. Wright of 'Marineland of the Pacific' observed two and collected one Chameleon Goby at Fish Harbor in Los Angeles Harbor, California – the first of its species to be taken in American waters (Hubbs & Miller, 1965; Brittan *et al.*,

1970). Subsequent collection of numerous mature and spawning adults indicated the establishment of a naturalized population of the species in Los Angeles Harbor, and also in 1962 in the Redwood City docks in South San Francisco Bay (Lachner *et al.*, 1970; Miller & Lea, 1972; Moyle, 1976a,b; Haacker, 1979).

Courtenay *et al.* (1986) point out that although Moyle (1976b) suggested that the Chameleon Goby could be expected in fresh water in California and that 'it occurs in brackish Lake Merritt in Oakland . . .', Shapovalov *et al.* (1981) quoted Hubbs and Miller (1965) who indicated that although it connects with San Francisco Bay, Lake Merritt is a freshwater body.

Hubbs and Miller (1965) conjectured that the initial introduction of the Chameleon Goby into San Francisco Bay might have been as fertilized ova on the introduced Japanese Oyster *Crassostrea gigas*. The Goby's appearance in Los Angeles Harbor, further south, was most probably the result of transportation in ships' ballast water tanks (Courtenay, 1993).

In recent years the population of Chameleon Gobies has fallen dramatically, perhaps due in part to predation by Yellowfin Gobies *Acanthogobius flavimanus* (q.v.). Chameleon Gobies spawn 12–16 weeks after Yellowfin Gobies, and since the latter are partly piscivorous the eggs and larvae of the former are likely to be included in the diet of both adults and juveniles (Meng *et al.*, 1994).

Australasia

Australia

In May 1973, Chameleon Gobies were first collected in Sydney Harbour, New South Wales, where they subsequently became established. The means of entry to Australia was believed to have been, as in the earlier case of the Yellowfin Goby *Acanthogobius flavimanus* (q.v.), in the ballast tanks of trans-Pacific shipping from Japan (Friese, 1973; Hoese, 1973; Clements, 1988). This theory was confirmed by the discovery by Paxton and Hoese (1985) of a live Chameleon Goby in the ballast water of an Australian transoceanic vessel. The species' establishment in Sydney Harbour was confirmed by Pollard and Hutchings (1990).

Allied Species

Borisova (1972) recorded the establishment in 1961 of the Amur Goby *Rhinogobius similis*, accidentally introduced from the Yangtse River in China in a consignment of Chinese carp fry, in fish ponds near Tashkent, Uzbekistan, in the former USSR. As a result of its faster rate of growth and greater fecundity in Tashkent than in its native range, it has partially displaced several native species, and fears have been expressed that it could eventually gain access to the rivers of the Caspian Sea.

Payne (1987) refers to the establishment of an apparently unidentified *Glossogobius* sp. in Lake Lanao in the Philippines, where together with other introduced exotics it preys on native cyprinids.

Eldredge (1994) lists *Mugiligobius parva* as introduced to, and established in, coastal waters of the Hawaiian Islands since 1987.

ELEOTRIDAE

Hypseleotris swinhonis

Hypseleotris swinhonis is a small freshwater eleotrid of very little commercial value.

Natural Distribution

The basin of the Yangtze River in China, and Japan and Korea.

Naturalized Distribution

The former USSR.

Europe

USSR

According to Borisova (1972), *H. swinhonis* was accidentally introduced in 1961, together with several other unwanted exotic 'trash' species, from the Yangtze River in China to ponds of the Akkurgan Fish Combine near Tashkent in Uzbekistan with shipments of commercially valuable Chinese Grass Carp *Ctenopharyngodon idella* (q.v.) and Silver Carp *Hypophthalmichthys molitrix* (q.v.). Some 5 years later *H. swinhonis* (and some other exotics) escaped from the ponds, and became established in inlet and outlet ditches and in other neighbouring natural waters, from where they soon spread via feeder canals into the Syrdar'ya, Chirchik and Akhangaran Rivers, and thence into the Tuyabuguz River reservoir. *H. swinhonis* is now common (together with the similarly introduced Bitterling *Rhodeus sericeus* (q.v.), Korean Sharpbelly *Hemiculter eigenmanni* (q.v.) and Spotted Stead *Hemibarbus maculatus* (q.v.)) in pools of the Syrdar'ya and in the mouth of the Chirchik and Akhangaran Rivers. Because they grow rapidly and are more fecund than in their native range in the Far East, *H. swinhonis* and some of the other aliens in Uzbekistan have tended to displace local species and have formed new biocoenoses.

OSPHRONEMIDAE

Giant Gourami
Osphronemus gouramy

The Giant Gourami is a large (up to 60 cm) and thickset mainly vegetarian fish with a greatly extended pelvic fin. It possesses auxiliary breathing organs in its gill chambers which enable it to take in air at the surface, and often lives in oxygen-depleted and brackish waters. In its native range it is widely reared in aquaculture, and has acquired a high reputation as a human food-fish.

Natural Distribution

Java, Sumatra and Borneo in the Greater Sunda Islands, Indonesia.

Naturalized Distribution

Asia: India; Philippines; Sri Lanka. *Africa:* Madagascar. *South America:* Colombia. *Austra-* *lasia:* Papua New Guinea. *Oceania:* Mauritius; New Caledonia.

Asia
India

The Giant Gourami was first introduced to India in the early 19th century, when it was stocked in the Botanical Gardens in Calcutta; by 1841, however, the population had died out due, it was said, to 'want of proper attention' (Jones & Sarojini, 1952).

Sometime after 1865 (according to Gopinath (1942) in 1886) Sir William Denison, then Governor of Madras, imported some Giant Gouramis from the island of Mauritius (see below), some of which were placed in ponds in the grounds of Government House, Madras, while others were taken to the

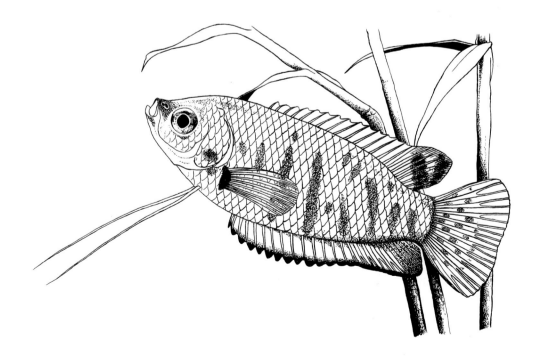

Nilgiris (Raj, 1916). In Madras the fish bred successfully, and some of the resulting fry were distributed to other waters in the then Presidency. The general condition of the fish was, however, apparently deemed not very healthy, and to augment the population 200 more were imported from Mauritius and Java. According to Sreenivasan (1989), quoting Anon. (1956) and Hora and Pillay (1962), Giant Gouramis were imported from Mauritius to Guindy Park, Tamil Nadu, by Francis Day in 1865, and from Java by Henry Wilson in 1916.

By 1937–1938, Giant Gouramis were breeding in Vellore and in the Chetpet fish farm, from where they were successfully transferred to Bombay (Kulkarni, 1946, 1947). By 1940, they were also reproducing in Sunkesula fish farms in the Madras Presidency, from where they were successfully transferred to Baroda in 1941 (Moses, 1944), Mysore in 1942 (Bhimichar *et al.*, 1944), Cochin and Hyderabad in 1945 (Jones & Sarojini, 1952), and unsuccessfully to the Punjab. In February 1942, 96 Giant Gourami fingerlings were transferred from the Madras Fishery Department to a rearing pond in the Markonahalli fish farm in Tumur, Karnataka, where within a year all but 25 had died. The survivors, however, were breeding by August 1943, although the species never became popular in the State (Chandrasekharaiah, 1989).

In 1950–1951 Giant Gouramis were stocked in the Chengalpattu fort moat, and in Webster moat, Thanjavoor, in 1958–1959, when they were being reared successfully in Poondi fish farms as well as in Buderi dam, near Poondi. They were later stocked unsuccessfully in Ooty Lake, where they failed to survive low winter temperatures.

In such reservoirs as Mettur in Tamil Nadu, and in ponds and reservoirs elsewhere in peninsular India, Giant Gouramis continue to be caught in small numbers, but because of their slow growth rate are of no commercial value (Sreenivasan, 1989). From Tamil Nadu, Giant Gouramis have been transferred to Kerala, Maharashtra, West Bengal, Orissa, Bombay, the Punjab and Andhra Pradesh (Natarajan & Ramachandra Menon, 1989; Shetty *et al.*, 1989; Yadav, 1993). (See also Vooren, 1968.)

Ecological Impact   In India the Giant Gourami is 'highly compatible with local fish and has no adverse impact on them' (Sreenivasan, 1991). 'Being phytophagous, it also serves to control aquatic weeds to some extent' (Shetty *et al.*, 1989).

Philippines

Giant Gouramis from Java were introduced to the Philippines in 1927, where they established naturally breeding populations in lakes, rice-paddies, and ponds. Although for a time they were also successfully reared in aquaculture they were eventually supplanted by other more popular and faster growing species (Juliano *et al.*, 1989).

Sri Lanka

In 1909 (Welcomme says 1939, presumably a misprint) Giant Gouramis from Indonesia were imported for aquacultural purposes to Ceylon (Sri Lanka), where they became established in a number of streams and impoundments (Ellepola & Fernando, 1968; Fernando, 1971; De Silva, 1987), including the lower reaches of the Mahaweli River (Welcomme, 1981), but are no longer regarded as an important food fish on the island.

Unsuccessful Introductions

In Asia, Welcomme (1981, 1988) lists unsuccessful introductions of Giant Gouramis to Thailand (Piyakarnchana (1989) does not mention this introduction). From Thailand, Gouramis were in 1956 introduced to Japan, where their present status is unknown (Chiba *et al.*, 1989).

Africa

Madagascar

In 1857, Giant Gouramis from Mauritius were introduced for aquacultural purposes to the island of Madagascar (Kiener, 1963; Moreau, 1979; Welcomme, 1981, 1988; Reinthal & Stiassny, 1991). Moreau (1979), and Moreau *et al.* (1988) who list the date of introduction as 1957 (presumably a misprint) and the source as Mauritius and the Far East, say Giant Gouramis have not been successfully cultured in captivity on Madagascar, but that escaped or released fish provide some 5% of the total fish catch where they have become established in the Canal des Pangalenes (see Glossary). The species was not apparently collected by Reinthal and Stiassny (1991) during their survey of Malagasy waters in 1988.

Unsuccessful Introductions

In Africa, Welcomme (1981, 1988) lists an unsuccessful attempt in 1957 to introduce Giant Gouramis to the Ivory Coast. Moreau *et al.* (1988) mention the species' failure to become established in Uganda following an introduction in 1963.

South America
Colombia

Welcomme (1988) says that as a result of the accidental escape of ornamental fish, Giant Gouramis are established and breeding in the wild in Colombia.

Australasia
Papua New Guinea

Van Pel (1956) initiated the introduction of Giant Gouramis to Papua New Guinea, where the first fish (30 in number) arrived from Malaya on 2 October of the following year and were released in a billabong at Bomana jail, where they subsequently reproduced. West and Glucksman (1976: table 7) give full particulars of further importations of Giant Gouramis to Papua New Guinea from Malaya, Jayapura, Irian Jaya, and Sydney, Australia, and internal translocations, between 1959 and 1971, to Dobel ponds; Balimo; Rabaul; a golf course pond on Mount Hagen; Dumpu (Bogia); Amanab; Wewak; Sirinumu impoundment; Lake Wanum; the Iarowari High School; Lake Puwan on Tong Island (Manus); the Sepik and Madang areas; and the Vudal Agricultural College. To this list Clements (1988) added Lake Imbia near Maprik and the Kerevat River. During this period a total of well over 1900 fish were imported to, or translocated within, Papua New Guinea.

Glucksman *et al.* (1976) listed Giant Gouramis as introduced in the following districts of Papua New Guinea: West Sepik; East Sepik; Western Highlands; Morobe; Madang; Western; East New Britain; and Central.

West and Glucksman (1976) describe the Giant Gourami in Papua New Guinea as 'of value in

extremely localized areas, and may prove to be so in others'. Allen (1991) says that most of the introductions of Giant Gouramis to the country 'appear to have been unsuccessful'.

Unsuccessful Introductions

Thomson (1922) describes briefly, and Clements (1988) in detail, unsuccessful attempts to introduce Giant Gouramis to New Zealand and Australia respectively.

Oceania
Mauritius

In 1761, the French imported Giant Gouramis from Batavia (Java, Dutch East Indies) to the island of Mauritius, with the intention of introducing them from there to France. (On the recommendation of Professor M. Coste, a pioneering French pisciculturist, some were successfully landed in Marseilles and Algeria via Suez before 1866 (Clements, 1988).) Welcomme (1988) gives the date of introduction as 1951 – presumably a misprint. Moreau *et al.* (1988), who use the old scientific name *O. olfax* for the fish introduced to Mauritius, list a second introduction to the island in 1944, where they imply that the species is tenuously established. Welcomme (1981, 1988), describes Giant Gouramis on Mauritius as 'of little value and low potential. Widespread but being replaced by tilapias'.

New Caledonia

Van Pel (1956), Devambez (1964) and Maciolek (1984) list the Giant Gourami as successfully naturalized in New Caledonia. When and from where it was introduced is apparently unknown.

Unsuccessful Introduction

Brock (1952, 1960), Kanayama (1968) and Maciolek (1984) list nine immature Giant Gouramis as introduced from the Philippines to Oahu in the Hawaiian Islands in 1950, where although they survived 'for a long time' (Brock, 1960) they never reproduced.

Pearl Gourami (Pearl Plasalid)
Trichogaster leeri

The Pearl Gourami is a popular warmwater aquarium fish and is also used in aquaculture. It is an extremely hardy species and is tolerant of low dissolved oxygen concentrations (Welcomme, 1988).

Natural Distribution

Borneo, Sumatra, Thailand and the Malay Peninsula.

Naturalized Distribution

Asia: Philippines. *South America:* Colombia.

Asia

Philippines

Juliano *et al.* (1989) list the Pearl Gourami as introduced in 1938 from Bangkok, Thailand, to lakes, rivers, rice-paddies, ponds and swamps in the Philippines for aquacultural purposes, where Welcomme (1981, 1988) says it occurs in the wild mainly in marshy and swampy localities, where it is 'regarded as a valuable addition to the local fauna'.

South America

Colombia

Pearl Gouramis which are believed to have escaped from aquarium fish farms have become successfully established in the wild in Colombia (Welcomme, 1988).

Unsuccessful Introductions

In 1940, a small number of Pearl Gouramis were released in Nuuanu reservoir on Oahu in the Hawaiian Islands, where they were never seen again (Brock, 1952, 1960; Kanayama, 1968; Maciolek, 1984). Courtenay and Stauffer (1990) and Courtenay *et al.* (1991) list the Pearl Gourami as having been collected, but not known to be reproducing, in Florida, USA.

Allied Species

The Moonlight Gourami *T. microlepis*, a native of Thailand, has become naturalized in waters of Colombia presumably as a result of escaping from aquarium fish farms (Welcomme, 1988).

Snakeskin Gourami (Snakeskin Plasalid; Sepat Siam)
Trichogaster pectoralis

In spite of its fairly small (25 cm) size, the Snakeskin Gourami is a popular food-fish in its native range. Its fast rate of growth and extreme hardiness have caused it to be widely used in aquaculture, particularly in South East Asia. It is also a favourite species among warmwater aquarists (Welcomme, 1988).

Natural Distribution

South Vietnam, Thailand, Cambodia and the Malay Peninsula.

Naturalized Distribution

Asia: Indonesia (Java, Borneo, Celebes); Philippines; Singapore; Sri Lanka. *South America:* Colombia. *Australasia:* Papua New Guinea. *Oceania:* New Caledonia.

Asia

Indonesia (Java, Borneo, Celebes)

Snakeskin Gouramis were first imported to Indonesia from Malaysia in 1934 by the Inland Fisheries

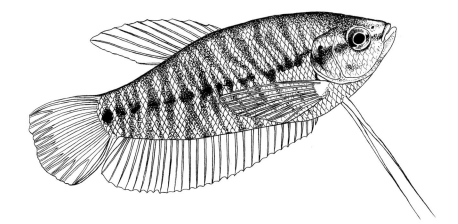

Department. Initially the fish proved hard to breed, but success was achieved in 1935 and by 1937 the species had begun to spread and was becoming economically important. It is currently widely cultured in captivity and has also successfully established self-maintaining populations in natural waters (Muhammad Eidman, 1989). Welcomme (1981, 1988) says that the Snakeskin Gourami occurs in Java, Borneo and Celebes.

Philippines

In 1938, Snakeskin Gouramis from Bangkok, Thailand, were introduced to lakes, rice-paddies and ponds in the Philippines, where they have become established in the wild; they are also reared in captivity, and have become an accepted food fish in inland areas where sea fish are unobtainable (Juliano *et al.*, 1989).

Singapore

According to Chou and Lam (1989), the Snakeskin Gourami is well established in the wild in Singapore.

Sri Lanka

In 1951, Snakeskin Gouramis from Malaysia were introduced to fill an empty niche in lagoons and marshes in Ceylon (Sri Lanka) (Fernando, 1971; De Silva, 1987), where Welcomme (1981) described them as 'very abundant in east coast lakes'. Jhingran and Gopalakrishnan (1974) found them to be economically insignificant on the island.

Unsuccessful Introductions

In Asia, Snakeskin Gouramis have failed to become naturalized in Hong Kong, India, Japan and Pakistan (Welcomme, 1981, 1988).

South America
Colombia

For details see under *T. leeri.*

Australasia
Papua New Guinea

West and Glucksman (1976) recorded the following introductions and translocations of Snakeskin Gouramis in Papua New Guinea between 1957 and 1963:

DATE	DISTRICT	SOURCE	LOCATION	REMARKS
2 October 1957	Central	Malaya	Bomana	300 imported
24 July 1959	West Highlands	Bomana	Dobel ponds	5 translocated; failed to reproduce
June 1960	Central	Malaya	Waigani swamp	Of 300 imported only 20 survived
21 December 1960	Central	Singapore	Waigani swamp	150 released
November 1962	?	Malaya	?	500 survived of 1000 shipped
13 January 1963	Central	Malaya	Bomana	120 released
21 May 1963	Central	Bomana	Sirinumu dam	700 released

Breeding experiments at Dobel showed that, as might have been expected, Snakeskin Gouramis were unsuited to upland regions, and subsequent releases were all made in lowland natural waters or in ponds.

By 1969, Snakeskin Gouramis were reported to be 'very rare' in the Sirinumu dam. Three years later they were discovered in eastern coastal areas of the Gulf Province as far west as Moviavi, although how they arrived there is unknown. This was the only area where West and Glucksman (1976) found that the species was fished for food. Elsewhere, it was occasionally cultured in sewage ponds and in manure–methane digestor systems. Berra *et al.* (1975) recorded the presence of Snakeskin Gouramis in the Laloki, Brown and Goldie Rivers.

According to Allen (1991), Snakeskin Gouramis were present in the Ajamaru Lakes in Irian Jaya in the 1950s, although the species' present status there, as elsewhere in Papua New Guinea, is uncertain. (See also West, 1973; Glucksman *et al.*, 1976; Clements, 1988.)

Oceania
New Caledonia

Van Pel (1956), Devambez (1964) and Maciolek (1984) list the Snakeskin Gourami as naturalized in New Caledonia. The date and source of the introduction are unknown.

Three-spot Gourami (Blue Gourami; Blue Plasalid)
Trichogaster trichopterus

Although not as popular as *T. pectoralis*, *T. trichopterus* is both cultured as a food fish and kept in aquaria (Welcomme, 1988).

Natural Distribution

Southern China, South Vietnam, Thailand, the Greater Sunda Islands (Java, Sumatra and Borneo) and the Malay Peninsula.

Naturalized Distribution

Asia: Philippines; ? Sri Lanka. ***South and Central America:*** Colombia; Hispaniola (Dominican Republic); ***Australasia:*** Papua New Guinea.

Asia
Philippines

For details see under *T. leeri.*

Sri Lanka

Welcomme (1981, 1988) lists the Three-spot Gourami as established but very rare in some dams in Sri Lanka, where he says it was introduced in 1948–1949. Since neither Fernando (1971) nor De

Silva (1987) include this species in their list of fish introduced to Sri Lanka, its presence there must be in doubt.

North America
Unsuccessful Introductions

Crossman (1984) lists the Three-spot Gourami as having been released in heated waters east of Cave Springs near Banff, Alberta, Canada, where it was said to have reproduced before dying out by 1981 when the flow of water was interrupted (Nelson, 1984). The species is not mentioned by McAllister (1969).

Courtenay and Stauffer (1990) and Courtenay *et al.* (1991) list the Three-spot Gourami as having been taken in the wild but not established in Florida, USA.

South and Central America
Colombia

For details see under *T. leeri.*

Hispaniola (Dominican Republic)

During hurricane 'David' in 1979, Three-spot Gouramis escaped, presumably from a tropical fish farm or a private aquarist, into the polluted waters of the Río Ozama in the Dominican Republic, where they are now abundant (Wetherbee, 1989).

Australasia

Papua New Guinea

In about 1970, Three-spot Gouramis from an apparently unrecorded source were introduced to the Central District of Papua New Guinea, where they were first recorded in the wild in the Laloki River near Port Moresby. In the shallow backwaters that it frequents, the Three-spot Gourami has become the dominant species, exceeding in numbers both native fish species and the introduced Mozambique Tilapia *Oreochromis mossambicus* (q.v.) and the Snakeskin Gourami *T. pectoralis* (q.v.). The presence of the Three-spot Gourami in natural waters of Papua New Guinea is believed to be the result of the accidental escape or deliberate release of aquarium stock (West, 1973). Allen (1991) recorded it as abundant in a few other streams near Port Moresby. (See also Berra *et al.*, 1975; Glucksman *et al.*, 1976; West & Glucksman, 1976; Clements, 1988.)

ANABANTIDAE

Climbing Perch
Anabas testudineus

The Climbing Perch is an exceptionally hardy fish which often lives in stagnant and poorly oxygenated water; this, combined with its small gills, forces it to obtain some of its oxygen supply from the air. Each gill chamber contains a labyrinthine organ which is well supplied with blood vessels. Air inhaled through the mouth passes through this auxiliary breathing organ, in which oxygen is absorbed. This also enables the fish to survive for lengthy periods (possibly for up to several days) out of water. The Climbing Perch's extremely mobile suboperculum and strong pectoral fin spines enable it to haul itself for considerable distances overland in search of new waters. It is a common warmwater aquarium species.

Natural Distribution

Southeastern Asia from India and Sri Lanka westward through Indonesia and the Philippines to southern China.

Naturalized Distribution

Australasia: Papua New Guinea.

North America
Unsuccessful Introduction

Courtenay and Robins (1973) and Courtenay *et al.* (1974) record the establishment of the Climbing Perch in natural waters in Manatee County, Florida, USA, following an accidental escape from an aquarium fish farm; it has since been eradicated (Courtenay *et al.*, 1991), probably by the exceptionally cold weather of January 1977 (Courtenay & Stauffer, 1990).

Australasia
Papua New Guinea

The Climbing Perch appeared in Irian Jaya, possibly in the Merauke area, at an apparently unrecorded date, from where it has spread eastwards, presumably naturally over land and water, as far as the Morehead River in Papua New Guinea, where it is now established (Allen, 1991). Since the Climbing Perch is not mentioned by West (1973), by Glucksman *et al.* (1976), nor by West & Glucksman (1976), it seems safe to assume that its introduction occurred after the mid-1970s.

Siamese Fighting Fish
Betta splendens

The tiny (6 cm) Siamese Fighting Fish often frequents stagnant and poorly oxygenated water, where it is forced to obtain some of its oxygen supply from the air through the use of auxiliary breathing organs situated inside its gill chambers. Notorious for their pugnacity, cock Siamese Fighting Fish compete with other males for supremacy and/or territory, although much of the contest takes the form of ritualized threat displays rather than actual combat. Mosquito larvae form a large part of the diet of Siamese Fighting Fish, which are thus sometimes regarded as an important controlling agent of these insects. The Siamese Fighting Fish

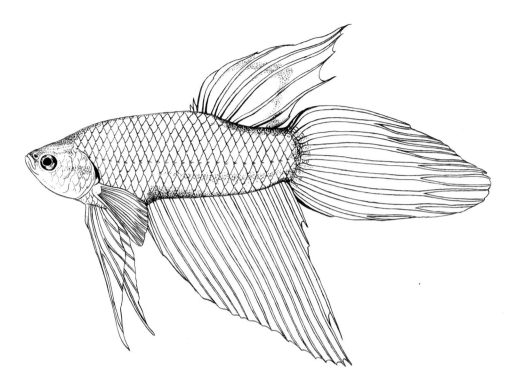

is an extremely popular warmwater aquarium species, and many forms with excessively long fins, especially in the males, have been bred in captivity.

Natural Distribution

Thailand and the Malay Peninsula.

Naturalized Distribution

Asia: Singapore. *Central and South America:* Brazil; Colombia; Hispaniola (Dominican Republic).

Asia

Singapore

Small self-perpetuating populations of the Siamese Fighting Fish have become naturalized in Jurong Lake and in some waterbodies in the Sembawang region of Singapore (Johnson, 1964, 1973; Chou & Lam, 1989).

North America

Unsuccessful Introduction

Courtenay and Robins (1973) and Courtenay *et al.* (1974) record the presence of Siamese Fighting Fish in a canal in Parkland, Broward County, Florida, USA, as the result of an accidental escape from a fish farm; the population was destroyed by cold in January 1977 (Courtenay & Stauffer, 1990; Courtenay *et al.*, 1991). Nelson (1984) recorded a local report that Siamese Fighting Fish had been released in heated waters east of Cave Springs near Banff, Alberta, Canada. Crossman (1984) could trace no confirmation of this report, and Nelson found none of these fish during his survey in 1981.

Central and South America

Brazil; Colombia

Siamese Fighting Fish that are believed to have escaped from an aquarium fish farm have become successfully established in the wild in Brazil and Colombia. In the former, Welcomme (1981) says that the species is 'presumably established near São Paulo in Amazonia'.

Hispaniola (Dominican Republic)

During hurricane 'David' in 1979, *B. splendens* and the Slim Fighting Fish *B. bellica* escaped, presumably from a tropical fish farm or a private aquarist, into the polluted waters of the Río Ozama in the

Dominican Republic, where they are now abundant (Wetherbee, 1989).

Allied Species

The Javan Mouthbrooding Fighting Fish or Java Betta **B. brederi**, a native of Java and Sumatra, is established in the wild on Guam in the Mariana Islands

(Best & Davidson, 1981; Shepard & Myers, 1981; Maciolek, 1984).

Unsuccessful Introduction

Natarajan and Ramachandra Menon (1989) list the unsuccessful introduction of an unidentified *Betta* sp. into Tamil Nadu, India.

Banded Gourami
Colisa fasciata

Thicklip Gourami
Colisa labiosa

Dwarf Gourami
Colisa lalia

These three small to medium-sized gouramis are all popular warmwater aquarium species.

Natural Distribution

C. fasciata: Burma, Thailand, Malay Peninsula. *C. labiosa*: southern Burma. *C. lalia*: India (Assam and Bengal).

Naturalized Distribution

North America: USA (*C. fasciata* and *C. lalia*). **South America:** Colombia (*C. fasciata*, *C. labiosa* and *C. lalia*).

North America
United States

Courtenay and Robins (1973), and Courtenay *et al.* (1974, 1991) record the Dwarf Gourami as having

been collected, but not known to be reproducing, in the wild in Florida. Courtenay *et al.* (1991) list the Banded Gourami as having the same status in Pennsylvania. Hoffman and Schubert (1984) reported the discovery of the parasite *Heteronocleidus gracilis* on the gills of Thicklip Gouramis (presumably in captivity) in California.

South America
Colombia

Banded Gouramis, Thicklip Gouramis and Dwarf Gouramis that are believed to have escaped from aquarium fish farms have become naturalized in the wild in Colombia (Welcomme, 1988).

Ornate Ctenopoma
Ctenopoma ansorgei

The Ornate Ctenopoma lives in both fast- and slow-flowing rivers, favouring in particular weedy forest streams and river backwaters.

Natural Distribution

Basin of the Zaire River.

Naturalized Distribution

Africa: Madascar.

Africa
Madagascar

In their survey in 1988 of natural waters on the island of Madagascar, Reinthal and Stiassny (1991) collected *C. ansorgei* from the River Mangoro, a large river with rocky and sandy habitats, low waterfalls, rapids, and some deep pools, lying at an altitude of around 850 m asl in the northeast central rainforest east from the town of Antananarivo. Reinthal and Stiassny could trace no record for the introduction of this species to Madagascar.

Kissing Gourami
Helostoma temmincki

The Kissing Gourami is a popular fish for aquacultural purposes both within and without its natural range in South East Asia, where its tolerance of low dissolved oxygen makes it ideal for rearing in waters unsuitable for many other species. It is also a favourite ornamental fish among warmwater aquarists (Welcomme, 1988).

Natural Distribution

The Greater Sunda Islands (Java, Sumatra and Borneo), Thailand and the Malay Peninsula.

Naturalized Distribution

Asia: ? Indonesia (Bali and Lombok); Philippines; ? Sri Lanka. *South America:* Colombia. *Oceania:* ? South Pacific.

Asia

Indonesia (Bali and Lombok)

Welcomme (1981, 1988) records the successful introduction for aquacultural purposes of Kissing Gouramis from Java to the neighbouring islands of

Bali and Lombok. This introduction is not mentioned by Muhammad Eidman (1989), and it is possible that these islands are within the species' natural range.

Philippines

Juliano *et al.* (1989) recorded the following introductions of Kissing Gouramis to the Philippines, where they became established in ponds and other natural waters: from Bangkok, Thailand, in 1948 and 1950; and from Bojor, Java, in July 1952 and February 1956.

Sri Lanka

De Silva (1987) listed the Kissing Gourami as introduced from Thailand in 1951 for experimental stocking in Sri Lanka, where Welcomme (1981) reported the species to be established in the wild in the Allai/Kanjalai area, where it was 'once abundant but receding'. De Silva (1987) had 'no data available on the present status of these introductions'.

Unsuccessful Introduction

In Asia, Welcomme (1981, 1988) reported the unsuccessful introduction of Kissing Gouramis from Borneo to Celebes.

North America

Unsuccessful Introduction

The Kissing Gourami has been collected, but is not known to be established, in natural waters of Florida, USA, having accidentally escaped from a tropical fish farm (Courtenay & Robins, 1973; Courtenay *et al.*, 1974, 1984, 1991; Courtenay & Stauffer, 1990; Courtenay, 1993).

South America

Colombia

Kissing Gouramis which are believed to have escaped from an aquarium fish farm have become established in natural waters in Colombia (Welcomme, 1988).

Oceania

South Pacific

The Kissing Gourami is said by Van Pel (1958) to be established in an unspecified location in the South Pacific.

Allied Species

Chiba *et al.* (1989) record the introduction from Korea in 1914 of the Round-tailed Paradisefish *Macropodus chinensis* to Japan where it is currently 'established as self reproducing populations'.

Welcomme (1988) lists an unidentified *Macropodus* sp. from South East Asia as naturalized in Colombia, following escapes from an aquarium fish farm.

Croaking Gourami (Talking Gourami)
Trichopsis vittata

The Croaking Gourami is another popular warm-water aquarium species.

Natural Distribution

The Greater Sunda Islands (Java, Sumatra and Borneo), Malaya, Thailand, Laos, Cambodia and Vietnam.

Naturalized Distribution

North America: USA.

North America
United States

The Croaking Gourami has been established since at least the late 1970s in a small area on the south side of Lake Worth Drainage District canal L-36, Delray Beach, Palm Beach County Florida, having probably escaped from a nearby aquarium fish farm. It successfully survived the exceptionally cold weather of January 1977 and several later cold winters, such as that of 1980–1981, and the population persists (Shafland, 1979; Courtenay *et al.*, 1986, 1991; Courtenay & Stauffer, 1990; Courtenay, 1993).

PLEURONECTIDAE

European Flounder
Platichthys flesus

The European Flounder is the only European flatfish known to enter and live in fresh water.

Natural Distribution

Marine coastal and brackish estuarine waters and freshwaters of lowland rivers of the Mediterranean and much of Eastern Europe, especially the Atlantic, the southern and eastern coastal North Sea, and the Baltic, as far north as the White Sea. It also occurs in the Black Sea.

Naturalized Distribution

North America: ? Canada; ? USA.

North America
Canada; United States

The first two specimens of the European Flounder in North America were caught in Lake Erie (between Ontario, Canada, and Ohio, USA), in July 1974 and January 1976 (Emery & Teleki, 1978). No further records of the species in the Great Lakes were received until 1981, when another specimen was taken in Lake Superior (between Ontario and Michigan, USA) (Crossman, 1984). The means of the species' arrival in the Great Lakes is believed to have been the ballast water carried by trans-Atlantic vessels (Emery & Teleki, 1978).

Evidence on the current status of the European Flounder in the Great Lakes is contradictory. According to Emery and Teleki (1978), the species 'cannot reproduce' in fresh water, and there is thus 'little likelihood of a population becoming established . . .'. Welcomme (1988), however, who lists the species as 'reproducing' in both Canada and the USA, says that some of the specimens in Lakes Erie and Superior 'were ready to breed and it is assumed that self-sustaining populations have become established'. Even were breeding to be proven, the establishment of self-maintaining populations seems unlikely. Courtenay *et al.* (1991) more realistically include the European Flounder in their list of 'exotic fishes collected but not known to be established in waters of the continental United States and Canada'.

ADDENDUM

From 25 to 29 March 1996 an international symposium and workshop on 'Stocking and Introduction of Fish in Freshwater and Marine Ecosystems' was held in Hull, England, under the auspices of the University of Hull International Fisheries Institute in collaboration with the European Inland Fisheries Advisory Commission. Attended by delegates from 33 nations, this was the most important conference on the subject ever to be convened.

The proceedings of this symposium will be published under the following citation:

Cowx, I. G. (Ed.), 1997. *Stocking and Introductions of Fish.* Oxford: Fishing News Books.

The following are adapted extracts taken from those papers presented at the conference that bear directly on the subject of the present work.

Ali, A.B. & Fatt, C.L. Impacts of exotic introduction on indigenous fish populations and fisheries in Malaysia

The introduction of exotic fishes into Malaysia has been through a combination of accidental and deliberate importations by fish farmers and governmental agencies, primarily for aquaculture. One of the earliest introductions for this purpose was the **Snakeskin Gourami** *Trichogaster pectoralis*, which was imported from Thailand to the rice-paddies of North Kerian. This species adapted well to the swampy and acidic habitat, and has replaced the native Three-spot Gourami *T. trichopterus* as the dominant herbivore in the rice-fields of peninsular Malaysia. Another early introduction, the **Mozambique Tilapia** *Oreochromis mossambicus*, was imported to Malaysia from Indonesia during the 1940s, but has become naturalized only locally, principally in water treatment ponds and in coastal lagoons and swamps. The **Nile Tilapia** *O. niloticus* was introduced at a much later date, and occurs in the wild only near centres of aquaculture. Yet another early introduction was the **Giant Gourami** *Osphronemus gouramy* from Indonesia, which is now established in major rivers and reservoirs in much of peninsular Malaysia. The **Guppy** *Poecilia reticulata* was introduced to control mosquito larvae, and is now virtually ubiquitous throughout the region.

Chinese and Indian major carps (including the Common Carp *Cyprinus carpio*, Grass Carp *Ctenopharyngodon idella*, Bighead Carp *Aristichthys nobilis*, Silver Carp *Hypophthalmichthys molitrix*, Mrigal *Cirrhinus mrigala*, Catla *Catla catla*, and Rohu *Labeo rohita*), and the Swai or Striped Catfish *Pangasius sutchi* from the basin of the Mekong River in southeast Asia, have been deliberately introduced to Malaysia for aquacultural purposes, as have, accidentally, the Tambaqui (Portuguese) or Cachama (Spanish) *Colossoma macropomum* from the catchments of the Amazon and Orinoco Rivers, and the **African Catfish** *Clarias gariepinus* from the Rivers Nile and Niger. Exotics imported by aquarists include the Green Swordtail *Xiphophorus helleri* from southeastern Mexico and Guatemala, the Angelfish *Pterophyllum scalare* from the Amazon, the Discus *Symphysodon discus*, and **Armored Catfish** *Plecostomus punctatus*. With the exception of the last-named species and the African Catfish, none has become established in the wild due to their inability either to reproduce or to adapt successfully to local conditions, and their failure to compete with native species.

Ecological and Socio-economic Impact

Only the Snakeskin Gourami and the Mozambique and Nile Tilapias seem to have had noticeable ecological and socio-economic impacts in Malaysia.

The replacement of the Three-spot Gourami by the Snakeskin Gourami was welcomed because the latter proved a more acceptable food-fish with local people. It has become established in lakes, reservoirs and large rivers, where it tends to occupy mainly the vegetated littoral zones. The species' population is limited by its slow growth rate.

Although both the tilapias tend to become the dominant species in some habitats, such as eutrophic and saline waters, they occur mainly in vacant niches unoccupied by indigenous species. Tilapia in Malaysia have not become established in natural lakes or in artificial impoundments, and have thus not contributed to artisanal fisheries as they have done in Sri Lanka. Moreover, neither species of tilapia flourishes in habitats with extreme physio-chemical parameters such as rice-paddies, owing to their inability to tolerate low dissolved oxygen and high water temperature.

The Armored Catfish, which escaped from aquarium fish farms, has become naturalized locally in polluted rivers, where it grazes on aufwuchs (periphyton organisms attached or clinging to the stems and leaves of plants or other objects projecting above the bottom sediments of freshwater ecosystems) encouraged by organic pollution.

In recent years, the introduction to Malaysia of species closely related to indigenes, with which they are able to hybridize, has caused some concern. Thus the African Catfish has interbred with the native *Clarias macrocephalus*, and has become probably the most important freshwater fish in the aquaculture of both Malaysia and Thailand. The ability of the species to hybridize, albeit with a lower success rate, with the native Walking Catfish *C. batrachus*, has raised concern for the continued genetic integrity of indigenous catfish species. Although early studies do not indicate that introgression is taking place, these studies are continuing in order to determine the genetic implication, if any, of this introduction to the gene pool of the native species. Moreover, the more rapid rate of growth and larger size of *C. gariepinus* could cause the displacement of local species from their present habitats.

Amarsinghe, U.S. **How effective are the stocking strategies for the management of reservoir fisheries of Sri Lanka?**

Exotic African cichlids and, to a much lesser extent, Chinese and Indian major carps, have been stocked by the Ministry of Fisheries since the 1950s in Sri Lanka's principal reservoirs, where the former are now the main component of commercial fisheries. In small seasonal village reservoirs, which may dry out

for between 2 and 6 months a year, fingerling fish are stocked annually.

Socio-economic Impact

When stocking densities of from 2000 to 3500 fingerlings per hectare of five species of major carps are made, yields ranging from 220 to 2300 kg ha^{-1} (mean 892 kg ha^{-1}) can be obtained in seasonal reservoirs in the year of release. It has been found that even in perennial impoundments annual restocking is necessary in order to establish viable fisheries.

Berg, S. **Rainbow Trout (*Oncorhynchus mykiss*) stocked as a biomanipulation tool in a eutrophic brackish lake**

The improvement of water quality in eutrophic lakes by the manipulation of the fish community, for example by the introduction of predatory species to feed on planktivores, is a useful conservation strategy. To this end, in the spring of 1992, 1993 and 1994, **Rainbow Trout** were introduced into Ferring Lagoon, a shallow brackish lake of some 320 ha in west Jutland, Denmark. Due to high phosphorus loading in the 1960s and 1970s, this lagoon has altered from clearwater with submerged macrophytes supporting 11 fish species in 1971 to a turbid water with high densities of phytoplankton containing large populations of only two species, the Three-spined Stickleback *Gasterosteus aculeatus* and Nine-spined Stickleback *Pungitius pungitius*. Two sizes of Rainbow Trout were stocked, small (mean length 17 cm) intended to prey on a mysid shrimp, *Neomysis integer*, which can have a marked negative impact on zooplankton, and a larger one (mean length 32 cm) designed to eat sticklebacks.

Ecological Impact

It was found, after stocking, that the most important prey species overall was the Three-spined Stickleback, which comprised 58.4% by weight of the stomach contents of Rainbow Trout. The most important invertebrate prey (by weight) were the Freshwater Shrimp *Gammarus pulex* (10.1%) and *N. integer* (7.9%). The most numerous prey species eaten were *N. integer* and a small snail, Jenkins' Spire Shell

Potamopyrgus jenkinsi. As might be expected, the diet of Rainbow Trout changed from invertebrates to fish and the total weight of the stomach contents increased with the growth of the introduced species. Although Rainbow Trout are able to tolerate the high pH conditions which occur in the lagoon every summer, their rate of growth was adversely affected; this, together with high mortality as a result of illegal angling, caused lower than anticipated pressure on the sticklebacks and mysid shrimps.

Bianco, P.G. Freshwater fish transfers in Italy: history, local modifications of fish composition and a prediction on the future of native populations

About 94 freshwater fish species have been recorded in the wild in Italian waters, of which 45 are native (30 having been translocated outside their natural range) and 49 are exotic, at least 26 of which are now naturalized. The two most recent records are the African catfish *Clarias anguillaris* and the Barbel *Barbus barbus.*

Binali, W. Dam restocking programme: fish transportation and stocking between catchment areas in Zimbabwe

Since the severe drought of 1992, several indigenous species have been transplanted within the 10 catchment areas of Zimbabwe. In addition, further introductions of three exotics, the **Nile Tilapia** *Oreochromis niloticus*, the **Common Carp** *Cyprinus carpio* and the **Largemouth Bass** *Micropterus salmoides* have been made because of their rapid rate of growth and popularity with anglers. Largemouth Bass are now widely naturalized throughout Zimbabwe, and there are no restrictions on releasing the species in uncolonized waters. Nile Tilapia now occur in the Zambezi and Lake Kariba systems.

Chifamba, P.C. The status of *Oreochromis niloticus* and *Macrobrachium rosenbergi* in Lake Kariba: an illustration of problems in preventing escapes from aquaculture production units

The first fish farm on the shores of Lake Kariba, Zimbabwe, was started in 1983, but was later flooded and abandoned. A second farm operated from 1985 and raceways opened in 1989, none of which had direct drainage into the lake. A series of ponds built in 1992 did, however, drain into the lake, and it is from these that the **Nile Tilapia** *Oreochromis niloticus* in Lake Kariba are derived.

Between September 1993 and July 1994, a survey was conducted to discover whether the Nile Tilapia had become naturalized in Lake Kariba. It was found that the species was well established and that no stunting was occurring, but that there was a degree of hybridization. The species is still restricted to those areas near its places of origin, but an increase in abundance and distribution is to be expected, as is competition with native species. An exotic freshwater prawn, *Macrobrachium rosenbergi*, has also become recently established in Lake Kariba.

Elvira, B. Impact of introduced fish on the native freshwater fish fauna of Spain: an overview and a programme for monitoring the spread of aliens

The freshwater ichthyofauna of the Iberian peninsula, which is characterized by a low diversity of species but a high degree of endemism, is currently threatened by the naturalization of introduced exotics, of which at least 19 species of 10 families have become successfully established in Spanish waters. Thus, some 37% of the present freshwater fish fauna (51 species, excluding diadromous ones) are aliens, and the continued importation of new exotic species, mostly for angling purposes, has increased exponentially during the 20th century. Although many of the allochthonous species have become only locally established in the wild, some are more widely distributed.

Ecological Impact

The introduction of alien predators and competitors has almost certainly adversely affected native species. Preliminary results from a monitoring programme to assess the progressive degradation of indigenous fish communities, concentrating especially on trophic (mainly predator–prey) relationships between introduced and native fish species, has shown that alien species such as **Northern Pike** *Esox lucius* and **Largemouth Bass** *Micropterus salmoides* have had a negative impact on indigenous prey species.

Garcia-Berthou, E. Fish introductions into Banyoles Lake (Catalonia, Spain) during the last 85 years: present state and feeding relationships of the fish assemblage

Before 1910, the fish community of Banyoles Lake, Catalonia, Spain comprised six native species – the European Eel *Anguilla anguilla*, the Mediterranean Barbel *Barbus meridionalis*, the Freshwater Blenny *Blennius fluviatilis*, the Three-spined Stickleback *Gasterosteus aculeatus*, the Chub *Leuciscus cephalus*, and (perhaps an early introduction) the Tench *Tinca tinca*.

Ecological Impact

By the mid-1990s, two of these species – the Tench and the Three-spined Stickleback – seem to have died out, and the surviving species, mainly Eels and Barbels, had become uncommon. The fish assemblage in the lake is now dominated by such North American aliens as the **Largemouth Bass** *Micropterus salmoides*, **Pumpkinseed** *Lepomis gibbosus* and **Mosquitofish** *Gambusia affinis holbrooki*, and such European species as **Rudd** *Scardinius erythrophthalmus*, **Roach** *Rutilus rutilus*, **Common Carp** *Cyprinus carpio* and **Perch** *Perca fluviatilis*.

Research has shown that Chub, Eels and Largemouth Bass consume relatively small numbers of a few larger prey species. Roach, which feed mainly on zooplankton, have the lowest feeding diversity, with a low mean weight but a high number of prey items. Feeding diversity is highest among the smallest littoral species, the alien Mosquitofish and

native Freshwater Blenny; adults of the latter species may feed on larger prey.

Hauser, L. & Carvalho, G.R. Morphological and genetic diversity of the clupeid (*Limnothrissa miodon*) introduced into African Lakes

Although much work has been done to assess the ecological and socio-economic impact that alien fishes have had on native ecosystems and local human populations, comparatively little research has been conducted on the effects on the introduced species themselves, including their genetic variability and thus their ecological adaptability.

Investigations have, however, been made into the genetic effects of introducing the **Kapenta** *Limnothrissa miodon* into Lakes Kivu (Rwanda and Zaire) and Kariba (Zambia and Zimbabwe), using morphometric, allozyme and mtDNA analyses to compare morphological and genetic variation of introduced and source populations. Although no reduction in genetic diversity was found at the allozyme level (nuclear DNA), there was strong supporting evidence for a reduction in mtDNA diversity in Lake Kivu, suggesting that even the introduction of 57400 fry may in some cases be too small a founder stock to transfer completely the genetic variability of a species. A high rate of mortality, in conjunction with poor reproductive success, may have reduced the population of effective founders to well below the number of fish actually released. No evidence was found of differentiation within the lakes where Kapentas have been introduced, probably because of the relatively recent dates of release (1959 and 1967 respectively). In Lake Kariba, however, individuals taken from adjacent areas were morphologically similar, and distinct from those collected in other locations. Although morphological differentiation can be environmentally induced, and thus cannot prove reproductive isolation, it can identify discrete cohorts of fish where genetic markers are inapplicable. It is apparent that the introduction of a fish species can result in a loss of that species' genetic variability, with possible effects on its ecological adaptability to the new environment. While the precise relationship between genetic variability and ecological adaptability remains only poorly understood, evidence exists for some reduction in genetic diversity affecting the fitness of populations.

Leveque, C. **Fish species introductions in African fresh waters**

Since the mid-19th century numerous alien fish species have been introduced into African fresh waters for a variety of reasons – for example to provide sport and commercial fisheries, for aquaculture, and for the biological control of disease vectors (such as the aquatic snail which harbours *Schistosomiasis mansoni*, the causative agent of schistosomiasis (bilharzia)) and excessive growths of aquatic vegetation. Some exotics, principally the pelagic **Kapenta** *Limnothrissa miodon* and tilapias (*Oreochromis* and *Saratherodon*) have thrived in their new environments, and contribute significantly to local artisanal and subsistence fisheries.

Ecological Impact

Introductions to impoverished waters can result in trophic cascades. Thus, in Lake Nakuru in the Rift Valley of Kenya the introduction of a tilapiine caused a considerable increase in biodiversity by the expansion of the food chain to piscivorous birds. In Lakes Kyoga (Uganda) and Victoria (Kenya, Uganda and Tanzania), on the other hand, the introduction of a top predator, the **Nile Perch** *Lates niloticus* in conjunction with such environmental changes as eutrophication and increased fishing pressure, may be associated with the disappearance of numerous endemic haplochromine species.

The introduction to Lake Victoria of the alien Water Hyacinth *Eichhornia crassipes*, whose vast mats now pervade much of the lake's surface, has had a marked negative impact on native fish species.

Löffler, H. **The pikeperch (*Stizostedion lucioperca* L.) in Lake Constance (Bodensee–Obersee) – an example of successful introduction?**

The **Pikeperch** or **Zander** *Stizostedion lucioperca*, a native of eastern Europe (especially the Danube River system) and western Asia, did not occur in the Bodensee in the River Rhine system after the last

glaciation. In an attempt to increase the total fish yield, in 1882 some 200 Zander from Galicia, then a province of Austria, were introduced, followed by others in 1884, 1885, and from 1887 to 1890, after which the species was considered to be naturalized.

Elsewhere in mainland Europe, Zander were introduced from Germany to Denmark between 1879 and 1913, from Sweden to Denmark from 1915 to 1936, to France between 1912 and 1958, from France to Italy in 1964–1966, and elsewhere.

Ecological and Socio-economic Impact

The successful establishment of the Zander in the Bodensee does not appear to have adversely affected any of the 50 or more autochthonous species, but nor did it succeed in its objective of increasing the overall fish yield.

Mamcarz, A. **Introduction of fishes in Poland: a review**

Since the 12th century no fewer than 32 species and subspecies of fish from 10 families (Acipenseridae, Salmonidae, Coregondiae, Thymallidae, Umbridae, Cyprinidae, Catostomidae, Ictaluridae, Centrarchidae and Eleotridae) have been introduced mostly from Germany and the USSR, principally for rearing in aquaculture and, to a lesser extent, to provide commercial fisheries and sport-angling in lakes. Of these, 16 species were unsuccessful, eight breed only in fish farms, and eight – the **Rainbow Trout** *Oncorhynchus mykiss*, **Brook Trout** *Salvelinus fontinalis*, **Peled** *Coregonus peled*, **Mudminnow** *Umbra krameri*, **Black Bullhead** *Ictalurus melas*, **Brown Bullhead** *I. nebulosus*, **Pumpkinseed** *Lepomis gibbosus* and **Silver Carp** *Hypophthalmichthys molitrix* are established in the wild.

Muli, J. **An appraisal on stocking and introduction of fish in Lake Victoria (East Africa)**

The early introduction of exotic fish species to Lake Victoria (Kenya, Uganda and Tanzania) originated in the urgent requirement for increased protein to feed the human populations in the three countries surrounding the lake, together with an increasing demand for **Nile Perch** *Lates niloticus* fillets in the markets of Europe, the Middle East and Australia.

Paaver, T. **Introduction of foreign fish species to Estonia**

The introduction of alien fish species to Estonia can be divided into three distinct phases. In the late 19th/early 20th century, Common Carp *Cyprinus carpio*, Rainbow Trout *Oncorhynchus mykiss*, Brook Trout *Salvelinus fontinalis*, and Sterlets *Acipenser ruthenus* were imported by landowners for rearing in fish farms. During the Stalinist era of the 1950s and 1960s, when there was an urgent need to increase the supply of food for the human population, many alien species were released in both fresh and marine waters in the hope that they would establish self-maintaining populations. Sturgeons *Acipenser* spp. and Pacific Salmon *Oncorhynchus* spp. were released in the Baltic Sea, and coregonids and cyprinids were planted in lakes; most of these introductions were unsuccessful. Finally, since the 1970s many exotics have been introduced for aquacultural purposes, again without noticeable success.

The following are recorded introductions of exotic freshwater fish species to Estonia:

SPECIES	YEAR	SOURCE	PURPOSE	REMARKS
Rainbow Trout *Oncorhynchus mykiss*	1896	Germany & USSR	Aquaculture	Occasional reproduction in the wild. Aquacultural success
	1952	Germany & USSR		
	1970s	USA, Japan & Finland		
	1980s	USA & Finland		
	1990s	Finland		
Coho Salmon *O. kisutch*	1975	USSR	Establishment in the Baltic Sea and fish farms	Unsuccessful
Pink Salmon *O. gorbuscha*	1972	USSR	Establishment in the Baltic Sea	Unsuccessful
Chum Salmon *O. keta*	1969	USSR	Establishment the Baltic Sea	Unsuccessful
Gegarkuni Trout *Salmo ischan*	1973	USSR	Aquaculture	Unsuccessful
Brook Trout *Salvelinus fontinalis*	1896	USSR	Aquaculture	Natural reproduction in one fish farm until the 1980s
Arctic Charr *S. alpinus*	1995	Finland	Aquaculture	?
Peled *Coregonus peled*	1959–1962	USSR	Establishment in lakes	One or two naturalized populations probably established. Aquacultural success
Broad Whitefish *C. nasus*	1964	USSR	Establishment in lakes	Unsuccessful
Inconnu *Stenodus leucichthys*	1959–1961	USSR	Establishment in lakes	Unsuccessful
Goldfish *Carassius auratus*	1949	USSR	Angling	Naturalized in some lakes and ponds
Common Carp *Cyprinus carpio*	1893	Latvia	Aquaculture and establishment in natural waters	Naturalized in some waters. Aquacultural success
	1947–1952	USSR		
	1966	USSR		
	1977 & 1995	Germany		
Grass Carp *Ctenopharyngodon idella*	1969	USSR	Aquatic weed control	Unsuccessful
Silver Carp *Hypophthalmichthys molitrix*	1969	USSR	Aquatic weed control	Unsuccessful
Khramulya *Varicorchinus capoeta*	1961–1962	USSR	Establishment in lakes	Unsuccessful
Catfish *Ictalurus* sp.	1980s	USSR	Warmwater aquaculture	Abandoned

Thus, only the **Rainbow Trout** *Oncorhynchus mykiss*, **Peled** *Coregonus peled*, **Goldfish** *Carassius auratus*, and **Common Carp** *Cyprinus carpio* are, or may be, currently naturalized in Estonian fresh waters.

Ramussen, G. Stocking of fish in Denmark

The stocking of fish (and crayfish) in fresh and salt waters in Denmark, either to try to increase the total yield or to re-establish a declining species, began in the early 20th century and continues today. All stockings are made with artificially reared individuals which are either the offspring of captured wild fish or are bred from stock which have been held in captivity for several generations. A total of nine native species are stocked, as well as three exotics – the **Rainbow Trout** *Oncorhynchus mykiss*, **Zander** *Stizostedion lucioperca* and **Common Carp** *Cyprinus carpio*.

Schiller, C. Impact of carp (*Cyprinus carpio*) introductions on Australian freshwater fish species

In contrast to most other authorities quoted in the main text of this book, Schiller considers that **Common Carp** in Australia are a serious ecological pest ('aquatic rodents'). They are seldom fished for by anglers, occur virtually ubiquitously in vast numbers, and attempts are currently being made to reduce the population.

Ecological Impact

Schiller lists the negative effects which can be caused by Common Carp in Australian waters as follows: an increase in turbidity; an increase of algal blooms; loss of aquatic vegetation; damage to water banks; and a reduction in the numbers and diversity of native fish species.

Although the presence of macrophytes increases turbidity naturally, Common Carp may well be a contributory factor, depending on their density and the geology, hydrology and depth of water, though the evidence is inconclusive and contradictory. Most turbidity caused by Common Carp occurs in shallow waters.

Common Carp are usually associated with the destruction of submerged shallow-rooted aquatic plants with soft and fragile leaves, whose uprooting causes banks to collapse. *Vallisneria* species are those most frequently and seriously affected.

Over the past 30 years there has been a dramatic decline in the size and distribution of native fish species in Australian waters. Concomitant with this decline has been an increase in numbers and range of Common Carp, especially in New South Wales, northern Victoria and southern Queensland. In the Murray and Murrumbidgee Rivers in New South Wales, Common Carp comprise nearly the entire total catch.

Although predation by Common Carp on native fishes is insignificant – indeed they may themselves be a prey species – they probably compete with some natives for food and spawning sites.

One of the main reasons for the successful establishment of the Common Carp in Australia is that the regulation of rivers by man favours them over native species; these regulatory factors include alterations in the volume of water, a reduction in the amount of floodwater, and thermal pollution downstream of impoundments.

Smith, P.A., Leah, R.T. & Eaton, J.W. A review of Zander (*Stizostedion lucioperca*) in the UK

In some of the waters in which **Zander** *Stizostedion lucioperca* are established in England, mostly eutrophic shallow ones in rivers and flood-relief channels in eastern England and, through deliberate translocations, in some east Midland counties, their presence is associated with a decline in local fisheries. In contrast to the situation in Europe, where the Zander is a popular sporting fish wherever it occurs, in England, other than for some specialist Zander anglers, native cyprinids are preferred, and attempts have been made to reduce Zander populations in order to preserve native species.

Ecological Impact

At the present time not enough is known about the impact of Zander on aquatic ecosystems to devise and evaluate management techniques regarding the use of fisheries populated by them. For this to be done, it will be necessary (i) for an estimation to be made of the numbers of fish present with sufficient accuracy to detect any spatial and temporal alterations; (ii) for a determination of the direct impact of Zander on prey species; (iii) for a description and quantification of the basics that control the abundance of prey species, and thus the assessment of their response to predation; and (iv) for the modelling of a variety of management options. Although the two species tend to live allopatrically, it is suspected that there is some competition for food between Zander and native Northern Pike *Esox lucius*.

Staras, M. & Năvodaru, I.
Known and unknown impacts of exotic introductions on Danube Delta Fisheries, Romania

The following alien fish species are presently naturalized in Romanian waters: **Prussian Carp** *Carassius auratus gibelio*, **? Grass Carp** *Ctenopharyngodon idella*, **? Silver Carp** *Hypophthalmichthys molitrix*, **Bighead Carp** *Aristichthys nobilis*, **Stone Moroko** *Pseudorasbora parva*, and **Pumpkinseed** *Lepomis gibbosus*. None were deliberately introduced, but escaped from aquarists or aquacultural centres.

Prussian Carp

The Prussian Carp was first recorded in Romanian waters in 1938, although it had probably occurred there since about 1920. Since around 1970 the species has become established in the basin and delta of the River Danube, where it has impinged on local fisheries based on native species.

The Prussian Carp is an example of an invasive species able to maximize its reproductive potential and trophic level efficiency. It has a filtering system which is extremely effective in eutrophic waters, and is highly tolerant of poor water quality. The fact that it occurred in the Danube long before its invasive explosion and subsequent establishment shows that environmental factors and the status of the indigenous fish assemblage played a major, possibly the principal, role in its success.

The invasion of the Danube by the Prussian Carp was assisted by the general weakness of native fishes and by three acts of human interference on habitats

essential for the survival of native cyprinids; these were the damming of the flood-basin upstream of the Danube catchment, the spawning area for semi-migratory species such as the Common Carp *Cyprinus carpio*; the damming of the delta itself, which caused a decline in non-migratory species; and nitrogen and phosphorus pollution of the Danube catchment which created turbidity, resulting in a decline of Crucian Carp *Carassius carassius*, Tench *Tinca tinca*, Northern Pike *Esox lucius*, and Perch *Perca fluviatilis*.

Ecological Impact

The ecological effects of Prussian Carp in the River Danube include competition for habitats with native species; alterations in the composition of the avifauna; and predation on the eggs of native fishes, which has contributed to changes in the fish community and a resulting decrease in biodiversity. On the positive side, the presence of Prussian Carp has increased the total fish biomass.

Socio-economic Impact

Because of their relatively small size, Prussian Carp are less valuable in fisheries than some native species, and have forced fishermen to decrease the mesh size of their nets, thus causing the death of undersized native fishes; they have also gained access to local fish farms, forcing aquaculturists to take expensive preventative measures. Positively, Prussian Carp have in some places increased total catch weights and added value to weak fisheries. The maximum catches of Prussian Carp, between 1973 and 1978, amounted to between 3500 and 4000 t per annum; since then there has been a decline and apparent stabilization to the current level of 1500–2000 t annually, suggesting a new equilibrium in the fish community.

Chinese carps

Silver Carp and Grass Carp have been in the Danube since around 1970, where they appear to be in the course of becoming established. The populations of neither species, however, are yet stabilized, and large fluctuations of both stock size and age structure occur. Eutrophication and increasing turbidity have curtailed the Grass Carp's expansion, but have favoured Silver Carp, which show a higher rate of recruitment when the water level rises in June.

Neither species seems to compete with natives when adult, but both may perhaps do so in the larval stages. Concern has been expressed about the future of Silver Carp in the Danube where the habitat seems suitable for the species' expansion, although successful spawning has so far only been recorded when climatic and hydrological conditions meet the species' specific requirements.

Ecological Impact

The ecological effects of Chinese carps in the Danube are as yet unknown, but competition for food with Prussian Carp seems a possibility. They may also eventually provide competition for food and nesting-sites with native and migratory species, cause a decline in macrophytes and zooplankton, and alterations in the fish community. On the positive side, they fill a vacant trophic niche and help to improve water quality.

Socio-economic Impact

Because of their numerous small bones, Chinese carps are in general less valuable as a food than native Danubian species, and local fishermen are put to the expense of changing tackle and their mode of fishing. There is also some consumer resistance to eating the new species. Positively, Chinese carps increase the weight of fish catches, add value to weak fisheries, and are of a size suitable for commercialization and factory processing, thus helping to maintain or even increase local employment.

Wanink., J.H. & Witte, F. **Effects of introduced Nile perch on some cyprinids of Lake Victoria**

Although much has been written about the decline of haplochromine cichlids in Lake Victoria (Kenya, Uganda and Tanzania) following the introduction of **Nile Perch** *Lates niloticus*, surprisingly little attention has been paid to the fate of the 16 species of native cyprinids which dominated the lake's ichthyofauna before the explosion in the Nile Perch population.

Ecological Impact

The numbers of Dagaa *Rastrineobola argentea*, the only cyprinid closely studied, increased despite predation by Nile Perch. A relatively high rate of fecundity of Dagaa in comparison to that of haplochromine cichlids has been suggested as a possible reason for their success. If fecundity was a key factor, however, one would expect a high rate of survival among other cyprinids with a similar sublittoral distribution, such as the barbels *Barbus radiatus profundus* and *B. magdalenae*. The former species, however, disappeared after the rise in the Nile Perch population, while the numbers of the latter decreased. The barbels seem to be more benthic than the Dagaa, although *B. magdalenae*, which lives in shallower water than *B. r. profundus*, is occasionally caught near the surface. In common with juvenile Dagaa, juveniles of *B. magdalenae* probably live comparatively safe from Nile Perch in shallow inshore waters. Sympatry with Nile Perch is probably most common with *B. r. profundus* and rarest with Dagaa. Two littoral barbel species, *B. jacksonii* and *B. palludinosus*, were still regularly caught after the rise in the Nile Perch population. *Labeo victorianus*, a species that was grossly overfished by the simple expedient of damming rivers during its upstream spawning migration, is now common around rocky islands which are avoided by Nile Perch. The survival of cyprinid fishes in Lake Victoria after the rise in the numbers of Nile Perch has depended mainly on their degree of sympatry with the introduced predator.

Winfield, I.J., Dodge, D.P., & Rosch, R. **Introductions of the ruffe to new areas of Europe and to North America: history, the present situation and management implications**

Recent extensions of the European distribution of the **Ruffe** *Gymnocephalus cernuus* include in Italy and Norway; to Lake Geneva (Switzerland) in 1980; to Lake Constance (Austria, Germany and Switzerland) in 1987; to lakes in the Midland and western counties of England, to Bassenthwaite Lake in north-western England, and to Loch Lomond in Scotland in the 1980s and 1990s. The species' accidental introduction to the Laurentian Great Lakes of North America (first to Lake Superior in 1987) in the ballast-water of transoceanic vessels, has raised concern about the possible adverse effects of this hitherto harmless and valueless species, which by August 1995 had spread to the mouth of Thunder Bay River, near Alpena, Michigan, in Lake Huron.

Ecological Impact

The greatest potential threat of introduced and translocated Ruffe seems likely to be to commercial fisheries and the long-term survival of native species, though so far the evidence is inconclusive.

In Lake Constance, which is eutrophic tending to mesotrophic, the main fishery is for Common Whitefish *Coregonus lavaretus* and Perch *Perca fluviatilis*. By 1991 (4 years after their arrival) Ruffe were ubiquitous, and are now the most common fish inshore, where they prey on the eggs of inshore-spawning Whitefish.

In parts of Lake Superior, where the most important fishery is for salmonids and Yellow Perch *P. flavescens*, Ruffe were by 1993 (6 years after their arrival) the most abundant inshore fish, and it is feared that they may be competing with Yellow Perch for food. Populations of several native fishes, including the cyprinids Spot-tail Shiner *Notropis hudsonius* and Emerald Shiner *N. atherinoides*, which are both important prey species, have declined as the number of Ruffe increased.

In the Great Lakes the main management options are the enhancement of populations of predators such as Walleye *Stizostedion vitreum*, Muskellunge *Esox masquinongy* and Northern Pike *Esox lucius*, and the prohibition of the discharging of ballast-water by visiting vessels.

Recent references include:

Selgeby, J.H. 1994. Status of ruffe in the St Louis River Estuary, 1993, with emphasis on predator-prey relations. *Minutes of the Lake Superior Committee, Great Lakes Fishery Commission, Ann Arbor, Michigan,* USA: 8 pp.

Selgeby, J.H. & Edwards, A.J. 1995. Status of the ruffe in the St Louis River Estuary, 1994, with emphasis on predator-prey relations. *Minutes of the Lake Superior Committee, Great Lakes Fishery Commission, Ann Arbor, Michigan,* USA: 8 pp.

APPENDIX I

Fishes which are or may be naturalized in major African lakes (adapted from Pitcher & Hart, 1995)

LAKE	COUNTRY	SPECIES	DATE	REMARKS
Bangweulu	Zambia	*Limnothrissa miodon*	?	Moreau *et al.*, 1988
Cahora Bassa	Mozambique	*Limnothrissa miodon*	1975	Natural diffusion down the Zambezi River
Itezhi-Tezhi (Kafue Gorge)	Zambia	*Limnothrissa miodon*	1992	No fishery yet
Kariba	Zambia/Zimbabwe	*Limnothrissa miodon*	1964–1968	From Lake Tanganyika
		Oreochromis niloticus	*c.* 1993	Binali; Chifamba (Addendum)
Kivu	Zaire/Rwanda	*Limnothrissa miodon*	1958–1960	From Lake Tanganyika
Kyoga	Uganda	*Lates niloticus;* tilapias	Since 1951 (tilapias)	See note 1
Victoria	Kenya, Uganda, Tanzania	*Lates niloticus;* tilapias	1962 (*Lates*); Since 1951 (tilapias)	See note 2

Notes
1. Since 1951 four alien tilapias, *Oreochromis leucostictus, O. niloticus, Tilapia rendalli* and *T. zillii* have been introduced from Lake Albert (Zaire and Uganda). *O. leucostictus*, which lives in pools and shallows behind papyrus beds, does not seem to compete with native species in open waters. *T. zillii* and *O. niloticus* had displaced endemic *O. variabilis* and *O. esculentus* by 1970. *O. niloticus* competes advantageously with endemic *Oreochromis* spp., and has a broader diet, is more fecund, grows more rapidly and to a larger size, and has greater longevity. Hybridization may occur between *O. niloticus* and *O. variabilis*.
2. Individuals of *Lates niloticus* were caught at Bugungu just above the Ripon Falls in May 1960 and in Waigala Hannington Bay in November of the same year, following unofficial releases in the 1950s. The first official stocking of *Lates niloticus* took place in May 1962, when 35 fish from Lake Albert (Zaire/Uganda) were released off Entebbe pier, Uganda. In September 1963, 339 *Lates* from Lake Rudolph (Turkana), Kenya were freed near Kisumu, Kenya. For tilapias, see Note 1.

APPENDIX II

Countries and island groups in which introduced fishes are, or may be, naturalized

Europe

Albania

Oncorhynchus mykiss Rainbow Trout

Austria

Oncorhynchus mykiss Rainbow Trout
Salvelinus alpinus Arctic Charr
Salvelinus fontinalis Brook Trout
Aristichthys nobilis Bighead Carp
Pseudorasbora parva Stone Moroko
Ictalurus melas Black Bullhead
Ictalurus nebulosus Brown Bullhead
Gambusia affinis Mosquitofish
Micropterus salmoides Largemouth Bass

Belgium

Salvelinus fontinalis Brook Trout
Coregonus lavaretus Common Whitefish
Coregonus nasus Broad Whitefish
Umbra pygmea Eastern Mudminnow
Pimephales promelas Fathead Minnow
Ictalurus melas Black Bullhead
Ictalurus nebulosus Brown Bullhead
Lepomis gibbosus Pumpkinseed
Micropterus dolomieui Smallmouth Bass
Micropterus salmoides Largemouth Bass

British Isles

Oncorhynchus mykiss Rainbow Trout
Salvelinus fontinalis Brook Trout
Esox lucius Northern Pike
Carassius auratus Goldfish
Cyprinus carpio Common Carp
Leucaspius delineatus Sunbleak
Leuciscus idus Orfe
Leuciscus leuciscus Dace

Rhodeus sericeus Bitterling
Rutilus rutilus Roach
Tinca tinca Tench
Ictalurus melas Black Bullhead
Ictalurus nebulosus Brown Bullhead
Ictalurus punctatus Channel Catfish
Silurus glanis Wels
Poecilia reticulata Guppy
Ambloplites rupestris Rock Bass
Lepomis gibbosus Pumpkinseed
Micropterus salmoides Largemouth Bass
Stizostedion lucioperca Zander
Tilapia zillii Redbelly Tilapia

Bulgaria

Oncorhynchus mykiss Rainbow Trout
Salvelinus fontinalis Brook Trout
Aristichthys nobilis Bighead Carp
Carassius auratus Goldfish
Gambusia affinis Mosquitofish

Corfu

Gambusia affinis Mosquitofish

Corsica

Gambusia affinis Mosquitofish

Cyprus

Salvelinus alpinus Arctic Charr
Alburnus alburnus Bleak
Blicca bjoerkna Silver Bream
Carassius auratus Goldfish
Carassius carassius Crucian Carp
Cyprinus carpio Common Carp
Rutilus rutilus Roach
Ictalurus punctatus Channel Catfish

Clarias gariepinus African Catfish
Gambusia affinis Mosquitofish
Micropterus salmoides Largemouth Bass
Perca fluviatilis Common Perch
Oreochromis aureus Blue Tilapia
Oreochromis niloticus Nile Tilapia

Czechoslovakia

Oncorhynchus mykiss Rainbow Trout
Salvelinus fontinalis Brook Trout
Coregonus lavaretus Common Whitefish
Coregonus peled Peled
Aristichthys nobilis Bighead Carp
Ctenopharyngodon idella Chinese Grass Carp
Hypophthalmichthys molitrix Silver Carp
Pseudorasbora parva Stone Moroko
Ictalurus melas Black Bullhead
Ictalurus nebulosus Brown Bullhead
Lepomis gibbosus Pumpkinseed
Micropterus salmoides Largemouth Bass

Denmark

Oncorhynchus mykiss Rainbow Trout
Salvelinus fontinalis Brook Trout
Cyprinus carpio Common Carp
Ictalurus melas Black Bullhead
Ictalurus nebulosus Brown Bullhead
Stizostedion lucioperca Zander

Finland

Oncorhynchus gorbuscha Pink Salmon
Oncorhynchus keta Chum Salmon
Tinca tinca Tench

France

Oncorhynchus kisutch Coho Salmon
Oncorhynchus mykiss Rainbow Trout
Salvelinus alpinus Arctic Charr
Salvelinus fontinalis Brook Trout
Umbra pygmea Eastern Mudminnow
Carassius auratus Goldfish
Pimephales promelas Fathead Minnow
Ictalurus melas Black Bullhead
Ictalurus nebulosus Brown Bullhead
Gambusia affinis Mosquitofish
Ambloplites rupestris Rock Bass
Lepomis gibbosus Pumpkinseed
Micropterus dolomieui Smallmouth Bass

Micropterus salmoides Largemouth Bass
Stizostedion lucioperca Zander

Germany

Oncorhynchus mykiss Rainbow Trout
Salvelinus alpinus Arctic Charr
Salvelinus fontinalis Brook Trout
Salvelinus namaycush Lake Trout
Coregonus lavaretus Common Whitefish
Umbra pygmea Eastern Mudminnow
Aristichthys nobilis Bighead Carp
Carassius auratus Goldfish
Cyprinus carpio Common Carp
Pseudorasbora parva Stone Moroko
Pimephales promelas Fathead Minnow
Ictalurus melas Black Bullhead
Ictalurus nebulosus Brown Bullhead
Gambusia affinis Mosquitofish
Lepomis cyanellus Green Sunfish
Lepomis gibbosus Pumpkinseed
Micropterus salmoides Largemouth Bass

Greece

Oncorhynchus mykiss Rainbow Trout
Salvelinus fontinalis Brook Trout
Gambusia affinis Mosquitofish

Hungary

Oncorhynchus mykiss Rainbow Trout
Aristichthys nobilis Bighead Carp
Carassius auratus Goldfish
Ctenopharyngodon idella Chinese Grass Carp
Hypophthalmichthys molitrix Silver Carp
Pseudorasbora parva Stone Moroko
Ictalurus melas Black Bullhead
Ictalurus nebulosus Brown Bullhead
Gambusia affinis Mosquitofish
Lepomis gibbosus Pumpkinseed
Micropterus salmoides Largemouth Bass

Italy

Oncorhynchus mykiss Rainbow Trout
Salvelinus alpinus Arctic Charr
Salvelinus fontinalis Brook Trout
Coregonus lavaretus Common Whitefish
Carassius auratus Goldfish
Cyprinus carpio Common Carp
Ictalurus melas Black Bullhead

Ictalurus nebulosus Brown Bullhead
Ictalurus punctatus Channel Catfish
Gambusia affinis Mosquitofish
Odontesthes bonariensis Pejerrey
Lepomis auritus Redbreast Sunfish
Lepomis gibbosus Pumpkinseed
Micropterus salmoides Largemouth Bass
Stizostedion lucioperca Zander

Luxembourg

Oncorhynchus mykiss Rainbow Trout

Malta

Oreochromis mossambicus Mozambique Tilapia

Monaco

Gambusia affinis

The Netherlands

Oncorhynchus mykiss Rainbow Trout
Umbra pygmea Eastern Mudminnow
Carassius auratus Goldfish
Pseudorasbora parva Stone Moroko
Vimba vimba
Ictalurus melas Black Bullhead
Ictalurus nebulosus Brown Bullhead
Poecilia reticulata Guppy
Lepomis gibbosus Pumpkinseed
Micropterus salmoides Largemouth Bass
Stizostedion lucioperca Zander

Norway

Oncorhynchus gorbuscha Pink Salmon
Oncorhynchus keta Chum Salmon
Salvelinus fontinalis Brook Trout
Tinca tinca Tench
Ictalurus melas Black Bullhead
Ictalurus nebulosus Brown Bullhead

Poland

Salvelinus fontinalis Brook Trout
Coregonus peled Peled
Umbra krameri Mudminnow
Ictalurus melas Black Bullhead
Ictalurus nebulosus Brown Bullhead
Lepomis gibbosus Pumpkinseed

Portugal

Oncorhynchus mykiss Rainbow Trout
Carassius auratus Goldfish
Gambusia affinis Mosquitofish
Micropterus salmoides Largemouth Bass
Cichlasoma facetum Chanchito

Romania

Oncorhynchus mykiss Rainbow Trout
Salvelinus fontinalis Brook Trout
Aristichthys nobilis Bighead Carp
Ctenopharyngodon idella Chinese Grass Carp
Hypophthalmichthys molitrix Silver Carp
Pseudorasbora parva Stone Moroko
Ictiobus cyprinellus Bigmouth Buffalo
Ictiobus niger Black Buffalo
Ictalurus melas Black Bullhead
Ictalurus nebulosus Brown Bullhead
Gambusia affinis Mosquitofish
Leopomis gibbosus Pumpkinseed

Sardinia

Micropterus salmoides Largemouth Bass

Spain

Hucho hucho Huchen
Oncorhynchus mykiss Rainbow Trout
Salvelinus fontinalis Brook Trout
Esox lucius Northern Pike
Alburnus alburnus Bleak
Carassius auratus Goldfish
Cyprinus carpio Common Carp
Gobio gobio Gudgeon
Rutilus rutilus Roach
Scardinius erythropthalmus Rudd
Tinca tinca Tench
Ictalurus melas Black Bullhead
Silurus glanis Wels
Fundulus heteroclitus Common Killifish
Gambusia affinis Mosquitofish
Lepomis gibbosus Pumpkinseed
Micropterus salmoides Largemouth Bass
Perca fluviatilis Common Perch
Stizostedion lucioperca Zander
Cichlasoma facetum Chanchito

Sweden

Oncorhynchus mykiss Rainbow trout

Salvelinus fontinalis Brook Trout
Carassius auratus Goldfish
Cyprinus carpio Common Carp
Micropterus dolomieui Smallmouth Bass

Switzerland

Oncorhynchus mykiss Rainbow Trout
Salvelinus fontinalis Brook Trout
Salvelinus namaycush Lake Trout
Ictalurus melas Black Bullhead
Ictalurus nebulosus Brown Bullhead
Lepomis gibbosus Pumpkinseed
Micropterus salmoides Largemouth Bass

Yugoslavia

Oncorhynchus mykiss Rainbow Trout
Salvelinus alpinus Arctic Charr
Salvelinus fontinalis Brook Trout
Ctenopharyngodon idella Chinese Grass Carp
Hypophthalmichthys molitrix Silver Carp
Pseudorasbora parva Stone Moroko
Ictalurus melas Black Bullhead
Ictalurus nebulosus Brown Bullhead
Gambusia affinis Mosquitofish
Lepomis gibbosus Pumpkinseed
Micropterus salmoides Largemouth Bass

Asia

Afghanistan

Oncorhynchus mykiss Rainbow Trout
Cyprinus carpio Common Carp
Gambusia affinis Mosquitofish

Bangladesh

Oreochromis mossambicus Mozambique Tilapia
Oreochromis niloticus Nile Tilapia

Borneo (Brunei)

Gambusia affinis Mosquitofish

Borneo (Kalimantan)

Oreochromis mossambicus Mozambique Tilapia
Trichogaster pectoralis Snakeskin Gourami

China

Rhodeus ocellatus Rosy Bitterling

Clarias batrachus Walking Catfish
Clarias gariepinus African Catfish
Gambusia affinis Mosquitofish
Oreochromis mossambicus Mozambique Tilapia
Oreochromis niloticus Nile Tilapia

Hong Kong

Gambusia affinis Mosquitofish
Micropterus dolomieui Smallmouth Bass
Oreochromis mossambicus Mozambique Tilapia
Oreochromis niloticus Nile Tilapia

India

Oncorhynchus mykiss Rainbow Trout
Salmo trutta Brown Trout
Barbus javanicus/gonionotus Tawes
Carassius carassius Crucian Carp
Cyprinus carpio Common Carp
Hypophthalmichthys molitrix Silver Carp
Tinca tinca Tench
Gambusia affinis Mosquitofish
Poecilia reticulata Guppy
Etroplus suratensis Pearl Spot
Oreochromis mossambicus Mozambique Tilapia
Oreochromis niloticus Nile Tilapia
Osphronemus gouramy Giant Gourami

Indonesia (Bali)

Oreochromis mossambicus Mozambique Tilapia
Helostoma temmincki Kissing Gourami

Indonesia (Banka)

Oreochromis mossambicus Mozambique Tilapia

Indonesia (Celebes)

Barbus javanicus/gonionotus Tawes
Clarias batrachus Walking Catfish
Oreochromis mossambicus Mozambique Tilapia
Trichogaster pectoralis Snakeskin Gourami

Indonesia (Java)

Oreochromis mossambicus Mozambique Tilapia
Trichogaster pectoralis Snakeskin Gourami

Indonesia (Lombok)

Oreochromis mossambicus Mozambique Tilapia
Helostoma temmincki Kissing Gourami

Indonesia (Amboina, Molucca Islands)

Oreochromis mossambicus Mozambique Tilapia

Indonesia (Banjak, Pulau Pulau Islands)

Oreochromis mossambicus Mozambique Tilapia

Indonesia (Sumatra)

Oreochromis mossambicus Mozambique Tilapia

Indonesia (unspecified localities)

Cyprinus carpio Common Carp
Tinca tinca Tench
Channa striata Snakehead

Iran

Oncorhynchus mykiss Rainbow Trout
Gambusia affinis Mosquitofish

Iraq

Gambusia affinis Mosquitofish

Israel

Oncorhynchus mykiss Rainbow Trout
Cyprinus carpio Common Carp
Gambusia affinis Mosquitofish
Odontesthes bonariensis Pejerrey
Oreochromis mossambicus Mozambique Tilapia
Oreochromis niloticus Nile Tilapia

Japan

Oncorhynchus mykiss Rainbow Trout
Salmo trutta Brown Trout
Salvelinus fontinalis Brook Trout
Carassius auratus Goldfish
Rhodeus ocellatus Rosy Bitterling
Gambusia affinis Mosquitofish
Channa maculata
Lepomis macrochirus Bluegill
Micropterus salmoides Largemough Bass
Oreochromis mossambicus Mozambique Tilapia
Oreochromis niloticus Nile Tilapia
Tilapia sparrmani Banded Tilapia
Tilapia zillii Redbelly Tilapia
Macropodus chinensis Round-tailed Paradisefish

Korea

Oncorhynchus mykiss Rainbow Trout

Carassius auratus Goldfish
Rhodeus ocellatus Rosy Bitterling
Lepomis cyanellus Green Sunfish
Lepomis macrochirus Bluegill
Micropterus salmoides Largemouth Bass

Laos

Oreochromis mossambicus Mozambique Tilapia

Lebanon

Oncorhynchus mykiss Rainbow Trout

Malaysia (unspecified localities)

Cyprinus carpio Common Carp
Clarias gariepinus African Catfish
Micropterus salmoides Largemouth Bass
Etroplus suratensis Pearl Spot
Oreochromis mossambicus Mozambique Tilapia
Oreochromis niloticus Nile Tilapia

Pakistan

Salmo trutta Brown Trout
Oreochromis mossambicus Mozambique Tilapia

Philippines

Barbus javanicus/gonionotus Tawes
Carassius carassius Crucian Carp
Catla catla Catla
Cyprinus carpio Common Carp
Labeo rohita Rohu
Misgurnus anguillicaudatus Oriental Weatherfish
Clarias batrachus Walking Catfish
Clarias gariepinus African Catfish
Gambusia affinis Mosquitofish
Poecilia latipinna Sailfin Molly
Lepomis cyanellus Green Sunfish
Lepomis macrochirus Bluegill
Micropterus salmoides Largemouth Bass
Oreochromis aureus Blue Tilapia
Oreochromis mossambicus Mozambique Tilapia
Oreochromis niloticus Nile Tilapia
Glossogobius sp.
Osphronemus gouramy Giant Gourami
Trichogaster leeri Pearl Gourami
Trichogaster pectoralis Snakeskin Gourami
Trichogaster trichopterus Three-spot Gourami
Helostoma temmincki Kissing Gourami

Saudi Arabia

Gambusia affinis Mosquitofish
Oreochromis mossambicus Mozambique Tilapia

Singapore

Barbus semifasciolatus Half-banded Barb
Gambusia affinis Mosquitofish
Poecilia latipinna Sailfin Molly
Poecilia reticulata Guppy
Poecilia sphenops Liberty Molly
Oreochromis aureus Blue Tilapia
Oreochromis mossambicus Mozambique Tilapia
Trichogaster pectoralis Snakeskin Gourami
Betta splendens Siamese Fighting Fish

Sri Lanka

Oncorhynchus mykiss Rainbow Trout
Salmo trutta Brown Trout
Carassius carassius Crucian Carp
Cyprinus carpio Common Carp
Poecilia reticulata Guppy
Xiphophorus helleri Green Swordtail
Oreochromis mossambicus Mozambique Tilapia
Oreochromis niloticus Nile Tilapia
Tilapia rendalli Redbreast Tilapia
Osphronemus gouramy Giant Gourami
Trichogaster pectoralis Snakeskin Gourami
Trichogaster trichopterus Three-spot Gourami
Helostoma temminck Kissing Gourami

Syria

Gambusia affinis Mosquitofish

Taiwan

Oncorhynchus mykiss Rainbow Trout
Carassius auratus Goldfish
Cyprinus carpio Common Carp
Clarias batrachus Walking Catfish
Gambusia affinis Mosquitofish
Oreochromis aureus Blue Tilapia
Oreochromis mossambicus Mozambique Tilapia
Oreochromis niloticus Nile Tilapia
Tilapia rendalli Redbreast Tilapia

Thailand

Gymnocorymbus ternetzi Black Tetra
Carassius auratus Goldfish

Cyprinus carpio Common Carp
Gambusia affinis Mosquitofish
Cichlasoma octofasciatum Jack Dempsey
Oreochromis mossambicus Mozambique Tilapia
Oreochromis niloticus Nile Tilapia
Tilapia rendalli Redbreast Tilapia

Turkey

Oncorhynchus mykiss Rainbow Trout
Gambusia affinis Mosquitofish
Stizostedion lucioperca Zander

USSR (Europe and Asia)

Aristichthys nobilis Bighead Carp
Hemibarbus maculatus Spotted Steed
Hemiculter eigenmanni Korean Sharpbelly
Hypophthalmichthys molitrix Silver Carp
Opsariichthys uncirostris
Pseudorasbora parva Stone Moroko
Rhodeus sericeus Bitterling
Ictiobus cyprinellus Bigmouth Buffalo
Ictiobus niger Black Buffalo
Ictalurus melas Black Bullhead
Ictalurus nebulosus Brown Bullhead
Gambusia affinis Mosquitofish
Micropterus salmoides Largemouth Bass
Oreochromis aureus Blue Tilapia
Oreochromis mossambicus Mozambique Tilapia
Rhinogobius similis Amur Goby
Hypseleotris swinhonis

Vietnam

Carassius auratus Goldfish
Micropterus dolomieui Smallmouth Bass
Oreochromis mossambicus Mozambique Tilapia
Oreochromis niloticus Nile Tilapia

Yemen

Gambusia affinis Mosquitofish

Africa
Algeria

Salmo trutta Brown Trout
Esox lucius Northern Pike
Gambusia affinis Mosquitofish
Micropterus salmoides Largemouth Bass
Oreochromis macrochir Longfin Tilapia

Oreochromis mossambicus Mozambique Tilapia

Benin

Oreochromis macrochir Longfin Tilapia
Oreochromis mossambicus Mozambique Tilapia
Oreochromis niloticus Nile Tilapia

Botswana

Micropterus salmoides Largemouth Bass

Burundi

Oreochromis macrochir Longfin Tilapia
Oreochromis niloticus Nile Tilapia
Tilapia rendalli Redbreast Tilapia

Cameroun

Clarias gariepinus African Catfish
Astatoreochromis alluaudi
Oreochromis macrochir Longfin Tilapia
Oreochromis niloticus Nile Tilapia

Central African Republic

Heterotis niloticus Bony Tongue
Gambusia affinis Mosquitofish
Astatoreochromis alluaudi
Oreochromis macrochir Longfin Tilapia
Oreochromis niloticus Nile Tilapia

Congo

Astatoreochromis alluaudi
Oreochromis macrochir Longfin Tilapia

Egypt

Gambusia affinis Mosquitofish

Ethiopia

Oncorhynchus mykiss Rainbow Trout
Salmo trutta Brown Trout
Esox lucius Northern Pike
Ctenopharyngodon idella Chinese Grass Carp
Cyprinus carpio Common Carp
Hypopthalmichthys molitrix Silver Carp
Gambusia affinis Mosquitofish
Tilapia zillii Redbelly Tilapia

Gabon

Heterotis niloticus Bony Tongue
Clarias gariepinus African Catfish
Oreochromis macrochir Longfin Tilapia

Ghana

Oreochromis macrochir Longfin Tilapia

Guinea

Oreochromis niloticus Nile Tilapia

Ivory Coast

Heterotis niloticus Bony Tongue
Ctenopharyngodon idella Chinese Grass Carp
Clarias gariepinus African Catfish
Astronotus ocellatus Oscar
Oreochromis macrochir Longfin Tilapia
Oreochromis niloticus Nile Tilapia
Oreochromis urolepis hornorum Wami Tilapia

Kenya

Oncorhynchus mykiss Rainbow trout
Salmo trutta Brown Trout
Salvelinus fontinalis Brook Trout
Barbus amphigramma
Cyprinus carpio Common Carp
Gambusia affinis Mosquitofish
Poecilia reticulata Guppy
Lates niloticus Nile Perch
Micropterus salmoides Largemouth Bass
Oreochromis leucostictus
Oreochromis macrochir Longfin Tilapia
Oreochromis niloticus Nile Tilapia
Tilapia rendalli Redbreast Tilapia
Tilapia zillii Redbelly Tilapia

Lesotho

Oncorhynchus mykiss Rainbow Trout
Salmo trutta Brown Trout
Micropterus salmoides Largemouth Bass

Madagascar

Heterotis niloticus Boney Tongue
Oncorhynchus mykiss Rainbow Trout
Salmo trutta Brown Trout

Carassius auratus Goldfish
Cyprinus carpio Common Carp
Tanichthys albonubes White Cloud Mountain Minnow
Gambusia affinis Mosquitofish
Poecilia reticulata Guppy
Xiphophorus helleri Green Swordtail
Xiphophorus maculatus Southern Platyfish
Channa striata Snakehead
Micropterus salmoides Largemouth Bass
Oreochromis macrochir Longfin Tilapia
Oreochromis mossambicus Mozambique Tilapia
Oreochromis niloticus Nile Tilapia
Oreochromis spilurus
Tilapia rendalli Redbreast Tilapia
Tilapia sparrmani Banded Tilapia
Tilapia zillii Redbelly Tilapia
Osphronemus gouramy Giant Gourami
Ctenopoma ansorgei Ornate Ctenopoma

Malawi

Oncorhynchus mykiss Rainbow Trout
Phalloceros caudimaculatus One-spot Livebearer
Micropterus salmoides Largemouth Bass

Mali

Oreochromis niloticus Nile Tilapia

Morocco

Hucho hucho Huchen
Esox lucius Northern Pike
Barbus barbus Barbel
Cyprinus carpio Common Carp
Rutilus rutilus Roach
Scardinius erythrophthalmus Rudd
Tinca tinca Tench
Gambusia affinis Mosquitofish
Lepomis cyanellus Green Sunfish
Lepomis gibbosus Pumpkinseed
Lepomis macrochirus Bluegill
Lepomis microlophus Redear Sunfish
Micropterus salmoides Largemouth Bass
Stizostedion lucioperca Zander

Mozambique

Limnothrissa miodon Tanganyika Sardine
Micropterus salmoides Largemouth Bass

Namibia

Cyprinus carpio Common Carp
Poecilia reticulata Guppy
Xiphophorus helleri Green Swordtail
Micropterus salmoides Largemouth Bass
Oreochromis mossambicus Mozambique Tilapia

Nigeria

Xiphophorus maculatus Southern Platyfish

Rwanda

Limnothrissa miodon Tanganyika Sardine
Oreochromis esculentus Ngege
Oreochromis macrochir Longfin Tilapia
Oreochromis niloticus Nile Tilapia
Tilapia rendalli Redbreast Tilapia
Haplochromis sp.

Sierra Leone

Oreochromis niloticus Nile Tilapia

South Africa

Oncorhynchus mykiss Rainbow Trout
Salmo trutta Brown Trout
Salvelinus fontinalis Brook Trout
Carassius auratus Goldfish
Cyprinus carpio Common Carp
Tinca tinca Tench
Gambusia affinis Mosquitofish
Poecilia reticulata Guppy
Xiphophorus helleri Green Swordtail
Morone saxatilis Striped Bass
Lepomis cyanellus Green Sunfish
Lepomis macrochirus Bluegill
Lepomis microlophus Redear Sunfish
Micropterus dolomieui Smallmouth Bass
Micropterus punctulatus Spotted Bass
Micropterus salmoides Largemouth Bass
Perca fluviatilis Common Perch
Oreochromis aureus Blue Tilapia
Oreochromis niloticus Nile Tilapia
Serranochromis robustus Banded Largemouth
Tilapia zillii Redbelly Tilapia

Sudan

Oncorhynchus mykiss Rainbow Trout

Ctenopharyngodon idella Chinese Grass Carp
Gambusia affinis Mosquitofish

Swaziland

Oncorhynchus mykiss Rainbow Trout
Salmo trutta Brown Trout
Lepomis cyanellus Green Sunfish
Lepomis macrochirus Bluegill
Micropterus salmoides Largemouth Bass
Serranochromis robustus Banded Largemouth

Tanzania

Oncorhynchus mykiss Rainbow Trout
Salmo trutta Brown Trout
Lates niloticus Nile Perch
Micropterus salmoides Largemouth Bass
Oreochromis andersoni Three-spot Tilapia
Oreochromis leucostictus
Oreochromis mossambicus Mozambique Tilapia
Oreochromis niloticus Nile Tilapia
Tilapia rendalli Redbreast Tilapia
Tilapia zillii Redbelly Tilapia

Tunisia

Oncorhynchus mykiss Rainbow Trout
Esox lucius Northern Pike
Cyprinus carpio Common Carp
Scardinius erythrophthalmus Rudd
Tinca tinca Tench
Oreochromis mossambicus Mozambique Tilapia
Oreochromis niloticus Nile Tilapia

Uganda

Oncorhynchus mykiss Rainbow Trout
Salmo trutta Brown Trout
Esox lucius Northern Pike
Cyprinus carpio Common Carp
Poecilia reticulata Guppy
Lates niloticus Nile Perch
Micropterus salmoides Largemouth Bass
Oreochromis mossambicus Mozambique Tilapia
Oreochromis niloticus Nile Tilapia
Oreochromis spilurus
Tilapia rendalli Redbreast Tilapia
Tilapia zillii Redbelly Tilapia

Zaire

Limnothrissa miodon Tanganyika Sardine

Heterotis niloticus Bony Tongue
Clarias gariepinus African Catfish
Astatoreochromis alluaudi
Oreochromis leucostictus
Oreochromis macrochir Longfin Tilapia
Oreochromis mossambicus Mozambique Tilapia
Oreochromis niloticus Nile Tilapia
Oreochromis spilurus

Zambia

Limnothrissa miodon Tanganyika Sardine
Oncorhynchus mykiss Rainbow Trout
Lepomis cyanellus Green Sunfish
Astatoreochromis alluaudi
Oreochromis macrochir Longfin Tilapia

Zanzibar

Oreochromis urolepis hornorum Wami Tilapia

Zimbabwe

Limnothrissa miodon Tanganyika Sardine
Oncorhynchus mykiss Rainbow Trout
Salmo trutta Brown Trout
Salvelinus fontinalis Brook Trout
Barbus holubi Smallmouth Yellowfish
Barbus natalensis Natal Yellowfish
Carassius auratus Goldfish
Cyprinus carpio Common Carp
Labeo altivelis
Neobola brevianalis
Tinca tinca Tench
Gambusia affinis Mosquitofish
Lepomis macrochirus Bluegill
Micropterus dolomieui Smallmouth Bass
Micropterus punctulatus Spotted Bass
Micropterus salmoides Largemouth Bass

North America
Canada

Oncorhynchus aquabonita Golden Trout
Salmo trutta Brown Trout
Carassius auratus Goldfish
Cyprinus carpio Common Carp
Scardinius erythrophthalmus Rudd
Tinca tinca Tench
Noturus insignis Margined Madtom

Gambusia affinis Mosquitofish
Poecilia latipinna Sailfin Molly
Gymnocephalus cernuus Ruffe
Hemichromis bimaculatus/letourneauxi African Jewelfish
Neogobius melanostomus Round Goby
Proterorhinus marmoratus Tubenose Goby
Platichthys flesus European Flounder

Mexico

Salvelinus fontinalis Brook Trout
Barbus conchonius Rosy Barb
Barbus titteya Cherry Barb
Carassius auratus Goldfish
Carpoides cyprinus Quillback
Ctenopharyngodon idella Chinese Grass Carp
Cyprinus carpio Common Carp
Misgurnus anguillicaudatus Oriental Weatherfish
Ictalurus melas Black Bullhead
Fundulus zebrinus Plains Killifish
Poecilia reticulata Guppy
Morone saxatilis Striped Bass
Morone chrysops White Bass
Ambloplites rupestris Rock Bass
Lepomis auritus Redbreast Sunfish
Lepomis gulosus Warmouth
Lepomis microlophus Redear Sunfish
Micropterus dolomieui Smallmouth Bass
Pomoxis annularis White Crappie
Pomoxis nigromaculatus Black Crappie
Oreochromis aureus Blue Tilapia
Oreochromis mossambicus Mozambique Tilapia
Oreochromis niloticus Nile Tilapia
Oreochromis urolepis hornorum Wami Tilapia
Tilapia zillii Redbelly Tilapia
Acanthogobius flavimanus Yellowfin Goby

United States

Salmo trutta Brown Trout
Hypomesus nipponensis Wakasagi
Aristichthys nobilis Bighead Carp
Brachydanio rerio Zebra Danio
Carassius auratus Goldfish
Ctenopharyngodon idella Chinese Grass Carp

Cyprinus carpio Common Carp
Hypophthalmichthys molitrix Silver Carp
Leuciscus idus Orfe
Rhodeus sericeus Bitterling
Scardinius erythrophthalmus Rudd
Tinca tinca Tench
Misgurnus anguillicaudatus Oriental Weatherfish
Clarias batrachus Walking Catfish
Hypostomus spp. Suckermouth Catfishes
Pterygoplichthys multiradiatus Radiated Ptero
Rivulus harti Giant Rivulus
Belonesox belizanus Pike Killifish
Poecilia latipunctata Broadspotted Molly
Poecilia mexicana Shortfin Molly
Poecilia petenensis Swordtail Molly
Poecilia reticulata Guppy
Poecilia sphenops Liberty Molly
Poeciliopsis gracilis Porthole Livebearer
Xiphophorus helleri Green Swordtail
Xiphophorus maculatus Southern Platyfish
Xiphophorus variatus Variable Platyfish
Gymnocephalus cernuus Ruffe
Bairdiella icistia Bairdiella
Cynoscion xanthulus Orangemouth Corvina
Astronotus ocellatus Oscar
Cichla ocellaris Tucunare
Cichlasoma bimaculatum Black Acara
Cichlasoma citrinellum Midas Cichlid
Cichlasoma managuense Jaguar Guapote
Cichlasoma meeki Firemouth Cichlid
Cichlasoma nigrofasciatum Convict Cichlid
Cichlasoma octofasciatum Jack Dempsey
Cichlasoma urophthalmus Mayan Cichlid
Geophagus surinamensis Redstriped Eartheater
Hemichromis bimaculatus/letourneauxi African Jewelfish
Oreochromis aureus Blue Tilapia
Oreochromis urolepis hornorum Wami Tilapia
Oreochromis mossambicus Mozambique Tilapia
Sarotherodon melanotheron Blackchin Tilapia
Tilapia mariae Spotted Tilapia
Tilapia rendalli Redbreast Tilapia
Acanthogobius flavimanus Yellowfin Goby
Neogobius melanostomus Round Goby
Proterorhinus marmoratus Tubenose Goby
Tridentiger bifasciatus
Tridentiger trigonocephalus Chameleon Goby
Colisa fasciata Banded Gourami
Colisa lalia Dwarf Gourami
Trichopsis vittata Croaking Gourami
Platichthys flesus European Flounder

South and Central America

West Indies

Antigua

Oreochromis aureus Blue Tilapia
Tilapia rendalli Redbreast Tilapia
Tilapia zillii Redbelly Tilapia

Barbados

Oreochromis mossambicus Mozambique Tilapia

Cuba

Cyprinus carpio Common Carp
Ictiobus cyprinellus Bigmouth Buffalo
Ictiobus niger Black Buffalo
Oryzias latipes Japanese Rice Fish
Poecilia reticulata Guppy
Lepomis sp.
Micropterus salmoides Largemouth Bass
Cichlasoma managuense Jaguar Guapote
Oreochromis aureus Blue Tilapia
Oreochromis mossambicus Mozambique Tilapia
Oreochromis niloticus Nile Tilapia
Oreochromis urolepis hornorum Wami Tilapia
Tilapia rendalli Redbreast Tilapia

Dominica

Oreochromis mossambicus Mozambique Tilapia

Grenada

Oreochromis mossambicus Mozambique Tilapia

Hispaniola (Dominican Republic)

Cyprinus carpio Common Carp
Ictalurus punctatus Channel Catfish
Gambusia affinis Mosquitofish
Poecilia reticulata Guppy
Xiphophorus helleri Green Swordtail
Xiphophorus maculatus Southern Platyfish
Micropterus salmoides Largemouth Bass
Cichla ocellaris Tucunare
Oreochromis aureus Blue Tilapia
Oreochromis mossambicus Mozambique Tilapia
Oreochromis niloticus Nile Tilapia
Trichogaster trichopterus Three-spot Gourami
Betta splendens Siamese Fighting Fish

Betta bellica Slim Fighting Fish

Hispaniola (Haiti)

Gambusia affinis Mosquitofish
Poecilia reticulata Guppy
Xiphophorus helleri Green Swordtail
Xiphophorus maculatus Southern Platyfish
Micropterus salmoides Largemouth Bass
Oreochromis aureus Blue Tilapia
Oreochromis mossambicus Mozambique Tilapia
Oreochromis niloticus Nile Tilapia

Jamaica

Poecilia reticulata Guppy
Xiphophorus helleri Green Swordtail
Xiphophorus maculatus Southern Platyfish
Oreochromis mossambicus Mozambique Tilapia

Lesser Antilles

Poecilia reticulata Guppy

Martinique

Oreochromis mossambicus Mozambique Tilapia

Puerto Rico

Dorosoma petenense Threadfin Shad
Barbus conchonius Rosy Barb
Ictalurus catus White Catfish
Ictalurus nebulosus Brown Bullhead
Ictalurus punctatus Channel Catfish
Rivulus marmoratus Mangrove Rivulus
Gambusia affinis Mosquitofish
Poecilia reticulata Guppy
Xiphophorus helleri Green Swordtail
Xiphophorus maculatus Southern Platyfish
Lepomis gulosus Warmouth
Lepomis macrochirus Bluegill
Lepomis microlophus Redear Sunfish
Micropterus coosae Redeye Bass
Micropterus salmoides Largemouth Bass
Cichla ocellaris Tucunare
Oreochromis aureus Blue Tilapia
Oreochromis mossambicus Mozambique Tilapia
Oreochromis niloticus Nile Tilapia
Oreochromis urolepis hornorum Wami Tilapia
Tilapia rendalli Redbreast Tilapia

St Lucia

Oreochromis mossambicus Mozambique Tilapia

Trinidad

Oreochromis mossambicus Mozambique Tilapia

Central America
Belize

Micropterus dolomieui Smallmouth Bass

Costa Rica

Oncorhynchus mykiss Rainbow Trout
Cyprinus carpio Common Carp
Micropterus salmoides Largemouth Bass
Oreochromis aureus Blue Tilapia
Oreochromis mossambicus Mozambique Tilapia
Oreochromis niloticus Nile Tilapia
Oreochromis urolepis hornorum Wami Tilapia

El Salvador

Micropterus salmoides Largemouth Bass
Cichlasoma managuense Jaguar Guapote
Cichlasoma guttulatum Mojara Azul
Oreochromis aureus Blue Tilapia
Oreochromis mossambicus Mozambique Tilapia
Oreochromis niloticus Nile Tilapia
Oreochromis urolepis hornorum Wami Tilapia
Tilapia rendalli Redbreast Tilapia

Guatemala

Lepomis gibbosus Pumpkinseed
Micropterus salmoides Largemouth Bass
Pomoxis nigromaculatus Black Crappie
Cichlasoma managuense Jaguar Guapote
Cichlasome motaguense
Oreochromis aureus Blue Tilapia
Oreochromis mossambicus Mozambique Tilapia
Oreochromis niloticus Nile Tilapia
Oreochromis urolepis hornorum Wami Tilapia

Honduras

Micropterus salmoides Largemouth Bass
Cichlasoma managuense Jaguar Guapote
Oreochromis mossambicus Mozambique Tilapia
Oreochromis niloticus Nile Tilapia
Oreochromis urolepis hornorum Wami Tilapia

Nicaragua

Oreochromis aureus Blue Tilapia
Oreochromis mossambicus Mozambique Tilapia
Oreochromis niloticus Nile Tilapia
Oreochromis urolepis hornorum Wami Tilapia

Panama

Oncorhynchus mykiss Rainbow Trout
Salmo trutta Brown Trout
Cyprinus carpio Common Carp
Lepomis macrochirus Bluegill
Lepomis microlophus Redear Sunfish
Lepomis humilis Orangespotted Sunfish
Micropterus salmoides Largemouth Bass
Pomoxis nigromaculatus Black Crappie
Cichla ocellaris Tucunare
Cichlasoma managuense Jaguar Guapote
Oreochromis aureus Blue Tilapia
Oreochromis mossambicus Mozambique Tilapia
Oreochromis niloticus Nile Tilapia
Oreochromis urolepis hornorum Wami Tilapia

South America
Argentina

Oncorhynchus gorbuscha Pink Salmon
Oncorhynchus mykiss Rainbow Trout
Salmo salar Atlantic Salmon
Salmo trutta Brown Trout
Salvelinus fontinalis Brook Trout
Salvelinus namaycush Lake Trout
Carassius auratus Goldfish
Cyprinus carpio Common Carp
Gambusia affinis Mosquitofish

Bolivia

Oncorhynchus mykiss Rainbow Trout
Salmo trutta Brown Trout
Salvelinus fontinalis Brook Trout
Carassius auratus Goldfish
Cyprinus carpio Common Carp
Gambusia affinis Mosquitofish
Odontesthes bonariensis Pejerrey
Micropterus salmoides Largemouth Bass
Oreochromis mossambicus Mozambique Tilapia
Oreochromis niloticus Nile Tilapia

Brazil

Oncorhynchus mykiss Rainbow Trout
Carassius auratus Goldfish
Cyprinus carpio Common Carp
Ictalurus punctatus Channel Catfish
Lepomis cyanellus Green Sunfish
Lepomis macrochirus Bluegill
Micropterus salmoides Largemouth Bass
Oreochromis aureus Blue Tilapia
Oreochromis mossambicus Mozambique Tilapia
Oreochromis niloticus Nile Tilapia
Oreochromis urolepis hornorum Wami Tilapia
Tilapia rendalli Redbreast Tilapia
Betta splendens Siamese Fighting Fish

Chile

Oncorhynchus gorbuscha Pink Salmon
Orcorhynchus keta Chum Salmon
Oncorhynchus kisutch Coho Salmon
Oncorhynchus mykiss Rainbow Trout
Salmo salar Atlantic Salmon
Salmo trutta Brown Trout
Salvelinus fontinalis Brook Trout
Coregonus clupeaformis Lake Whitefish
Cyprinus carpio Common Carp
Tinca tinca Tench
Ictalurus melas Black Bullhead
Ictalurus nebulosus Brown Bullhead
Gambusia affinis Mosquitofish
Odontesthes bonariensis Pejerrey
Lepomis gibbosus Pumpkinseed
Cichlasoma facetum Chanchito

Colombia

Oncorhynchus mykiss Rainbow Trout
Gymnocorymbus ternetzi Black Tetra
Barbus conchonius Rosy Barb
Barbus gelius Golden Barb
Barbus oligolepis Island Barb
Barbus titteya Cherry Barb
Brachydanio rerio Zebra Danio
Carassius auratus Goldfish
Cyprinus carpio Common Carp
Danio malabaricus Giant Danio
Rasbora trilineata Three-lined Rasbora
Tanichthys albonubes White Cloud Mountain Minnow
Poecilia latipinna Sailfin Molly
Poecilia reticulata Guppy

Poecilia velifera Yucatan Sailfin Molly
Xiphophorus helleri Green Swordtail
Xiphophorus maculatus Southern Platyfish
Xiphophorus variatus Variable Platyfish
Micropterus salmoides Largemouth Bass
Cichlasoma meeki Firemouth Cichlid
Oreochromis mossambicus Mozambique Tilapia
Oreochromis niloticus Nile Tilapia
Oreochromis urolepis hornorum Wami Tilapia
Tilapia rendalli Redbreast Tilapia
Osphronemus gouramy Giant Gourami
Trichogaster leeri Pearl Gourami
Trichogaster pectoralis Snakeskin Gourami
Trichogaster trichopterus Three-spot Goruami
Colisa fasciata Banded Gourami
Colisa labiosa Thicklip Gourami
Colisa lalia Dwarf Gourami
Helostoma temmincki Kissing Gourami
Macropodus sp.

Ecudaor

Oncorhynchus mykiss Rainbow Troup
Salmo trutta Brown Trout
Micropterus salmoides Largemouth Bass
Oreochromis niloticus Nile Tilapia

Guyana

Oreochromis mossambicus Mozambique Tilapia

Paraguay

Oreochromis niloticus Nile Tilapia

Peru

Oncorhynchus mykiss Rainbow Trout
Salmo trutta Brown Trout
Salvelinus fontinalis Brook Trout
Carassius auratus Goldfish
Cyprinus carpio Common Carp
Gambusia affinis Mosquitofish
Poecilia reticulata Guppy
Odontesthes bonariensis Pejerrey
Oreochromis aureus Blue Tilapia
Oreochromis mossambicus Mozambique Tilapia
Oreochromis niloticus Nile Tilapia
Oreochromis urolepis hornorum Wami Tilapia
Tilapia rendalli Redbreast Tilapia

Uruguay

Carassius auratus Goldfish
Cyprinus carpio Common Carp

Venezuela

Oncorhynchus mykiss Rainbow Trout
Salmo trutta Brown Trout
Salvelinus fontinalis Brook Trout
Cyprinus carpio Common Carp
Poeciliopsis gracilis Porthole Livebearer
Lepomis macrochirus Bluegill
Oreochromis mossambicus Mozambique Tilapia
Oreochromis niloticus Nile Tilapia

Australasia

Australia

Oncorhynchus mykiss Rainbow Trout
Salmo trutta Brown Trout
Salvelinus fontinalis Brook Trout
Barbus conchonius Rosy Barb
Barbus tetrazona Tiger Barb
Carassius auratus Goldfish
Cyprinus carpio Common Carp
Rutilus rutilus Roach
Tinca tinca Tench
Misgurnus anguillicaudatus Oriental Weatherfish
Gambusia affinis Mosquitofish
Phalloceros caudimaculatus One-spot Livebearer
Poecilia latipinna Sailfin Molly
Poecilia reticulata Guppy
Xiphophorus helleri Green Swordtail
Xiphophorus maculatus Southern Platyfish
Perca fluviatilis Common Perch
Cichlasoma nigrofasciatum Convict Cichlid
Cichlasoma octofasciatum Jack Dempsey
Geophagus brasiliensis Pearl Cichlid
Hemichromis bimaculatus/letourneauxi African Jewelfish
Oreochromis mossambicus Mozambique Tilapia
Tilapia mariae Spotted Tilapia
Acanthogobius flavimanus Yellowfin Goby
Tridentiger bifasciatus
Tridentiger trigonocephalus Chameleon Goby

New Zealand

Oncorhynchus mykiss Rainbow Trout
Oncorhynchus nerka Sockeye Salmon
Oncorhynchus tshawytscha Chinook Salmon

Salmo salar Atlantic Salmon
Salmo trutta Brown Trout
Salvelinus fontinalis Brook Trout
Salvelinus namaycush Lake Trout
Carassius auratus Goldfish
Cyprinus carpio Common Carp
Scardinius erythrophthalmus Rudd
Tinca tinca Tench
Ictalurus nebulosus Brown Bullhead
Gambusia affinis Mosquitofish
Poecilia latipinna Sailfin Molly
Poecilia reticulata Guppy
Xiphophorus helleri Green Swordtail
Perca fluviatilis Common Perch

Papua New Guinea

Oncorhynchus mykiss Rainbow Trout
Salmo trutta Brown Trout
Carassius auratus Goldfish
Cyprinus carpio Common Carp
Clarias batrachus Walking Catfish
Gambusia affinis Mosquitofish
Poecilia reticulata Guppy
Xiphophorus helleri Green Swordtail
Channa striata Snakehead
Oreochromis mossambicus Mozambique Tilapia
Tilapia rendalli Redbreast Tilapia
Osphronemus gouramy Giant Gourami
Trichogaster pectoralis Snakeskin Gourami
Trichogaster trichopterus Three-spot Gourami
Anabas testudineus Climbing Perch

Oceania
Atlantic Ocean

Azores

Carassius auratus Goldfish
Carassius carassius Crucian Carp
Cyprinus carpio Common Carp
Rutilus macrolepidotus
Micropterus salmoides Largemouth Bass
Perca fluviatilis Common Perch

Bermuda

Gambusia affinis Mosquitofish

Canary Islands

Poecilia reticulata Guppy

Falkland Islands

Salmo trutta Brown Trout
Salvelinus fontinalis Brook Trout

Indian (and Southern) Oceans

Îles Crozet

Salmo trutta Brown Trout
Salvelinus fontinalis Brook Trout

Îles Kerguelen

Salmo trutta Brown Trout
Salvelinus fontinalis Brook Trout

Maldive Islands

Gambusia affinis Mosquitofish

Mauritius

Carassius auratus Goldfish
Cyprinus carpio Common Carp
Hypophthalmichthys molitrix Silver Carp
Labeo rohita Rohu
Poecilia reticulata Guppy
Channa striata Snakehead
Lepomis cyanellus Green Sunfish
Lepomis macrochirus Bluegill
Lepomis microlophus Redear Sunfish
Micropterus salmoides Largemouth Bass
Oreochromis macrochir Longfin Tilapia
Oreochromis niloticus Nile Tilapia
Tilapia rendalli Redbreast Tilapia
Tilapia zillii Redbelly Tilapia
Osphronemus gouramy Giant Gourami

Réunion

Oncorhynchus mykiss Rainbow Trout
Oreochromis niloticus Nile Tilapia

Pacific Ocean

Caroline Islands

Oreochromis mossambicus Mozambique Tilapia

Cook Islands

Gambusia affinis Mosquitofish
Poecilia reticulata Guppy
Oreochromis mossambicus Mozambique Tilapia

Oreochromis niloticus Nile Tilapia

Fiji Islands

Barbus javanicus/gonionotus Tawes
Carassius auratus Goldfish
Cyprinus carpio Common Carp
Pseudorasbora parva Stone Moroko
Rhodeus ocellatus Rosy Bitterling
Gambusia affinis Mosquitofish
Poecilia mexicana Shortfin Molly
Poecilia reticulata Guppy
Xiphophorus helleri Green Swordtail
Channa striata Snakehead
Micropterus dolomieui Smallmouth Bass
Micropterus salmoides Largemouth Bass
Oreochromis mossambicus Mozambique Tilapia
Oreochromis niloticus Nile Tilapia

Gilbert Islands

Oreochromis mossambicus Mozambique Tilapia

Hawaiian Islands

Monopterus albus Swamp Eel
Dorosoma petenense Threadfin Shad
Oncorhynchus mykiss Rainbow Trout
Barbus semifasciolatus Half-banded Barb
Barbus filamentosus Black-spot Barb
Carassius auratus Goldfish
Cyprinus carpio Common Carp
Misgurnus anguillicaudatus Oriental Weatherfish
Ictalurus punctatus Channel Catfish
Clarias fuscus Chinese Catfish
Corydoras aenus Bronze Corydoras
Hypostomus spp. Suckermouth Catfishes
Pterygoplichthys multiradiatus Radiated Ptero
Ancistrus sp. Bristle-nosed Catfish
Strongylura kreffti Freshwater Long Tom
Xenentodon cancila Asian Needlefish
Gambusia affinis Mosquitofish
Poecilia latipinna Sailfin Molly
Poecilia mexicana Shortfin Molly
Poecilia reticulata Guppy
Poecilia sphenops Liberty Molly
Poecilia vittata Cuban Limia
Xiphophorus helleri Green Swordtail
Xiphophorus maculatus Southern Platyfish
Xiphophorus variatus Variable Platyfish

Channa striata Snakehead
Lepomis macrochirus Bluegill
Micropterus dolomieui Smallmouth Bass
Micropterus salmoides Largemouth Bass
Astronotus ocellatus Oscar
Cichla ocellaris Tucunare
Cichlasoma meeki Firemouth Cichlid
Cichlasoma nigrofasciatum Convict Cichlid
Cichlasoma spilurum Blue-eyed Cichlid
Hemichromis bimaculatus/letourneauxi African Jewelfish
Oreochromis macrochir Longfin Tilapia
Oreochromis mossambicus Mozambique Tilapia
Pelvicachromis pulcher Rainbow Krib
Sarotherodon melanotheron Blackchin Tilapia
Tilapia rendalli Redbreast Tilapia
Tilapia zillii Redbelly Tilapia
Mugiligobius parva

Line Islands (Fanning Atoll & Washington Island)

Gambusia affinis Mosquitofish
Oreochromis mossambicus Mozambique Tilapia

Mariana Islands (Guam)

Cyprinus carpio Common Carp
Clarias batrachus Walking Catfish
Gambusia affinis Mosquitofish
Poecilia latipinna Sailfin Molly
Poecilia reticulata Guppy
Xiphophorus helleri Green Swordtail
Astronotus ocellatus Oscar
Cichla ocellaris Tucunare
Oreochromis mossambicus Mozambique Tilapia
Tilapia zillii Redbelly Tilapia
Betta brederi Javan Mouthbrooding Fighting Fish

Marshall Islands (Jaluit)

Gambusia affinis Mosquitofish

Micronesia (unspecified islands)

Gambusia affinis Mosquitofish

Nauru Island

Oreochromis mossambicus Mozambique Tilapia

New Caledonia

Poecilia reticulata Guppy

Channa striata Snakehead
Micropterus salmoides Largemouth Bass
Oreochromis mossambicus Mozambique Tilapia
Tilapia zillii Redbelly Tilapia
Osphronemus gouramy Giant Gourami
Trichogaster pectoralis Snakeskin Gourami

New Hebrides

Gambusia affinis Mosquitofish

Palau Islands (Babelthuap)

Barbus sealei
Misgurnus anguillicaudatus Oriental Weatherfish
Poecilia reticulata Guppy
Xiphophorus maculatus Southern Platyfish

Samoa (American and Western)

Carassius auratus Goldfish
Gambusia affinis Mosquitofish
Poecilia mexicana Shortfin Molly
Poecilia reticulata Guppy
Oreochromis mossambicus Mozambique Tilapia

Society Islands (Tahiti)

Gambusia affinis Mosquitofish
Poecilia mexicana Shortfin Molly
Poecilia reticulata Guppy
Lates calcarifer Barramundi
Oreochromis mossambicus Mozambique Tilapia

Solomon Islands

Gambusia affinis Mosquitofish
Oreochromis mossambicus Mozambique Tilapia

South Pacific (unspecified island)

Helostoma temmincki Kissing Gourami

Tonga Islands

Oreochromis mossambicus Mozambique Tilapia

Tuvalu Islands

Oreochromis mossambicus Mozambique Tilapia

Vanuatu Islands

Poecilia reticulata Guppy
Oreochromis mossambicus Mozambique Tilapia

Wallis Island

Oreochromis macrochir Longfin Tilapia
Oreochromis mossambicus Mozambique Tilapia
Tilapia rendalli Redbreast Tilapia

GLOSSARY

(For terms associated with the introduction of living organisms see Preface)

Abiotic   Non-living or devoid of life.

Aerobic   Growing only in the presence of oxygen.

Aestivate   The survival of long periods of hot weather by sleeping in a shaded location. Some species of freshwater fish burrow into soft mud at the onset of the dry season.

Agonistic   Behaviour between two individuals of the same species that may involve aggression, threat, appeasement or avoidance.

Alevin   Yolk-bearing larvae of salmonids.

Alga   Vernacular name for simple type of plant that is never differentiated into leaves, stem, and root. Most are aquatic. (Pl. algae).

Allopatric   Species that occupy separate habitats.

Amphidromous   Fish that migrate between the sea and fresh water for other than breeding purposes.

Anadromous   Fish that spend most of their lives at sea but migrate to fresh water to spawn.

Anaerobic   Growing only in the absence of oxygen.

Anoxic   Difficiency of oxygen in tissues.

Anthropogenic   Originated by man.

Benthic   Species that live on or near the sea bottom.

Billabong   Australasian river branch that forms a backwater or small stagnant oxbow (q.v.) pool.

Biocoenoses   The living components of a biome (q.v.) in an ecosystem (q.v.), comprising the phytocoenosis (primary producers), the zoocoenosis (secondary producers or consumers), and the microbiocoenosis (decomposer organisms).

Biomass   The total weight of the living components in an ecosystem (q.v.).

Biome   The biological subdivision of the surface of the Earth which reflects the ecological and physiognomic character of the vegetation.

Biosphere   That part of the environment in which living organisms are found.

Biota   Animals and plants collectively of a given area.

Canal des Pangalenes   A series of natural swampy coastal lagoons with barrier islands on the central east coast of the island of Madagascar, formed by soil erosion and joined by man-made canals to provide a habitat for fishes and a navigable waterway for small boats.

Cestode   Endoparasitic (q.v.) tapeworm.

Chironomid   Non-biting midges of the order Diptera.

Compensation level   The depth at which light penetration in aquatic ecosystems is so reduced that oxygen production barely balances consumption.

Congener   Closely related species.

Conspecific   Individuals of the same species.

Cordillera   System of mountain ranges.

Cut-Off   See Oxbow.

Demography   The study of life in communities.

Detritivore   Animal that feeds largely on dead material.

Diadromous   Fish that regularly migrate, for any reason, between marine and fresh waters.

DNA   Deoxyribonucleic acid. The genetic material of organisms, its sequence of paired bases constituting the genetic code.

Dystrophic   Waterbody that has become so shallow (through organic and inorganic sedimentation) and so oxygen-depleted (through the aerobic (q.v.) bacterial decomposition of the organic matter), that bog and peat begin to form.

Ecad   Organism modified by its environment.

Ecological niche   Functional position of an organism within a community.

Ecology   Scientific study of the inter-relationships among organisms and between organisms and all aspects, biotic (q.v.) and abiotic (q.v.), of their environment.

Ecosystem   The interdependence of species in the biota (q.v.) with one another and with their abiotic (q.v.) environment.

Ectoparasite   Parasite (q.v.) living on its host externally.

Endemic   Species or any other taxonomic group that is a native only of the geographic area under discussion.

Endoparasite   Parasite (q.v.) living within the body of its host.

Epilimnion   Upper level of circulating warm water in a thermally stratified lake in summer.

Euryhaline   Able to tolerate a wide range of degrees of salinity.

Eurythermal   Able to tolerate a wide range of temperatures.

Eutrophic   Nutrient-rich water with high primary productivity.

Exotherm   See Poikilotherm.

Folivore   Animal whose food consists mainly of leaves and other foliage.

Gene   Fundamental physical unit of heredity.

Gillraker   Fairly stiff tooth-like processes (q.v.) on the inner side of the gill arch of most bony fishes which strain the water flowing over them.

Guild Ecological group (of animals or plants) distinguished from others by specific characteristics.

Halophyte Plant adapted morphologically (q.v.) and/or physiologically (q.v.) to growing in a saline environment.

Herbivore An animal that feeds on primary producers, usually green plants.

Homologue Animal that corresponds to another in a different part of the world, probably through their descent from a common ancestor and thus of the same evolutionary origin.

Hydric An extremely wet environment.

Hydrology Science of the study of water.

Hydrophyte Plant adapted morphologically (q.v.) and/or physiologically (q.v.) to growing in water or a wet environment.

Hypolimnion The deeper and colder non-circulating water in a thermally heated lake in summer.

Lacustrine Living or growing in lakes.

Lentic Freshwater habitat dominated by standing water.

Leucistic Of pale (or white) coloration.

Limnetic zone The area in large and deep lakes which lies between the compensation level (q.v.) and littoral zone (q.v.).

Limnology Study of freshwater ecosystems.

Littoral zone The area in shallow waters around lake shores where light can penetrate to the bottom.

Lotic Freshwater habitat dominated by flowing water.

Macrophyte Member of the macroscopic plant life in a body of water.

Mesotrophic Providing only a moderate amount of nutrition.

Microbial Pertaining to micro-organisms.

Morphology Form and structure of individual organisms.

Nauplius The first, free-swimming, planktonic larva of most marine and some freshwater crustaceans.

Nematode Class of worms, including eelworms, roundworms and threadworms.

Neonatal Newly born.

Oligotrophic Water low in nutrient and of poor primary productivity.

Ontogeny The development of an individual from early stages to maturity.

Operculum (gill cover) Hard but flexible cover which in bony fishes forms the outer wall of the gill chamber.

Ovum Strictly an unfertilized egg cell. Frequently used in the more general sense of any egg. (Pl. ova).

Oxbow A section of a river channel that no longer carries the main flow.

Parr Juvenile freshwater stage of a salmon before becoming a smolt (q.v.).

Parasite Animal or plant living on or within another species, from which it derives nutriment.

Pelagic Marine organism living in open water.

Physiology Science of functions and phenomena of living organisms.

Phytophagous Feeding on plants. See also Polyphagous.

Phytophilous See Phytophagous.

Phytoplankton Plant plankton (q.v.) and primary producers (of organic material) of an aquatic ecosystem.

Piscivorous Fish-eating.

Plankton Minute aquatic organisms that drift with water movement, usually without locomotive organs.

Poikilotherm Organisms, such as fish, that regulate their body temperature by behavioural means, such as burrowing and/or basking.

Polyphagous Phytophagous (q.v.) species that feeds on a wide variety of plant material.

Potamodromous Fish that make regular migrations within large freshwater systems.

Processes Natural appendages, outgrowths or protuberances.

Redd Gravelly spawning depression or nest in a river bed made by salmonids.

Refugia Small isolated areas where animals and plants find refuge from newly created unfavourable conditions.

Rheophilic Stream-loving.

Roil Make water turbid by disturbance.

Sessile Attached to a substrate (q.v.) and thus non-motile.

Sigmoid growth curve An S-shaped growth curve, in which a population initially increases fairly rapidly and then declines, often with equal rapidity.

Slough Quagmire, marsh or swamp.

Smolt A young salmon migrating to the sea for the first time.

Spat Spawn of shellfish, especially oysters.

Stenographic Having extremely specialized feeding requirements.

Stenothermal Adapted to only slight variations in temperature.

Stunt The tendency for fishes whose numbers have increased explosively to produce populations in which the size of mature individuals is much reduced.

Substrate Any object or material on which an organism grows or to which it is attached. An underlying layer or substance.

Sympatric Two or more species that occupy similar habitats or whose habitats overlap.

Taxon Group of organisms of any taxonomic rank.

Trematode Parasitic (q.v.) flatworm, such as a fluke, of the class Trematoda.

Trophic level Step in transfer of food or energy within a chain.

Unimodal Having a single form or mode (e.g. of size or age).

Wallace's Line Zoogeographical division in the Malay archipelago separating the Oriental and Notogaean (Australasian) faunal regions.

Zooplankton See Plankton.

REFERENCES

Acere, T.O. 1988. The controversy over Nile perch *Lates niloticus* in Lake Victoria, east Africa. *Naga Report* **11**: 3–5.

Achieng, A.P. 1990. The effects of the introduction of the Nile perch *Lates niloticus* (L.) with special reference to the haplochromines of Lake Victoria. In: *The Biology and Conservation of Rare Fish.* Papers from an international symposium of The Fisheries Society of the British Isles.

Ahuja, S.K. 1964. Salinity tolerance of *Gambusia affinis*. *Indian Journal of Experimental Biology* **2**: 9–11.

Akaboshi, T. 1959. Blackbass. *Miyazaki-ken Tansui-gyogyos hidojyo*: 1–71.

Akekio, S. 1880. Memorandum on fish-culture in Japan with a notice of experiments in breeding the California trout. *Report of the United States Commissioner for Fish and Fisheries* **7**: 645–648.

Akihito, Emperor of Japan & Sakamoto, K. 1989. Reexamination of the status of the striped goby. *Japanese Journal of Ichthyology* **36**: 100–112.

Aksiray, F. 1961. About Sudat (*Lucioperca lucioperca* L.) introduced into some lakes in Turkey. *Proceedings of the General Fisheries Council of the Mediterranean* **6**: 335–343.

Albrecht, A.B. 1964. Some observations on factors associated with survival of striped bass eggs and larvae. *California Fish and Game* **50**: 100–113.

Al-Daham, N.K., Huq, M.F. & Sharma, K.P. 1977. Notes on the ecology of fishes of the genus *Aphanius* and *Gambusia affinis* in southern Iraq. *Freshwater Biology* **7**: 245–251.

Alessio, G. 1984. Le Black-bass *Micropterus salmoides* (Lacep.) dans les eaux italiennes. Un antagoniste du brochet? *Bulletin Français de Pisciculture* **292**: 1–17.

Alikunhi, K.H. 1966. Synopsis of biological data on the common carp *Cyprinus carpio* (Linnaeus, 1758) (Asia and the Far East). *FAO Fisheries Synopsis* **(31.1)**: 73 pp.

Aliyev, D.S., Verigina, I.A. & Svetovidova, A.A. 1963. The species of fish imported together with grass carp and silver carp from China. *Problemy rybokhozyay-stvennogo ispol'-zovaniya rastitel'noyadnykh ryb v vodoyemakh SSSR.* Ashkhabad. [In Russian.]

Allen, A.W. 1980. *Cyprinus carpio.* In: Lee, D.S., Gilbert, C.R., Hocutt, C.H., Jenkins, R.E., McAllister, D.E. & Stauffer Jr., J.R. (Eds), *Atlas of North American Freshwater Fishes.* Raleigh: North Carolina State Museum.

Allen. G.R. 1989. *Freshwater Fishes of Australia.* Brookvale, NSW: T.F.H. Publications.

Allen, G.R. 1991. *Freshwater Fishes of New Guinea.* Madang, PNG: Christensen Research Institute.

Allen, S. 1984. Occurrence of juvenile weatherfish *Misgurnus anguillicaudatus* (Pisces: Cobitidae) in the Yarra River. *Victorian Naturalist* **101**: 240–242.

Alletson, D.J. 1985. Observations on some piscivorous birds in a trout fishing area of Natal. *Lammergeyer* **35**: 41–46.

Allport, C. 1874. A report on the fish breeding ponds of the Acclimatisation Society of Victoria and the introduction of fish into the rivers of the colony. *Proceedings of the Zoological & Acclimatization Society of Victoria* **3**: 37–40.

Allport, M. 1870a. Brief history of the introduction of salmon (*Salmo salar*) and other Salmonidae to the waters of Tasmania. *Proceedings of the Zoological Society of London* 14–30.

Allport, M. 1870b. Additional notes on the introduction of Salmonidae into Tasmania. *Proceedings of the Zoological Society*: 750–752.

Allport, M. 1875. Some further notes on the introduction of the salmon into Tasmanian waters. *Proceedings of the Royal Society of Tasmania*, 1874: 12–18.

Allport, M. 1880. Present stage of the Salmon experiment in Tasmania. *Report of the United States Fisheries Commission*, 1878: 819–823.

Alm, G. 1920. Resultaten av fiskinplanteringar i Sverige. *Meddelingen Lantbruksstyrelsen* **226**: 1–108.

Alm, G. 1929. Niagra ord om blackbassen. *Svensk Fiskeri Tidskrift* **38**: 61–62.

Alvarez, J. 1959. Nota preliminar sobre la ictiofauna del estado de San Luis Potosi. *Acta Científica Potosina* **3**: 71–88.

Alvarez, J. & Cortés, M.T. 1962. Ictiologia Michoacána. I. Claves y catálogo de las especies conocidas. *Anales Escuela Nacional Ciencias Biológicas, Mexico* **9**: 85–142.

Alvarez, J. & Navarro, L. 1957. Los peces del Valle de México. Secretaria de Marina. Comissión para el Fomento de la Piscicultura Rural, México.

Amatayakul, C. 1957. Gold fish. *Thai Fisheries Gazette* **10**: 169.

Anderson, I. 1995. Trout in trouble as carp cross Bass Strait. *New Scientist* **146**: 6.

Andrews, C.W. 1966. Early importation and distribution of exotic fresh water fish species in Newfoundland. *Canadian Fish-Culturist* **36**: 35–36.

Andrews, S. 1985. Aquatic species introduced to Fiji. *Domodomo* **111**: 67–82.

Ang, K.J., Gopinath, R. & Chua, T.E. 1989. The status of introduced fish species in Malaysia. In: De Silva, S.S. (Ed.), *Exotic Aquatic Organisms in Asia. Asian Fisheries Society Special Publication* **3**: 71–82.

Anon. 1887. Fish culture in Scotland. *Bulletin of the United States Fish Commission* **6**: 408.

Anon. 1897. Fish represented in colored plates – the steel-head. *Annual Report of the Commissioner of Fisheries & Game Forests for the State of New York* **3**: 241–243.

Anon. 1908. Further proof of the successful introduction of the Pacific Coast Quinnat Salmon (*Oncorhynchus tschawytscha*) into New Zealand. *New Zealand Fishing Gazette* **56**: 21.

Anon. 1909. Report of the Committee on Foreign Relations. *Transactions of the American Fisheries Society* **39**: 174–182.

Anon. 1944. *Cape Provincial Administration Inland Fisheries Department Report* **1**: 1–5.

Anon. 1948. Surveys and stocking of local waters. *Cape Provincial Administration Report* **5**: 25–31.

Anon. 1956. Fish farming. *Madras Government Fisheries Information Bureau*: 25 pp.

Anon. 1958. New introduction of edible pondfish from Philippines. *Quarterly Bulletin of the South Pacific Commission* **8**: 19.

Anon. 1960. Survey of dams in the Northern Cape. CPA Department of Nature Conservation. Report **17**: 55–56.

Anon. 1962–1963. From the Cape Agricultural Journal, 26 July 1894. Acclimation of trout. *Piscator* **56**: 109–110.

Anon. 1963. Black bass success in New Caledonia. *Quarterly Bulletin of the South Pacific Commission* **13**: 54.

Anon. 1971. Fish Introductions. FAO Aquaculture Bulletin **3**: 14–15.

Anon. 1974. A strange find. *Fauna and Flora* **25**: 17.

Anon. 1988. Monster fish may be innocent of ecological crimes. *New Scientist* **119**: 34.

Anon. 1992. Ruffe in the Great Lakes: a threat to North American fisheries. Great Lakes Fishery Commission Ruffe Task Force Report. Ann Arbor, Michigan: Great Lakes Fishery Commission.

Appleton, C.C. 1974. Check-list of the flora and fauna of the Gladdespruit, Nelspruit district, Eastern Transvaal. *Newsletter of the Limnological Society of Southern Africa* **22**: 49–58.

Arbocco, G. 1966. I pesci d'acqua dolce della Liguria. *Estratto Dagli Annali del Museo Civico Di Storia Naturale Di Genova* **76**: 149–150.

Arenburg, P. 1911. Acclimation du blac-bass en France (*Micropterus salmoides*). *Bulletin de la Société d'Acclimatation* **58**: 533–535.

Arentz, A.F. 1966. Acclimatisation history covers 100 years. *The Fisherman*: 8–10.

Armistead, J.J. 1895. *An Angler's Paradise, and How to Obtain It.* London.

Arrowsmith, E. & Pentelow, F.T.K. 1965. The introduction of trout and salmon to the Falkland Islands. *Salmon and Trout Magazine* **174**: 119–129.

Arthington, A.H. 1986. Introduced cichlid fish in Australian inland waters. *Monographiae Biologicae* **61**: 239–248.

Arthington, A.H. 1989. Impacts of introduced and translo-cated freshwater fishes in Australia. In: De Silva, S.S.

(Ed.), *Exotic Aquatic Organisms in Asia. Asian Fisheries Society Special Publication* **3**: 7–20.

Arthington, A.H. 1991. Ecological and genetic impacts of introduced and translocated freshwater fishes in Australia. *Canadian Journal of Fisheries and Aquatic Sciences* **48** (Suppl. 1): 33–43.

Arthington, A.H. & Blühdorn, D.R. 1994. Distribution, genetics, ecology and status of the introduced cichlid, *Oreochromis mossambicus*, in Australia. *Mitteilungen Internationale Vereinigung für Theoretische und Angewandte Limnologie* **24**: 53–62.

Arthington, A.H. & Lloyd, L.N. 1989. Introduced poeciliids in Australia and New Zealand. In: Meffe, G.K. & Snelson, F.F. Jr (Eds), *Ecology and Evolution of Livebearing Fishes (Poeciliidae)*: 333–348. Prentice Hall.

Arthington, A.H., McKay, R.J. & Milton, D. 1981. Consultancy on ecology and interactions of exotic and endemic freshwater fishes in southeastern Queensland streams. *Report* **1**. Canberra: Australian National Parks & Wildlife Service: 96 pp.

Arthington, A.H. & Milton, D.A. 1986. Reproductive biology, growth and age composition of the introduced *Oreochromis mossambicus* (Cichlidae) in two reservoirs, Brisbane, Australia. *Environmental Biology of Fishes* **16**: 257–266.

Arthington, A.H., Milton, D.A. & McKay, R.J. 1983. Effects of urban development and habitat alterations on the dis-tribution of native and exotic freshwater fish in the Brisbane region, Queensland. *Australian Journal of Ecology* **8**: 87–101.

Arthington, A.H., McKay, R.J., Russell, D.J. & Milton, D.A. 1984. Occurrence of the introduced cichlid *Oreochromis mossambicus* (Peters) in Queensland. *Australian Journal of Marine and Freshwater Research* **35**: 267–272.

Arthur, W. 1879 and 1884. On the brown trout introduced into Otago. *Transactions of the New Zealand Institute* **11**: 271–290; **16**: 467–512.

Arthur, W. 1881. History of fish culture in New Zealand. *Transactions of the New Zealand Institute* **14**: 1–79.

Arthur, W. 1882. History of fish culture in New Zealand. *Transactions of the New Zealand Institute* **14**: 180–210.

Artom, C. 1924. La specie di Gambusia acclimata in Italia (*Gambusia holbrooki* Grd) in relazione colla stabilita del carattere del gonopodio. *Atti della Reale Academia Nazionale dei Lincei*, Series 5, **33**: 278–282.

Ashton, P.J., Appleton, C.C. & Jackson, P.B.N. 1986. Ecological impacts and economic consequences of alien or invasive organisms in southern African aquatic ecosys-tems. In: Macdonald, I.A.W., Kruger, F.J. & Ferrar, A.A. (Eds), *The Ecology and Management of Biological Invasions in Southern Africa*: 247–260. Cape Town: Oxford University Press.

Atton, F.M. 1959. The invasion of Manitoba and Saskatchewan by carp. *Transactions of the American Fisheries Society* **88**: 203–205.

Atton, F.M. & Johnson, R.P. 1955. First records of eight species of fishes in Saskatchewan. *Canadian Field Naturalist* **69**: 82–84.

Atz, J.M. 1954. The perigrinating *Tilapia*. *Animal Kingdom* **57**: 148–155.

Ayson, D.F. 1910. Introduction of American fishes into New Zealand. *Bulletin of the Bureau of Fisheries*, Washington (1908) **28**: 969–975.

Bade, E. 1926. The central European bitterling found in the States. *Bulletin of the New York Zoological Society* **29**: 188: 205–206.

Baigun, C.R.M. & Quiros, R. 1985. Introduction de peces exoticos en la Republica de Argentina. *Informe Tecnico Deptmento Aquas Instituto Nacional de Investigación Desarrollo Pesquero Argentina* (2): 5 vols.

Baird, S.F. 1879. The carp. In: *Report of the United States Fish Commission* (1876–1877): 40–44. Washington, DC: US Government Printing Office.

Baird, S.F. 1893. Report of the Commissioner. In: *Report of the United States Fish Commission (1890–1891)*: 1–96. Washington, DC: US Government Printing Office.

Bakshtansky, E.L. 1980. The introduction of pink salmon into the Kola Peninsula. In: Thorpe, J.E. (Ed.), *Salmon Ranching*: 245–260. London: Academic Press.

Balarin, J.D. 1979. *Tilapia: A Guide to their Biology and Culture in Africa*. Stirling, Scotland: University of Stirling.

Baldwin, N.S. & Saalfield, R.W. 1962. Commercial fish production in the Great Lakes, 1867–1960. *Great Lakes Fisheries Technical Report* **3**: 166 pp.

Balon, E.K. 1974. Domestication of the carp *Cyprinus carpio* L. *Royal Ontario Museum, Life Sciences Miscellaneous Publications*: 37 pp.

Balon, E.K. 1995. Origin and domestication of the wild carp, *Cyprinus carpio*: from Roman gourmets to the swimming flowers. *Aquaculture* **129**: 3–48.

Bănărescu, P. 1964. *Pisces – Osteichthyes. Fauna Republicii Populare Romine, XIII*. Bucuresti: Academiei RPR.

Banerji, S.R. & Satish, M.S. 1989. Changes brought about by the introduction of exotic fishes in Bihar. In: Joseph, M.M. (Ed.), *Exotic Aquatic Species in India. Asian Fisheries Society Special Publication* **1**: 97–99.

Barajas, M.L. & Contreras, S. 1986. Variación estacional y morfologia de los peces de la Presa Marte R. Gomez, noreste de México. IV Congreso Nacional de Zoológica, México.

Bard, J. 1960. Introduction du black bass dans l'ouest Cameroon. Paper at symposium on problems of major lakes, Lusaka, 14–18 August. CTA/CSA; Ms. 1 p.

Bard, J. 1977. The unforeseen results of pike–perch (*Lucioperca lucioperca* L.) introduction in an Anatolian lake. *Journal of Pisciculture of France* **49**: 37–39.

Barel, C.D.N., Dorit, R., Greenwood, P.H., Fryer, G., Hughes, N., Jackson, P.B.N., Kawanabe, H., Lowe-McConnell, R.H., Nagoshi, M., Ribbink, A.J., Trewavas, E., Witte, F. & Yamaoka, K. 1985. Destruction of fisheries in Africa's lakes. *Nature* **315**: 19–20.

Barlow, C.G. & Lisle, A. 1987. Biology of the Nile perch, *Lates niloticus* (Pisces: Centropomidae) with reference to its proposed role as a sport fish in Australia. *Biological Conservation* **39**: 269–289.

Barron, J.C. 1964. Reproduction and apparent over-winter survival of the sucker-mouth armoured catfish, *Plecostomus* sp., in the headwaters of the San Antonio River. *Texas Journal of Science* **16**: 449.

Barros, V.R. 1931. Introducion de un nuevo salmon en Chile. *Revista Chilena de Historia Natural* **35**: 57–62.

Barrow, I. 1971. Freshwater fish. In: *The Swartkops Estuary: An Ecological Survey*: 33–34. Port Elizabeth: Zwartkops Trust.

Bartley, D.M. 1993. An application of international codes of practice on introductions of aquatic organisms: assessment of a project on the use of Chinese carps in Mozambique. *Food and Agricultural Organization of the United Nations Fisheries Circular* **863**: 21 pp.

Bartley, D.M. & Subasinghe, R.P. 1995. Historical aspects of international movement of living aquatic species. Paper presented at the OIE International Conference on Problems in Control of Diseases in Aquatic Animals Due to International Trade. Paris, France, 7–9 June 1995: 13 pp. mimeo.

Baskin, Y. 1992. Africa's troubled waters. Fish introductions and a changing physical profile: muddy Lake Victoria's future. *Bioscience* **42**: 476–481.

Bastedo, S.T. 1904. Report of the Deputy Commissioner of Fisheries for the year 1904. *Annual Report of the Ontario Department of Fisheries* **6**: 104 pp.

Batchelor, G.R. 1974. An ecological investigation of the Doorndraai Dam, Sterk River, Transvaal, with special reference to fish management. MSc dissertation, Rand Africaans University: 131 pp.

Batchelor, G.R. 1978. Aspects of the biology of *Tilapia rendalli* in the Doorndraai Dam, Transvaal. *Journal of the Limnological Society of Southern Africa* **4**: 65–68.

Bauch, G. 1955. *Die einheimischen Süsswasserfische*. Radebeul, Berlin: Neumann.

Baugh, T.M., Deacon, J.E. & Withers, D. 1985. Conservation efforts with the Hiko White River springfish. *Journal of Aquaculture and Aquatic Sciences* **4**: 49–53.

Baughman, J.L. 1947. The tench in America. *Journal of Wildlife Management* **11**: 197–204.

Baughman, J. 1985. A brief history of the golden trout. *Fisheries* **10**: 2.

Baxter, A.F., Vallis, S.L. & Hume, D.J. 1985. The predation of recently released rainbow trout fingerlings (*Salmo gairdneri*) by redfin (*Perca fluviatilis*) in Lake Burrumbeet, October–December 1983. Arthur Rylah Institute for Environmental Research, Technical Report Services **16**: 1–24.

Bayoumi, A.R. 1969. Notes on the occurrence of *Tilapia zillii* (Pisces) in Suez Bay. *Marine Biology* **4**: 255–256.

Bean, J.H. 1895. The rainbow trout. *Annual Report of the Commissioners for Fisheries and Game Forests in the State of New York* **1**: 135–140.

Beckman, W.C. 1974. *Guide to the Fishes of Colorado.* Boulder, Colorado: University of Colorado Museum.

Begg, G.W. 1976. Some notes on the Sandvlei fish fauna, Muizenberg, Cape. *Piscator* **96**: 5–14.

Bell-Cross, G. & Bell-Cross, B. 1971. Introduction of *Limnothrissa miodon* and *Limnocaridina tanganicae* from Lake Tanganyika to Lake Kariba. *Fisheries Research Bulletin of Zambia* **5**: 207–214.

Bennion, B. 1923. *The Angler in South Africa.* Johannesburg: Hortors.

Belshe, J.F. 1961. Observations of an introduced tropical fish (*Belonesox belizanus*) in southern Florida. Unpublished MSc thesis, University of Miami: 71 pp.

Ben-Tuvia, A. 1981. Man-induced changes in the freshwater fish fauna of Israel. *Fisheries Management* **12**: 139–148.

Berg, L.S. 1964–1965. *Freshwater Fishes of the USSR and adjacent countries.* Jerusalem: Israel Programme for Scientific Translations.

Berg, M. 1961. Pink Salmon (*Oncorhynchus gorbuscha*) in northern Norway in the year 1960. *Acta Borealis* **17A**: 1–24.

Berg, M. 1977. Pink Salmon *Oncorhynchus gorbuscha* (Walbaum) in Norway. *Report of the Institute of Freshwater Research* **56**: 12–17.

Berra, T.M., Moore, R. & Reynolds, L.F. 1975. The freshwater fishes of the Laloki River system of New Guinea. *Copeia* **1975**: 316–325.

Besana, G. 1910. American fishes in Italy. *Bulletin of the United States Bureau of Fisheries* **28**: 949–954.

Besana, G. 1915. *Coregonus maraena* im Manatesee. *Allgemeine Fischerei-Zeitung* **40**: 121.

Best, B. & Davidson, C. 1981. Inventory and atlas of the inland aquatic ecosystems of the Marianas Archipelago. *University of Guam, Marine Laboratory Technical Report* **75**.

Bhimichar, B.S., David, A. & Muniappa, B. 1944. Observations on the acclimatisation, nesting habits and early development of *Osphronemus goramy* (Lacep.). *Proceedings of the Indian Academy of Sciences*, Bangalore **20**: 88–100.

Bird, E.A. 1960. *Fishing off Puerto Rico.* New York: A.S. Barnes.

Blanc, M., Bănărescu, P., Gaudet, J-L., & Hureau, J-C. 1971. *European Inland Water Fish, a multilingual catalogue.* London: Fishing News Books.

Blühdorn, D.R. & Arthington, A.H. 1989. Somatic characteristics of an Australian population of *Oreochromis mossambicus* (Pisces: Cichlidae). *Environmental Biology of Fishes* **29**: 277–291.

Blühdorn, D.R. & Arthington, A.H. 1990. The incidence of stunting in Australian populations of the introduced cichlid *Oreochromis mossambicus* (Peters). In: Hirano, R. & Hanyu, I. (Eds), *The Second Asian Fisheries Forum.* Manila: Asian Fisheries Society.

Blühdorn, D.R., Arthington, A.H., & Mather, P.B. 1990. The introduced cichlid, *Oreochromis mossambicus*, in Australia: a review of distribution, population, genetics, ecology, management issues and research priorities. In: Pollard, D. (Ed.), *Introduced and Translocated Fishes and their Ecological Effects. Bureau of Rural Resources Proceedings* **8**: 83–92. Canberra: Australian Government Printing Service.

Bond, C.E. 1973. Keys to Oregon freshwater fishes. Agricultural Experimental Station, Oregon State University, *Technical Bulletin* **58**: 1–42.

Bonnet, L.C. 1941. Desarrollo y planes para la pesca de agua dulce en Puerto Rico. Revista de Agricultura, Industria y Comercio. *Departmento de Agricultura Puerto Rico* **4**: 339–346.

Borgstrom, G. 1966. Our experiments with bass. *Napp och Nytt*: 18–22.

Borgstrom, G. 1978. The contribution of freshwater fish to human food. In: Gerking, S.D. (Ed.), *Ecology of Freshwater Fish Production*: 469–491. Oxford: Blackwell Scientific.

Borisova, A.T. 1972. Accidental introductions of fishes into the waters of Uzbekistan. *Journal of Ichthyology* **12**: 41–45.

Bourquin, O. & Mathias, I. 1984. The vertebrates of Oribi Gorge Nature Reserve: 1. *Lammergeyer* **33**: 35–44.

Bourquin, O. & Sowler, S.G. 1980. The vertebrates of Vernon Crookes Nature Reserve: 1. *Lammergeyer* **28**: 20–32.

Bourquin, O. & Van Rensburg, J. 1984. Vertebrates of Vernon Crookes Nature Reserve: 1: additional and confirming records. *Lammergeyer* **34**: 59–61.

Bourquin, O., Pike, T., Johnston, D., Rowe-Rowe, D. & Appleton, C.C. 1984. Alien animal species. Internal Report of the Natal Parks, Game and Fish Preservation Board, Pietermaritzburg: 36 pp.

Boyd, C.E. 1971. The limnological role of aquatic macrophytes and their relationship to reservoir management. In: Hall, G.E. (Ed.), *Reservoir Fisheries and Limnology Special Publication* **8**: 153–166. American Fisheries Society.

Braack, H.H. 1981. Lower vertebrates of the Bontebok National Park. *Koedoe* **24**: 67–77.

Branson, B.A., McCoy, C.J. & Sisk, M.E. 1960. Notes on the freshwater fishes of Sonora with an addition to the known fauna. *Copeia* **1960**: 217–220.

Brenner, T. 1984. The introduction of Arctic charr (*Salvelinus alpinus*) in Nordhein Westfalen (Federal Republic of Germany). In: Johnson, L. & Burns, B. (Eds), *Biology of the Arctic Charr. Proceedings of the International Symposium on Arctic Charr.* University of Manitoba Press: 293–301.

Bright, G. 1979. The inland waters of Palau, Caroline

Islands. Report, Office of the Chief Conservationist, Trust Territory of the Pacific Islands, Koror, Palau.

Bright, G. & June, J. 1981. The freshwater fishes of Palau, Caroline Islands. *Micronesica* 17: 107–111.

Brittan, M.R., Albrecht, A.B. & Hopkirk, J.D. 1963. An oriental goby collected in the San Joaquin River Delta near Stockton, California. *California Fish and Game* 49: 302–304.

Brittan, M.R., Hopkirk, J.D., Conners, J.D. & Martin, M. 1970. Explosive spread of the oriental goby *Acanthogobius flavimanus* in the San Francisco Bay–Delta region of California. *Proceedings of the California Academy of Sciences* 38: 207–214.

Brock, V.E. 1952. A history of the introduction of certain aquatic animals to Hawaii. *Report of the Board of Agriculture and Forestry for the Territory of Hawaii,* 1952: 114–123.

Brock, V.E. 1954. A note on the spawning of *Tilapia mossambica* in sea water. *Copeia* 1954: 72.

Brock, V.E. 1960. The introduction of aquatic animals into Hawaiian waters. *Internationale Revue der Gesamten Hydrobiologie und Hydrographie* 45: 463–480.

Brock, V.E. & Takata, M. 1956. A limnological survey of Fena Reservoir, Guam, Marianas Islands. Territory of Hawaii, Board of Commissioners of Agriculture, Honolulu, Hawaii.

Brock, V.E. & Yamaguchi, Y. 1955. A limnological survey of Fena River Reservoir, Guam, Marianas Islands. Territory of Hawaii, Board of Commissioners of Agriculture and Forestry, Honolulu, Hawaii.

Brouwer, A.B. 1925. Het invoeren van vreemde vischsoorten I. *Onze Zoetwatervisscherij* 21: 17–18.

Brown, C.J.D. 1971. *Fishes of Montana.* Boseman, Montana: Big Sky Books.

Brown, C.J.D. & Fox, A.C. 1966. Mosquito fish (*Gambusia affinis*) in a Montana pond. *Copeia* 1966: 614–616.

Brown, W.H. 1961. First record of the African mouthbreeder *Tilapia mossambica* Peters in Texas. *Texas Journal of Science* 13: 352–354.

Brumley, A.R. 1991. Cyprinids of Australasia. In: Winfield, I.J. & Nelson, J.S. (Eds), *Cyprinid Fishes: systematics, biology and exploitation*: 264–283. London: Chapman & Hall.

Brunn, A. & Pfaff, J.R. 1950. *List of Danish Vertebrates – Fishes.* Copenhagen.

Bruton, M.N. 1985. The effects of suspensoids on fish. *Hydrobiologia* 125: 221–241.

Bruton, M.N. 1990. The conservation of the fishes of Lake Victoria, Africa: an ecological perspective. *Environmental Biology of Fishes* 27: 161–175.

Bruton, M.N. & Boltt, R.E. 1975. Aspects of the biology of *Tilapia mossambica* Peters (Pisces: Cichlidae) in a natural freshwater lake (Lake Sibaya, South Africa). *Journal of Fish Biology* 7: 423–445.

Bruton, M.N. & Merron, S.V. 1985. *Alien and Translocated Aquatic Animals in Southern Africa: a general introduction,*

checklist and bibliography. South African National Scientific Programmes Report 113: 71 pp.

Bruton, M.N. & Van As, J.G. 1986. Faunal invasions of aquatic ecosystems in southern Africa, with suggestions for their management. In: Macdonald, I.A.W., Kruger, F.J. & Ferrar, A.A. (Eds), *The Ecology and Management of Biological Invasions in Southern Africa*: 47–61. Cape Town: Oxford University Press.

Buck, D.H. 1956. Effects of turbidity on fish and fishing. *Transactions of the North American Wildlife Conference* 21: 249–261.

Buckow, E. 1969. Exotics: new threat to U.S. waters. *Field and Stream* (**May**): 16; 18; 20; 22; 24; 28.

Bundy, A. & Pitcher, T.J. 1995. An analysis of species changes in Lake Victoria: did the Nile perch act alone? In: Pitcher, T.J. & Hart, P.J.B. (Eds), *The Impact of Species Changes in African Lakes. Fish and Fisheries Series* 18: 111–136. London: Chapman & Hall.

Buntz, J. & Manooch, C.S. III. 1968. *Tilapia aurea* Steindachner, a rapidly spreading exotic in south central Florida. *Proceedings of the Annual Conference of the South Eastern Association of Game and Fish Commissioners* 22: 495–501.

Buntz, J. & Manooch, C.S. III. 1969a. Fisherman utilization of *Tilapia aurea* (Steindachner) in Lake Parker, Lakeland, Florida. *Proceedings of the Annual Conference of the South Eastern Association of Game and Fish Commissioners* 23: 312–319.

Buntz, J. & Manooch, C.S. III. 1969b. A brief summary of the cichlids in the south Florida region. Mimeo. report, Florida Game and Fresh Water Fish Commission: 3 pp.

Burchmore, J.J. & Battaglene, S. 1990. Introduced fishes in Lake Burrinjuck, New South Wales, Australia. In: *The Biology and Conservation of Rare Fish.* Papers from an international symposium of The Fisheries Society of the British Isles.

Burchmore, J., Faraghar, R. & Thorncraft, G. 1990. Occurrence of the introduced oriental weather loach (*Misgurnus anguillicaudatus*) in the Wincecarribee River, New South Wales. In: Pollard, D. (Ed.), *Introduced and Translocated Fishes and their Ecological Effects. Bureau of Rural Resources Proceedings* 8: 38–46. Canberra: Australian Government Publishing Service.

Burgess, G.H. 1980. *Dorosoma petenense, Poecilia latipinna, Morone chrysops,* and *Morone saxatilis.* In: Lee, D.S., Gilbert, C.R., Hocutt, C.H., Jenkins, R.E., McAllister, D.E. & Stauffer, J.R. Jr. (Ed), *Atlas of North American Freshwater Fishes.* Raleigh: North Carolina State Museum.

Burgess, G.H., Gilbert, C.R. & Taphorn, D.C. 1977. Distributional notes on some north Florida freshwater fishes. *Florida Scientist* 40: 33–41.

Burgess, J.E. 1958. The fishes of Six Mile Creek, Hillsborough County, Florida, with particular reference to the presence of exotic species. *Proceedings of the Annual*

Conference of the Southeastern Association of Game and Fish Commissioners **12**: 1–8.

Burkhead, N.M. & Williams, J.D. 1991. An inter-generic hybrid of a native minnow, the golden shiner, and an exotic minnow, the rudd. *Transactions of the American Fisheries Society* **120**: 781–795.

Burstall, P-J. 1980. The introduction of freshwater fish into Rotorua lakes. In: Stafford, D., Steele, R. & Boyd, J. (Eds), *Rotorua 1880–1980*: 115–121. Rotorua: Rotorua and District Historical Society.

Butcher, A.D. 1962. The implications of the introduction of European carp into Victoria waters. Publications on Fisheries & Wildlife, Victoria: 73 pp.

Cadwallader, P.L. 1977. Introduction of rudd *Scardinius erythrophthalmus* into New Zealand. Part 1. Review of the ecology of rudd and the implications of its introduction into New Zealand. *New Zealand Ministry of Agriculture and Fisheries, Fisheries Technical Report* **147**: 1–18.

Cadwallader, P.L. 1978a. Some causes of the decline in range and abundance of the native fish in the Murray-Darling river system. *Proceedings of the Royal Society of Victoria* **90**: 211–224.

Cadwallader, P.L. 1978b. Acclimatisation of rudd *Scardinius erythrophthalmus* (Pisces: Cyprinidae) in the North Island of New Zealand. *New Zealand Journal of Marine and Freshwater Research* **12**: 81–82.

Cadwallader, P.L. 1995. Review of the impacts of introduced salmonids on Australian fauna. Australian Nature Conservation Agency, Project No. FPP 82: 61 pp.

Cadwallader, P.L. & Backhouse, G.N. 1983. *A Guide to the Freshwater Fish of Victoria*. Melbourne: Victorian Government Printing Office.

Cadwallader, P.L. & Gooley, G.J. 1984. Past and present distributions and translocations of Murray Cod, *Maccullochella peeli* and Trout Cod *M. macquariensis* (Pisces: Percichthyidae) in Victoria. *Proceedings of the Royal Society of Victoria* **96**: 33–43.

Cadwallader, P.L. & Tilzey, R.D.J. 1980. The role of trout as sport fish in impounded waters of Victoria and New South Wales. *Bulletin of the Australian Society for Limnology* **7**: 17–29.

Cadwallader, P.L., Backhouse, G.N. & Fallu, R. 1980. Occurrence of exotic tropical fish in the cooling pondage of a power station in temperate south-eastern Australia. *Australian Journal of Marine and Freshwater Research* **31**: 541–546.

Cahn, A.R. 1927. An ecological study of southern Wisconsin fishes, the brook silversides (*Labidesthes sicculus*) and the cisco (*Leucichthys artedi*) in their relations to the region. *Illinois Biological Monograph* **2**: 1–151.

Calderon Andreu, E.G. 1955. Acclimatation du brochet en Espagne. *Verhandlungen der Internationalen Vereinigung für Theoretische und Angewandte Limnologie* **12**: 536–542.

Cambray, J.A. & Stuart, C.T. 1985. Aspects of the biology of a rare redfin minnow, *Barbus burchelli* (Pisces: Cyprinidae) from South Africa. *South African Journal of Zoology* **20**: 155–165.

Cambray, J.A., Hanhdiek, S. & Hanhdiek, Q. 1977. A survey of fishes in Grassridge Dam. *Research Report of the Department of Nature and Environmental Conservation, Cape Provincial Administration*: 107–124.

Campos Cereda, H. 1970. Introducción de especias exoticas y su relación con los peces de agua dulce de Chile. *Noticiario Mensual Museo Nacional de Historia Natural, Santiago, Chile*: 3–9.

Carl, C.G. & Clemens, W.A. 1953. *The Fresh-water Fishes of British Columbia*. Victoria, BC: British Columbia Provincial Museum (Department of Education); Handbook **5**: 136 pp.

Carl, G.C. & Guiguet, C.J. 1958. *Alien Animals in British Columbia*. Victoria, BC: British Columbia Provincial Museum (Department of Education); Handbook **14**: 94 pp.

Carl, G.C., Clemens, W.A. & Lindsay, C.C. 1967. *The Fresh-Water Fishes of British Columbia*. Victoria, BC: British Columbia Provincial Museum (Department of Education) Handbook **55**: 132 pp.

Carson, J. & Handlinger, J. 1988. Virulence of the aetiological agent of goldfish ulcer disease in Atlantic salmon, *Salmo salar* L. *Journal of Fish Diseases* **11**: 471–479.

Castelnau, De F. 1861. *Memoire Sur Les Poissons de L'Afrique Australe*. Paris: Libraries de L'Academie Imperiale de Medecine.

Catt, J. 1950. Some notes on brown trout with particular reference to their status in New Brunswick and Nova Scotia. *Canadian Fish-Culturist* **4**: 15–18.

Cawkwell, C. & McAngus, J. 1976. Spread of the zander. *Angler's Mail*, **3 March**: 12–13.

Cazemier, W.G. & Heesen, M.J. 1989. First record of *Vimba vimba* (Linnaeus, 1758) (Pisces: Cyprinidae) in The Netherlands. *Bulletin of the Zoological Museum, University of Amsterdam* **12**: 97–100.

Chacko, P.J. 1945. Acclimatization of mirror carp in Nilgiris. *Journal of the Bombay Natural History Society* **45**: 244–247.

Chacko, P.J. 1948. Development of fisheries in the Periyar Lake. *Proceedings of the 35th Indian Scientific Congress*, Part 3, Abstracts: 204–205.

Chandrasekharaiah, H.N. 1989. Status of exotic fish culture in Karnataka. In: Joseph, M.M. (Ed.), *Exotic Aquatic Species in India. Asian Fisheries Society Special Publication* **1**: 75–77.

Chapman, W.M. 1942. Alien fishes in the waters of the Pacific northwest. *California Fish & Game* **28**: 9–15.

Chaudhuri, H. 1964. Introduction of exotic species of fish and their effects on the culture of indigenous species of India. In: *Proceedings of the Seminar on Inland Fisheries Development, Lucknow*: 83–92.

Chaudhuri, H. & Hla Tin, U. 1971. Preliminary observations of the survival and growth of Chinese carp in Burma. *Occasional Papers of the Indo-Pacific Fisheries Council* **71**: 1–16.

Chen, T.P. 1976. *Aquaculture Practices in Taiwan.* Farnham, England: Fishing News Books.

Chiba, K., Taki, Y., Sakai, K. & Oozeki, Y. 1989. Present status of aquatic organisms introduced into Japan. In: De Silva, S.S. (Ed.), *Exotic Aquatic Organisms in Asia. Asian Fisheries Society Special Publication* **3**: 63–70.

Chilton, E.W. & Muoneke, M.I. 1992. Biology and management of grass carp (*Ctenopharyngodon idella*, Cyprinidae) for vegetation control: a north American perspective. *Reviews in Fish Biology and Fisheries* **2**: 283–320.

Chimits, P. 1957. The tilapias and their culture. *Fisheries Bulletin of the Food and Agriculture Organization* **10**: 1–24.

Chou, L.M. & Lam, T.J. 1989. Introduction of exotic aquatic species in Singapore. In: De Silva, S.S. (Ed.), *Exotic Aquatic Organisms in Asia. Asian Fisheries Society Special Publication* **3**: 91–97.

Clark, F.N. 1885. Report of operations at the Northville and Alpena Stations during the season of 1884–85. *Report of the United States Commissioner of Fish and Fisheries* **12**: 156–157.

Clark, F.N. 1887. Report on the distribution of fish and eggs from Northville and Alpena stations for the season of 1885–1886. *Bulletin of the United States Fish Commission* (1886) **6**: 395.

Clay, D. 1972. Introduction of exotic fish to Swaziland. *Piscator* **85**: 74–77.

Clemens, W.A., Rawson, D.S. & McHugh, J.L. 1939. A biological survey of Okanagan Lake, British Columbia. *Bulletin of the Fisheries Research Board of Canada* **61**: 70 pp.

Clements, J. 1988. *Salmon at the Antipodes: a history and review of trout, salmon and char and introduced coarse fish in Australasia.* Australia: Author.

Coates, D. 1987. Considerations of fish introductions into the Sepik River, Papua New Guinea. *Aquaculture & Fisheries Management* **18**: 231–241.

Coates, G.D. & Turner, A.S. 1977. Introduction of rudd *Scardinus erythrophthalmus* into New Zealand. 2. First survey of a rudd population in New Zealand. *New Zealand Ministry of Agriculture and Fisheries, Fisheries Technical Report* **147**: 19–24.

Cochrane, K.L. 1983. Hartbeespoort Dam: are the fish surviving? *Tight Lines*: 32–33; 35–36.

Coetzee, P.W. 1977. The history and distribution of freshwater fish in Albany: largemouth bass. *Newsletter of the Albany Freshwater Angling Association* **32**: 3–4.

Cole, L.J. 1905a. The German carp in the United States. *Report of the United States Bureau of Fisheries* **1904**: 525–641.

Cole, L.J. 1905b. The status of carp in America. *Transactions of the American Fisheries Society* **34**: 201–207.

Conner, J.V., Gallagher, R.P. & Chatry, M.F. 1980. Larval evidence for natural reproduction of the grass carp (*Ctenopharyngodon idella*) in the lower Mississippi River. *Proceedings of the 4th Annual Larval Fish Conference* **43**: 1–19.

Contreras, S. 1967. Lista de peces del Estado de Nuevo León. *Cuadernos del Instituto de Investigaciones Científicas, Universidad de Nuevo León* **11**: 1–12.

Contreras, S. 1969. Perspectives de la ictiofauna en las zonas aridas del norte de México. Memorias del Simposio Internacional sobre el Aumento de Producción de Alimentos Zonas Aridas. *International Center for Arid Lands Studies Publications* **3**: 293–304.

Contreras, S. 1975. Cambios de composición de especies en comunidades de en zonas semiáridas de México. Publicaciones Biológicas del Instituto de Investigaciones Cientificas. *Universidad Autónoma de Nuevo León, México* **1**: 181–194.

Contreras, S. 1978. Speciation aspects and man-made community composition changes in Chihuahua desert fishes. In: Wauer, R.H. & Riskind, D.H. (Eds), *Transactions of the Symposium on Biological Resources of the Chihuahuan Desert Region, U.S. and Mexico. National Park Service Transactions and Proceedings, Section* **3**: 405–431.

Contreras, A.J. & Contreras, S. 1985. Los peces de aquascalientes, México. In: *Proceedings of the 3rd Congreso Nacional de Zoologica, México*, 1979.

Contreras, S. & Escalante, M.A. 1984. Distribution and known impacts of exotic fishes in Mexico. In: Courtenay, W.R. Jr. & Stauffer, J.R. Jr. (Eds), *Distribution, Biology and Management of Exotic Fishes*: 102–130. Baltimore: Johns Hopkins University Press.

Contreras, S., Landa, V., Villegas, T. & Rodriguez, G. 1976. Peces, piscicultura, presas, polución, planificación pesquera, y monitoreo en México, o la danza de las P. *Memorias del Simposio de Pesquerías en Aguas Continentales* **1**: 315–346.

Cooke, B.W.C. 1977. A threat to the Atlantic Salmon: should the Coho be introduced to Britain? *Country Life*: 752.

Cooper, J., Crafford, J.E. & Hecht, T. 1992. Introduction and extinction of brown trout (*Salmo trutta* L.) in an impoverished subantarctic stream. *Antarctic Science* **4**: 9–14.

Corral, J.I. 1936. Nuevos peces Cubanos de aqua dolce. *Revista Geographica* **9**: 22–27.

Cottiglia, M. 1968. La distribuzione della ittiofauna dulciacquiecola in Sardegna. *Revista di Idrobiologia (Revista di Biologia) Dell'Universita di Perugia Dirette da Aldo Spirito, Publicacione* **7**: 63–113.

Courtenay, W.R. Jr 1970. Florida's walking catfish. *Ward's Natural Science Bulletin* **10**: 1 & 6.

Courtenay, W.R. Jr 1975. Status of introduced Walking Catfish in Florida. *Bulletin of the International Union for Conservation of Nature*, n.s. No. **6**: 23–24.

Courtenay, W.R. Jr 1978. Additional range expansion in Florida of the introduced walking catfish. *Environmental Conservation* **5**: 273–276.

Courtenay, W.R. Jr 1979a. Continued range expansion in Florida of the walking catfish. *Environmental Conservation* **6**: 20.

Courtenay, W.R. Jr 1979b. The introduction of exotic organisms. In: Brokaw, H.P. (Ed.), *Wildlife and America*: 237–252. Washington, DC: US Government Printing Office.

Courtenay, W.R. Jr 1989. Fish introductions and translocations and their impacts in Australia. *Proceedings of the Department of Primary and Industrial Energy Bureau of Rural Resources* 8: 171–179.

Courtenay, W.R. Jr 1992. A summary of fish introductions in the United States. In: DeVoe, M.R. (Ed.), *Introductions and Transfers of Marine Species: achieving a balance between economic development and resource protection. Proceedings of a conference and workshop on Hilton Head Island, South Carolina, October 30–November 2*, **1991**: 9–15.

Courtenay, W.R. Jr 1993. Biological pollution through fish introductions. In: McKnight, B.N. (Ed.), *Biological Pollution: the control and impact of invasive exotic species. Proceedings of a Symposium at the Indiana University – Purdue University of Indianapolis on 25–26 October* **1991**: 35–61.

Courtenay, W.R. Jr 1995. The case for caution with fish introductions. *American Fisheries Society Symposium* **15**: 413–424.

Courtenay, W.R. Jr & Deacon, J.E. 1982. Status of introduced fishes in certain spring systems in southern Nevada. *Great Basin Naturalist* **42**: 361–366.

Courtenay, W.R. Jr & Deacon, J.E. 1983. Fish introductions in the American southwest: a case history of Rogers Spring, Nevada. *Southwestern Naturalist* **28**: 221–224.

Courtenay, W.R Jr & Hensley, D.A. 1979. Range expansion in southern Florida of the introduced spotted tilapia, with comments on its environmental impress. *Environmental Conservation* **6**: 149–151.

Courtenay, W.R. Jr & Hensley, D.A. 1980a. Special problems associated with monitoring exotic species. In: Hocutt, C.H. & Stauffer, J.R. Jr (Eds), *Biological Monitoring of Fish*: 281–307. Lexington, Massachusetts: Lexington Books.

Courtenay, W.R. Jr & Hensley, D.A. 1980b. Survey of introduced non-native fishes. Phase 1 Report. Introduced exotic fishes in North America: status, 1979. Department of Biological Sciences, Florida Atlantic University.

Courtenay, W.R. Jr & Miley, W.W. II. 1975. Range expansion and environmental impress of the introduced catfish in the United States. *Environmental Conservation* **2**: 145–148.

Courtenay, W.R. Jr & Moyle, P.B. 1992. Crimes against biodiversity: the lasting legacy of fish introductions. *Transactions of the 57th North American Wildlife and Natural Resources Conference, Special Session 6. Biological Diversity in Aquatic Management*: 365–372.

Courtenay, W.R Jr & Ogilvie, V.E. 1971. Species pollution. *Animal Kingdom* **74**: 22–28.

Courtenay, W.R. Jr & Robins, C.R. 1972. The grass carp enigma. *Bioscience* **22**: 210.

Courtenay, W.R. Jr & Robins, C.R. 1973. Exotic aquatic organisms in Florida with emphasis on fishes: a review and recommendations. *Transactions of the American Fisheries Society* **102**: 1–12.

Courtenay, W.R. Jr & Stauffer, J.R. Jr 1990. The introduced fish problem and the aquarium fish industry. *Journal of the World Aquaculture Society* **21**: 145–159.

Courtenay, W.R. Jr., Sahlman, H.F., Miley, W.W. II. & Herrema, D.J. 1974. Exotic fishes in fresh and brackish waters of Florida. *Biological Conservation* **6**: 292–302.

Courtenay, W.R. Jr, Hensley, D.A., Taylor, J.N. & McCann, J.A. 1984. Distribution of exotic fish in the continental United States. In: Courtenay, W.R. Jr & Stauffer, J.R. Jr (Eds), *Distribution, Biology and Management of Exotic Fishes*: 41–47. Baltimore: Johns Hopkins University Press.

Courtenay, W.R., Hensley, D.A., Taylor, J.N. & McCann, J.A. 1986. Distribution of exotic fishes in North America. In: Hocutt, C.H. & Wiley, E.O. (Eds), *Zoogeography of North American Fishes*: 675–698. New York: Wiley.

Courtenay, W.R. Jr, Robins, C.R., Bailey, R.M. & Deacon, J.E. 1987. Records of exotic fishes from Idaho and Wyoming. *Great Basin Naturalist* **47**: 523–526.

Courtenay, W.R. Jr, Jennings, D.P. & Williams, J.D. 1991. Exotic fishes. In: Robins, C.R., Bailey, R.M., Bond, C.E., Brooker, J.R., Lachmer, E.A., Lea, R.N. & Scott, W.B. (Eds), *Common and Scientific Names of Fishes from the United States and Canada*, 5th edn. *American Fisheries Society Special Publication* **20**: 97–107.

Courtright, A.M. 1970. Fresh water fishing adds new sport to Guam scene. *Pacific Daily News*, **16 April**: 16.

Craig, J.F. 1992. Human-induced changes in the composition of fish communities in the African Great Lakes. *Reviews on Fish Biology and Fisheries* **2**: 93–124.

Crass, R.S. 1964. *Freshwater Fishes of Natal*. Pietermaritzburg: Shuter & Shuter.

Crass, R.S. 1966. Features of freshwater fish distribution in Natal and a discussion on controlling factors. *Newsletter of the Limnological Society of Southern Africa* **7**: 31–35.

Crass, R.S. 1968. Kamloops trout in Natal. *Piscator* **72**: 14–18.

Crass, R.S. 1969. The effects of land use on freshwater fish in South Africa, with particular reference to Natal. *Hydrobiologia* **34**: 38–56.

Crittenden, E. 1962. Status of *Tilapia nilotica* in Florida. *Proceedings of the Annual Conference of the Southeastern Association of Game and Fish Commissioners* **16**: 257–262.

Crossman, E.J. 1968. Changes in the Canadian freshwater fish fauna. In: *Introductions of Exotic Species: Ontario Department of Lands and Forest Research Report* **82**: 1–20.

Crossman, E.J. 1984. Introduction of exotic fishes into Canada. In: Courtenay, W.R. Jr & Stauffer, J.R. Jr (Eds), *Distribution, Biology and Management of Exotic Fishes*: 78–101. Baltimore: Johns Hopkins University Press.

Crossman, E.J. 1991. Introduced freshwater fishes: a review of the North American perspective with emphasis on Canada. *Canadian Journal of Fisheries and Aquatic Sciences* **48** (Suppl.1): 46–57.

Crossman, E.J., Holm, E., Cholmondeley, R. & Tuininga, K. 1992. First records for Canada of the rudd, *Scardinius erythrophthalmus* and the round goby, *Neogobius melanostomus*. *Canadian Field Naturalist* **106**: 206–209.

Crowl, T.A., Townsend, C.R. & McIntosh, A.R. 1992. The impact of introduced brown and rainbow trout on native fish: the case of Australasia. *Reviews in Fish Biology and Fisheries* **2**: 217–241.

Cuerrier, J.P., Fry, F.E.J. & Prefontaine, G. 1946. Liste preliminaire de la region de Montreal et du lac Sainte-Pierre. *Le Naturaliste* **73**: 17–23.

Curtis, B. 1942. The general situation and the biological effects of the introduction of the alien fishes into California waters. *California Fish and Game* **28**: 2–8.

Dahl, J. 1962. The importance and profits from zander fishing in the cultivation of Danish lakes. *Zhurnal Fischereiwelt* **8/10**: 689–695.

Dahlberg, M.D. & Scott, D.C. 1971. Introductions of freshwater fishes in Georgia. *Bulletin of the Georgia Academy of Science* **29**: 245–252.

Dannevig, G.M. 1885. Salmonidae in Australia. *Bulletin of the United States Fish Commission* **5**: 440–442.

d'Aubenton, F., Daget, J. & Prin, P. 1983. Nouvelle station en France (Argonne) pour *Umbra pygmaea*. *Revue Française Aquariologie et Herpetologie* **10**: 13–16. [In French with English summary.]

Davaine, P. & Beall, E. 1982a. Introductions des salmonides dans les terres australes et antarctiques françaises. *CNFRA Publication* **51**, 1981: 289–300.

Davaine, P. & Beall, E. 1982b. Acclimatation de la truite commune, *Salmo trutta* L., en milieu subantarctique (îles Kerguelen). II. Stratégie adaptive. *Colloque sur les Ecosystèmes Subantarctiques*. Paimpont: *CNFRA* **51**: 399–412.

Davidoff, E.B. & Chervinski, J. 1984. Comments on fish introductions into Lake Kinneret. *Fisheries Management* **15**: 27–28.

Davidson, F.A. & Hutchinson, S.J. 1938. The geographic distribution and environmental limitations of the Pacific Salmon (Genus *Oncorhynchus*). *Bulletin of the Bureau of Fisheries* **48**: 667–692.

Day, F. 1876. On the introduction of trout and tench into India. *Journal of the Linnean Society of London* **12**: 562–565.

Day, L.A. 1932a. The introduction of Salmonidae into Natal rivers, with special reference to *Salmo irideus*. *South African Journal of Science* **29**: 473–479.

Day, L.A. 1932b. The introduction of trout in Natal. *Salmon and Trout Magazine* **69**: 345–352.

Deacon, J.E. 1979. Endangered and threatened species of the West. In: *The Endangered Species. Great Basin Naturalist* **3**: 41–64.

Deacon, J. 1988. The endangered woundfin and water management in the Virgin River, Arizona. *Fisheries* **13**: 18–29.

Deacon, J.E., Hubbs, C. & Zahuranec, B.J. 1964. Some effects of introduced fishes on the native fish fauna of southern Nevada. *Copeia* **1964**: 384–388.

De Buen, F. 1941. El *Micropterus (Huro) salmoides* y los resultado de su aclimatación en el Lago Patzcuaro. *Revista de la Sociedad Méxicana de Historia Natural* **2**: 69–78.

De Buen, F. 1958. Ictiologia. La familia Ictaluridae nueva para la fauna aclimatada de Chile y algunas consideraciones sobre los Siluroidei indigenas. *Investigaciones Zoologicas Chilenas* **4**: 146–158.

De Buen, F. 1959. Los Peces exoticos en las aquas dulces de Chile. *Investigaciones Zoologicas Chilenas* **5**: 103–137.

De Bunsen, J.M. 1962. The American Brook Trout in Britain. Unpublished memorandum.

De Groot, S.J. 1985. Introductions of non-indigenous fish species for release and culture in The Netherlands. *Aquaculture* **46**: 237–257.

De Iongh, H.H. & Van Zon, J.C.J. 1993. Assessment of impact of the introduction of exotic fish species in northeast Thailand. *Aquaculture and Fisheries Management* **24**: 279–289.

De Iongh, H.H., Spliethoff, P.C. & Roest, F. 1995. The impact of an introduction of sardine into Lake Kivu. In: Pitcher, T.J. & Hart, P.J.B. (Eds), *The Impact of Species Changes in African Lakes. Fish and Fisheries Series* **18**: 277–297. London: Chapman & Hall.

DeKay, J.E. 1842. *Zoology of New York: IV. Fishes*. Albany, New York: W. & A. White and J. Visscher.

Delachaux, A. 1901. L'acclimatation et l'elevage de l'omblechevalier americain dans un lac des Alpes. *Bulletin of the Société Centrale d'Aquiculture et de Pêche* **13**: 41–54.

Delfin, F.T. 1901. *Ictiologia Chilena. Catálogo de los Peces de Chile*. Valparaíso.

Delgadillo, M.S. 1976. La Estación de Temascal como factor de desarrollo en la acuacultura de la cuenca del Papaloapan. *Memorias del Simposio de Pesquerías en Aguas Continentales* **1**: 55–86.

De Man, E. 1983. Walking catfish. *Oceans* **16**: 41.

De Moor, I.J. & Bruton, M.N. 1988. Atlas of Alien and Translocated Indigenous Aquatic Animals in Southern Africa. *South African National Scientific Programmes Report* **144**: 310 pp.

Dence, W.A. 1925. Bitter carp (*Rhodeus amarus*) from New York State waters. *Copeia* **142**: 33.

De Plaza, M.L.F. & De Plaza, J.C. 1949. Salmonicultura. Ministry of Agriculture of Argentina, Yanderia. *Miscellaneous Publication* **321**.

De Silva, S.S. 1987. Impact of exotics on the inland fishery resources of Sri Lanka. *Ergebnisse der Limnologie* **28**: 273–293.

De Silva, S.S. & Fernando, C.H. 1980. Recent trends in the fishery of Parakrama Samudra, an ancient man-made lake in Sri Lanka. In: Furtado, J. (Ed.), *Tropical Ecology and Development*: 927–937. Kuala Lumpur: University of Malaya Press.

Devadas, D.D.P. & Chacko, P.I. 1953. Introduction of the exotic cichlid fish *Tilapia mossambica* Peters in Madras State. *Current Science*. 22–29.

Devambez, L.C. 1960. American game fish for New Caledonia. *South Pacific Bulletin* **10**: 25; 38.

Devambez, L.C. 1964. Tilapia in the south Pacific. *South Pacific Bulletin* **14**: 27; 28; 52.

Devick, W.S. 1991. Patterns of introductions of aquatic organisms to Hawaiian freshwater habitats. In: *New Directions in Research, Management, and Conservation of Hawaiian Freshwater Stream Ecosystems. Proceedings of the 1990 Symposium on Freshwater Stream Biology and Management, Hawaii*: 189–213.

De Vos, L., Snoeks, J. & van den Audenaerde, D.T. 1990. The effects of *Tilapia* introductions in Lake Luhondo, Rwanda. *Environmental Biology of Fishes* **27**: 303–308.

Dial, R.S. & Wainright, S.C. 1983. New distributional records for non-native fishes in Florida. *Florida Scientist* **46**: 8–15.

Diamond, J.M. 1984. The avifaunas of Rennell and Bellona Islands. Natural History of Rennell Island, British Solomon Islands 8 (*Zoology*): 127–168.

Dixon, J.E.W. & Blom, M.J. 1974. Some aquatic vertebrates from the Namib Desert, South West Africa. *Madoqua* **2**: 31–32.

Donne, T.E. 1927. *Rod Fishing in New Zealand Waters*. London: Seeley Service.

Donnelly, B.G. 1965. The first record of rainbow trout breeding in the Port Elizabeth area. *Piscator* **64**: 91–92.

Doroshev, S.I. 1970. Biological features of the eggs, larvae and young of the striped bass (*Roccus saxatilis* Walbaum) in connection with the problem of its acclimatization in the USSR. *Journal of Ichthyology* **10**: 235–248.

Dottrens, E. 1955. Acclimatation et hybridation de *Coregones*. *Revue Suisse de Zoologie* **62**: 101–118.

Douglas, P.A. 1953. Survival of some fishes recently introduced into the Salton Sea, California. *California Fish and Game* **39**: 264–265.

Drapkin, E.I. 1968. Penetration of the Kubanj pikeperch (*Lucioperca lucioperca* L.) into the eastern Black Sea (Caucasian coast). *Voprosy Ikhtiologii* **8**: 749–751. [In Russian.]

Druett, J. 1983. *Exotic Intruders: the introduction of plants and animals into New Zealand*. Auckland: Heinemann.

Dukravets, G.M. 1972. White amur in the Ili River basin. *Izvestiya Akademii Nauk Kazakhskoy SSR Series Biologia* **1**: 52–57. [In Russian.]

Dumont, H.J. 1986. The Tanganyikan sardine in Lake Kivu: another ecodisaster for Africa. *Environmental Conservation* **13**: 143–148.

Du Plessis, S.S. 1961. Trout pioneer F.C. Braun (1875–1959). *Fauna and Flora* **12**: 17–23.

Du Plessis, S.S. & Le Roux, P.J. 1965. Sport fisheries in river development with references to the Orange River scheme. *South African Journal of Science* **61**: 137–146.

Dymond, J.R. 1936. Some fresh-water fishes of British Columbia. *Report of the Commissioner for Fisheries, British Columbia*, **1935**: 60–73.

Dymond, J.R. 1947. A list of the freshwater fishes of Canada east of the Rocky Mountains. *Royal Ontario Museum of Zoology, Miscellaneous Publication* **1**: 1–36.

Dymond, J.R. 1955. The introduction of foreign fishes in Canada. *Proceedings of the International Association of Theoretical and Applied Limnology* **12**: 543–553.

Eigenmann, C.H. 1927. The freshwater fishes of Chile. *Memoirs of the National Academy of Sciences* **12** (2): 80 pp.

El Bolock, A.R. & Labib, W. 1967. Carp culture in the U.A.R. *FAO Fisheries Reports* **44**: 165–174.

Eldredge, L.G. 1988. Case studies of the impacts of introduced animal species on renewable resources in the US-affiliated Pacific Islands. In: Smith, B.D. (Ed.), *Topic Reviews in Insular Resource Development and Management in the Pacific US-affiliated islands. University of Guam Marine Laboratory Technical Report* **88**: 91–121.

Eldredge, L.G. 1994. *Perspectives in Aquatic Exotic Species Management in the Pacific Islands*. Vol. I *Introductions of Commercially Significant Organisms to the Pacific Islands*. Noumea, New Caledonia: South Pacific Commission: 127 pp.

Ellepola, W.B. & Fernando, C.H. 1968. Hatching and rearing of gouramy (*Osphronemus goramy* Lacepede), in the Polonnaruwa nursery. *Bulletin of the Fisheries Research Station of Ceylon* **19**: 1–3.

Elliott, J.M. 1989. Wild brown trout *Salmo trutta*: an important national and international resource. *Freshwater Biology* **21**: 1–5.

Elton, C.S. 1958. *The Ecology of Invasions by Animals and Plants*. London: Chapman & Hall.

Elvira, B. 1995a. Native and exotic freshwater fishes in Spanish river basins. *Freshwater Biology* **33**: 103–108.

Elvira, B. 1995b. Freshwater fishes introduced in Spain and relationships with autochthonous species. In: Voigtlander, C.W. (Ed.), *The State of the World's Fisheries Resources. Proceedings of the World Fisheries Congress, Theme* **3**: 261–264. New Delhi: Oxford & IBH Publishing Co. Pvt.

Emery, A.R. & Teleki, G. 1978. European flounder (*Platichthys flesus*) captured in Lake Erie, Ontario, Canada. *Canadian Field Naturalist* **92**: 89–91.

Erdman, D.S. 1947. Aquarium fishes in Puerto Rico. *Revista de Agricultura* **28**: 90–91.

Erdman, D.S. 1972. Inland game fishes of Puerto Rico. Departmento de Agricultura de Puerto Rico **4**: 1–96.

Erdman, D.S. 1984. Exotic fishes in Puerto Rico. In: Courtenay, W.R. Jr & Stauffer, J.R. Jr (Eds), *Distribution, Biology and Management of Exotic Fishes*: 162–176. Baltimore: Johns Hopkins University Press.

Erencin, Z., Baran, I. & Ergueven, H. 1972. Culture of Baligi, the rainbow trout, *Salmo gairdneri irideus* (Richardson 1836). On the supply of rainbow trout eggs and the first establishment of these fish in Turkey. *Journal of the Faculty of Veterinary Medicine of the University of Ankara* **19**: 12–20.

Escalante, M.A. & Contreras, S. 1985. Distribución de algunas especies exóticas (transfaunadas de sus esosistemis nativos) en la República Méxicana. In: *Proceedings of the 3rd Congreso Nacional do Zoologica, México, 1979.*

Euzenat, G. & Forunel, F. 1982. Les saumons du Pacifique en France. *Peche Maritime* **61**: 391–395.

Everett, G.V. 1971. The rainbow trout of Lake Titicaca and the fisheries of Lake Titicaca. Report to the Government of the Republic of Peru. Mimeo: 180 pp.

Everett, G.V. 1973. The rainbow trout *Salmo gairdneri* (Rich.) fishery of Lake Titicaca. *Journal of Fish Biology* **5**: 429–440.

Farman, R. 1984. The fate of a North American game fish in the South Pacific. *South Pacific Commission Fisheries Newsletter* **31**: 25–30.

Farr-Cox, F., Leonard, S., Wheeler, A. In press. The status of the recently introduced fish *Leucaspius delineatus* (Cyprinidae) in Britain. *Fisheries Management & Ecology.*

Fehlmann, W. 1930. Der forellenbarsch. *Allgemeine Fischerei-Zeitung* **55**: 80.

Ferguson, A.D. 1915. Extending the range of the golden trout. *Transactions of the Pacific Fishing Society*: 65–70.

Ferguson, F.F. 1978. The role of biological control agents in the control of schistosome-bearing snails. Atlanta, Georgia: US DHEW Public Health Service, Center for Disease Control.

Fernando, C.H. 1971. The role of introduced fish on fish production in Ceylon's freshwaters. In: Duffay, E.B. & Watt, A.S. (Eds), *The Scientific Management of Animal and Plant Communities for Conservation*: 295–310. Oxford: Blackwell.

Ffennell, H. 1885. Hatching American fish at South Kensington, and their introduction to English waters. *Bulletin of the United States Fish Commission* **5**: 273–275.

Fickling, N.J. & Lee, R.L.G. 1983. A review of the ecological impact of the introduction of the zander (*Stizostedion lucioperca* L.) into waters of the Eurasian mainland. *Fishery Management* **14**: 151–155.

Finucane, J.H. & Rinckey, G.R. 1964. A study of the African cichlid *Tilapia heudeloti* Dumeril in Tampa Bay, Florida. *Proceedings of the Annual Conference of the Southeastern Association of Game and Fish Commissioners* **18**: 259–269.

Fish, G.R. 1966. An artificially maintained trout population in a Northland lake. *New Zealand Journal of Science* **9**: 200–210.

Fishelson, L. 1962. Tilapia hybrids. *Fisherman's Bulletin* (Haifa) **4**: 14–19. [In Hebrew with English abstract.]

Fitter, R.S.R. 1959. *The Ark in Our Midst.* London: Collins.

FitzGerald, W.J. Jr. & Nelson, S.G. 1979. Development of aquaculture in an island community (Guam, Mariana Islands). *Proceedings of the World Maricultural Society* **10**: 39–50.

Flain, M. 1982. The history of New Zealand's salmon fishery. *Occasional Publications of the Fisheries Research Division, Wellington* **30**: 8–10.

Fletcher, A.R. 1986. Effects of introduced fish in Australia. *Monographaie Biologicae* **61**: 231–238.

Fletcher, A.R., Morison, A.K. & Hume, D.J. 1985. Effects of carp, *Cyprinus carpio* L., on communities of aquatic vege-tation and turbidity of waterbodies in the Lower Goulburn River Basin. *Australian Journal of Marine and Freshwater Research* **36**: 311–327.

Follett, W.I. 1960. The fresh-water fishes – their origins and affinities. In: Symposium on the biogeography of Baja California and adjacent seas. *Systematic Zoology* **9**: 212–232.

Foote, K.J. 1977. Blue tilapia investigations. Study 1. Preliminary status investigations. Mimeo. report, Florida Game and Fresh Water Fish Commission: 71pp.

Forester, F. 1850. *Fish and Fishing.* New York: Stringer & Townsend.

Forester, T.S. & Lawrence, J.M. 1978. Effects of grass carp and carp in populations of bluegill and largemouth bass in ponds. *Transactions of the American Fisheries Society* **107**: 172–175.

Fowke, P. 1938. Trout culture in Ceylon. *Ceylon Journal of Science, Section C (Fisheries)* **6**: 78 pp.

Fowler, H.W. 1925. Fishes of Guam, Hawaii, Samoa, and Tahiti. *Bernice P. Bishop Museum Bulletin* **22**: 1–38.

Fowler, H.W. 1932a. Fishes obtained at Samoa in 1929. *Occasional Papers of the Bernice P. Bishop Museum* **9**: 1–16.

Fowler, H.W. 1932b. Fresh-water fishes from the Marquesas and Society Islands. *Occasional Papers of the Bernice P. Bishop Museum* **9**: 4–11.

Francis, F. 1879. Trout in the Antipodes. *The Field*: 359.

Frank, V.G. 1977. Transplant sardines thrive in Lake Kivu. *Fish Farming International* **4**: 31–32.

Friese, U.E. 1973. Another Japanese goby in Australian waters; what next? *Proceedings of the Royal Zoological Society of New South Wales* **2**: 5–7.

Frost, N. 1938. Trout and their conservation. *Newfoundland Department of Natural Resources, Service Bulletin* **6** (Fisheries): 1–16.

Frost, W.E. 1940. Rainbows of a peat lough on Arranmore. *Salmon & Trout Magazine* **100**: 234–240.

Frost, W.E. 1974. A survey of the rainbow trout (*Salmo gairdneri*) in Britain and Ireland. London: Salmon and Trout Association: 36 pp.

Frost, W.E. & Brown, M.E. 1967. *The Trout.* London: Collins.

Fryer, G. 1960. Concerning the proposed introduction of Nile perch into Lake Victoria. *East African Agricultural Journal* **25**: 267–270.

Fryer, G. & Iles, T.D. 1972. *The Cichlid Fishes of the Great Lakes of Africa: their biology and evolution.* London: Oliver & Boyd.

Fuchs, J. 1987. Growth of introduced larvae and fingerlings of sea bass (*Lates calcifer*) in Tahiti. In: *Management of Wild and Cultured Seabass/Barramundi. ACIAR Monograph* **20**: 189–192.

Fuster de Plaza, M.L. & Plaza, J.C. 1954. Salmonicultura. *Publicaciones Misceláneas Ministerio de Agricultura y Ganadería, Republica Argentina.*

Gaigher, C.M. 1973. The Clanwilliam River: it is not yet too late. *Piscator* **88**: 75–78.

Gaigher, C.M. 1981. Tiny jewels of the Cedarberg. *African Wildlife* **35**: 25–29.

Gaigher, I.G. 1975a. The ecology of a minnow, *Barbus trevelyani* (Pisces; Cyprinidae) in the Tyume River, Eastern Cape. *Annals of the Cape Provincial Museum* **11**: 1–19.

Gaigher, I.G. 1975b. The Hardap dam: an angler's paradise in South West Africa. *Piscator* **93**: 4–8.

Gaigher, I.G. 1978. The distribution, status and factors affecting the survival of indigenous freshwater fishes in the Cape Province. Cape Provincial Administration Department of Nature and Environmental Conservation, Internal Report: 56 pp.

Gaigher, I.G. 1979. Overgrazing endangers fish species. *African Wildlife* **33**: 41.

Gaigher, I.G., Hamman, K.C.D. & Thorne, S.C. 1980. The distribution, conservation status and factors affecting the survival of indigenous freshwater fishes in the Cape Province. *Koedoe* **23**: 57–88.

Garcia Avila, I., Koldenkova, L., Santamarina Mijares, A. & Gonzalez Broche, R. 1991. Introduccion del pez larvivoro *Poecilia reticulata* (Peters, 1895), agente biorregulador de culicidos en lagunas de oxidacion y zanjas contaminadas en la Isla de la Juventud. *Revista Cubana Tropicana* **43**: 45–49. [In Spanish with English and French summaries.]

George, T.T. 1976. Introduction and transplantation of cultivable species into Africa. In: Supplement 1 to the Report of the Symposium on Aquaculture in Africa; Accra, Ghana, 1975. Reviews and experience papers, CIFA. *Technical Paper* **4** (Suppl.) 408–432.

Gephard, S.R. 1977. Status of trout populations in Lesotho. *Piscator* **100**: 71–74.

Gerberich, J.B. & Laird, M. 1968. Bibliography of papers relating to the control of mosquitoes by the use of fish. An annotated bibliography for the years 1901–1966. *FAO Fisheries Paper* **75**: 70 pp.

Germany, R.D. 1977. Population dynamics of the blue tilapia and its effects on fish populations in Trinidad Lake, Texas. Unpub. PhD dissertation, Texas A & M University: 85 pp.

Giard, A. 1894. La truite arc-en-ciel: son acclimatation en Allemagne et en Belgique. *Buletin Centrale Agriculture et de Pêche 2e partie.* 14.

Gibbs, E.J. 1981. A review of the Atlantic salmon in New Zealand with notes on current status and management. In: Hopkins, C.L. (Ed.), *Proceedings of the Salmon Symposum. New Zealand Ministry of Agriculture and Fisheries, Fisheries Research Division Occasional Publications* **30**: 55–64.

Giblin, W.E. 1936. *Gambusia affinis* as a factor in mosquito elimination in Papua. Annexure A, in the *Territory of Papua Annual Report, 1934–1935*: 67 pp. Canberra: Government Printer.

Gilhen, J. 1974. *The Fishes of Nova Scotia Lakes and Streams.* Nova Scotia Museum: 49 pp.

Gilmore, K.S. 1978. A fishery survey of Shashi Dam. Unpublished report of the Department of Agriculture and Fisheries, Ministry of Agriculture, Botswana: 26 pp.

Glucksman, J., West, G. & Berra, T.M. 1976. The introduced fishes of Papua New Guinea, with special reference to *Tilapia mossambica*. *Biological Conservation* **9**: 37–44.

Godby, M.H. 1934. Atlantic salmon in New Zealand: the evidence of the Lake Coleridge fish. *Salmon & Trout Magazine* **75**: 173–176.

Goldner, H.J. 1967. 'n Populasiestudie van varswatervisse in Baberspan, Wes-Transvaal. MSc thesis, Potchefstroom University for CHO (Unpublished).

Goll, H. 1887. The American brook trout recommended for Swiss waters. *Bulletin of the United States Fish Commission* **1886**: 206–208.

Golusda, P. 1927. Aclimatación y cultivo de especies salmonideas en Chile. *Boletín de la Sociedad de Biologia de Concepción* **1**: 80–100.

González, G.R. 1988. La familia Centrarchidae (Pisces: Perciformes) en la Republica de Panama. *Brenesia* **29**: 7–14. [In Spanish with English summary.]

Goode, G.B. 1903. *American Fishes.* Boston, Massachusetts: Dana Estes.

Gophen, M., Drenner, R.W. & Vinyard, G.G. 1983. Fish introductions into Lake Kinneret – call for concern. *Fisheries Management* **14**: 43–45.

Gopinath, K. 1942. Acclimatisation of foreign fish in Travencore. *Journal of the Bombay Natural History Society* **43**: 267–271.

Gore, R. & Doubilet, D. 1976. Noah's Ark for exotic newcomers. *National Geographic Magazine* **150**: 538–559.

Gotschlich, B. 1913. Llanquihue y Valdivia. *Boletin Museo Nacional (Chile)*, **VI**(I): 197–204.

Goubier, J. 1975. Biogeographie, biometric et biologie du sandre. DSc thesis, University of Claude-Bernard, Lyon, France: 259 pp.

Goubier, J., Hoestlandt, H. & Goubier, M. 1983. Recherches biologiques sur la peche (*Perca fluviatilis* L.) de Sao Miguel (Açores). *Cybium* (Series 3) **7**: 25–49.

Graham, M. 1929. *The Victoria Nyanza and its fisheries. A report on the fishing survey of Lake Victoria 1927–1928, and appendices.* London: Crown Agents for the Colonies.

Grande, M. 1964. A study of the brook trout (*Salvelinus fontinalis*) in Telemark. *Fauna* **1**: 32–33.

Greeley, J.R. 1937. Fishes of the area with annotated list. In: *A Biological Survey of the Lower Hudson Watershed. Supplement to the 26th Annual Report of the New York State Conservation Department (1936)*, **II**: 45–103.

Greene, C.W. 1935. *The Distribution of Wisconsin Fishes.* Madison, Wisconsin: Wisconsin Conservation Commission.

Gressitt, J.L. 1961. Terrestrial fauna. In: Blementhal, D.I. (Ed.), *A Report on Typhoon Effects upon Jaluit Atoll. Atoll Research Bulletin* **75**: 69–73.

Grimaldi, E. 1972. Lago Maggiore: effects of exploitation and introductions on the salmonid community. *Journal of the Fisheries Research Board of Canada* **29**: 777–785.

Grimås, V. & Nilsson, N-A. 1962. Nahrungsfauna und Kadanische Seeforelle in Berner Gebirgsseen. *Schweizerische Zeitschrift für Hydrologie* **24**: 49–75.

Guillory, V. & Gasaway, R.D. 1978. Zoogeography of the grass carp in the United States. *Transactions of the American Fisheries Society* **107**: 105–112.

Guinther, E.B. 1971. Ecologic observations on an estuarine environment at Fanning Atoll. *Pacific Science* **25**: 249–259.

Haack, H. 1893. Zander im Bodensee. *Allgemeine Fischerei-Zeitung* **18**: 177–178.

Haacker, P.L. 1979. Two Asiatic gobiid fishes, *Tridentiger trigonocephalus* and *Acanthogobius flavimanus*, in southern California. *Bulletin of the Southern California Academy of Sciences* **78**: 56–61.

Haas, R. & Pal, R. 1984. Mosquito larvivorous fishes. *Bulletin of the Entomological Society of America* **30**: 17–25.

Hahn, D.E. 1966. An introduction of *Poecilia mexicana* (Osteichthyes: Poeciliidae) into Colorado. *Southwestern Naturalist* **11**: 307–308.

Halliwell, N. 1992. Lake Victoria – a nightmare come true. *Cichlidae* **13**: 53–58.

Hamman, K.C.D. 1980. Post-impoundment trends in the fish populations of the Hendrik Verwoerd Dam, South Africa. *Journal of the Limnological Society of Southern Africa* **6**: 101–108.

Hanek, G. (Comp.). 1982. Peru. La pesquería en el lago Titicaca (Peru): presente y futuro. Un informe preparado por el Proyecto de Investigacíon de los Recursos Hidrobiologia de Aguas Continental. Rome: FAO, FI: DP/PER/76/022. Documento de campo **1**: 58 pp.

Hanström, B. & Johnels, A.G. 1962. Fishar (Part 2). *Djurens Värld* **6**: 94–95.

Hardy, C.J. (Ed.). 1972. South Island Council of Acclimatization Societies proceedings of the quinnat salmon fisheries symposium, 2–3 October 1971. Fisheries Technical Report, New Zealand and Marine Department **83**: 1–298.

Hardy, C.J. 1983. Origins of NZ's sockeye. *Freshwater Catch (New Zealand)* **18**: 11–13.

Hardy, E. 1954. The Bitterling in Lancashire. *Salmon & Trout Magazine* **142**: 548–553.

Hardy, E. 1961. New salmon in Europe and research leading to the introduction of Pacific *Oncorhynchus*. *Salmon & Trout Magazine* **163**: 132–138.

Harris, C. 1978. Tilapia: Florida's alarming foreign menace. *Florida Sportsman* **9**: 12, 15, 17–19.

Harris, C.K., Wiley, D.S. & Wilson, D.C. 1995. Socio-economic impacts of introduced species in Lake Victoria. In: Pitcher, T.J. & Hart, P.J.B. (Eds), *The Impact of Species Changes in African Lakes. Fish and Fisheries Series* **18**: 215–242. London: Chapman & Hall.

Harrison, A.C. 1934. Black bass in the Cape Province. Report on the early history of the acclimatization of American largemouth black bass (*Micropterus salmoides*, Lacepede) in the Cape Province. Fisheries and Marine Biological Survey Division, Union of South Africa. *Investigational Report* **4**: 92 pp.

Harrison, A.C. 1936. Black bass in the Cape Province: second report on the progress of the American largemouth black bass. Department of Commerce and Industries, Fisheries and Marine Biological Survey Division, Union of South Africa. *Investigational Report* **7**: 119 pp.

Harrison, A.C. 1938. Fish of the rivers of the Laneberge, south-western Cape. *Cape Piscatorial Society, Freshwater Fisheries Circular* **30**: 2–5.

Harrison, A.C. 1940. The acclimation of freshwater game fish in the Cape province and its relation to forest areas. *Journal of the South African Forestry Association* **4**: 5–14.

Harrison, A.C. 1948. Report on the rivers of Maclear and East Griqualand. *Cape Provincial Administration Inland Fisheries Department Report* **5**: 15–19.

Harrison, A.C. 1949. *Freshwater Fishing in the Cape South-Western Districts, South Africa.* Cape Town: Cape Peninsula Publicity Association and Cape Piscatorial Society.

Harrison, A.C. 1952. Introduction of exotic fishes to the Cape Province. Section III: black bass and bluegills. *Piscator* **22**: 57–64; **23**: 92–95; **24**: 117–128; **25**: 12–15.

Harrison, A.C. 1953. The acclimatization of the smallmouth bass. *Piscator* **27**: 89–96.

Harrison, A.C. 1954. Spotted bass in the Buffalo River. *Piscator* **30**: 39–40.

Harrison, A.C. 1956. Carp from the nature conservation angle. *Department of Nature Conservation, Cape Provincial Administration Report* **13**: 75–80.

Harrison, A.C. 1959. The menace of carp. *Piscator* **46**: 46.

Harrison, A.C. 1963. The Olifants/Doorn River system and its fishes. *Piscator* **57**: 25–39.

Harrison, A.C. 1964. Steenbras reservoir: trout plantings and recovery. *Piscator* **60**: 29–46.

Harrison, A.C. 1964/1965a. Spotted bass. *Piscator* **62**: 127–128.

Harrison, A.C. 1966a. Trout in the peat-stained waters of the southern Cape. *Piscator* **66**: 22–23.

Harrison, A.C. 1966b. Early references to carp in the Cape Colony. *Piscator* **66**: 23.

Harrison, A.C. 1966–1967. Rainbow trout in Wemmershoek Reservoir. *Piscator* **68**: 126–134.

Harrison, A.C. 1976. The early transactions of the Cape Piscatorial Society. Part III. *Piscator* **97**: 67–72.

Harrison, A.C. 1977. The early transactions of the Cape Piscatorial Society. Part V. *Piscator* **99**: 14–18.

Harrison, A.C. & Lewis, D.C. 1968–1969. Twenty-one years of trout fishing in Liesbeeck. *Piscator* **74**: 128–144.

Harrison, A.C., Shortt-Smith, K.E., Yates, J.H., Jubb, R.A.,

Rushby, G. & Flamwell, C.T. 1963. *Freshwater Fish and Fishing in Africa*. Thomas Nelson.

Harrison, J.L. & Tham, A.K. 1973. The exploitation of animals. In: Chuang, S.H. (Ed.), *Animal Life and Nature in Singapore*. 251–259. Singapore: Singapore University Press.

Harrison, K., Crimmen, O., Travers, R., Maikweki, J. & Mutoro, D. 1989. Balancing the scales in Lake Victoria. *Biologist* **36**: 189–191.

Hart, J.L. 1934. Black crappies in British Columbia. *Canadian Field-Naturalist* **48**: 103–104.

Hauser, W.J., Legner, E.F., Medved, R.A. & Platt, S. 1976. *Tilapia* – a management tool for biological conservation of aquatic weeds and insects. *Fisheries* **1**: 15–16.

Healey, M. 1984. Fish predation on aquatic insects. In: Resh, V.H. & Rosenberg, D.M. (Eds), *The Ecology of Aquatic Insects*. 255–289. New York: Praeger.

Heard, H.W. & King, M. 1982. The fish populations and ecology of the Baakens River. *The Naturalist* **26**: 21–24.

Heeg, J. 1983. Secondary production – utilization of invertebrates. In: Breen, C.M. (Ed.), *Limnology of Lake Midmar*. *South African National Scientific Programmes Report* **78**: 113–114. Pretoria: CSIR.

Hefford, A.E. 1926. The rainbow trout of New Zealand: a problem of fish classification. *New Zealand Journal of Science and Technology* **8**: 102–106.

Heinz, K. & Lorenz, H. 1955. Ein Zuchtstamm von *Cristivomer namaycush* in der Schweiz. *Schweizerisches Fische* **11**: 288–290; **12**: 312–314.

Hemsen, J. 1964. The adoption of American salmon, especially of rainbow trout in the last century – a historical survey. In: *Die Regenbogenforelle als Neüburger Unserer Gewässer, Osterreichs Fischcherei* **17**: 180–183.

Hendrickson, D.A., Minckley, W.L., Miller, R.R., Siebert, D.J. & Minckley, P.H. 1980. Fishes of the Río Yaqui basin, Mexico and United States. *Journal of the Arizona Nevada Academy of Sciences* **15**: 65–106.

Hensley, D.A. & Courtenay, W.R. Jr 1980. *Carassius auratus, Misgurnus anguillicaudatus, Poecilia reticulata, Xiphophorus helleri, Xiphophorus maculatus, Xiphophorus variatus, Tilapia aurea, Tilapia mossambica, & Tilapia zillii*. In: Lee, D.S., Gilbert, C.R., Hocutt, C.H., Jenkins, R.E., McAllister, D.E. & Stauffer, J.R. Jr (Eds), *Atlas of North American Freshwater Fishes*. Raleigh: North Carolina State Museum.

Herald, E.S. 1961. *Living Fishes of the World*. New York: Doubleday.

Herre, A.W. 1953. Checklist of Philippine fishes. *United States Fish and Wildlife Service, Research Report* **20**: 977 pp.

Hervey, G.F. & Hems, J. 1968. *The Goldfish*. London: Faber & Faber.

Hessel, R. 1874. The salmon of the Danube, or the hucho (*Salmo hucho*) and its introduction into American waters. *Report of the United States Fish Commission* **1872–1873**: 161–165.

Hessel, R. 1878. The carp and its culture in rivers and lakes and its introduction into America. *Report of the United States Commission of Fisheries* **1875–1876**: 865–900.

Hessel, R. 1884. The carp, *Cyprinus carpio*. Fishery Industry of the United States, Section 1: 618–627. *Natural History of Aquatic Animals*.

Hestand, R.S. & Carter, C.C. 1978. Comparative effects of grass carp and selected herbicides on macrophyte and phytoplankton communities. *Journal of Aquatic Plant Management* **16**: 43–50.

Heusser, A. 1964. Chile. In: McClane, A.J. (Ed.), *McClane's Fishing Encyclopedia*. New York: Holt, Rinehart & Winston.

Hey, D. 1947. The fertility of brown trout eggs at the Jonkershoek Inland Fish Hatchery. *Transactions of the American Fisheries Society* **77**: 26–31.

Hey, S.A. 1926. Fisheries survey 1925–1926: inland waters. *Department of Mines and Industries Report* **2**.

Hickley, P. 1987. Invasion by zander and the management of fish stocks. In: Kornberg, H. & Williamson, M.H. (Eds). *Quantitative Aspects of the Ecology of Biological Invasions*. 571–582. London: Royal Society.

Hicks, B.J. & Watson, N.R.N. 1983. Quinnat salmon (*Oncorhynchus tshawytscha*) spawning in the Rangitikei River. *New Zealand Journal of Marine and Freshwater Research* **17**: 17–19.

Hida, T.S. & Thomson, D.A. 1962. Introduction of the threadfin shad to Hawaii. *Progressive Fish-Culturist* **24**: 159–163.

Hikita, T. 1960. On the chinook salmon eggs, *Oncorhynchus tschawytscha* (Walbaum), first planted in Hokkaido, Japan. *Scientific Reports of the Hokkaido Salmon Hatchery* **15**: 1–6. [In Japanese with English summary.]

Hildebrand, S.F. 1934. An investigation of the fishes and fish culture possibilities of the fresh waters of Puerto Rico, with recommendations. Mimeo. report to the Commissioner of Agriculture and Commerce: 38 pp.

Hildebrand, S.F. 1935a. An annotated list of fishes of the freshwaters of Puerto Rico. *Copeia* **1935**: 49–56.

Hildebrand, S.F. 1935b. Trout fishing in the tropics, rainbow trout in the Rio Chiriqui Viejo, Panama. *Pan American Union Bulletin* (**October**): 763–767.

Hildebrand, S.F. 1938. A new catalog of the freshwater fishes of Panama. *Field Museum of Natural History, Publications in Zoology*, Series 22(4), **425**: 215–359.

Hinds, V.T. 1969. *A Fisheries Reconnaissance to Wallis Island*. Noumea, New Caledonia: South Pacific Commission.

Hinks, D. 1943. *The Fishes of Manitoba*. Winnipeg: Manitoba Department of Mines and Natural Resources.

Hobbs, D.R. 1937. Natural reproduction of quinnat salmon, brown and rainbow trout in certain New Zealand waters. *Fisheries Bulletin of the New Zealand Marine Department, Wellington* **6**: 1–104.

Hobbs, D.R. 1948. Trout fisheries in New Zealand, their

development and management. *New Zealand Marine Department of Fisheries Bulletin* **9**: 3–10.

Hocutt, C.H. & Skelton, P.H. 1983. Fishes of the Sak River, South Africa. *JLB Smith Institute of Ichthyology Special Publication* **32**: 11 pp.

Hodgkiss, I.J. & Man, H.S.H. 1977. Stock density and mortality assessment of *Sarotherodon mossambicus* (Cichlidae) in Plover Cove Reservoir, Hong Kong. *Environmental Biology of Fishes* **1**: 171–180.

Hodgkiss, I.J. & Man, H.S.H. 1978. Reproductive biology of *Sarotherodon mossambicus* (Cichlidae) in Plover Cove Reservoir, Hong Kong. *Environmental Biology of Fisheries* **3**: 287–292.

Hoese, D.F. 1973. The introduction of the gobiid fishes *Acanthogobius flavimanus* and *Tridentiger trigonocephalus* into Australia. *Koolewong* **2**: 3–5.

Hoffman, G.L. & Schubert, G. 1984. Some parasites of exotic fishes. In: Courtenay, W.R. Jr. & Stauffer, J.R. Jr. (Eds), *Distribution, Biology and Management of Exotic Fishes*: 233–261. Baltimore: Johns Hopkins University Press.

Hofstede, A.E. & Botke, F. 1950. *Tilapia mossambica* Peters as a factor in malaria control in the vicinity of Djakarta. *Landbouwetenskap Documentatie* **22**: 453–468.

Hogg, R.G. 1974. Environmental hazards posed by cichlid fish species newly established in Florida. *Environmental Conservation* **1**: 176.

Hogg, R.G. 1976a. Established exotic cichlid fishes in Dade County, Florida. *Florida Scientist* **39**: 97–103.

Hogg, R.G. 1976b. Ecology of the fishes of the family Cichlidae introduced into the fresh waters of Dade County, Florida. PhD dissertation, University of Miami: 142 pp.

Holčik, J. 1970. Standing crop abundance, production and some ecological aspects of fish populations in some inland waters of Cuba. *Věstnik Ceskoslovenské Společnosti Zoologické* **33**: 184–201.

Holmes, S. 1954. Report on the possibility of using *Tilapia mossambica* as human food. *Fiji Agricultural Journal* **25**: 79.

Hoover, F.G. & St. Amant, J.A. 1970. Establishment of *Tilapia mossambica* Peters in Bard Valley, Imperial County, California. *California Fish and Game* **56**: 70–71.

Hora, S.K. & Pillay, T.V.R. 1962. Hand-book on fish culture in the Indo-Pacific region. *FAO Fisheries Biology Technical Paper*.

Hornel, J. 1935. *Report on the Fisheries of Palestine*. Crown Agents for the Colonies: 106 pp.

Howard, G.V. & Godfrey, E.R. 1950. *Fishery Research and Educational Institutions in North and South America*. FAO Special Publication. Washington, DC: UN.

Howell, G.C.L. 1916. The making of a Himalayan trout water. *Journal of the Bombay Natural History Society* **24**: 317–328.

Hubault, E. 1955. Introduction of species into lakes in the eastern part of France. *Verhandlungen der Internationalen Vereinigung für Theroetische und Angewandte Limnologie* **12**: 515–519.

Hubbs, C. 1972. A checklist of Texas freshwater fishes. *Texas Parks and Wildlife Department, Technical Series* **1**: 1–11.

Hubbs, C. 1976. A revised checklist of Texas freshwater fishes. *Texas Parks and Wildlife Department, Technical Series* **2**: 1–11.

Hubbs, C. 1982. Occurrence of exotic fishes in Texas waters. *Pearce–Sellards Series, Texas Memorial Museum* **36**: 1–19.

Hubbs, C. & Deacon, J.E. 1964. Additional introductions of tropical fishes into southern Nevada. *Southwestern Naturalist* **9**: 249–251.

Hubbs, C.L. & Miller, R.R. 1965. Studies of cyprinodont fishes. XXII. Variation in *Lucania parva*, its establishment in western United States, and description of a new species from an interior basin in Coahuila, México. *Miscellaneous Publications of the Museum of Zoology, University of Michigan* **127**: 1–11.

Hubbs, C., Miller, R.R., Edwards, R.J., Thompson, K.W., Marsh, E., Garrett, G.P., Powell, G.L., Morris, D.J. & Zerr, R.W. 1977. Fishes inhabiting the Río Grande, Texas and Mexico, between El Paso and the Pecos confluence. In: Importance, preservation and management of riparian habitat. *USDA Forest Service Technical Report* **RM -43**: 91–97.

Hubbs, C., Lucier, T., Garrett, G.P., Edwards, R.J., Dean, S.M., & Marsh, E. 1978. Survival and abundance of introduced fishes near San Antonio, Texas. *Texas Journal of Science* **30**: 369–376.

Hubbs, C.L., Follett, W.I. & Dempster, L.J. 1979. List of the fishes of California. *Occasional Papers of the California Academy of Sciences* **133**: 1–51.

Hunter, J.S. 1915. Introduced game in New Zealand. *California Fish and Game* **1**: 41–44.

Hurlbert, S.H., Zedler, J. & Fairbanks, D. 1972. Ecosystem alteration by mosquitofish (*Gambusia affinis*) predation. *Science* **175**: 639–641.

Hutchinson, R.T. 1975. Atlantic salmon. *Wildlife: A Review* **6**: 6–10.

Hutton, J.A. 1927. A note on the scales. *Salmon & Trout Magazine* **48**: 240–252.

Idyll, C.P. 1969. New Florida resident, the walking catfish. *National Geographic Magazine* **135**: 846–851.

Ikebe, K. 1939. Records of the introduction of top minnows into the south sea islands. *South Sea Fishery News* **3**: 7–10. [In Japanese: translation by the Southwest Fisheries Center, National Marine Fisheries Service, Honolulu, Hawaii.]

Imai, S. 1980. Tilapia. In: Kawai T. (Ed.), *Nihon no Tansuiseibutsu (Freshwater aquatic species in Japan)*: 124–132. Tokyo, Japan.

Infante, O.C. 1985. Bio-ecological aspects of tilapia *Sarotherodon mossambicus* (Peters) 1852 (Teleostei, Perciformes, Cichlidae) of Lake Valencia, Venezuela. *Acta Científica Venezolana*, **36**: 68–76. [In Spanish.]

Iñigo, F. 1949. Introduction of North American fishes into Puerto Rico. *Revista de Agricultura de Puerto Rico* **40**: 107.

Jackson, P.B.N. 1960. On the desirability or otherwise of introducing fishes to waters that are foreign to them. *Publications Conseil Scientifique pour l'Afrique au sud du Sahara* **63**: 157–164.

Jackson, P.B.N. 1976. Water resources and freshwater fishes in southern Africa. In: Baker, G. (Ed.), *Resources of Southern Africa, Today and Tomorrow.* 196–207. Johannesburg: Association of Scientific and Technical Societies.

Jackson P.B.N., 1982. Fish in the Buffalo River catchment system. In: Hart, R.C. (Ed.), *Water Quality in the Buffalo River Catchment: a synthesis:* 119–132. Grahamstown: Rhodes University.

Jackson, P.B.N., Cambray, J.A., Eccles, D.H., Hamman, K.C.D., Tomasson, T. & White, P.N. 1983. Distribution, structure and relative abundance of fish populations. In: Allanson, B.R. & Jackson, P.B.N. (Eds), *Limnology and fisheries potential of Lake Le Roux. South African National Scientific Programmes Report* **77**: 77–107, Pretoria: CISR.

Jackson, P.D. 1981. Trout introduced into southeastern Australia: their interaction with native fishes. *Victorian Naturalist* **98**: 18–24.

Jackson, P.D. & Davies, W.D. 1983. Survey of the fish fauna in the Grampians region, south western Victoria. *Proceedings of the Royal Society of Victoria* **95**: 39–51.

Jackson, P.D. & Williams, W.D. 1980. Effects of brown trout, *Salmo trutta* L., on the distribution of some native fishes in three areas of southern Victoria. *Australian Journal of Marine and Freshwater Research* **31**: 61–67.

Jara, H. 1945. Poblaciones y repoblaciones piscicolas efectuadas en el pais 1936 a Julio de 1945. *Memoria de la Secretaria de Marina 1944–1945*, México, DF: 302–308.

Jasinski, A. 1981. [The char *Salvelinus lepechini* in Polish waters.] *Wszechswiat* **1981**: 125–126. [In Polish.]

Jensen, F. 1987. Introductions of non-native freshwater fish to Danish streams. *Fauna och Flora* **93**: 105–110. [In Danish with English summary.]

Jernelov, A. 1985. [Grass carp – a new fish in Sweden.] *Fauna och Flora* **80**: 146–148. [In Swedish with English summary.]

Jhingran, A.G. 1989. Status of exotic fishes in capture fishery waters of India. In: Joseph, M.M. (Ed.), *Exotic Aquatic Species in India. Asian Fisheries Society Special Publication* **1**: 19–23.

Jhingran, V.G. & Gopalakrishnan, V. 1974. A catalogue of cultivated aquatic organisms. *FAO Fisheries Technical Paper* **130**: 83 pp.

Jhingram, V.G. & Pullin, R.S.V. 1985. A hatchery manual for common Chinese and Indian major carps. *ICLARM Studies and Reviews* **11**: 191 pp.

Johnson, D.S. 1964. *An Introduction to the Natural History of Singapore.* Singapore: Rayirath (Raybooks).

Johnson, D.S. 1973. Freshwater life. In: Chuang, S.H. (Ed.), *Animal Life and Nature in Singapore* 103–127. Singapore: Singapore University Press.

Johnson, M.C. 1954. Preliminary experiments of fish culture in brackish water ponds. *Progressive Fish Culturist* **16**: 131–133.

Johnson, R.S. 1915. The distribution of fish and fish eggs during the fiscal year 1915. *United States Bureau of Fisheries Report for 1915 (Appendix I)*, Document **828**: 138 pp.

Johnston, R.M. 1883. On the fishes of Tasmania. *Proceedings of the Royal Society of Tasmania* **1882**: 53–144.

Johri, V.K. & Prasad, R. 1989. Impact of introduction of *Salmo trutta fario* and *Salmo gairdneri* in Garhwal Himalayas. In: Joseph, M.M. (Ed.), *Exotic Aquatic Species in India. Asian Fisheries Society Special Publication* **1**: 83–84.

Jones, S. & Sarojini, K.K. 1952. History of transplantation and introduction of fishes in India. *Journal of the Bombay Natural History Society* **50**: 594–609.

Jonez, A. & Sumner, R.C. 1954. Lakes Mead and Mojave investigations. Final report, Nevada Fish and Game Commission (D–J Project, F-1-R): 186 pp. (Mimeograph).

Jordan, D.S. & Evermann, B.W. 1905. Aquatic resources of the Hawaiian Islands. Part 1. The shore fishes. *Bulletin of the United States Fish Commission* **23**: 1–574.

Joubert, P. 1984. Daar's baars in the Vaal. *Stywe Lyne* (**January**): 5–7.

Jubb, R.A. 1959a. Carp, mud-mullet and eels in Eastern Cape Rivers. *Piscator* **46**: 56–60.

Jubb, R.A. 1959b. The plump redfin (rooivlerk) *Barbus afer* Boulenger, 1911. *Piscator* **45**: 34–35.

Jubb, R.A. 1961. *An Illustrated Guide to the Freshwater Fishes of the Zambezi River, Lake Kariba, Pungwe, Sabi, Lundi and Limpopo Rivers.* Bulawayo: Stuart Manning (PVT).

Jubb, R.A. 1965. Freshwater fishes of the Cape Province. *Annals of the Cape Provincial Museum* **4**: 72 pp.

Jubb, R.A. 1967. *Freshwater Fishes of Southern Africa.* Cape Town: Balkema.

Jubb, R.A. 1971. Some interesting Eastern Cape freshwater fishes. *The Eastern Cape Naturalist* **43**: 7–9.

Jubb, R.A. 1972. The Hendrik Verwoerd Dam and Orange River fishes. *Piscator* **84**: 22–26.

Jubb, R.A., 1973, 1976, 1977. Notes on exotic fishes introduced into South African waters. *Piscator* **87**: 9–12; **88**: 62–64; **98**: 132–134.

Jubb, R.A. 1978. The distribution of the common carp *Cyprinus carpio* in southern Africa. *Piscator* **103**: 63–65.

Jubb, R.A. 1979. Some endangered freshwater fish species. *The Eastern Cape Naturalist* **68**: 24–26.

Jubb, R.A. & Petrick, F.O. 1970. *Tilapia mossambica* Peters from Australia. *Annals of the Cape Provincial Museum* **8**: 67–71.

Jude, D.J., Reider, R.H. & Smith, G.R. 1992. Establishment of Gobiidae in the Great Lakes Basin. *Canadian Journal of Fisheries and Aquacultural Science* **49**: 416–421.

Jude, D.J., Janssen, J. & Crawford, G. In press. Ecology, distribution and impact of the newly introduced Round and Tubenose Gobies on the Biota of the St Clair and Detroit Rivers. In: Munawar, M., Edsall, T. & Leach, J. (Eds), *The Lake Huron Ecosystem: Ecology, Fisheries and Management*. Ecovision World Monograph Series. The Netherlands: Academic Publishing.

Juliano, R.O., Guerrero, R. III. & Ronquillo, I. 1989. The introduction of exotic aquatic species in the Philippines. In: De Silva, S.S. (Ed.), *Exotic Aquatic Organisms in Asia. Asian Fisheries Society Special Publication* **3**: 83–90.

Kailola, P. 1981. Gambusia is a pest. *SAFIC* **5**: 11.

Kami H.T., Ikehara, I.I. & DeLeon, F.P. 1968. Checklist of Guam fishes. *Micronesica* **4**: 95–131.

Kanayama, R.K. 1968. Hawaii's aquatic animal introductions. *Proceedings of the Annual Conference of the Western Association of State Game and Fish Commissioners* **47**: 123–131.

Karenge, L. & Kolding, J. 1995. Inshore fish population and species changes in Lake Kariba, Zimbabwe. In: Pitcher, T.J. & Hart, P.J.B. (Eds), *The Impact of Species Changes in African Lakes. Fish and Fisheries Series* **18**: 245–275. London: Chapman & Hall.

Kawanabe, H. 1980. Nijimasu (Rainbow Trout). In: Kawai, T. (Ed.), *Nihon no Tansuiseibutsu (Freshwater aquatic species in Japan)*. Tokyo, Japan.

Keam, R. 1994. *The Impact of Trout in Australian Waters*. Victoria: Australian Trout Foundation.

Kear, J. 1972. The Blue Duck in New Zealand. *Living Bird* **11**: 175–192.

Kear, J. 1990. *Man and Wildfowl*. London: T. & A.D. Poyser.

Kear, J. & Williams, G. 1978. Waterfowl at risk. *Wildfowl* **29**: 5–21.

Keleher, J.J. 1956. The northern limits of distribution in Manitoba for cyprinid fishes. *Canadian Journal of Zoology* **34**: 263–266.

Keleher, J.J. & Kooyman, B. 1957. Supplement to Hinks' *The Fishes of Manitoba*. Winnipeg: Department of Mines and Natural Resources.

Kell, L.T. 1985. The impact of an alien piscivore the zander (*Stizostedion lucioperca* L.) on a freshwater fish community. Doctoral dissertation, University of Liverpool: 442 pp.

Kennedy, M. 1969. Irish pike investigations. 1. Spawning and early life history. *Irish Fisheries Investigations, Series* A, **5**: 4–33.

Kennedy, M. & Fitzmaurice, P. 1970. The biology of the Tench *Tinca tinca* (L.) in Irish waters. *Proceedings of the Royal Irish Academy*, Series B, **69**: 31–82.

Kendall, W.C. 1910. American catfishes: habits, culture, and commercial importance. *Report of the United States Fish Commission* **733**: 1–39.

Kiener, A. 1961. Documentation piscicole: liste des poissons introduits à Madagascar jusqu'en juillet 1960. *Bulletin de Madagascar* **176**: 1–12.

Kiener, A. 1963. Poissons pêche et pisciculture à Madagascar. *Publication 24 du Centre Technique Forestier Tropical.*

Kikukawa, T. 1980. Blackbass. In: Kawai, T. (Ed.), *Nihon no Tansuiseibutsu (Freshwater aquatic species in Japan)*: 20–29. Tokyo, Japan.

King, M.J. & Bok, A.H. 1984. Report on the ichthyofauna and ecology of the Baakens River. Internal Report, Cape Department of Nature and Environmental Conservation: 9 pp.

Kirk, J.T.O. 1977. Attenuation of light in natural waters. *Australian Journal of Marine and Freshwater Research* **28**: 497–508.

Klee, C. 1981. An assessment of the contribution made by zander to the decline of fisheries in the lower Great Ouse area. *Proceedings of the 2nd British Freshwater Fisheries Conference*, **1981**: 80–93.

Kleijn, L.J.K. 1968. Identificatie van de in Nederland voorkomende soort van het genus *Umbra* [*pygmaea*] Walbaum, 1792 (Honsvissen). *Natuurhistorisch Maandblad* **57**: 35–40.

Kleynhans, C.J. 1983. A checklist of the fish species of the Mogol and Palala Rivers (Limpopo system) of the Transvaal. *Journal of the Limnological Society of Southern Africa* **9**: 29–32.

Kleynhans, C.J. 1984. Die verspreiding en status van sekere seldsame vissorte van die Transvaal en die ekologie van sommige spesies. DSc thesis, University of Pretoria (unpublished).

Klindt, R. 1990–1991. Distribution of rudd in the St Lawrence River: 1989–1991. *Report to the Great Lakes Fishery Commission, St Lawrence River Subcommittee.* 71–72.

Knaggs, E.H. 1977. Status of the genus *Tilapia* in California's estuarine and marine waters. *California–Nevada Wildlife Transactions* **1977**: 60–67.

Koch, B.B. & Schoonbee, H.J. 1980. A fish mark–recapture study in Boskop Dam, Western Transvaal. *Water S.A.* **6**: 149–155.

Koster, W.J. 1957. *Guide to the Fishes of New Mexico*. Albuquerque, New Mexico: University of New Mexico Press.

Koura, R. & El-Bolock, A.R. 1960. Acclimatization and growth of mirror carp in Egyptian ponds. *Notes and Memoirs of the Hydrobiological Department, Ministry of Agriculture, UAR* **51**: 1–15.

Kreitman, L. 1929. L'acclimatation du lavaret du Bourget dans la lac Leman et sa relation avec la systematique des coregones. *Actes du IVe Congrès Internationale de Limnologie Pure et Appliqué* 415–433.

Krueger, C.C. & May, B. 1991. Ecological and genetic effects of salmonid introductions in North America. *Canadian Journal of Fisheries and Aquatic Science* **48** (Suppl. 1): 66–77.

Kruger, E.J. 1971. 'n Ondersoek na die ekologiese voedselstruktuur van sekere visspesies in Loskopdam. Unpublished DSc thesis, Potchefstroom University for CHO: 131 pp. (unpublished).

Krumholz, L.A. 1948. Reproduction in the western mosqui-tofish, *Gambusia affinis affinis* (Baird & Girard), and its use in mosquito control. *Ecological Monographs* **18**: 1–43.

Kudersky, L.A. 1982. Acclimatization of American Channel catfish in Cherepets. *Sbornik Nauch Trudov Gosniokh Nauchno-Issled* **187**: 219–232. [In Russian with English summary on p. 294.]

Kudhongania, A.W. & Chitamwebwa, D.B.R. 1995. Impact of environmental change, species introductions and eco-logical interactions on the fish stocks of Lake Victoria. In: Pitcher, T.J. & Hart, P.J.B. (Eds), *The Impact of Species Changes in African Lakes. Fish and Fisheries Series* **18**: 19–32. London: Chapman & Hall.

Kukowski, G.E. 1972. Southern range extension for the yel-lowfin goby, *Acanthogobius flavimanus* (Temminck & Schlegel). *California Fish and Game* **58**: 326–327.

Kulkarni, C.V. 1946. Gourami culture. *Indian Farming* **7**: 565–571.

Kulkarni, C.V. 1947. Notes on freshwater fishes of Bombay and Salsette Island. *Journal of the Bombay Natural History Society* **47**: 319–326.

Kushlan, J.A. 1972. The exotic fish (*Aequidens portalegrensis*) in the Big Cypress Swamp. *Florida Naturalist* **45**: 29.

La Bastille, A. 1991. *Mama Poc: the account of a species' extinc-tion*. New York: Norton.

Lachner, E.A., Robins, C.R. & Courtenay, W.R. Jr 1970. Exotic fishes and other aquatic organisms introduced into North America. *Smithsonian Contributions to Zoology* **59**: 1–29.

Ladiges, W. & Vogt, D. 1965. *Die Süsswasserfische Europas*. Hamburg & Berlin: Paul Parey.

Laird, M. 1956. Studies of mosquitoes and freshwater ecol-ogy of the South Pacific. *Bulletin of the Royal Society of New Zealand* **6**: 1–288.

Lake, J.S. 1957. Trout populations and habitats in New South Wales. *Australian Journal of Marine and Freshwater Research* **4**: 414–450.

Lake, J.S. 1959. The freshwater fishes of New South Wales. *New South Wales State Fisheries Research Bulletin* **5**: 19 pp.

Lamarque, P., Therézien, Y. & Charlon, N. 1975. Etudes des conditions de pêche à l'électricité dans les eaux tropi-cales. I. Etudes conduites à Madagascar et en Zambie (1972). *Bulletin du Centre d'Etudes et de Recherches Scientifiques* **10**: 403–554; 575–665.

Lampman, B.H. 1949. *The Coming of the Pond Fishes*. Portland, Oregon: Binfords & Mort.

Langdon, J.S. 1986. A new viral disease of redfin perch. *Australian Fish* **45**: 35–36.

Langdon, J.S. 1988. Diseases of introduced Australian fish. In: *Fish Diseases*: 225–276. Sydney: Post-graduate Committee in Veterinary Science.

Langdon, J.S. 1989. Experimental transmission and patho-genicity of epizootic haematopoietic necrosis virus (EHNV) in redfin perch, *Perca fluviatilis* L., and 11 other teleosts. *Journal of Fish Diseases* **12**: 295–310.

Langdon, J.S. 1990. Disease risks of fish introductions and translocations. In: Pollard, D.A. (Ed.), *Introduced and Translocated Fishes and their Ecological Effects. Australian Bureau of Rural Resources Proceedings* **8**: 98–107. Canberra: Australian Government Publishing Service.

La Rivers, I. 1962. *Fishes and Fisheries of Nevada*. Nevada State Fish and Game Commission.

Larsen, K. 1945. The liberation of salmon and trout fry in Denmark. *Report of the Danish Biological Station to the Board of Agriculture*. **47**: 17–24.

Larsen, K. 1972. New trends in planting trout in lowland streams. The result of some controlled Danish libera-tions. *Aquaculture* **1**: 137–171.

Laycock, G. 1966. *The Alien Animals: the story of imported wildlife*. New York: Ballantine Books.

Laycock, G. 1992. Aliens. *Wildlife Conservation* **95**: 60–67.

Leach, G.C. & James, M.C. 1935. Propagation and distribu-tion of food fishes, fiscal year 1935. *United States Bureau of Fisheries Report for 1935* (Appendix IV), *Administrative Report No. 22*: 401–427.

Lee, D.S. 1980. *Notemigonus crysoleucas, Lepomis auritus, Lepomis cyanellus, Lepomis gulosus, Lepomis macrochirus, Lepomis microlophus, Lepomis punctatus, Micropterus dolomieui, Micropterus salmoides, Pomoxis annularis*. In: Lee, D.S., Gilbert, C.R., Hocutt, C.H., Jenkins, R.E., McAllister, D.E. & Stauffer, J.R. Jr (Eds), *Atlas of North American Freshwater Fishes*. Raleigh: North Carolina State Museum.

Lefebvre, A. 1883. The proposed introduction of American catfish into the rivers of Belgium. *Bulletin of the United States Fish Commission* **1882**: 153–154.

Legner, E.F. 1979. Considerations in the management of *Tilapia* for biological aquatic weed control. *Proceedings of the California Mosquito Control Association* **47**: 44.

Legner, E.F. & Pelsue, F.W. 1977. Adaptations of *Tilapia* to *Culex* and chironomid midge ecosystems in south California. *Proceedings of the California Mosquito Control Association* **45**: 95–97.

Legner, E.F., Fisher, T.W. & Medved, R.A. 1973. Biological control of aquatic weeds in the lower Colorado River Basin. *Proceedings of a Conference of the California Mosquito Control Association* **41**: 115–117.

Legner, E.F., Hauser, W.J., Fisher, T.W. & Medved, R.A. 1975. Biological aquatic weed control by fish in the lower Sonoran Desert of California. *California Agriculture News* **29**: 8–10.

Legner, E.F., Medved, R.A. & Pelsue, F. 1980. Changes in chironomid breeding patterns in a paved river channel following adaptation of cichlids of the *Tilapia mossambica – hornorum* complex. *Annals of the Entomological Society of America* **73**: 293–299.

Leith, A., Nelson, S.G., & Gates, P. 1984. Mass mortality of *Oreochromis mossambicus* (Pisces: Cichlidae) in Fena Lake, Guam, associated with a *Pseudomonas* infection. *University of Guam Marine Laboratory Technical Report* **85**: 12 pp.

Lema, R., Giaoom, B. & Ibrahim, K.H. 1975. Observations on the introduction of *Tilapia andersonii* (Castelnau) into Tanzania from Zambia. FAO/CIFA Symposium on Aquaculture in Africa, 30 September 1975, Accra, Ghana. FAO/CIFA, 75/SE-3: 6 pp.

Léon, F.J.I. 1966. Piscicultura rural en Venezuela. *Bulletin of the Office of International Epizootiology* 65: 1127–1134.

Le Roux, P. & Steyn, L. 1968. *Fishes of the Transvaal.* Capetown: Cape and Transvaal Printers.

Lesel, R., Therézien, Y. & Vibert, R. 1971. Introduction des Salmoides aux îles Kerguelen. I. Premiers résultats et observations préliminaires. *Annals of Hydrobiology* 2: 275–304.

Lever, C. 1977. *The Naturalized Animals: of the British Isles.* London: Hutchinson.

Lever, C. 1987. *Naturalized Birds of the World.* London: Longman.

Lever, C. 1992. *They Dined on Eland: the story of the acclimatization societies.* London: Quiller Press.

Lever, C. 1994. *Naturalized Animals: the ecology of successfully introduced species.* London: T. & A.D. Poyser.

Levine, D.S., Krummrich, J.T. & Shafland, P.L. 1979. Renovation of a borrow pit in Levy County, Florida, containing Jack Dempseys (*Cichlasoma octofasciatum*). Mimeo. report, Non-Native Fish Laboratory, Florida Game and Fresh Water Fish Commission: 6 pp.

Liao, I.C. & Chen, T.P. 1983. Status and prospects of tilapia culture in Taiwan. In: Fishelson, L. & Yaron, Z. (Eds), *Proceedings of the International Symposium on Tilapia in Aquaculture,* 588–598. Nazareth, Israel.

Liao, I-Chiu & Liu, Hsi-Chiang. 1989. Exotic aquatic species in Taiwan. In: De Silva, S.S. (Ed.), *Exotic Aquatic Organisms in Asia. Asian Fisheries Society Special Publication* 3: 101–118.

Liebenberg, D.P. 1967–1968. Trout transfer in Lesotho Highlands. *Piscator* 71: 106–112.

Ligtvoet, W. 1989. The Nile perch in Lake Victoria: a blessing or a disaster? *Koninklijk Museum Voor Midden-Afrika Tervuren Belgïe, Annalen Zoologische Wetenschappen* 257: 151–156.

Linfield, R.S.J. 1981. The current status of the major coarse fisheries in Anglia. *Proceedings of the Second British Freshwater Fisheries Conference, University of Liverpool.* 67–79.

Linfield, R.S.J. 1984. The impact of zander (*Stizostedion lucioperca* L.) in the United Kingdom and the future management of the affected fisheries in the Anglian region. In: *EIFAC Technical Paper* 42 (Suppl. 2): 353–361. Rome: FAO.

Linfield, R.S.J. & Rickards, R.B. 1979. The zander in perspective. *Fisheries Management* 10: 1–16.

Lintermans, M., Rutzou, T. & Kukolic, K. 1990. Introduced fish of the Canberra region – recent range expansions. In: Pollard, D. (Ed.), *Introduced and Translocated Fishes and their Ecological Effects.* Australian Bureau of Rural Resources Proceedings 8: 50–60. Canberra: Australian Government Publishing Service.

Llewellyn, L.C. 1983. The distribution of fish in New South Wales. *Australian Society for Limnology, Special Publication* 7: 23 pp.

Lloyd, L.N. 1986. An alternative to insect control by 'Mosquitofish', *Gambusia affinis.* In: St George, T.D., Kay, B.H. & Blok, J. (Eds), *Arbovirus Research in Australia. Proceeding of the Fourth Australian Arbovirus Symposium, Brisbane*: 156–163.

Lloyd, L.N. 1987. Ecology and distribution of small native fish of the lower River Murray, South Australia and their interactions with the exotic mosquitofish, *Gambusia affinis holbrooki.* MSc thesis, University of Adelaide.

Lloyd, L. 1990. Ecological interactions of *Gambusia holbrooki* with Australian native fishes. In: Pollard, D. (Ed.), *Introduced and Translocated Fishes and their Ecological Effects. Australian Bureau of Rural Resources Proceedings* 8: 94–97. Canberra: Australian Government Publishing Service.

Lloyd, L. & Tomasov, J. 1985. The status of the genus *Gambusia* in Australia. *Australian Journal of Marine and Freshwater Research* 36: 1–6.

Lloyd, L.N., Arthington, A.H. & Milton, D.A. 1986. The mosquitofish – a valuable mosquito-control agent or a pest: 7–25. In: Kitching, R.L. (Ed.), *The Ecology of Exotic Animals and Plants: some Australian case histories.* Queensland: John Wiley.

Lobel, P.S. 1980. Invasion by the Mozambique tilapia (*Sarotherodon mossambicus*; Pisces; Cichlidae) of a Pacific atoll marine ecosystem. *Micronesica* 16: 349–355.

Loftus, W.F. 1987. Possible establishment of the Mayan cichlid, *Cichlasoma urophthalmus* (Günther) (Pisces: Cichlidae), in Everglades National Park. *Florida Scientist* 50: 1–6.

Loftus, W.F. 1989. Distribution and ecology of exotic fishes in Everglades National Park. In: Thomas, L.K. (Ed.), *Proceedings of the Conference on Science in the National Parks,* Vol. 5. *Management of Exotic Species in Natural Communities*: 24–34. Washington, DC: George Wright Society and US National Park Service.

Looyen, A.J.L. 1948. De Nederlandsche Heidemaatschappij en haar bemoeiingen op visserijgebied. In: *De Nederlandse Heidemaatschappij 60 jaar.* 152–168. Arnhem: De Nederlandsche Heidemaatschappij.

Louw, J.J.R. 1979. The Wemmershoek Reservoir. *The Cape Angler* 6: 5–6.

Lowe-McConnell, R.H. 1958. Observations on the biology of *Tilapia nilotica* L. in East Africa. *Revue de Zoologie et de Botanique Africaines* 57: 129–170.

Lowe-McConnell, R.H. 1982. Tilapias in fish communities. In: Pullin, R.S.V. & Lowe-McConnell, R.H. (Eds), *The Biology and Culture of Tilapias. Proceedings of the Conference of the International Center for Living Aquatic Resources Management, Manila, Philippines* 7: 83–113.

Ludbrook, J.V. 1974. Feeding habits of the largemouth bass (*Micropterus salmoides* Lacepede 1802) in Lake Kyle, Rhodesia. *Arnoldia (Rhodesia)* 6: 1–2.

Luton, J.R. 1985. The first introductions of brown trout, *Salmo trutta*, in the United States. *Fisheries* **10**: 10–13.

Lynch, D.D. 1970. Species of Salmonidae in Tasmania. *Tasmanian Year Book* **1970**: 84–96.

Maar, A. 1960. The introduction of carp in Africa south of the Sahara. Troisième Colloque sur L'Hydrobiologie et les Pedies en Eau Donnée, Oisaka. *CCTA/CSA Publication CSA* **63**: 204–210.

McAllister, D.E. 1969. Introduction of tropical fishes into a hotspring near Banff, Alberta. *Canadian Field-Naturalist* **83**: 31–35.

McClane, A.J. 1965. *McClane's Standard Fishing Encyclopaedia and International Fishing Guide.* Toronto: Holt, Rinehard & Winston.

McClaren, P. 1980. Is carp an established asset? *Fisheries* **5**: 31–32.

MacCrimmon, H.R. 1956. *Fishing in Lake Simcoe.* Publications of the Ontario Department of Lands and Forests: 69–74. Ottowa, Ontario (Canada).

MacCrimmon, H.R. 1968. The carp in Canada. *Bulletin of the Fisheries Research Board of Canada* **165**: 93 pp.

MacCrimmon, H.R. 1971. World distribution of the rainbow trout (*Salmo gairdneri*). *Journal of the Fisheries Research Board of Canada* **28**: 663–704.

MacCrimmon, H.R. 1972. World distribution of the rainbow trout (*Salmo gairdneri*): further observations. *Journal of the Fisheries Research Board of Canada* **29**: 1788–1791.

MacCrimmon, H.R. & Scott Campbell, J. 1969. World distribution of the brook trout, *Salvelinus fontinalis. Journal of the Fisheries Research Board of Canada* **26**: 1699–1725.

MacCrimmon H.R., Marshall, T.L. and Gots, B.L. 1970. World distribution of brown trout, *Salmo trutta*: further observations. *Journal of the Fisheries Research Board of Canada* **27**: 811–818.

MacCrimmon, H.R., Gots, B.L. & Scott Campbell, J. 1971. World distribution of brook trout, *Salvelinus fontinalis*: further observations. *Journal of the Fisheries Research Board of Canada* **28**: 452–456.

MacCrimmon, H.R. & Marshall, T.L. 1968. World distribution of brown trout, *Salmo trutta. Journal of the Fisheries Research Board of Canada* **25**: 2527–2548.

McCulloch, A.R. 1922. *Check List of the Fishes and Fish-like Animals of New South Wales.* Sydney: Royal Zoological Society of NSW.

McDowall, R.M. 1968. Interactions of the native and alien faunas of New Zealand and the problems of fish introductions. *Transactions of the American Fisheries Society* **97**: 1–11.

McDowall, R.M. 1979. Exotic fishes of New Zealand – dangers of illegal releases. *New Zealand Ministry of Agriculture & Fisheries (Fisheries Research Division) Information Leaflet* **9**: 1–17.

McDowall, R.M. (Ed.). 1980. *Freshwater Fishes of South-Eastern Australia.* Sydney: A.H. & A.W. Reed.

McDowall, R.M. 1984. Exotic fishes: the New Zealand experience. In: Courtenay, W.R. Jr & Stauffer, J.R. Jr (Eds), *Distribution, Biology and Management of Exotic Fishes*: 200–214. Baltimore: Johns Hopkins University Press.

McDowall, R.M. 1987. Impacts of exotic fishes on the native fauna. *Bulletin of the New Zealand Department of Scientific and Industrial Research* **241**: 333–347.

McDowall, R.M. 1990a. Filling in the gaps – the introduction of exotic fishes into New Zealand. In: Pollard, D. (Ed.), *Introduced and Translocated Fishes and their Ecological Effects. Australian Bureau of Rural Resources Proceedings* **8**: 69–82. Canberra: Australian Government Printing Service.

McDowall, R.M. 1990b. *New Zealand Freshwater Fishes – a natural history and guide.* Auckland: Heinemann.

McDowall, R.M. 1990c. When galaxiid and salmonid fishes meet – a family reunion in New Zealand. *Journal of Fish Biology* **37** (Suppl. A): 35–43.

McDowall, R.M. 1991. Freshwater fisheries research in New Zealand: processes, projects and people. *New Zealand Journal of Marine and Freshwater Research* **25**: 393–413.

McDowall, R.M. 1994a. The origins of New Zealand's Chinook Salmon, *Oncorhynchus tshawytscha. Marine Fisheries Review* **56**: 1–7.

McDowall, R.M. 1994b. *Gamekeepers for the Nation: the story of New Zealand's acclimatisation societies, 1861–1990.* Christchurch: Canterbury University Press.

McDowall, R.M. & Tilzey, R.D.J. 1980. Family Salmonidae. In: McDowall, R.M. (Ed.), *Freshwater Fishes of South-Eastern Australia.* Sydney: A.H. & A.W. Reed.

Maciolek, J.A. 1984. Exotic fishes in Hawaii and other islands of oceania. In: Courtenay, W.R. Jr & Stauffer, J.R. Jr (Eds), *Distribution, Biology and Management of Exotic Fishes*: 131–161. Baltimore: Johns Hopkins University Press.

Maciolek, J.A. & Timbol, A.S. 1980. Electroshocking in tropical insular streams. *Progressive Fish-Culturist* **42**: 57–58.

Maciolek, J.A. & Yamada, R. 1981. Vai Lahi and other lakes of Tonga. *Proceedings of the International Association of Theoretical and Applied Limnology* **21**: 693–698.

MacKay, H.H. 1963. *Fish of Ontario.* Publications of the Ontario Department of Lands and Forests: 166–175. Ottawa, Ontario (Canada).

McKay, R.J. 1977. The Australian aquarium fish industry and the possibility of the introduction of exotic fish species and diseases. *Fisheries Paper* **5**, Department of Primary Industry: 36 pp.

McKay, R.J. 1978. The exotic freshwater fishes of Queensland. Report to the Australian National Parks and Wildlife Service, Canberra: 144 pp.

McKay, R.J. 1984. Introduction of fishes in Australia. In: Courtenay, W.R Jr & Stauffer, J.R. Jr. (Eds), *Distribution, Biology and Management of Exotic Fishes*: 177–179. Baltimore: Johns Hopkins University Press.

McKay, R. 1986–1987. It's your problem too! Parts 3 and 4. The Australian introductions. *Aquarium Life Australia* **2**: 37–39; 39–40.

McKay, R.J. 1989. Exotic and Translocated Freshwater Fishes in Australia. In: De Silva, S.S. (Ed.), *Exotic Aquatic Organisms in Asia. Asian Fisheries Society Special Publication* **3**: 21–34.

MacKay, W.S. 1945. Trout of Travancore. *Journal of the Bombay Natural History Society* **45**: 352–373; 542–557.

MacKinnon, M. 1987. Review of inland fisheries – Australia. *FAO Fisheries Report* **371** (Suppl.): 1–37.

McNally, T. 1959. World's strangest trout fishing. *Outdoor Life* **124**: 29–31; 70–71; 113–115.

McVeigh, S.J. 1984. The sunfish syndrome. *Tight Lines* **25**: 26–27.

Mail, G.A. 1954. The mosquito fish, *Gambusia affinis* (Baird and Girard), in Alberta. *Mosquito News* **14**: 120.

Maitland, P.S. 1969. A preliminary account of the mapping of the distribution of freshwater fish in the British Isles. *Journal of Fish Biology* **1**: 45–58.

Maitland, P.S. 1977. *The Hamlyn Guide to Freshwater Fishes of Britain and Europe*. London: Hamlyn.

Maitland, P.S. & Price, C.E. 1969. *Urocleidus principalis* (Mizelle 1936), a North American monogenetic trematode new to the British Isles, probably introduced with the largemouth bass *Micropterus salmoides* (Lacepede 1802). *Journal of Fish Biology* **1**: 17–18.

Man, H.S.H. & Hodgkiss, I.J. 1977. Studies on the ichthyofauna in Plover Cove Reservoir, Hong Kong. *Journal of Fish Biology* **10**: 493–503; **11**: 1–13.

Mann, F.G. 1954. *Vida de los Peces en Aguas dulces Chilenas*. Santiago, Chile.

Marini, T.L. 1942. El landlocked salmon en la Republica Argentina. *Publicaciones Misión Buenos Aires* **117**: 1–20.

Marshall, B.E. 1979. Fish populations and the fisheries potential of Lake Kariba. *South African Journal of Science* **75**: 485–488.

Marshall, B.E. 1991. The impact of the introduced sardine *Limnothrissa miodon* on the ecology of Lake Kariba. *Biological Conservation* **55**: 151–165.

Marshall, B.E. 1995. Why is *Limnothrissa miodon* such a successful introduced species and is there anywhere else we should put it? In: Pitcher, T.J. & Hart, P.J.B. (Eds), *The Impact of Species Changes on African Lakes. Fish and Fisheries Series* **18**: 527–545. London: Chapman & Hall.

Marshall, B.E. & Junor, F.J.R. 1981. The decline of *Salvinia molesta* in Lake Kariba. *Hydrobiologica* **83**: 477–484.

Marshall, R.D. 1972. A preliminary fish survey of the Caledon River system. *The Civil Engineer in South Africa* **14**: 94–95.

Marshall, T.L. & Johnson, R.P. 1971. History and results of fish introduction in Saskatchewan 1900–1969. *Department of Natural Resources, Fisheries and Wildlife Branch, Fisheries Report* **8**: 31 pp.

Matern, S.A. & Fleming, K.J. In press. Invasion of a third Asian goby, *Tridentiger bifasciatus*, into California. *California Fish and Game*.

Mather, F. 1879. Account of trip to Europe with eggs of the Quinnat salmon. *Report of the United States Fish Commission* **1877**: 811–816.

Mather, F. 1889. Brown trout in America. *Bulletin of the United States Fish Commission* (1887) **7**: 21–22.

Mather, P.B. & Arthington, A.H. 1991. An assessment of genetic differentation among feral Australian tilapia populations. *Australian Journal of Marine and Freshwater Research* **42**: 721–728.

Matsui, Y. 1962. On the rainbow trout (*Salmo gairdneri irideus*) in Lake Titicaca, South America. *Bulletin of the Japanese Society of Scientific Fisheries* **28**: 497–498. [In Japanese with English summary.]

Mayekiso, M. 1986. Some aspects of the ecology of the Eastern Cape rocky *Sandelia bainsii* (Pisces: Anabantidae) in the Tyumie River, Eastern Cape, South Africa. MSc thesis, Rhodes University, Grahamstown.

Mazzola, A. 1992. Allochthonous species and aquaculture. *Bollettino Dei Musei e degli Instituti Biologici dell'Università di Genova* **56–57**: 235–246.

Meadows, B.S. 1968. Guppy in the River Lee. *Essex Naturalist* **32**: 186–189.

Mearns, A.J. 1975. *Poeciliopsis gracilis* (Heckel), a newly introduced poecillid fish in California. *California Fish and Game* **61**: 251–253.

Meehean, O.L., Douglass, E.J. & Duncan, L.M. 1948. Propagation and distribution of food fishes for the calendar years 1944–1948. *US Fish and Wildlife Service Statistical Digest* **24**: 82 pp.

Meek, S.E. 1904. The freshwater fishes of Mexico north of the Isthmus of Tehuantepec. *Field Columbian Museum of Zoology, Series* **5**: 1–252.

Melancan, C. 1936. *Les Poissons de Nos Eaux*: 111–113.

Mendis, A.J. 1977. The role of man-made lakes in the development of fisheries in Sri Lanka. *Proceedings of the 17th Indo-Pacific Fish Council*: 24 pp.

Meng, L., Moyle, P.B. & Herbold, B. 1994. Changes in abundance and distribution of native and introduced fishes in Suisun Marsh. *Transactions of the American Fisheries Society* **123**: 498–507.

Merrick, J.R. & Schmida, G.E. 1984. *Australian Freshwater Fishes*. North Ryde, NSW: Macquarie University Press.

Meynell, P.J. 1973. A hydrobiological survey of a small Spanish river grossly polluted by oil refinery and petrochemical wastes. *Freshwater Biology* **3**: 503–520.

Michel, C. 1986. A clinical case of enteric redmouth in minnows (*Pimephales promelas*). *Bulletin of the European Association of Fish Pathologists* **1986**: 1–6.

Middleton, J.J. 1982. The oriental goby, *Acanthogobius flavimanus*, an introduced fish in the coastal waters of New South Wales, Australia. *Journal of Fish Biology* **21**: 513–523.

Miley, W.W. II. 1978. Ecological impact of the pike killifish, *Belonesox belizanus* Kner (Poeciliidae), in southern

Florida. Unpublished MSc thesis, Florida Atlantic University: 55 pp.

Miller, D.J. 1989. Introductions and extinctions of fish in the African great lakes. *Trends in Ecological Evolution* **4**: 56–59.

Miller, D.L. & Lea, R.N. 1972. Guide to the coastal marine fishes of California. *Fisheries Bulletin of the California Department of Fish and Game* **157**: 1–235.

Miller, R.B. 1949. The status of the hatchery. *Canadian Fish Culturist* **4**: 5.

Miller, R.R. 1943. The introduced fishes of Nevada. *Transactions of the American Fisheries Society* **23**: 181.

Miller, R.R. 1963. Extinct, rare and endangered American freshwater fishes. *XVI International Congress of Zoology* **8**: 4–11.

Miller, R.R. & Chernoff, B. 1980. Status of the endangered Chihuahua chub, *Gila nigrescens*, in New Mexico and Mexico. *Proceedings of the Desert Fishes Council* **1979**: 74–84.

Mills, E.L., Leach, J.H., Carlton, J.T. & Secor, C.L. 1993. Exotic species in the Great Lakes: a history of biotic crises and anthropogenic introductions. *Journal of Great Lakes Research* **19**: 1–54.

Mills, E.L., Leach, J.H., Carlton, J.T. & Secor, C.L. 1994. Exotic species and the integrity of the Great Lakes. *Bioscience* **44**: 666–676.

Minckley, W.L. 1973. *The Fishes of Arizona.* Phoenix: Arizona Game and Fish Department.

Mitchell, F.J. 1918. How trout were introduced into Kashmir. *Journal of the Bombay Natural History Society* **26**: 295–299.

Miyada, R. 1991. Up the creek. *Hawaiian Fishing News* (**October**) **16**: 23.

Moe, M.A. Jr. 1964. Survival potential of piranhas in Florida. *Quarterly Journal of the Florida Academy of Science* **27**: 197–210.

Mohsin, A.K.M. & Ambak, M.A. 1983. *Freshwater Fishes of Peninsular Malaysia.* Kuala Lumpur: Universiti Pertanian Malaysia Press.

Molesworth, C. & Bryant, J.F. 1921. Trout culture in the Nilgiris. *Journal of the Bombay Natural History Society* **27**: 898–910.

Monod, T. 1949. Equilibres hydrobiologiques. *Proceedings of the International Union for Conservation of Nature Conference, Lake Success*: 421–423. IUCN: Morges (Switzerland).

Moreau, J. 1979. Introductions d'especes etrangeres dans les eaux continentales africaines. *Society of International Limnologists*: pag. var.

Moreau, J., Arrignon, J. & Jubb, R.A. 1988. Introduction of foreign fishes in African inland waters: suitability and problems. In: *Biology and Ecology of African Freshwater Fishes*: 395–425. *Travaux et Documents* **216**. Paris: Editions Orstom. [In French.]

Moreira Da Silva, A. 1977. A pesca desportiva nas aguas interiores da Ilha de S. Miguel. I. Sua evoluçào. II. Perspectivas para o seu melhoramento. Circunscrçào

Florestal de Ponta Delgade. *Estudos, Experimentaçào e Divulgaçào* **7**: 25 pp.

Morison, A. & Hume, D. 1990. Carp (*Cyprinus carpio* L.) in Australia. In: Pollard, D. (Ed.), *Introduced and Translocated Fishes and their Ecological Effects. Australian Bureau of Rural Resources Proceedings* **8**: 110–113. Canberra: Australian Government Printing Service.

Morrissy, N.W. 1967. The ecology of trout in South Australia. Unpublished PhD thesis, University of Adelaide: 374 pp.

Moses, S.T. 1944. Report of the Department of Fisheries of Baroda State, 1942–1943.

Motobar, M. 1978. Larvivorous fish, *Gambusia affinis* – a review. *World Health Organization Vector Biology and Control* **78**: 703.

Moyle, P.B. 1976a. *Inland Fishes of California.* Berkeley: University of California Press.

Moyle, P.B. 1976b. Fish introductions in California: history and impact on native fishes. *Biological Conservation* **9**: 101–118.

Moyle, P.B. 1984. America's carp. *Natural History, New York* **93**: 42–51.

Moyle, P.B. & Leidy, R.A. 1992. Loss of biodiversity in aquatic ecosystems: evidence from fish faunas. In: Fielder, P.L. & Jain, S.K. (Eds), *Conservation Biology: the theory and practice of nature conservation, preservation and management*: 129–168. New York: Chapman & Hall.

Muchiri, S.M. & Hickley, P. 1991. The fishery of Lake Naivasha, Kenya. In: Cowx, I.G. (Ed.), *Catch Effort Sampling Strategies: their application in freshwater fisheries management*: 382–392. Oxford: Blackwell Scientific Publications.

Muchiri, S.M., Hart, P.J.B. & Harper, D.M. 1995. The persistence of two introduced tilapia species in Lake Naivasha, Kenya, in the face of environmental variability and fishing pressure. In: Pitcher, T.J. & Hart, P.J.B. (Eds), *The Impact of Species Changes in African Lakes. Fish and Fisheries Series* **18**: 299–319. London: Chapman & Hall.

Muentyan, S. 1963. On the reproduction of the pink salmon in the Moochka River (a tributary into Teriberka River, the Kola Peninsula). ICES, C.M. 1963. *Salmon and Trout Committee Document* **67**: 4pp.

Muhammad Eidman, H. 1989. Exotic aquatic species introduction into Indonesia. In: De Silva, S.S. (Ed.), *Exotic Aquatic Organisms in Asia. Asian Fisheries Society Special Publication* **3**: 57–62.

Mukhachev, I.S. 1965. Acclimatization and breeding of *Coregonus peled* (Gmelin) in the water basins of the Chelybinsk District. *Voprosy Ichtiologii* **5**: 630–638. [In Russian.]

Mulder, P.F.S. 1973. Aspects on the ecology of *Barbus kimberleyensis* and *B. holubi* in the Vaal River. *Zoologica Africana* **8**: 1–14.

Mulier, W. 1900. *Vischkweekterij en Instandhouding van den Vischstand*. Haarlem: De Erven Loosjes.

Muus, B.J. & Dahlstrøm, P. 1968. *Guide des Poissons d'eau Douce et Pêche*. Neuchâtel, Switzerland: Delachaux & Niestlé.

Myers, O. 1925. Introduction of the European bitterling (*Rhodeus*) in New York and the rudd (*Scardinius*) in New Jersey. *Copeia* **140**: 20–21.

Myers, G.S. 1965. *Gambusia*, the fish destroyer. *Tropical Fish Hobbyist* **13**: 31–32. (See also *Australian Zoologist* **13**: 102.)

Mylius, W. 1894. Zander in der Mosel. *Allgemeine Fischerei-Zeitung* **19**: 211–212.

Nakamura, M. 1955. (Some exotic fishes appeared at the Kanto Plains). *Nihon-seibutsu-chiri-gakkaiho* **16/19**: 333–337.

Natarajan, A.V. 1989. Ecological and aquacultural roles of exotic fishes in aquatic productivity in India. In: Joseph, M.M. (Ed.), *Exotic Aquatic Species in India. Asian Fisheries Society Special Publication* **1**: 1–3.

Natarajan, M.V. & Ramachandra Menon, V. 1989. Introduction of exotic fishes in Tamil Nadu. In: Joseph, M.M. (Ed.), *Exotic Aquatic Species in India. Asian Fisheries Society Special Publication* **1**: 63–65.

Navas, J.R. 1987. Los vertebrados exoticos introducidos en la Argentina. *Revista del Museo Argentino de Ciencias Naturales 'Bernardino Rivadavia' Instituto Nacional de Investigacion de las Ciencias Naturales (Zoologica)* **14**: 7–38.

Neave, F. & Carl, G.C. 1940. The brown trout on Vancouver Island. *Proceedings of the 6th Pacific Science Congress* **3**: 341–343.

Needham, P.R. 1949. A fisheries survey of the streams of Kauai and Maui with special reference to rainbow trout (*Salmo gairdneri*). *Special Bulletin* **1**, Territory of Hawaii, Division of Fish and Game.

Needham, P.R. & Welsh, J.P. 1953. Rainbow trout (*Salmo gairdneri*) in the Hawaiian Islands. *Journal of Wildlife Management* **17**: 233–255.

Negonovskaya, I.T. 1981. On the results and prospects of the introduction of phytophagous fishes into waters of the USSR. *Journal of Ichthyology* **20**: 101–111.

Neil, E.H. 1966. Observations on the behaviour of *Tilapia mossambica* (Pisces, Cichlidae) in Hawaiian ponds. *Copeia* **1966**: 50–56.

Nelson, J.S. 1965. Effects of fish introductions and hydro-electric development on fishes in the Kananaskis River system, Alberta. *Journal of the Fisheries Research Board of Canada* **22**: 721–753.

Nelson, J.S. 1984. The tropical fish fauna in Cave and Basin Hotsprings drainage, Banff National Park. *Canadian Field-Naturalist* **9**: 255–261.

Nelson, S.G. 1988. Development of aquaculture in the US-affiliated islands of Micronesia. In: Smith, B.D. (Ed.), Topic Reviews in Insular Resource Development and Management in the Pacific US-affiliated Islands. *University of Guam Marine Laboratory Technical Report* **88**: 61–83.

Nelson, S.G. & Eldredge, L.G. 1991. Distribution and status of introduced cichlid fishes of the genera *Oreochromis* and *Tilapia* in the islands of the South Pacific and Micronesia. *Asian Fisheries Science* **4**: 11–22.

Nelson, S.G. & Hopper, D. 1989. The freshwater fishes of Yap. In: Inland aquatic habitats of Yap. *University of Guam Marine Laboratory Technical Report* **92**: 10–20.

Nezdolic, V.K. & Mitrofanov, V.P. 1975. Natural reproduction of the grass carp, *Ctenopharyngodon idella*, in the Ili River. *Journal of Ichthyology* **15**: 927–933.

Nichols, J.T. 1929. The fishes of Porto Rico and the Virgin Islands. Scientific Survey of Porto Rico and the Virgin Islands. *New York Academy of Sciences* **10**: 161–295.

Nicholls, A.G. 1957–1958. The Tasmanian trout fishery. *Australian Journal of Marine and Freshwater Research* **8**: 451–475; **9**: 167–190; 19–59; **12**: 17–53.

Nicols, A. 1882. *The Acclimatisation of the Salmonidae at the Antipodes: its history and results*. London: Sampson, Low, Marston, Searle & Rivington.

Nijssen, H. & De Groot, S.J. 1974. Catalogue of fish species of the Netherlands. *Beaufortia* **21**: 173–207.

Nikolsky, G.V. 1958. Salmon investigations in USSR. *Journal du Conseil Permanent International pour l'Exploration de la Mer* **23**: 434–439.

Nikolsky, G.V. 1961. *Special Ichthyology*. Jerusalem: Israel Program for Scientific Translations.

Nikolsky, G.V. 1963. *The Ecology of Fishes*. New York: Academic Press.

Nilsson, N.A. 1972. Effects of introductions of salmonids into barren lakes. *Journal of the Fisheries Research Board of Canada* **29**: 693–697.

Nilsson, N.A. & Svärdson, G. 1968. Some results of the introduction of lake trout (*Salvelinus namaycush* Walbaum) into Swedish lakes. *Report of the Institute of Freshwater Research, Drottningholm* **48**: 5–16.

Noble, R.L., Germany, R.D. & Hall, C.R. 1975. Interactions of blue tilapia and largemouth bass in a power plant cooling reservoir. *Proceedings of the Annual Conference of the Southeastern Association of Game and Fish Commissioners* **29**: 247–251.

Nomura, H. 1976. Present development and future perspective in the intensive and extensive culture of fish in São Paulo State. *Ciencia e Cultura* **28**: 1097–1107.

Nomura, H. 1977. Principales espiecies de peces cultivadas en el Brazil. *FAO Fisheries Report* **159**: 211–219.

Nomura, M. & Furuta, Y. 1977. Gunma-ken niokeru koshyo kasen no koudo gyogyou-riyou ni-kansuru chyosha-hokokushyo. *Gunma Prefecture.* 40–43.

Obregón, F. 1960. Cultivo de la carpa seleccionada en México. Secretaria de Agricultura y Ganaderia, México: 63 pp.

Ochumba, P.B.O. 1995. Limnological changes in Lake Victoria since the Nile perch introduction. In: Pitcher, T.J. & Hart, P.J.B. (Eds), *The Impact of Species Changes in African Lakes. Fish and Fisheries Series* **18**: 33–44. London: Chapman & Hall.

Ogari, J. 1990. Introduction and the transfer of fish species: a case study of the exotic species found in Lake Victoria (Kenya waters). In: *Fisheries of the African Great Lakes. Occasional Papers of the Wageningen International Agricultural Centre* **3**: 55–57.

Ogilvie, V.E. 1966. Report on the Peacock Bass Project including Venezuelan trip report and a description of the five cichla species. Florida Game and Freshwater Fish Commission: 62 pp. mimeo.

Ogutu-Ohwayo, R. 1988. Reproductive potential of the Nile perch *Lates niloticus* (L.) and the establishment of the species in Lakes Kyoga and Victoria (East Africa). *Hydrobiologia* **162**: 193–200.

Oguto-Ohwayo, R. 1990a. The decline of the native fishes of Lakes Victoria and Kyoga (east Africa) and the impact of introduced species, especially the Nile perch, *Lates niloticus*, and the Nile tilapia, *Oreochromis niloticus*. *Environmental Biology of Fishes* **27**: 81–96.

Ogutu-Ohwayo, R. 1990b. The reduction in fish species diversity in lakes Victoria and Kyoga (East Africa) following human exploitation and introduction of non-native fishes. In: *The Biology and Conservation of Rare Fish.* Papers from an international symposium of The Fisheries Society of the British Isles.

Ogutu-Ohwayo, R. 1995. Diversity and stability of fish stocks in Lakes Victoria, Kyoga and Nabugabo after establishment of introduced species. In: Pitcher, T.J. & Hart, P.J.B. (Eds), *The Impact of Species Changes in African Lakes. Fish and Fisheries Series* **18**: 59–82. London: Chapman & Hall.

Okada, Y. 1960. *Studies On the Freshwater Fishes of Japan, Mie Tsu (Mie Prefecture).* Mie Tsu: Prefectural University: 860 pp.

Okada, Y. 1966. *Fishes of Japan.* Tokyo: Uno Shoten.

Oliver, S.C. 1949. Catálogo de los peces fluviales de la provincia de Concepción. *Boletín de la Sociedad de Biologia de Concepción*, Chile **24**: 51–60.

Oliver-Gonzalez, J. 1946. The possible role of the guppy, *Lebistes reticulatus*, on the biological control of *Schistosoma mansoni. Science* **104**: 605.

Ortiz-Carrasquillo, W. 1980. Resumen historico de la introducción de los peces de agua dulce en los lagos artificiales de Puerto Rico desde 1915 hasta 1975. *Science-Ciencia* **7**: 95–107.

Ortiz-Carrasquillo, W. 1981. Notas sobre los crustaceos y peces des Río Matrullas, Orocovis, P.R. *Science-Ciencia* **8**: 9–13.

Osborne, P.L. 1993. Biodiversity and conservation of freshwater wetlands in Papua New Guinea. In: Beehler, B.M. (Ed.), *Papua New Guinea Conservation Needs Assessment.* Vol. 2. Washington, DC Biodiversity Support Program.

Otto, R.G. 1973. Temperature tolerance of the mosquitofish *Gambusia affinis* (Baird and Girard). *Journal of Fish Biology* **5**: 575–585.

Pagan-Font, F.A. 1973. Potential for cage culture of the cichlid fish *Tilapia aurea. 10th Meeting of the Association of Island Marine Laboratories of the Caribbean*: 52pp.

Panikkar, N.K. & Tampi, P.R.S. 1954. On the mouth brooding cichlid *Tilapia mossambica* Peters. *India Journal of Fisheries* **1**: 217–230.

Parent, A.W. 1950. Uitheemse vissen in de Nederlandse. Fauna. *Aquarium* **21**: 37–39.

Patchell, C.J. 1977. Studies on the biology of the catfish *Ictalurus nebulosus* Le Sueur in the Waikata region. Hamilton, New Zealand: Unpublished MSc thesis, University of Waikata: 144 pp.

Paxton, J.R. & Hoese, D.F. 1985. The Japanese sea bass, *Lateolabrax japonicus* (Pisces: Percichthyidae), an apparent marine introduction into eastern Australia. *Japanese Journal of Ichthyology* **31**: 369–372.

Payne, I. 1987. A lake perched on piscine peril. *New Scientist* **115**: 50–54.

Pechlaner, R. 1984. Historical evidence for the introduction of Arctic charr into high-mountain lakes in the Alps by man. In: Johnson, L. & Burns, B. (Eds), *Biology of the Arctic Charr*: 549–557. Winnipeg, Manitoba: University of Manitoba Press.

Pelren, D.W. & Carlander, K.D. 1971. Growth and reproduction of yearling *Tilapia aurea* in Iowa ponds. *Proceedings of the Iowa Academy of Sciences* **78**: 27–29.

Pelzman, R.J. 1972. Evaluation of introduced *Tilapia sparrmanii* into California. *California Department of Fish and Game, Inland Fisheries Administration Report* **72**: 7 pp.

Pelzman, R.J. 1973. A review of the life history of *Tilapia zillii* with a reassessment of its desirability in California. *California Department of Fish and Game, Inland Fisheries Administration Report* **74**: 1–9.

Peng, S.H. 1962. An experiment on the propagation of rainbow trout in the Korean People's Democratic Republic. *Sbornik Dokladov Na II Plenume Komis Rybone Issled Zap Chasti Tikhogo Okeana Moscow Pishche Promizdat*: 216–225. (Biological Abstracts 45 (19) No. 80932 (1964.)

Pennekamp, G.A. 1905. De Vischkweekerji van de Nederlandse Heidemaatschappij te Vaassen. *Tijdschrift voor Nederlandsche Heidemaatsche* **17**: 337–349.

Peterson, E.T. & Drews, R.A. 1957. Some historical aspects of the carp. *Michigan Department of Conservation, Fisheries Division Pamphlet* **23**: 5pp.

Petr, T. 1987a. Food fish as vector control and strategies for their use in agriculture. In: Effects of agricultural development on vector-borne diseases. Working papers read to the 7th Annual Meeting of the joint WHO/FAO/UNEP panel of experts on environmental management of vector control, Rome, 7–11 September 1987. Rome: FAO (AGL/MISC/12/87): 87–92.

Petr, T. 1987b. Fish, fisheries, aquatic macrophytes and water quality in inland fisheries. *Water Quality Bulletin* **12** (3): pag. var.

Pflieger, W. 1978. Distribution and status of the grass carp (*Ctenopharyngodon idella*) in Missouri streams. *Transactions of the American Fisheries Society* **107**: 113–118.

Pflieger, W.L. 1989. Natural reproduction of bighead carp (*Hypophthalmichthys nobilis*) in Missouri. *Introduced Fish Section, American Fisheries Society* **9**: 9–10.

Philippart, J-Cl. & Ruwet, J-Cl. 1982. Ecology and distribution of tilapias. In: Pullin, R.S.V. & Lowe-McConnell, R.H. (Eds), *The Biology and Culture of Tilapias. Proceedings of the Conference of the International Center for Living Aquatic Resources Management, Manila, Philippines* **7**: 15–59.

Phillips, B. 1883. Holland carp put into Hudson River about 1830. *Report of the United States Fish Commission* Vol. II **1882**: 25.

Phillipps, W.J. 1923. An attempted acclimatization of rainbow trout in Samoa. *New Zealand Journal of Science and Technology* **6**: 168.

Pienaar, U. de V. 1978. Undesirable immigrants in the Kruger National Park. *Custos* **7**: 6–7; 14–15.

Pierce, B.A. 1983. Grass carp status in the United States: a review. *Environmental Management* **7**: 151–160.

Pigg, J. 1978. The tilapia *Sarotherodon aurea* (Steindachner) in the North Canadian River in central Oklahoma. *Proceedings of the Oklahoma Academy of Science* **58**: 111–112.

Pike, T. 1980. An historical review of freshwater fish hatcheries in Natal. *Piscator* **106**: 49–53.

Pike, T. & Tedder, A.J. 1973. Rediscovery of *Oreodaimon quathlambae* (Barnard). *Lammergeyer* **19**: 9–15.

Pike, T. & Wright, C.W. 1972. Brown and rainbow trout in Highmoor Dam. *Newsletter of the Limnological Society of Southern Africa* **19**: 25–26.

Pillai, T.G. 1972. Pests and predators in coastal aquaculture: systems of the Indo-Pacific Region: In: Pillay, T.V.R. (Ed.), *Coastal Aquaculture in the Indo-Pacific Region*: 456–470. Surrey, England: Fishing News Books.

Pinter, K. 1976. Silver crucian *Carassius auratus gibelio* Bloch. *Halászat* **22**: Suppl. 3–4. [In Hungarian.]

Pinter, K. 1978. Largemouth Bass (*Micropterus salmoides*). *Halászat* **24**: Suppl. 1–4. [In Hungarian.]

Pitcher, T.J. & Bundy, A. 1995. Assessment of the Nile perch fishery in Lake Victoria. In: Pitcher, T.J. & Hart, P.J.B. (Eds), *The Impact of Species Changes in African Lakes. Fish and Fisheries Series* **18**: 163–180. London: Chapman & Hall.

Pitcher, T.J. & Hart, P.J.B. (Eds) 1995. *The Impact of Species Changes in African Lakes. Fish and Fisheries Series* **18**. London: Chapman & Hall.

Piyakarnchana, T. 1989. Exotic aquatic species in Thailand. In: De Silva, S.S. (Ed.), *Exotic Aquatic Organisms in Asia. Asian Fisheries Society Special Publication* **3**: 119–124.

Plumstead, E.E., Prinsloo, J.F. & Schoonbee, H.J. 1985. A survey of the fish fauna of Transkei estuaries. Part 1. The Kei River estuary. *South African Journal of Zoology* **20**: 213–220.

Poll, M. 1949. L'introduction en Belgique et acclimatation dans la nature d'un poisson américan supplementaire, *Umbra pygmaea*. *Bulletin de l'Institut Royal des Sciences Naturelles de Belgique* **25**: 1–11.

Pollard, D.A. & Hutchings, P.A. 1990. A review of exotic marine organisms introduced to the Australian region. 1. Fishes. *Asian Fisheries Science* **3**: 205–221.

Poppe, R.A. 1880. The introduction and culture of the carp in California. *Report of the United States Fish Commission* **1878**: 661–666.

Popper, D. & Lichatowich, T. 1975. Preliminary success in predator control of *Tilapia mossambica*. *Aquaculture* **5**: 213–214.

Pott, R. McC. 1981. The Treur River barb: a rare fish in good company. *African Wildlife* **35**: 29–31.

Prashad, B. & Hora, S.L. 1936. A general review of the probably larvivorous fishes of India. *Record of Malarial Survey of India* **6**: 642.

Pratt, D.M., Blust, W.H. & Selgeby, J.H. 1992. Ruffe, *Gymnocephalus cernuus*: newly introduced in North America. *Canadian Journal of Fisheries and Aquatic Science* **49**: 1616–1618.

Preciado, A.S. 1955. La pesca en el Lago de Patzcuaro, Mich. y su importancia económica regional. Mexico, Secretaría de la Marina, Direccíon General de Pesca: 58 pp.

Preston, G.L. 1990. Brackish-water aquaculture in Tahiti. *SPC Fisheries Newsletter* **55**: 36–40.

Provine, W.C. 1975. The grass carp. Texas Parks & Wildlife Department (Inland Fisheries Research). Mimeo report: 51 pp.

Pullin, R.S.V. (Ed.). 1988. *Tilapia Genetic Resources for Aquaculture. Proceedings of the Workshop on Tilapia Genetic Resources for Aquaculture, Bangkok, Thailand*. Manila (Philippines): International Center for Living Aquatic Resources Management.

Quijada, B.B. 1913. Catálago ilustrado i descriptivo de la colección de peces chilenas i extranjeros. Santiago de Chile. *Boletin Museo Nacional* V (I): 139 pp.

Quinn, T.P. & Unwin, M.J. 1993. Life history patterns of New Zealand chinook salmon (*Oncorhynchus tshawytscha*) populations. *Canadian Journal of Fisheries and Aquacultural Science* **50**: 1414–1421.

Radforth, I. 1944. Some consideration of the distribution of fishes in Ontario. *Royal Ontario Museum of Zoology, Contribution* **25**: 116 pp.

Raidt, E.H. (Ed.). 1971. *Description of the Cape of Good Hope, with matters concerning it, by F. Valentyn, Amsterdam, 1727*. Cape Town: Van Riebeeck Society.

Raj, B.S. 1916. Notes on the freshwater fish of Madras. *Records of the Indian Museum* **12**: 249–294.

Raminosa, N.R. 1987. Ecologie et biologie d'un poisson teleosteen: *Ophiocephalus striatus* (Bloch, 1793), introduit à Madagascar. University of Madagascar thesis; unpublished mimeo: 225 pp.

Randall, J.E. 1960. New fishes for Hawaii. *Sea Frontiers* **6:** 33–43.

Randall, J.E. 1980. New records of fishes from the Hawaiian Islands. *Pacific Science* **34:** 211–232.

Randall, J.E. & Kanayama, R.K. 1972. Hawaiian fish immigrants. *Sea Frontiers* **18:** 144–153.

Ranoemihardjo, B.S. 1981. Nauru: eradication of tilapia from fresh- and brackishwater lagoons and ponds with a view to promoting milkfish culture. A report prepared for the tilapia eradication project. FI: DP/NAU/78/001. *Field Document* **1:** 15 pp. Rome:FAO.

Raquel, P.F. 1988. Report of the chameleon goby, *Tridentiger trigonocephalus*, from the Sacramento-San Joaquin delta. *California Fish and Game* **74:** 60–61.

Ravenel, W. de C. 1898. Report on the propagation and distribution of food-fishes, 1897–98 – steelhead and rainbow trout. *Report of the United States Commissioner for Fish and Fisheries* **24:** CX–CXIV.

Raveret-Wattel, C. 1900. *Poissons D'Eau Douce.* Paris: Librairie des Sciences Naturelles.

Ravichandra Reddy, S., Nijaguna, G.M. & Shakuntala, K. 1990. The impact of the introduction of *Oreochromis mossambicus* (Peters) on the native carps. In: *The Biology and Conservation of Rare Fish.* Papers from an international symposium of The Fisheries Society of the British Isles.

Ravichandra Reddy, S.R. & Pandian, T.J. 1972. Heavy mortality of *Gambusia affinis* reared on diet restricted to mosquito larvae. *Mosquito News* **32:** 108–110.

Redeke, H.C. 1941. *Pisces (Cyclostomi–Euichthyes), Fauna van Nederland,* Part 10. Leiden: A.W. Sijthoff.

Reed, H.D. & Wright, A.H. 1909. The vertebrates of the Cayuga Lake basin, NY. *Proceedings of the American Philosophical Society* **48:** 370–459.

Reich, K. 1978. Lake Kinneret fishing and its development. *Bamidgeh Bulletin of Fish Culture in Israel* **30:** 37–64.

Reinthal, P.N. & Stiassny, M.L.J. 1991. The freshwater fishes of Madagascar: a study of an endangered fauna with recommendations for a conservation strategy. *Conservation Biology* **5:** 231–243.

Renault, R. 1951. *Le Black Bass ou Achigan, ses Moeurs – ses Pêches.* Paris: Bornemann Co.

Reynolds, J.E. & Gréboval, D.F. 1988. Socio-economic effects of the evolution of the Nile perch fisheries of Lake Victoria: a review. *CIFA Technical Paper* **17:** 148 pp.

Reynolds, J.E., Gréboval, D.F. & Mannini, P. 1995. Thirty years on: the development of the Nile perch fishery in Lake Victoria. In: Pitcher, T.J. & Hart, P.J.B. (Eds), *The Impact of Species Changes in African Lakes. Fish and Fisheries Series* **18:** 181–214. London: Chapman & Hall.

Reyntjens, D. 1982. Bijdrage tot de Limnologie van het Kivu meer. Unpublished MSc thesis, University of Gand, Belgium: 95 pp.

Ribbink, A.J. 1987. African lakes and their fishes: conservation scenarios and suggestions. *Environmental Biology of Fishes* **19:** 3–26.

Ribbink, T. 1991. Devastation in Lake Victoria. *Buntbarsche Bulletin* **138:** 2–3.

Rickards, R.B. & Fickling, N. 1979. *Zander.* London: A & C Black.

Riedel, D. 1965. Some remarks on the fecundity of tilapia (*T. mossambica* Peters) and its introduction into middle Central America (Nicaragua) together with a first contribution towards the limnology of Nicaragua. *Hydrobiologia* **25:** 357–388.

Rincón, P.A., Velasco, J.C., González-Sánchez, N. & Pollo, C. 1990. Fish assemblages in small streams in western Spain: the influence of an introduced predator. *Archiv für Hydrobiologie* **118:** 81–91.

Ringuelet, R.A., Aramburu, R.E. & Alonso de Arumburu, A. 1967. *Los Peces Argentinos de Agua Dulce.* La Plata: Comisión de Investigación Científica.

Ritchie, J. 1988. *The Australian Trout: its introduction and acclimatisation in Victorian waters.* Melbourne: Victoria Fly-Fishers' Association.

Rivas, L.R. 1965. Florida freshwater fishes and conservation. *Quarterly Journal of the Florida Academy of Science* **28:** 255–258.

Rivera-González, J.E. 1979. Estudio de los poblaciónes pisicolas en los lagos Loiza y Guatjataca. *Tercer Symposio del Departmento de Recursos Naturales,* **1976:** 1–10.

Rivero, L.H. 1936. The introduced largemouth bass, a predator upon native Cuban fishes. *Transactions of the American Fisheries Society* **66:** 367–368.

Robbins, W.H. & MacCrimmon, H.R. 1974. *The Blackbass in America and Overseas.* Sault Sainte Marie, Ontario, Canada: Biomanagement Research Enterprises.

Rogan, P.L. 1982. Australian chinook salmon fishery. *Occasional Publications of the Fisheries Research Division, Wellington* **30:** 78–82.

Rolls, E.C. 1969. *They All Ran Wild: the story of pests on the land in Australia.* Sydney: Angus & Robertson.

Romero, H. 1967. Catálogo sistemático de los peces del Alta Lerma con descripción de una nueva especie. *Anales de la Escuela Nacional de Ciencies Biológicas, México* **14:** 47–77.

Rondorf, D.W. 1976. New locations of *Oreodaimon quathlambae* (Barnard) (Pisces, Cyprinidae) populations. *South African Journal of Science* **72:** 150–151.

Roots, C. 1976. *Animal Invaders.* Newton Abbot: David & Charles.

Rosas, M. 1976a. Reproducción natural de la carpa herbivora en México, *Ctenopharyngodon idellus* Cyprinidae. *Memorias del Simposio sobre Pesquerías en Aguas Continentales* **1:** 1–28.

Rosas, M. 1976b. Peces dulceacuícolas que se explotan en México y datos sobre su cultivo. *Centro de Estudios Económicos y Sociales del Tercer Mundo* **2:** 1–135.

Rosen, D.E. & Bailey, R.M. 1963. The poeciliid fishes (Cyprinodontiformes), their structure, zoogeography, and systematics. *Bulletin of the American Museum of Natural History* **126**: 1–176.

Rosenthal, H. 1978. Bibliography on transplantation of aquatic organisms and its consequences on aquaculture and ecosystems. *Special Publication of the European Mariculture Society* **3**: 146 pp.

Ross, W. 1984. Mystery fish of Saudi Arabia. *Aquarist and Pondkeeper* **49**: 21–23.

Rossi, R. 1991. Specie alloctone per l'acquacoltura italiana. *Il Pesce* **1**: 28–41.

Roughley, T.C. 1951. *Fish and Fisheries of Australia.* Sydney: Angus & Robertson.

Roule, L. 1922. Un cas particular d'acclimatation de la truite arc-en-ciel dans le bassin du Rhone. *Bulletin Société Centrale d'Aquiculture et de Pêche* **29**: 65–66.

Roule, L. 1925. *Les Poissons des Eaux Douces de la France.* Paris.

Royer, Ch. 1902. Observations sur le Saumon de Californie et sur son acclimatation en France. *Bulletin Société Centrale d'Aquiculture et de Pêche* **14**: 17–21.

Rubec, P.J. & Coad, B.W. 1974. First record of the margined madtom (*Noturus insignis*) from Canada. *Journal of the Fisheries Research Board of Canada* **31**: 1430–1431.

Ryan, P.A. 1980. A checklist of the brackish and freshwater fish of Fiji. *South Pacific Journal of Natural Science* **1**: 58–73.

Ryder, R.M. 1956. Occurrence of carp on the north shore of Lake Superior, Port Arthur and Geraldton Districts. Ontario Department of Lands and Forests. *Fish and Wildlife Report* **31**: 1–2.

Sachs, T.R. 1878. Transportation of live pike-perch. *Land and Water* **25**: 476.

St Amant, J.A. 1966. Addition of *Tilapia mossambica* Peters to the California fauna. *California Fish and Game* **52**: 54–55.

St Amant, J.A. 1970. Addition of Hart's rivulus, *Rivulus harti* (Boulenger), to the California fauna. *California Fish and Game* **56**: 138.

St Amant, J.A. & Hoover, F.G. 1969. Addition of *Misgurnus anguillicaudatus* (Cantor) to the California fauna. *California Fish and Game* **55**: 330–331.

St Amant, J.A. & Sharp, I. 1971. Addition of *Xiphophorus variatus* (Meek) to the California fauna. *California Fish and Game* **57**: 128–129.

Sakuda, H.M. 1993. Freshwater fishing news. *Hawaii Fishing News* (**April**) **18**: 15.

San Feliu, J.M. 1973. Present state of aquaculture in the Mediterannean and south Atlantic coast of Spain. *GFCM Studies Review* **52**: 1–24.

Sarig, S. 1966. Synopsis of biological data on the common carp, *Cyprinus carpio* (Linnaeus, 1758). (Near East and Europe). *FAO Fisheries Synopsis* (**31.2**): 35 pp.

Sawara, Y. 1974. Reproduction of the mosquitofish, *Gambusia affinis affinis*, a freshwater fish introduced into Japan. *Japanese Journal of Ecology* **24**: 140–146.

Schindler, O. 1957. *Freshwater Fishes* (translated from German by Orkin, P.A.). London: Thames & Hudson.

Schmidt, R.E., Samaritan, J.M. & Pappantoniou, A. 1981. Status of the bitterling, *Rhodeus sericeus*, in southeastern New York. *Copeia* **1981**: 481–482.

Schrader, H.J. 1985. Invasive alien fishes of South West Africa/Namibia. *South African National Scientific Programme Report* **199**: 35–40.

Schroder, H. 1928. Zum import von *Gambusia*. *Blätter für Aquarien und Terrarienkunde* **39**: 115.

Schulte, T.S. 1974. An initial analysis of the fishery populations of five Puerto Rican reservoirs. *First Symposium, Department of Natural Resources of Puerto Rico, San Juan.*

Schultz, E.E. 1960. Establishment and early dispersal of a loach, *Misgurnus anguillicaudatus* (Cantor), in Michigan. *Transactions of the American Fisheries Society* **89**: 376–377.

Schuster, W.H. 1950. Comments on the importation and the transplantation of different species of fish into Indonesia. *Contributions of the General Agricultural Research Station, Bogor, Indonesia* **111**: 1–31.

Schuster, W.H. 1951. A survey of the inland fisheries of the Territory of New Guinea and Papua. *Australian Journal of Marine and Freshwater Research* **2**: 226–236.

Schuster, W.H. 1952. Discovery of *Tilapia mossambica* in East Java. *Special Publications of the Indo-Pacific Fisheries Council* **1**: 94–96; 103–107.

Schwartz, F.J. 1963. The freshwater minnows of Maryland. *Maryland Conservationist* **40**: 19–29.

Scott, D. 1964. The migratory trout (*Salmo trutta* L.) in New Zealand: the introduction of stocks. *Transactions of the Royal Society of New Zealand* (*Zoology*) **4**: 209–227.

Scott, D. 1984. Origin of the New Zealand sockeye salmon, *Oncorhynchus nerka* (Walbaum). *Journal of the Royal Society of New Zealand* **14**: 245–249.

Scott, D.A. 1993. *A Directory of Wetlands in Oceania.* Slimbridge, England: IWRB; & Kuala Lumpur, Malaysia: AWB.

Scott, D., Hewitson, I. & Fraser, J.C. 1978. The origin of rainbow trout *Salmo gairdneri* Richardson in New Zealand. *California Fish and Game* **64**: 210–218.

Scott, H.A. 1982. *The Olifants River System: a unique habitat for rare Cape fishes.* Cape Department of Nature and Environmental Conservation, Cape Conservation Series **2**.

Scott, W.B. 1967. *Freshwater Fishes of Eastern Canada.* Toronto: University of Toronto Press.

Scott, W.B. & Crossman, E.J. 1964. *Fishes Occurring in the Fresh Waters of Insular Newfoundland.* Canada: Department of Fisheries: 124 pp.

Scott, W.B. & Crossman, E.J. 1973. *Freshwater Fishes of Canada. Bulletin of the Fisheries Research Board of Canada* **184**: 966 pp.

Seager, P.S. 1889. A concise history of the acclimatisation of the Salmonidae in Tasmania. *Proceedings of the Royal Society of Tasmania* **1888**: 1–26.

Seale, A. 1915. The successful transference of blackbass to the Philippine Islands, with notes on the transporting of live fish long distances. *Philippine Journal of Science* **5**: 153–160.

Seale, A. 1917. The mosquitofish, *Gambusia affinis* (Baird & Girard) in the Philippine Islands. *Philippines Journal of Science* **12**: 177–187.

Sehgal, K.L. 1989. Present status of exotic coldwater fish species in India. In: Joseph M.M. (Ed.), *Exotic Aquatic Species in India. Asian Fisheries Society Special Publication* **1**: 41–47.

Serbétis, C.D. 1954. La station de pisciculture du Louros et la culture de la truite arc-en-ciel en Grèce. In: Conseil général des pêches pour la Mediterranée, debats et documents techniques. Rome: *FAO UN* **2**: 17–27.

Shafland, P.L. 1979. Non-native fish introductions with special reference to Florida. *Fisheries* **4**: 18–23.

Shafland, P.L. & Foote, K.J. 1979. A reproducing population of *Serrasalmus humeralis* Valenciennes in southern Florida. *Florida Scientist* **42**: 206–214.

Shapovalov, L. 1944. The tench in California. *California Fish and Game* **30**: 54–57.

Shapovalov, L., Cordone, A.J. & Dill, W.A. 1981. A list of the freshwater and anadromous fishes of California. *California Fish and Game* **61**: 4–38.

Shearer, K.D. & Mulley, J.C. 1978. The introduction and distribution of the carp, *Cyprinus carpio* Linnaeus, in Australia. *Australian Journal of Marine and Freshwater Research* **29**: 551–563.

Shelby, W.H. 1917. History of the introduction of food and game fishes into the waters of California. *California Fish and Game* **3**: 3–12.

Shelton, W.L. & Smitherman, R.O. 1984. Exotic fishes in warmwater aquaculture. In: Courtenay, W.R. Jr & Stauffer, J.R. Jr (Eds), *Distribution, Biology and Management of Exotic Fishes:* 262–301. Baltimore: Johns Hopkins University Press.

Shepard, J.W. & Myers, R.F. 1981. A preliminary checklist of the fishes of Guam and the southern Mariana Islands. In: A working list of marine organisms from Guam. *University of Guam Marine Laboratory Technical Report* **70**: 60–88.

Sherrin, R.A.A. 1886. *Handbook of the Fishes of New Zealand.* Auckland: Wilson & Horton.

Shetty, H.P.C., Nandeesha, M.C. & Jhingran, A.G. 1989. Impact of exotic aquatic species in Indian waters. In: De Silva, S.S. (Ed.), *Exotic Aquatic Organisms in Asia. Asian Fisheries Society Special Publication* **3**: 45–55.

Shireman, J.V. 1984. Control of aquatic weeds with exotic fishes. In: Courtenay, W.R. Jr & Stauffer, J.R Jr. (Eds), *Distribution, Biology and Management of Exotic Fishes:* 302–312. Baltimore: Johns Hopkins University Press.

Shireman, J.V. & Smith, C.R. 1983. Synopsis of biological data on the grass carp, *Ctenopharyngodon idella* (Cuvier and Valenciennes, 1844). *FAO Fisheries Synopsis* **135**: 86 pp.

Shuker, K. 1993. *The Lost Ark: new and rediscovered animals of the 20th century.* London: Collins.

Shutt, P. 1995. Landlocked but not forgotten. *Fish and Game New Zealand* **9**: 22–26.

Siegfried, W.R. 1962. Introduced vertebrates in the Cape Province. *Cape Provincial Administration, Department of Nature Conservation Report* **19**: 80–87.

Silverstein, A. & Silverstein, V. 1974. *Animal Invaders: the story of imported wildlife.* New York: Atheneum.

Simpson, J.C. & Wallace, R.L. 1978. *Fishes of Idaho.* Moscow, Idaho: University Press, Idaho.

Singh, M. & Kumar, K. 1989. Role of exotic fishes in the development of fisheries in Himachal Pradesh. In: Joseph, M.M. (Ed.), *Exotic Aquatic Organisms in Asia. Asian Fisheries Society Special Publication* **1**: 79–82.

Skelton, P.H. 1981. Fishes of the Aughrabies Falls National Park and the Lower Orange River. *Albany Museum Research Report*: 1–20.

Skelton, P.H. 1986. Fish of the Orange-Vaal system. In: Davies, B.R. & Walker, K.F. (Eds), *The Ecology of River Systems*: 143–161. Dordrecht: Kluwer Academic.

Skelton, P.H. 1987. South African Red Data Book. *Fishes. South African National Scientific Programmes Report* **137**. Pretoria: Council for Scientific and Industrial Research.

Skelton, P.H. & Merron, G.S. 1984. The fishes of Okavango River in South West Africa with reference to the possible impact of the Eastern National Water Carrier on fish distribution. *J.L.B. Smith Institute of Ichthyology, Investigational Report* **9**: 32 pp.

Skelton, P.H. & Skead, J. 1984. Early reports and paintings of freshwater fishes in the Cape Province. *Africana Notes and Records* **26**: 29–34.

Skinner, W.F. 1984. *Oreochromis aureus* (Steindachner: Cichlidae), an exotic fish species accidentally introduced to the lower Susquehanna River, Pennsylvania. *Proceedings of the Pennsylvania Academy of Science* **58**: 99–100.

Skrynski, W. 1967. Freshwater fishes of the Chatham Islands. *New Zealand Journal of Marine and Freshwater Research* **1**: 89–98.

Sloane, R.D. & French, G.C. 1991. Trout fishery management plan: Western Lakes – Central Plateau. Tasmanian World Heritage Area. Hobart: Department of Environment and Planning, Parks and Wildlife Service.

Smiley, C.W. 1883. The German carp and its introduction to the United States. *Bulletin of the United States Fish Commission* **3**: 333–336.

Smiley, C.W. 1884 & 1889a. Brief notes upon fish and fisheries. *Bulletin of the United States Fish Commission* **4**: 359–368; **7**: 33–47.

Smiley, C.W. 1885. Sending catfish to Europe. *Bulletin of the United States Fish Commission* **5**: 433–434.

Smiley, C.W. 1886. Some results of carp culture in the United States. *Report of the United States Fish Commission* **1884**: 657–890.

Smiley, C.W. 1889b. Loch Leven trout introduced into the United States. *Bulletin of the United States Fish Commission* **7**: 28–32.

Smith, D.L. 1960. The ability of the top minnow. *Gambusia affinis* (Baird and Girard), to reproduce and overwinter in an outdoor pond at Winnipeg, Manitoba, Canada. *Mosquito News* **20**: 55–56.

Smith, G.J. & Pribble, H.J. 1978. A review of the effects of carp (*Cyprinus carpio* L.) on aquatic vegetation and waterfowl. *Carp Program Publication* **4**: 1–16. Fisheries and Wildlife Division, Victoria.

Smith, H.M. 1892. Report on the investigation of the fisheries of Lake Ontario. *Bulletin of the United States Fish Commission* **15**: 379–472.

Smith, H.M. 1896. A review of the history and results of the attempts to acclimatize fish and other water animals in the Pacific states. *Bulletin of the United States Fish Commission* (1895) **15**: 379–475.

Smith, H.M. 1904. The status of carp in the Great Lakes. *Report of the Commissioners for Fish and Fisheries* (**1901–2**): 128–130.

Smith, W. 1983. Rietvlei – past and present. *Tight Lines* (**October**): 14–18.

Smith, W. 1984. Count down for SA bass team. *Tight Lines* (**March**): 19–23.

Smith-Vaniz, W.F. 1968. *Freshwater Fishes of Alabama.* Auburn, Alabama: Auburn University.

Sneed, K.E. 1972. The history of introduction and distribution of grass carp in the United States. Report of the United States Bureau of Sport Fishing and Wildlife; Mimeo: 5 pp.

Soler, P.J. 1951. A bibliography of the fishes and fisheries of Puerto Rico. *Proceedings of the Gulf and Caribbean Fisheries Institute* **3**: 143–149.

Soller, M., Shchori, Y., Moav, R., Wohlfarth, G. & Lehman, M. 1965. Carp growth in brackish waters. *Bamidgeh Bulletin of Fish Culture in Israel* **17**: 16–23.

Sommani, E. 1957. Caratteristiche eco-etologiche della trota iridea (*Salmo gairdneri* Rich.) in relazione ai ripopolamenti dei corsi d'acqua montani. *Bollettino di Pesca di Piscicoltura e di Idrobiologia* **12** (NS): 92–99.

Spillman, C.J. 1959. Un petit poisson americain, *Umbra pygmaea* (DeKay) acclimaté depuis 46 ans dans un etang du Bourbonnais. *Bulletin du Muséum d'Histoire Naturelle* **31**: 401–402.

Spillman, J. 1961. Poissons d'eau douce. *Faune de France* **65**: 303.

Spliethoff, P.C., De Iongh, H.H. & Frank, V.G. 1983. Success of the introduction of the fresh water clupeid *Limnothrissa miodon* (Boulenger) in Lake Kivu. *Fisheries Management* **14**: 17–31.

Springer, V.G. & Finucane, J.H. 1963. The African cichlid, *Tilapia heudeloti* Dumeril, in the commercial fish catch of Florida. *Transactions of the American Fisheries Society* **92**: 317–318.

Sreenivasan, A. 1967. *Tilapia mossambica*: its ecology and status in Madras State, India. *Madras Journal of Fisheries* **3**: 33–43.

Sreenivasan, A. 1989. Status of some exotic fish introductions in Tamil Nadu. In: Joseph, M.M. (Ed.), *Exotic Aquatic Organisms in India. Asian Fisheries Society Special Publication* **1**: 59–62.

Sreenivasan, A. 1991. Transfers of freshwater fishes into India. In: Ramakrishnan, P.S. (Ed.), *Ecology of Biological Invasions in the Tropics.* New Delhi: International Scientific Publications (Supplement No. 16 to *International Journal of Ecology and Environmental Sciences*): 131–138.

Sreenivasan, A. & Chandrasekaran, F. 1989. Status of *Tilapia* in Tamil Nadu. In: Joseph, M.M. (Ed.), *Exotic Aquatic Organisms in India. Asian Fisheries Society Special Publication* **1**: 67–74.

Stanton Hales, L. Jr 1991. Occurrence of an introduced Africa cichlid, the blue tilapia, *Tilapia aurea*, in a tidal creek of the Skidaway River, Georgia. *Brimleyana* **17**: 27–35.

Stanley, S.E., Moyle, P.B. & Shaffer, H.B. 1995. Allozyme analysis of Delta Smelt, *Hypomesus transpacificus* and Longfin Smelt, *Spirinchus thaleichthys* in the Sacramento–San Joaquin Estuary, California. *Copeia* **1995**: 390–396.

Stauffer, J.R. Jr 1984. Colonization theory relative to introduced populations. In: Courtenay, W.R. Jr & Stauffer, J.R. Jr (Eds), *Distribution, Biology and Management of Exotic Fishes*: 8–21. Baltimore: Johns Hopkins University Press.

Stead, D.G. 1908. *The Edible Fishes of New South Wales: their present importance and their potentialities.* Sydney: Government Printer.

Stead, D.G. 1929. Introduction of the great carp *Cyprinus carpio* into waters of New South Wales. *Australian Zoologist* **6**: 100–102.

Steinmetz, Ir. B. 1968. De merkactie van forel voor het Veerse Meer. *Visserij* **6**: 296–299.

Stephanidis, T. 1964. The influence of the antimosquitofish *Gambusia affinis* on the natural fauna of a Corfu lakelet. *Proceedings of the Hellenic Hydrobiological Institute* **9**: 3–6.

Stinson, D.W., Ritter, M.W. & Reichel, J.D. 1991. The Mariana common moorhen: decline in an island endemic. *Condor* **93**: 38–43.

Stokell, G. 1934. New light on New Zealand salmon. A comparison of Te Anau fish with Atlantic and quinnat salmon from Lake Coleridge. *Salmon & Trout Magazine* **76**: 260–276.

Stokell, G. 1951. The American lake char *Cristivomer namaycush. Transactions of the Royal Society of New Zealand* **79**: 213–217.

Stokell, G. 1955. *Freshwater Fishes of New Zealand.* Christchurch: Simpson & Williams.

Stokell, G. 1962. Pacific salmon in New Zealand. *Transactions of the Royal Society of New Zealand* (*Zoology*) **2**: 181–190.

Sukumaran, K.K. & Tripathi, J.D. 1989. *Tilapia mossambica* – a controversial exotic species in Indian waters. In: Joseph, M.M. (Ed.), *Exotic Aquatic Organisms in India. Asian Fisheries Society Special Publication* **1**: 91–95.

Svärdson, G. 1964. Rapportera förekomsten av vild bäckröding. *Svenskt Fiske* **9**: 4–5; 26.

Svärdson, G. 1968. *Regnbågen Särtryck ur Fiske (Sweden)*: 10–31.

Swartzmann, G.L. & Zaret, T.M. 1983. Modeling fish species introductions and prey extermination: the invasion of *Cichla ocellaris* to Gatun Lake, Panama. *Developments in Environmental Modelling* **5**: 361–371.

Swenson, R. & Matern, S.A. In Prep. Interactions between two estuarine gobies, the endangered tidewater goby (*Eucyclogobius newberryi*) and a recent Asian invader, the shimofuri goby (*Tridentiger bifasciatus*).

Swift, C.G., Haglund, T.R., Ruiz, M. & Fisher, R.N. 1993. The status and distribution of the freshwater fishes of southern California. *Bulletin of the Southern California Academy of Science* **92**: 101–167.

Swingle, H.S. 1957. A recessive factor controlling reproduction in fishes. *Proceedings of the 1953 Pacific Science Congress* **8**: 865–871.

Tal, S. & Shelubsky, M. 1951. Review of the fish-farming industry in Israel. *Transactions of the American Fisheries Society* **81**: 218–223.

Tanaka, M. & Shiraishi, Y. 1970. Studies on the effective stocking of salmonid fish. I. Survival of Himemasu (*Oncorhynchus nerka*) released in Lake Yunoko. *Bulletin of the Freshwater Fisheries Research Laboratory, Tokyo* **20**: 83–91. [In Japanese with English summary.]

Taylor, J.N., Courtenay, W.R. Jr & McCann, J.A. 1984. Known impacts of exotic fishes in the continental United States. In: Courtenay, W.R. Jr & Stauffer, J.R. Jr (eds), *Distribution, Biology and Management of Exotic Fishes*: 322–373. Baltimore: Johns Hopkins University Press.

Tedla, S. & Meskel, F.H. 1981. Introduction and transplantation of freshwater fish species in Ethiopia. *Sinet* **4**: 69–72.

Therézein, Y. 1960. L'introduction des poissons d'eau douce Madagascar, leur influence sur la modification du biotope. *Publications of the Scientific Council for Africa South of the Sahara* **63**: 145–156.

Thienemann, A. 1950. *Die Binnengewässer, Band 18, Verbreitungsgeschichte der Süsswassertierwelt Europas.* Stuttgart: Schweizerbart.

Thompson, D.H. 1939. Growth of the large-mouth black bass, *Huro salmoides*, in Lake Naivasha, Kenya. *Nature* **143**: 561–562.

Thompson, F.A. 1940. Salmonid fishes in the Argentine Andes. *Transactions of the American Fisheries Society (1939)* **69**: 279–284.

Thomson, G.M. 1922. *The Naturalisation of Animals and Plants in New Zealand.* Cambridge: Cambridge University Press.

Thomson, G.M. 1926. *Wildlife in New Zealand. 2. Introduced Birds and Fishes.* Wellington.

Thorpe, J.E. 1977. Morphology, physiology, behaviour and ecology of *Perca fluviatilis* L. and *Perca flavescens* Mitchill. *Journal of the Fisheries Research Board of Canada* **34**: 1504–1514.

Thys van den Audenaerde, D.F.E. 1992. The introduction of aquatic species into Zambian waters, and their importance for aquaculture and fisheries. Consultancy Report, ALCOM, Harare, Zimbabwe.

Tilzey, R.D.J. 1976. Observations on the interactions between indigenous Galaxiidae and introduced Salmonidae in the Lake Eucumbene catchment, New South Wales. *Australian Journal of Marine and Freshwater Research* **27**: 551–564.

Tilzey, R.D.J. 1977. Key factors in the establishment and success of trout in Australia. *Proceedings of the Ecological Society of Australia* **10**: 97–105.

Tilzey, R.D.J. 1980. Introduced Fish. In: Williams, W.D. (Ed.), *An Ecological Basis for Water Resource Management*: 271–279. Canberra: Australian National University Press.

Timmermans, J.A. 1978. La carpe herbivore; premières expériences en Belgique. *Travaux de la Station Recherches Eaux et Fôrets* **47**: 22 pp.

Toots, H. 1970. Exotic fishes in Rhodesia. *Rhodesia Agricultural Journal* **67**: 83–88.

Tortonese, E. 1967. Il pesce gatto. *Rivista Italiana di Piscicultura e Ittiopatologia* **2**: 46–47.

Trendall, J.T. & Johnson, M.S. 1981. Identification by anatomy and gel electrophoresis of *Phalloceros caudimaculatus*, previously mistaken for *Gambusia affinis holbrooki*. *Australian Journal of Marine and Freshwater Research* **32**: 993–996.

Treviño-Robinson, D. 1959. The ichthyofauna of the lower Río Grande, Texas and Mexico. *Copeia* **1959**: 253–256.

Trewavas, E. 1933. Scientific results of the Cambridge expedition to the East African lakes in 1930–1931. 2. The cichlid fishes. *Journal of the Linnean Society* **38**: 309–341.

Trewavas, E. 1965. *Tilapia aurea* and the status of *Tilapia nilotica exul, T. monodi* and *T. lemassoni. Israel Journal of Zoology* **14**: 258–276.

Trewavas, E. 1966. A preliminary review of fishes of the genus *Tilapia* in the eastward-flowing rivers of Africa, with proposals of two new scientific names. *Revue de Zoologie et de Botanique Africaines* **74**: 394–424.

Tubb, J.A. 1954. Introduction of *Tilapia* to Hong Kong. *Hong Kong University Fisheries Journal* **1**: 63–64.

Tulian, E.A 1910. Acclimatization of American fishes in Argentina. *Bulletin of the Bureau of Fisheries, Washington* **28**: 955–965.

Turnbull-Kemp, P. St J. 1957. Trout in Southern Rhodesia. *Rhodesia Agricultural Journal* **54**: 364–370; 438–449; 533–546.

Turner, J.S. 1981. Population structure and reproduction in

the introduced Florida population of the pike killifish, *Belonesox belizanus* (Pisces: Poeciliidae). MSc thesis, University of Central Florida: 56 pp.

Tweedie, M.W.F. 1952. Notes on Malayan freshwater fishes. *Bulletin of the Raffles Museum* **24**: 63–95.

Twongo, T. 1988. Recent trends in the fisheries of Lake Kioga, Uganda. *CIFA Occasional Papers* **15**: 140–151.

Twongo, T. 1995. Impact of fish species introductions on the tilapias of Lakes Victoria and Kyoga. In: Pitcher, T.J. & Hart, P.J.B. (Eds) *The Impact of Species Changes in African Lakes. Fish and Fisheries Series* **18**: 45–58. London: Chapman & Hall.

Twyford, G. 1991. *Australia's Introduced Animals and Plants.* Sydney: Reed.

Ulaiwi, K.N. 1990. The occurrence and spread of common carp, *Cyprinus carpio* (L.), in the Sepik River system, Papua New Guinea. *Second Asian Fisheries Forum*: 765–768.

Umeda, K., Matsumura, K., Okukawa, G., Sazawa, R., Honma, H., Arauchi, M., Kasahara, K. & Nara, K. 1981. On the coho salmon transplanted from North America into the Ichani River. *Scientific Reports of the Hokkaido Salmon Hatchery* **35**: 9–23. [In Japanese with English summary.]

Uwate, K.R., Kunatuba, P., Raobati, B. & Tenakonai, C. 1984. *A Review of Aquaculture Activities in the Pacific Islands Region.* East–West Center, Honolulu: Pacific Islands Development Program.

Vaas, K.F. & Hofstede, A.E. 1952. Studies on *Tilapia mossambica* (ikan mundjair) in Indonesia. *Pemberitaan Belai Penjelidikan Perikanan Darat* **1**: 1–68.

Van der Merwe, C.V. 1970. A short description of the Goukamma Nature Reserve. *Department of Nature Conservation, Cape Provincial Administration Investigational Report* **16**: 16 pp.

Van Dine, D.L. 1907. The introduction of top-minnow into the Hawaiian Islands. *Press Bulletin of the Hawaiian Agricultural Experiment Station* **20**: 10 pp.

Van Pel, H. 1956. Introduction of edible pond fish from the Philippines. *Quarterly Bulletin of the South Pacific Commission* **6**: 17–18.

Van Pel, H. 1958. Fresh water fish for the Pacific. *Quarterly Bulletin of the South Pacific Commission* **8**: 48–49, 52.

Van Pel, H. 1959. Fisheries in American Samoa, Fiji, and New Caledonia. *Quarterly Bulletin of the South Pacific Commission* **9**: 26–27.

Van Pel, H. 1961. S.P.C. fisheries investigations in Western Samoa. *South Pacific Bulletin* **11**: 20–22.

Van Rensburg, K.J. 1966. Die vis van die Olifantsrivier (weskus) met spesiale verwysing na die geelvis (*Barbus capensis*) en saagvin (*Barbus serra*). *Cape Provincial Administration Department of Nature Conservation Investigational* Report **10**: 14 pp.

Van Schoor, D.J. 1966. Studies on the culture and acclimatization of *Tilapia* in the Western Cape Province. *Cape Provincial Administration Department of Nature Conservation Investigational Report* **7**.

Van Schoor, D.J. 1969a. The growth of largemouth bass *Micropterus salmoides* at Jonkershoek Fish Hatchery without benefit of a forage fish. *Cape Provincial Administration Department of Nature Conservation Investigational Report* **11**.

Van Schoor, D.J. 1969b. The introduction of *Serranochromis robustus jallae* (Boulenger) to the Eerste River Basin, Western Cape Province. *Department of Nature Conservation, Cape Provincial Administration Investigational Report* **12**.

Vásárhelyi, I. 1963. Pisztrángtenyésztés. *Országos Erdészeti Föigazgatósag, Budapest*: 7–12.

Vergara de los Rios, M. 1976. Engorda de bagre (*Ictalurus punctatus*) en Jaulas. *Memorias del Simposio de Pesquerias en Aguas Continentales, Mexico* **1**: 89–97.

Viljoen, S. & Van As, J.G. 1985. Sessile peritrichs (Ciliophora: Peritricha) from freshwater fish in the Transvaal, South Africa. *South African Journal of Zoology* **20**: 79–96.

Villa, J. 1971. Presence of the cichlid fish *Cichlasoma managuense* Guenther in Lake Xiloa, Nicaragua. *Copeia* **1971**: 186.

Vivier, P. 1951a. Poissons et crustacés d'eau douce acclimatés en France en eaux libres depuis le début du siècle. *Terre et la Vie* **98**: 57–82.

Vivier, P. 1951b. Sur l'extension en France du Black-Bass à grande bouche (*Micropterus salmoides* Lacepède) et de l'Écrivisse américaine (*Cambarus affinis* Say). *Verhandlungen der Internationalen Vereinigung für Theoretische und Angewandte Limnologie* **11**: 430–436.

Vivier, P. 1955. Sur l'introduction des Salmonidés exotiques en France. *Travaux Association Internationale de Limnologie Théoretique et Appliquée* **12**: 527–535.

Vladykov, V.D. & McAllister, D.E. 1961. Preliminary list of marine fishes of Quebec. *Le Naturaliste Canadien* **88**: 55–78.

Von Behr, F. 1883. Five American Salmonidae in Germany. *Bulletin of the United States Fish Commission* (1882) **2**: 237–246.

Von dem Borne, M. 1885. Distribution of American fish and fish-eggs by the German Fishery Association. *Bulletin of the United States Fish Commission* **5**: 261–264.

Von dem Borne, M. 1888. *Der Schwarzbarsch und der Forellenbarsch, Black Bass, zwei Amerikanische Fische in Deutschland.* Neudamm: Verlag Journal Neumann: 35 pp.

Von dem Borne, M. 1890. *Sechs Amerikanische Salmoniden in Europa.* Neudamm: Verlag Journal Neumann: 38 pp.

Von dem Borne, M. 1892. *Der Schwarzbarsch und der Forellenbarsch, Black Bass, zwei Amerikanische Fische in Deutschland.* Berlin: Paul Parey.

Von dem Borne, M. 1894. *Teichwirtschaft.* Berlin: Paul Parey.

Von Pirko, F. 1910. Naturalization of American fishes in Austria. *Bulletin of the Bureau of Fisheries, Washington* (**1908**) **28**: 979–982.

Vooren, C.M. 1968. The influence of fish species introduced into natural waters: a literature survey. Unpublished MS report by the Netherlands State Institute for Field Biology Research for RIVON: 25 pp. [Translated from Dutch.]

Vooren, C.M. 1972. Ecological aspects of the introduction of fish species into natural habitats in Europe, with special reference to the Netherlands and literature survey. *Journal of Fish Biology* **4**: 565–583.

Vostradovsky, J. 1986. On the ichthyofauna and possibilities of fishery utilization of the Cahora Bassa Reservoir on the Zambezi River. *Prace VURH Vodany (Czechoslovakia)* **32**: 143–147.

Vutskits, G. 1913. A pisztrángsügér és a naphal meghonosodása a Drávában. *Pótfüzetek a Természettudományi Közlönyhöz* **45**: 748–749.

Wales, J.H. 1939. General report of investigations on the McCloud River drainage in 1938. *California Fish and Game* **25**: 272–309.

Wales, J.B. 1962. Introduction of pond smelt from Japan into California. *California Fish and Game* **48**: 141–142.

Walker, A.F. 1976. The American Brook Trout in Scotland. *Rod and Line* **16**: 24–26.

Walker, B.W., Whitney, R.R. & Barlow, G.W. 1961. The fishes of the Salton Sea. In: Walker, B.W. (Ed.), *The Ecology of the Salton Sea. California, in relation to the sport fishery.* California Department of Fish and Game, Fishery Bulletin **113**: 77–91.

Walker, C.E. & Patterson, C.S. 1898. *The Rainbow Trout.* London: Lawrence & Bullen.

Wallis, H.F. 1969. *Where to Fish – the Field Guide to the Fishing in Rivers and Lakes.* London: Harmsworth Press.

Walters, V. 1953. Notes on reptiles and amphibians from El Volcan de Chiriqui, Panama. *Copeia* **1953**: 126.

Watson, R. 1989. Nile perch in Lake Victoria. *Swara* **12**: 25–26.

Waugh, G.D. 1981. Salmon in New Zealand. In: Thorpe, J. (Ed.), *Salmon Ranching:* 277–303. London: Academic Press.

Weatherley, A.H. 1959. Some features of the biology of the tench, *Tinca tinca* (Linnaeus), in Tasmania. *Journal of Animal Ecology* **28**: 73–87.

Weatherley, A.H. 1962. Notes on distribution, taxonomy and behaviour of tench, *Tinca tinca* (Linnaeus), in Tasmania. *Annals and Magazine of Natural History* **4**: 713–719.

Weatherley, A.H. 1963. Zoogeography of *Perca fluviatilis* Linnaeus and *Perca flavescens* Mitchell with special reference to the effects of high temperature. *Proceedings of the Zoological Society of London* **141**: 557–576.

Weatherley, A.H. 1977. *Perca fluviatilis* in Australia. Zoogeographic expression of a life-cycle in relation to environment. *Journal of the Fisheries Research Board of Canada* **34**: 1464–1466.

Weatherley, A.H. & Lake, J.S. 1967. Introduced fish species in Australian inland waters. In: Weatherley, A.H. (Ed.), *Australian Inland Waters and their Fauna:* 217–239. Canberra: Australian National University Press.

Webster, D.A. 1941. The life histories of some Connecticut fishes. In: A fishery survey of important Connecticut Lakes. *State Geological and Natural History Survey of Connecticut, Bulletin* **63**: 122–227.

Weisinger, M. 1975. Acclimatization of goldfish in Hungary. *Akvarisztike* **112**: 1–18.

Welcomme, R.L. 1964. Notes on the present distribution and habits of the non-endemic species of *Tilapia* which have been introduced into Lake Victoria. *Report of the East African Freshwater Fisheries Research Organization* **1962–1963**: 36–39.

Welcomme, R.L. 1966. Recent changes in the stocks of *Tilapia* in Lake Victoria. *Nature* **212**: 52–54.

Welcomme, R.L. 1967. Observations on the biology of the introduced species of *Tilapia* in Lake Victoria. *Revue de Zoologie et de Botanique Africaines* **76**: 249–279.

Welcomme, R.L. 1981. Register of international transfers of inland fish species. *FAO Fisheries Technical Paper* **213**: 120 pp.

Welcomme, R.L. 1984. International transfers of inland fish species. In: Courtenay, W.R. Jr & Stauffer, J.R. Jr (Eds), *Distribution, Biology and Management of Exotic Fishes:* 22–40. Baltimore: Johns Hopkins University Press.

Welcomme, R.L. 1988. International introductions of inland aquatic species. *FAO Fisheries Technical Paper* **294**: 318 pp.

Welcomme, R.L. 1991. International introductions of freshwater fish species into Europe. *Finnish Fisheries Research* **12**: 11–18.

Welcomme, R.L. 1992. A history of international introductions of inland aquatic species. *ICES Marine Science Symposium* **194**: 3–14.

Went, A.E.J. 1950. Notes on the introduction of some freshwater fish into Ireland. *Journal of the Department of Agriculture, Dublin* **47**: 119–124.

Went, A.E.J. 1957. The pike in Ireland. *Irish Naturalists' Journal* **12**: 177–182.

West, G.J. 1973. The establishment of exotic freshwater aquarium fish in Papua New Guinea. *Papua and New Guinea Agricultural Journal* **24**: 30–32.

West, G.J. & Glucksman, J. 1976. Introduction and distribution of exotic fish in Papua New Guinea. *Papua and New Guinea Agricultural Journal* **27**: 19–48.

Wetherbee, D.K. 1989. *Guide to Freshwater Fishes Naturalized from Abroad to Hispaniola or to the West Indies.* Shelbourne, Massachusetts: Author.

Wharton, J.C.F. 1969. Trout liberation in Victorian streams and lakes from 1958 to 1967. *Fisheries and Wildlife Department, Victoria, Fisheries Circular* **19**: 357 pp.

Wharton, J.C.F. 1971. European carp in Victoria. *Fur, Feathers and Fin* **130**: 3–11.

Wharton, J.C.F. 1979. Impact of exotic animals, especially European carp *Cyprinus carpio*, on native fauna. *Fisheries & Wildlife (Victoria) Paper* **20**: 1–13.

Wheeler, A.C. 1974. *Changes in the Freshwater Fish Fauna of Britain. Systematics Association Special Volume* **6**. London: Academic Press.

Wheeler, A.C. 1978. *Ictalurus melas* (Rafinesque, 1820) and *I. nebulosus* (Lesueur, 1819): the North American catfishes in Europe. *Journal of Fish Biology* **12**: 435–439.

Wheeler, A. & Maitland, P.S. 1973. The scarcer freshwater fishes of the British Isles. I. Introduced Species. *Journal of Fish Biology* **5**: 49–68.

Whitley, G.P. 1951. Introduced fishes. *Australian Museum Magazine* **10**: 198–200; 234–239.

Whitehouse, F.C. 1946. *Sport Fisheries of Western Canada*. Vancouver, BC: Author.

Whitney, C.A. 1927. *Salmo salar* in Te Anau: its habits identical with Scottish salmon. *New Zealand Fishing and Shooting Gazette* **1**: 12–14.

Whittington, R.J. & Cullis, B. 1988. The susceptibility of salmonid fish to an atypical strain of *Aeromonas salmonicida* that infects goldfish *Carassius auratus* (L) in Australia. *Journal of Fish Diseases* **11**: 461–470.

Whitworth, W.R., Berrien, P.L. & Keller, W.T. 1968. Freshwater fishes of Connecticut. *Bulletin of the State Geological and Natural History Survey of Connecticut* **101**: 1–134.

Wiederholm, T. 1984. Responses of aquatic insects to environmental pollution. In: Resh, V.H. & Rosenberg, D.M. (Eds), *The Ecology of Aquatic Insects*: 508–557. New York: Praeger.

Wiggins, W.G.B. 1950. The introduction and ecology of the brown trout (*Salmo trutta* Linnaeus) with special reference to North America. MA thesis, University of Toronto; mimeo: 109 pp.

Wilson, B. 1995. Native fish and trout, a delicate balancing act. *Fish & Game New Zealand* **9**: 14–21.

Winterbourn, M.J. & Brown, T.J. 1967. Observations on the fauna of two warm springs in the Taupo thermal region. *New Zealand Journal of Marine and Freshwater Research* **1**: 38–50.

Witte, F., Goldschmidt, T., Goudswaard, P.C., Ligyvoet, W. Van Oijen, M.J.P. & Wanink, J.H. 1992. Species extinction and concomitant ecological changes in Lake Victoria. *Netherlands Journal of Zoology* **42**: 214–232.

Witte, F., Goldschmidt, T. & Wanink, J.H. 1995. Dynamics of the haplochromine cichlid fauna and other ecological changes in the Mwanza Gulf of Lake Victoria. In: Pitcher, T.J. & Hart, P.J.B. (Eds), *The Impact of Species Changes in African Lakes. Fish and Fisheries Series* **18**: 83–110. London: Chapman & Hall.

Wohlgemuth, E. & Sebela, M. 1987. [The stone moroko *Pseudorasbora parva* in the fauna of Europe.] *Ziva* **35**: 25–27. [In Czechoslovakian.]

Wolff, T. 1969. The fauna of Rennell and Bellona, Solomon Islands. *Transactions of the Royal Society* **B255**: 321–343.

Wood, J.W. 1970. Introduction del salmon del pacifico en Chile. Informe sobre investigaciones de piscicultura. Servicio Agricol y Gamadero de Pesca y Caza Departmento Desarrollo JW/IVP/ags: 8 pp.

Worth, S.G. 1893. Report on the propagation and distribution of food-fishes, 1892–93, for rainbow trout. *Report of the United States Commissioners for Fish and Fisheries* **19**: 130–132.

Worthington, E.B. 1932. *A report on the fisheries of Uganda, investigated by the Cambridge expedition to the East African lakes (1930–1931)*. London: Crown Agents for the Colonies.

Worthington, E.B. 1941. Rainbows: a report on attempts to acclimatize rainbow trout in Britain. *Salmon & Trout Magazine* **100**: 241–260; **101**: 62–99.

Worthington, E.B. 1973. The ecology of introductions – a case study from the African lakes. *Biological Conservation* **5**: 221–222.

Worthington, E.B. 1989. The Lake Victoria *Lates* saga. *Environmental Conservation* **16**: 266–267.

Wydoski, R.S. & Whitney, R.R. 1979. *Inland Fishes of Washington*. Seattle: Washington Press.

Yadav, B.N. 1993. *Fish and Fisheries*. Delhi: Daya Publishing House.

Yo-Jun, T. & He-Yi, T. 1989. The status of the exotic aquatic organisms in China. In: De Silva, S.S. (Ed.), *Exotic Aquatic Organisms in Asia. Asian Fisheries Society Special Publication* **3**: 35–39.

Youl, J.A. 1864. Who sent the trout ova to Tasmania? *Fisherman's Magazine and Review* **1**: 429.

Zaccagnini, M. 1965. I Black-bass de Lago di Santa Luce. *Pescare* (**Sept.**): 54.

Zamorano, R.M. 1991. Salmon farming in Chile. In: Cook, R.H. & Pennell, W. (Eds), *Proceedings of the Special Session on Salmonid Aquaculture, World Aquaculture Society, Los Angeles, 1989. Technical Report on Fisheries & Aquacultural Science* **1831**: 51–63.

Zaret, T.M. 1974. The ecology of introductions; a case study from a Central American lake. *Environmental Conservation* **1**: 308–309.

Zaret, T.M. 1982. The stability/diversity controversy: a test of hypotheses. *Ecology* **63**: 721–731.

Zaret, T.M. 1984. Central American limnology and Gatún Lake, Panama. In: Taub, F.B. (Ed.), *Ecosystems of the World* **23**. *Lakes and Reservoirs*: 447–465. Amsterdam: Elsevier.

Zaret, T.M. & Paine, R.T. 1973. Species introductions in a tropical lake. *Science, Washington, DC* **182**: 449–455.

INDEX OF ANIMALS AND PLANTS

Note: **Emboldened** names indicate naturalized fishes